T0213724

Communications in Computer and Information Science   631

Commenced Publication in 2007
Founding and Former Series Editors:
Alfredo Cuzzocrea, Dominik Ślęzak, and Xiaokang Yang

Editorial Board

Simone Diniz Junqueira Barbosa
　Pontifical Catholic University of Rio de Janeiro (PUC-Rio),
　Rio de Janeiro, Brazil
Phoebe Chen
　La Trobe University, Melbourne, Australia
Xiaoyong Du
　Renmin University of China, Beijing, China
Joaquim Filipe
　Polytechnic Institute of Setúbal, Setúbal, Portugal
Orhun Kara
　TÜBİTAK BİLGEM and Middle East Technical University, Ankara, Turkey
Igor Kotenko
　St. Petersburg Institute for Informatics and Automation of the Russian
　Academy of Sciences, St. Petersburg, Russia
Ting Liu
　Harbin Institute of Technology (HIT), Harbin, China
Krishna M. Sivalingam
　Indian Institute of Technology Madras, Chennai, India
Takashi Washio
　Osaka University, Osaka, Japan

More information about this series at http://www.springer.com/series/7899

Ana Fred · Jan L.G. Dietz
David Aveiro · Kecheng Liu
Joaquim Filipe (Eds.)

Knowledge Discovery, Knowledge Engineering and Knowledge Management

7th International Joint Conference, IC3K 2015
Lisbon, Portugal, November 12–14, 2015
Revised Selected Papers

 Springer

Editors

Ana Fred
Instituto de Telecomunicações/IST
Lisbon
Portugal

Jan L.G. Dietz
Delft University of Technology
Delft
The Netherlands

David Aveiro
University of Madeira
Funchal
Portugal

Kecheng Liu
University of Reading
Reading
UK

Joaquim Filipe
Polytechnic Institute of Setúbal/INSTICC
Setúbal
Portugal

ISSN 1865-0929   ISSN 1865-0937 (electronic)
Communications in Computer and Information Science
ISBN 978-3-319-52757-4   ISBN 978-3-319-52758-1 (eBook)
DOI 10.1007/978-3-319-52758-1

Library of Congress Control Number: 2017930149

© Springer International Publishing AG 2016
This work is subject to copyright. All rights are reserved by the Publisher, whether the whole or part of the material is concerned, specifically the rights of translation, reprinting, reuse of illustrations, recitation, broadcasting, reproduction on microfilms or in any other physical way, and transmission or information storage and retrieval, electronic adaptation, computer software, or by similar or dissimilar methodology now known or hereafter developed.
The use of general descriptive names, registered names, trademarks, service marks, etc. in this publication does not imply, even in the absence of a specific statement, that such names are exempt from the relevant protective laws and regulations and therefore free for general use.
The publisher, the authors and the editors are safe to assume that the advice and information in this book are believed to be true and accurate at the date of publication. Neither the publisher nor the authors or the editors give a warranty, express or implied, with respect to the material contained herein or for any errors or omissions that may have been made. The publisher remains neutral with regard to jurisdictional claims in published maps and institutional affiliations.

Printed on acid-free paper

This Springer imprint is published by Springer Nature
The registered company is Springer International Publishing AG
The registered company address is: Gewerbestrasse 11, 6330 Cham, Switzerland

Preface

The present book includes extended and revised versions of a set of selected papers from the 7th International Joint Conference on Knowledge Discovery, Knowledge Engineering and Knowledge Management (IC3K 2015), held in Lisbon, Portugal, during November 12–14, 2015. IC3K was sponsored by the Institute for Systems and Technologies of Information, Control and Communication (INSTICC) and was organized in cooperation with the AAAI (Association for the Advancement of Artificial Intelligence), ACM SIGMIS (ACM Special Interest Group on Management Information Systems), ACM SIGAI (ACM Special Interest Group on Artificial Intelligence) Associazione Italiana per l'Intelligenza Artificiale, APPIA (Portuguese Association for Artificial Intelligence) and ERCIM (European Research Consortium for Informatics and Mathematics) and technically co-sponsored by IEEE CS – TCBIS – IEEE Technical Committee on Business Informatics and Systems.

The main objective of IC3K is to provide a point of contact for scientists, engineers, and practitioners interested in the areas of knowledge discovery, knowledge engineering, and knowledge management.

IC3K is composed of three co-located complementary conferences, each specialized in one of the aforementioned main knowledge areas, namely: the International Conference on Knowledge Discovery and Information Retrieval (KDIR), the International Conference on Knowledge Engineering and Ontology Development (KEOD), and the International Conference on Knowledge Management and Information Sharing (KMIS)

The International Conference on Knowledge Discovery and Information Retrieval (KDIR) aims to provide a major forum for the scientific and technical advancement of knowledge discovery and information retrieval. Knowledge discovery is an interdisciplinary area focusing on methodologies for identifying valid, novel, potentially useful, and meaningful patterns from data, often based on underlying large data sets. A major aspect of knowledge discovery is data mining, i.e., applying data analysis and discovery algorithms that produce a particular enumeration of patterns (or models) over the data. Knowledge discovery also includes the evaluation of patterns and identification of which add to knowledge. Information retrieval (IR) is concerned with gathering relevant information from unstructured and semantically fuzzy data in texts and other media, searching for information within documents and for metadata about documents, as well as searching relational databases and the Web. IR can be combined with knowledge discovery to create software tools that empower users of decision support systems to better understand and use the knowledge underlying large data sets.

The purpose of the International Conference on Knowledge Engineering and Ontology Development (KEOD) is to provide a major meeting point for researchers and practitioners interested in the scientific and technical advancement of methodologies and technologies for knowledge engineering (KE) and ontology development, both theoretically and in a broad range of application fields. KE refers to all technical, scientific, and social aspects involved in building, maintaining, and using knowledge-based systems. KE is a multidisciplinary field, bringing in concepts and methods from several computer science domains such as artificial intelligence, databases, expert systems, decision support systems, and geographic information systems. Currently, KE is gradually more related to the construction of shared conceptual frameworks, often designated as ontologies. Ontology development (OD) aims at building reusable semantic structures that can be informal vocabularies, catalogs, glossaries, as well as more complex finite formal structures specifying types of entities and types of relationships relevant within a certain domain. A wide range of applications are emerging, especially given the current Web emphasis, including library science, ontology-enhanced search, e-commerce and business process design, and enterprise engineering.

The goal of the International Conference on Knowledge Management and Information Sharing (KMIS) is to provide a major meeting point for researchers and practitioners interested in the study and application of all perspectives of knowledge management (KM) and information sharing (IS). KM is a discipline concerned with the analysis and technical support of practices used in an organization to identify, create, represent, distribute, and enable the adoption and leveraging of good practices embedded in collaborative settings and, in particular, in organizational processes. Effective KM is an increasingly important source of competitive advantage, and a key to the success of contemporary organizations, bolstering the collective expertise of its employees and partners. IS is a term used for a long time in the information technology (IT) lexicon, related to data exchange, communication protocols, and technological infrastructures.

This book of selected papers from IC3K 2015 includes 25 papers, from a total of 280 paper submissions from 53 countries, representing an acceptance ratio of 9%.

We trust that this book will be of interest for all researchers in various fields involving knowledge extraction, knowledge discovery, knowledge engineering, and knowledge management.

September 2016

Ana Fred
Jan Dietz
David Aveiro
Kecheng Liu
Joaquim Filipe

Organization

Conference Chair

Joaquim Filipe Polytechnic Institute of Setúbal/INSTICC, Portugal

Program Co-chairs

KDIR

Ana Fred Instituto de Telecomunicações/IST, Portugal

KEOD

Jan Dietz Delft University of Technology, The Netherlands
David Aveiro University of Madeira/Madeira-ITI, Portugal

KMIS

Kecheng Liu University of Reading, UK

KDIR Program Committee

Sherief Abdallah British University in Dubai, United Arab Emirates
Muhammad Abulaish Jamia Millia Islamia, India
Amir Ahmad United Arab Emirates University, Saudi Arabia
Samad Ahmadi De Montfort University, UK
Mayer Aladjem Ben-Gurion University of the Negev, Israel
Francisco Martínez Álvarez Pablo de Olavide University of Seville, Spain
Eva Armengol IIIA CSIC, Spain
Zeyar Aung Masdar Institute of Science and Technology, United Arab Emirates

Vladan Babovic National University of Singapore, Singapore
Niranjan Balasubramanian University of Massachusetts Amherst, USA
Alberto Del Bimbo Università degli Studi di Firenze, Italy
Emanuele Di Buccio University of Padua, Italy
Maria Jose Aramburu Cabo Jaume I University, Spain
Rui Camacho Universidade do Porto, Portugal
Luis M. de Campos University of Granada, Spain
Maria Catalan Universitat Jaume I, Spain
Michelangelo Ceci University of Bari, Italy
Chien-Chung Chan University of Akron, USA
Keith C.C. Chan The Hong Kong Polytechnic University, Hong Kong, SAR China

Fei Chao	Xiamen University, China
Chien Chin Chen	National Taiwan University College of Management, Taiwan
Meng Chang Chen	Academia Sinica, Taiwan
Juan Manuel Corchado	University of Salamanca, Spain
Ingemar Cox	University of Copenhagen, Denmark
Bruce Croft	University of Massachusetts Amherst, USA
Roger Dannenberg	Carnegie Mellon University, USA
Kareem Darwish	Qatar Computing Research Institute, Qatar
Gianluca Demartini	University of Sheffield, UK
Thanh-Nghi Do	College of Information Technology, Can Tho University, Vietnam
Dejing Dou	University of Oregon, USA
Antoine Doucet	University of La Rochelle, France
Floriana Esposito	Università degli Studi di Bari, Italy
Iaakov Exman	The Jerusalem College of Engineering, Azrieli, Israel
Elisabetta Fersini	University of Milano-Bicocca, Italy
Philippe Fournier-Viger	University of Moncton, Canada
Fabrício Olivetti de França	Universidade Federal do ABC, Brazil
Ana Fred	Instituto de Telecomunicações/IST, Portugal
Ingo Frommholz	University of Bedfordshire, UK
Panorea Gaitanou	Ionian University, Greece
William Gasarch	University of Maryland, USA
Michael Gashler	University of Arkansas, USA
Susan Gauch	University of Arkansas, USA
Marco de Gemmis	Università degli Studi di Bari Aldo Moro, Italy
Rosario Girardi	UFMA, Brazil
Rosalba Giugno	University of Catania, Italy
Manuel Montes y Gómez	INAOE, Mexico
Nuno Pina Gonçalves	Superior School of Technology, Polytechnical Institute of Setúbal, Portugal
Cathal Gurrin	Dublin City University, Ireland
Yaakov Hacohen-Kerner	Jerusalem College of Technology (Machon Lev), Israel
Jianchao Han	California State University Dominguez Hills, USA
Jennifer Harding	Loughborough University, UK
Lynette Hirschman	The MITRE Corporation, USA
Frank Hopfgartner	University of Glasgow, UK
Xuanjing Huang	Fudan University, China
Beatriz de la Iglesia	University of East Anglia, UK
Bernard J. Jansen	Pennsylvania State University, USA
Szymon Jaroszewicz	Polish Academy of Sciences, Poland
Gareth Jones	Dublin City University, Ireland
Jae-Yoon Jung	Kyung Hee University, Republic of Korea
Mouna Kamel	IRIT, France
Mehmed Kantardzic	University of Louisville, USA
Ron Kenett	KPA Ltd., Israel

Claus-Peter Klas	GESIS Leibniz Institute for the Social Sciences, Germany
Natallia Kokash	Leiden University, The Netherlands
Donald Kraft	Colorado Technical University, USA
Steven Kraines	The University of Tokyo, Japan
Ralf Krestel	Hasso-Plattner-Institut, Germany
Nuno Lau	Universidade de Aveiro, Portugal
Anne Laurent	University of Montpellier 2, France
Carson K. Leung	University of Manitoba, Canada
Johannes Leveling	Elsevier, The Netherlands
Chun Hung Li	Hong Kong Baptist University, Hong Kong, SAR China
Lei Li	Florida International University, USA
Chun-Wei Lin	Harbin Institute of Technology Shenzhen Graduate School, China
Christina Lioma	University of Copenhagen, Denmark
Michel Liquiere	University of Montpellier II, France
Berenike Litz	Attensity Corporation, USA
Jun Liu	University of Ulster, UK
Rafael Berlanga Llavori	Jaume I University, Spain
Alicia Troncoso Lora	Pablo de Olavide University of Seville, Spain
Walid Magdy	Qatar Computing Research Institute, Qatar
Devignes Marie-Dominique	LORIA, CNRS, France
Sergio Di Martino	Università degli Studi di Napoli "Federico II", Italy
Edson T. Matsubara	UFMS, Brazil
Debora Medeiros	Universidade Federal do ABC, Brazil
Pietro Michiardi	EURECOM, France
Bamshad Mobasher	DePaul University, USA
Misael Mongiovi	Università di Catania, Italy
Stefania Montani	Piemonte Orientale University, Italy
Eduardo F. Morales	INAOE, Mexico
Yashar Moshfeghi	University of Glasgow, UK
Wolfgang Nejdl	L3S and University of Hannover, Germany
Raymond Ng	The University of British Columbia, Canada
Engelbert Mephu Nguifo	LIMOS, Université Blaise Pascal, France
Tae-Gil Noh	NEC Laboratories Europe, Germany
Giorgio Maria Di Nunzio	Università degli Studi di Padova, Italy
Mitsunori Ogihara	University of Miami, USA
Elias Oliveira	Universidade Federal do Espirito Santo, Brazil
Nicola Orio	Università degli Studi di Padova, Italy
Colm O'Riordan	NUI, Galway, Ireland
Rifat Ozcan	Turgut Ozal University, Turkey
Rui Pedro Paiva	University of Coimbra, Portugal
Krzysztof Pancerz	University of Rzeszow, Poland
Alberto Adrego Pinto	University of Porto, Portugal
Massimo Poesio	University of Essex, UK

Luigi Pontieri	National Research Council (CNR), Italy
Alfredo Pulvirenti	University of Catania, Italy
Marcos Gonçalves Quiles	Federal University of Sao Paulo, Brazil
Bijan Raahemi	University of Ottawa, Canada
Luis Paulo Reis	University of Minho, Portugal
Maria Rifqi	University Panthéon-Assas, France
Carolina Ruiz	WPI, USA
Ovidio Salvetti	National Research Council of Italy (CNR)s, Italy
Ismael Sanz	Jaume I University, Spain
Filippo Sciarrone	Open Informatica srl, Italy
Hong Shen	University of Adelaide, Australia
Zhongzhi Shi	Chinese Academy of Sciences, China
Fabricio Silva	FIOCRUZ - Fundação Oswaldo Cruz, Brazil
Andrzej Sluzek	Khalifa University, United Arab Emirates
Alan F. Smeaton	DCU, Ireland
Minseok Song	Pohang University of Science and Technology, Republic of Korea
Yu-Ru Syau	National Formosa University, Taiwan
Marcin Sydow	IPI PAN (and PJIIT), Warsaw, Poland
Andrea Tagarelli	University of Calabria, Italy
Kosuke Takano	Kanagawa Institute of Technology, Japan
Domenico Talia	University of Calabria and ICAR-CNR, Italy
Liang Tang	Florida International University, USA
Ulrich Thiel	Fraunhofer Gesellschaft, Germany
Kar Ann Toh	Yonsei University, Republic of Korea
Vicenc Torra	University of Skövde, Sweden
Juan-Manuel Torres-Moreno	Université d'Avignon et des Pays de Vaucluse, France
Manos Tsagkias	904Labs, The Netherlands
Theodora Tsikrika	Centre for Research and Technology Hellas, Greece
Evelyne Tzoukermann	The MITRE Corporation, USA
Domenico Ursino	University "Mediterranea" of Reggio Calabria, Italy
Jiabing Wang	South China University of Technology, China
Peiling Wang	The University of Tennessee Knoxville, USA
Harco Leslie Hendric Spits Warnars	Bina Nusantara University, Indonesia
Wouter Weerkamp	904Labs, The Netherlands
Theo van der Weide	Radboud University Nijmegen, The Netherlands
Leandro Krug Wives	Universidade Federal do Rio Grande do Sul, Brazil
Yinghui Wu	UCSB, USA
Yiyu Yao	University of Regina, Canada
Wai Gen Yee	Orbitz Worldwide, USA
Djamel Abdelkader Zighed	Lumière University Lyon 2, France

KDIR Additional Reviewers

Sabeur Aridhi	University of Trento, Italy
Pengwei Hu	The Hong Kong Polytechnic University, Hong Kong, SAR China
Roberto Interdonato	DIMES - Università della Calabria, Italy
Francesca Alessandra Lisi	Università degli Studi di Bari "Aldo Moro", Italy
Nicolás Rey-Villamizar	University of Houston, USA
Salvatore Romeo	UNICAL, Italy
Esaú Villatoro-tello	Universidad Autonoma Metropolitana Unidad Cuajimalpa, Mexico
Peiyuan Zhou	The Hong Kong Polytechnic University, Hong Kong, SAR China

KEOD Program Committee

Salah Ait-Mokhtar	Xerox Research Centre Europe, France
Raian Ali	Bournemouth University, UK
Carlo Allocca	FORTH Research Institute, University of Crete, Greece
Frederic Andres	Research Organization of Information and Systems, Japan
Francisco Antunes	Institute of Computer and Systems Engineering of Coimbra and Beira Interior University, Portugal
Chimay J. Anumba	Pennsylvania State University, USA
David Aveiro	University of Madeira/Madeira-ITI, Portugal
Panos Balatsoukas	University of Manchester, UK
Claudio de Souza Baptista	Universidade Federal de Campina Grande, Brazil
Jean-Paul Barthes	Université de Technologie de Compiègne, France
Teresa M.A. Basile	Università degli Studi di Bari, Italy
Sonia Bergamaschi	University of Modena and Reggio Emilia, Italy
Olivier Bodenreider	National Institutes of Health, USA
Patrick Brezillon	Pierre and Marie Curie (UPMC) University, France
Giacomo Bucci	Università degli Studi di Firenze, Italy
Gerhard Budin	University of Vienna, Austria
Vladimír Bureš	University of Hradec Kralove, Czech Republic
Ismael Caballero	Universidad de Castilla-La Mancha (UCLM), Spain
Caterina Caracciolo	Food and Agriculture Organization of the United Nations, Italy
Doina Caragea	Kansas State University, USA
Jin Chen	Michigan State University, USA
João Paulo Costa	Institute of Computer and Systems Engineering of Coimbra, Portugal
Constantina Costopoulou	Agricultural University of Athens, Greece
Christophe Cruz	Laboratoire Le2i, UMR 5158 CNRS, France
Sally Jo Cunningham	University of Waikato, New Zealand
Jan Dietz	Delft University of Technology, The Netherlands

Erdogan Dogdu   TOBB University of Economics and Technology,
                             Turkey
John Edwards   Aston University, UK
Nicola Fanizzi   Università degli studi di Bari "Aldo Moro", Italy
Catherine Faron-Zucker   University Nice Sophia Antipolis, France
Anna Fensel   STI Innsbruck, University of Innsbruck, Austria
Jesualdo Tomás   University of Murcia, Spain
  Fernández-Breis
Arnulfo Alanis Garza   Instituto Tecnologico de Tijuana, Mexico
James Geller   New Jersey Institute of Technology, USA
George Giannakopoulos   NCSR Demokritos, Greece
Rosario Girardi   UFMA, Brazil
Matteo Golfarelli   University of Bologna, Italy
Jane Greenberg   University of North Carolina at Chapel Hill, USA
Nicola Guarino   Institute of Cognitive Sciences and Technologies of the
                             Italian National Research Council (ISTC-CNR),
                             Italy
Choo Yun Huoy   Universiti Teknikal Malaysia Melaka, Malaysia
Sabina Jeschke   RWTH Aachen University, Germany
Achilles Kameas   Hellenic Open University, Greece
Dimitris Kanellopoulos   University of Patras, Greece
Sarantos Kapidakis   Ionian University, Greece
Pinar Karagoz   METU, Turkey
Kouji Kozaki   Osaka University, Japan
Deniss Kumlander   Tallinn University of Technology, Estonia
Antoni Ligeza   AGH University of Science and Technology, Poland
Adolfo Lozano-Tello   Universidad de Extremadura, Spain
Xudong Luo   Sun Yat-Sen University, China
Kun Ma   University of Jinan, China
Paulo Maio   Polytechnic of Porto, Portugal
Rocio Abascal Mena   Universidad Autónoma Metropolitana – Cuajimalpa,
                             Mexico
Riichiro Mizoguchi   Japan Advanced Institute of Science and Technology,
                             Japan
Andres Montoyo   University of Alicante, Spain
Claude Moulin   JRU CNRS Heudiasyc, University of Compiègne,
                             France
Ana Maria Moura   National Laboratory of Scientific Computing —
                             LNCC, Brazil
Azah Kamilah Muda   Universiti Teknikal Malaysia Melaka, Malaysia
Phivos Mylonas   Ionian University, Greece
Kazumi Nakamatsu   University of Hyogo, Japan
William Nelson   Devry University, USA
Erich Neuhold   University of Vienna, Austria
Jørgen Fischer Nilsson   Technical University of Denmark, Denmark
Matteo Palmonari   University of Milano-Bicocca, Italy

Laura Po   University of Modena and Reggio Emilia, Italy
Mihail Popescu   University of Missouri-Columbia, USA
Nives Mikelic Preradovic   University of Zagreb, Croatia
Violaine Prince   LIRMM-CNRS, France
Frank Puppe   University of Würzburg, Germany
Juha Puustjärvi   University of Helsinki, Finland
Amar Ramdane-Cherif   University of Versailles St. Quentin en Yvelines,
                                     France
Domenico Redavid   University of Bari, Italy
Thomas Risse   L3S Research Center, Germany
Genaina Rodrigues   University of Brasilia, Brazil
Colette Rolland   Université de Paris 1 Panthèon Sorbonne, France
M. Teresa Romá-Ferri   University of Alicante, Spain
Inès Saad   ESC Amiens, France
Fabio Sartori   University of Milano-Bicocca, Italy
Nuno Silva   Polytechnic of Porto, Portugal
Deborah Stacey   University of Guelph, Canada
Anna Stavrianou   Laboratoire Informatique de Grenoble, France
Cesar Augusto Tacla   Federal University of Technology in Parana, Brazil
Domenico Talia   University of Calabria and ICAR-CNR, Italy
Gheorghe Tecuci   George Mason University, USA
Orazio Tomarchio   University of Catania, Italy
George Tsatsaronis   Technische Universität Dresden, Germany
Shengru Tu   University of New Orleans, USA
Manolis Tzagarakis   University of Patras, Greece
Rafael Valencia-Garcia   Universidad de Murcia, Spain
Iraklis Varlamis   Harokopio University of Athens, Greece
Cristina Vicente-Chicote   Universidad de Extremadura, Spain
Bruno Volckaert   Ghent University, Belgium
Toyohide Watanabe   Nagoya Industrial Science Research Institute, Japan
Yue Xu   Queensland University of Technology, Australia
Gian Piero Zarri   Sorbonne University, France
Jinglan Zhang   Queensland University of Technology, Australia

KEOD Additional Reviewers

Fabio Benedetti   Unimore, Italy
Luca Gagliardelli   Università degli studi di Modena e Reggio Emilia, Italy
Fernanda Lima   Universidade de Brasilia, Brazil
Rohit Parimi   Motorola, USA
Rafael Peixoto   Grupo de Investigação em Engenharia do
                                     Conhecimento e Apoio à Decisão (GECAD),
                                     Portugal
Jean-Yves Vion-Dury   Xerox Research Centre Europe, France

KMIS Program Committee

Marie-Helene Abel	University of Compiègne, France
Shamsuddin Ahmed	University of Malaya, Bangladesh
Miriam C. Bergue Alves	Institute of Aeronautics and Space, Brazil
Rangachari Anand	IBM T.J. Watson Research Center, USA
Chimay J. Anumba	Pennsylvania State University, USA
Carlos Alberto Malcher Bastos	Universidade Federal Fluminense, Brazil
Sonia Bergamaschi	University of Modena and Reggio Emilia, Italy
Silvana Castano	Università degli Studi di Milano, Italy
Marcello Castellano	Politecnico di Bari, Italy
Xiaoyu Chen	School of Mathematics and Systems Science, China
Dickson K.W. Chiu	Dickson Computer Systems, Hong Kong, SAR China
Byron Choi	Hong Kong Baptist University, Hong Kong, SAR China
Ian Douglas	Florida State University, USA
Alan Eardley	Staffordshire University, UK
Joao Carlos Amaro Ferreira	ISEL, Portugal
Joan-Francesc Fondevila-Gascón	CECABLE (Centre d'Estudis sobre el Cable), UAO and UOC, Spain
Anna Goy	University of Turin, Italy
Francesco Guerra	University of Modena and Reggio Emilia, Italy
Renata Guizzardi	Federal University of Espirito Santo (UFES), Brazil
Jennifer Harding	Loughborough University, UK
Mounira Harzallah	LINA, France
Anca Daniela Ionita	University Politehnica of Bucharest, Romania
Nikos Karacapilidis	University of Patras and CTI, Greece
Radoslaw Katarzyniak	Wroclaw University of Technology, Poland
Helmut Krcmar	Technische Universität München, Germany
Dominique Laurent	Cergy-Pontoise University — ENSEA, France
Elise Lavoué	Université Jean Moulin Lyon 3, France
Kecheng Liu	University of Reading, UK
Heide Lukosch	Delft University of Technology, The Netherlands
Xiaoyue Ma	University of Xidian, China
Federica Mandreoli	University of Modena and Reggio Emilia Italy, Italy
Nada Matta	University of Technology of Troyes, France
Christine Michel	INSA-Lyon, Laboratoire LIRIS, France
Michele M. Missikoff	Institute of Sciences and Technologies of Cognition, ISTC-CNR, Italy
Owen Molloy	National University of Ireland, Galway, Ireland
Jean-Henry Morin	University of Geneva, Switzerland
Minh Nhut	Institute for Infocomm Research (I2R), A*STAR, Singapore
Augusta Maria Paci	National Research Council of Italy, Italy

Wilma Penzo	University of Bologna, Italy
José de Jesus Pérez-Alcázar	University of São Paulo (USP), Brazil
Milly Perry	The Open University, Israel
Erwin Pesch	University of Siegen, Germany
Filipe Portela	Centro ALGORITMI, University of Minho, Portugal
Arkalgud Ramaprasad	University of Illinois at Chicago, USA
Edie Rasmussen	University of British Columbia, Canada
Marina Ribaudo	Università di Genova, Italy
Colette Rolland	Université de Paris 1 Panthèon Sorbonne, France
Masaki Samejima	Osaka University, Japan
Marilde Terezinha Prado Santos	Federal University of São Carlos, Brazil
Conrad Shayo	California State University, USA
Paolo Spagnoletti	LUISS Guido Carli University, Italy
Malgorzata Sterna	Poznan University of Technology, Poland
Deborah Swain	North Carolina Central University, USA
Esaú Villatoro Tello	Universidad Autonoma Metropolitana (UAM), Mexico
Bhavani Thuraisingham	University of Texas at Dallas, USA
Shu-Mei Tseng	I-SHOU University, Taiwan
Martin Wessner	Darmstadt University of Applied Sciences, Germany
Uffe K. Wiil	University of Southern Denmark, Denmark
Leandro Krug Wives	Universidade Federal do Rio Grande do Sul, Brazil
Jie Yang	Shanghai Jiao Tong University, China

KMIS Additional Reviewers

| Fabio Benedetti | Unimore, Italy |
| Diego Magro | University of Turin, Italy |

Invited Speakers

Jan Vanthienen	KU Leuven, Belgium
João César das Neves	Universidade Católica Portuguesa (UCP), Portugal
Giancarlo Guizzardi	Federal University of Espirito Santo, Brazil/(LOA), Institute for Cognitive Science and Technology (CNR), Italy
Ralf Bogusch	Airbus Defence and Space, Germany

Contents

Invited Papers

Business Ethics as Personal Ethics

João César das Neves[✉]

Catolica Lisbon School of Business and Economics,
Universidade Católica Portuguesa, Palma de Cima, 1649-023 Lisbon, Portugal
jcn@ucp.pt

Abstract. Business ethics is a major issue for most, if not all companies. This paper attempts a clarification of the composite of forces and influences involved in the business ethics world. Its performance is then assessed, resulting in a relative failure of the huge efforts of the business world in recent decades, as the public doubts the moral sincerity of managers.

The diagnostics of this credibility disease points to a flawed trust on the mechanic and judicial approach to ethics, preferring rules and measurements to character and intentions. The solution presented is the use of a virtue-based ethics, returning to the personal elements of morality.

Keywords: Ethics · Business · Virtue

1 Introdution

Business ethics or corporate social responsibility is having a triumphant moment. Firms everywhere are supposed to comply with some form of its demands. Governments are ever tougher, with innumerous rules and regulations imposed on companies by the law. But, more than these, firms are also expected to follow some complex moral procedures, informally determined by society. The main question here is that, unlike laws, these tend do be diverse, vague and ambiguous, creating all sorts of doubts, debates, fights and ... yes, ethical questions. It's hard to be in business these days.

Economic ventures, like all human initiatives, have always been subject to morality. But this is very different from the complex procedures today implied by what is usually called «business ethics». If you are a manager trying to do the right thing with your life, a clarification is both urgent and hard.

The present paper will attempt an analysis of the ethical problems involved in contemporaneous business ethics. Section 1 will describe the main elements involved in the concept, attempting a portrayal rather than a definition. Section 2 will present the original difficulties involved in introducing ethics to the modern boardrooms, and the way this task was generally formulated. Section 3 draws the failure hypothesis, the alleged gap between rhetoric and reality in ethical behavior of firms. This is done in a clear-cut form, presenting some evidence to support it. More than generalized opinions, data from public surveys will be the cornerstone of the argument. The final section will briefly describe the classical virtue ethics approach to morality, pointing some guidelines derived from general personal ethics as a framework for business ethics.

© Springer International Publishing AG 2016
A. Fred et al. (Eds.): IC3K 2015, CCIS 631, pp. 3–12, 2016.
DOI: 10.1007/978-3-319-52758-1_1

2 What Is Business Ethics?

Business ethics is a very large and complex field. At least three very different realities are included under this classification, as one of the founders has stated some years ago: «The term business ethics is used in at least three different, although related, senses. (...) The primary sense of the term refers to recent developments and to the period, since roughly the early 1970s, when the term 'business ethics' came into common use in the United States. Its origin in this sense is found in the academy, in academic writings and meetings, and in the development of a field of academic teaching, research and publication. That is one strand of the story. As the term entered more general usage in the media and public discourse, it often became equated with either business scandals or more broadly with what can be called "ethics in business." In this broader sense the history of business ethics goes back to the origin of business, again taken in a broad sense, meaning commercial exchanges and later meaning economic systems as well. That is another strand of the history. The third strand corresponds to a third sense of business ethics which refers to a movement within business or the movement to explicitly build ethics into the structures of corporations in the form of ethics codes, ethics officers, ethics committees and ethics training.» [1].

There are thus three different elements to be considered inside the notion. The first aspect could be named «academic business ethics». This is to be evaluated scientifically, and this paper is a clear example of it. The second type, «behavioral business ethics», is the concrete performance of companies, to be evaluated ethically. An example is the topic of this paper. Finally we have «structural business ethics», to be evaluated managerially. This is the «industry of ethics», the enormous complex of institutions, documents, initiatives and efforts developed to make companies obey ethical instructions. As the first of the senses, pertaining to intellectual endeavors, is alien to everyday business life, the following considerations will be centered on the other two.

The role of business ethics in contemporaneous business life is easily identifiable, on either of these two strands, as managing companies require the second sense of the expression, using the third as instrument for that purpose. The main problem of current business ethics is that, while the third managerial sense of the expression is very prosperous and flourishing, many doubts remain about the operational relevancy of it at the second sense. All companies have an intense use of the industry of ethics, being very committed to the application of moral codes, commissions, auditing, prizes and many other similar mechanisms. Nevertheless, in spite of intense use of methods devised to implement good behavior for managers, strong misgivings linger about the actual performance of firms' staff. This creates a general climate of failure around one of the most remarkable movements in business history.

It is very easy to establish the first part of the above statement, as the affluence of ethical endeavors is very visible. In 2008 the reputed British magazine *The Economist* wrote: «THE CSR industry, as we have seen, is in rude health. Company after company has been shaken into adopting a CSR policy: it is almost unthinkable today for a big global corporation to be without one.» [14 p. 21].

This «rude health» has a very clear and imposing translation in corporate life. The same journal stated, eight years before the previous comment: «In America there is now a veritable industry, complete with consultancies, conferences, journals and "corporate conscience" awards. Accountancy firms such as PricewaterhouseCoopers offer to "audit" the ethical performance of companies. Corporate-ethics officers, who barely existed a decade ago, have become *de rigueur*, at least for big companies. (...) As many as one in five big firms has a full-time office devoted to the subject» [12].

This success is the result of a long evolution which may be summarized in another article of the same review: «Although the first course in business ethics was offered by Harvard Business School back in 1915, it is only since the mid-1980s that business schools have truly taken the subject to their hearts. Blame this renewed interest on a string of business scandals: Drexel Burnham Lambert, Guinness, Salomon Brothers, Robert Maxwell and Recruit, not forgetting Olivetti, Fiat and a big chunk of the rest of corporate Italy; the list can seem endless. Market-driven as ever, business schools have risen to the challenge. In America alone, on one estimate, more than 500 courses on business ethics are on offer; 90% of the country's business schools now teach the subject. Globally, more than 20 research units now study the topic, and business-ethics journal abound» [11].

This very simple description of a very complex evolution already points directly to the simple idea which underlies the second aspect of our main thesis: corporate scandals were crucial elements in the process leading to the development of business ethics. The moral breakdown of companies was the driving force behind their ethical awareness. But, as these crimes keep emerging, in spite of the large investments made by companies in ethical processes and instruments, it is not surprising the large skepticism about the relevancy of the industry of ethics.

Money corrupts. This is the basic concept underlying most of the ethical misgivings about business. It is a very old and influential notion, In Antiquity, economics was a matter for slaves. Citizens dealt with politics, war, art and philosophy. Buying, selling and producing were the slave's job. Progress and industrialization changed the cultural perception of economic endeavors, but still today the same corruption is visible, as most of the movie villains come from companies. It is not hard to find, in all countries and epochs, lots of examples to illustrate that money corrupts. Both the corporate scandals and the «robber barons», magnates accused of serious frauds, occupy large sections of every newspapers space.

Consequently, the disbelief about the real relevancy of the large movements towards business ethics is evident. Returning to the initial quote from *The Economist*, the reviewer, although observing the rude health of the industry of health, is unable to hide a strong cynicism: «All this is convoluted code for something simple: companies meaning (or seeming) to be good (...) because with a few interesting exceptions, the rhetoric falls well short of the reality.» [14 p. 3–4].

3   The Task of Business Ethics

What is the origin of such a paradoxical situation? In spite of all their efforts, why are firms unable to convince the public about the sincerity of their ethical accomplishments? There are good reasons for all this.

The original purpose of business ethics was to introduce moral sense in one of the toughest fields in human endeavor. Due to the ruthless nature of commerce, directing managers, entrepreneurs and business leaders towards a more ethical sense of life is something even today many consider impossible.

Markets are intense, merciless places. Competition is brutal, dramatic and unrepentant. Business Ethics knew it would never be able to change that. Ethics in management would have to proceed in spite of those conditions, not outside them. The only way managers could be made to pay attention was to be objective, pragmatic, scientific, and business-like. The only way business ethics was going to succeed was to be «Ethics without the sermon».

This last expression comes from one of the seminal contributions in this fields, the paper by Laura L. Nash of the same title in the Harvard Business Review in 1981 [8]. This article, reprinted in many of latter readings references of the area, presents the basic attitude of the then primitive practitioners. One of the collections, when introducing the paper, expressed this stance clearly: «When academic philosophers begin to discuss ethics with those who have more practical concerns (such as corporate executives), radically different styles, approaches and biases become quite apparent. Taking these differences as her cue, Laura L. Nash offers a set of twelve questions that, while free of philosophical abstracion, nonetheless embody the central concerns of ethical reasoning as it is applied in business» [6 p. 38].

The problem is thus that, from the start, business ethics had a complex and intricate relation with the philosophical field of ethics, of which it was suppose to be a part. Unlike any other of the many sector applications of morality, corporate ethics intended to be free from sermons. As Nash states in the mentioned paper: «Like some Triassic reptile, the theoretical view of ethics lumbers along in the far past of Sunday School and Philosophy 1, while the reality of practical business concerns is constantly measuring a wide range of competing claims on time and resources against the unrelenting and objective marketplace» [8 p. 80].

Thus, facing one of the most morally demanding fields of human activity, business ethics was also voluntarily alienated from the epistemological foundations of its own intellectual area. Business ethics had to be different from all other types of ethics. The reasons for this came, as we saw, from the specific characteristics of the same morally demanding field. Corporate executives were allegedly unable to dialog with academic philosophers.

Nobody ever explained why managers were so different from doctors, soldiers, and many other professionals which, also busy with practical concerns, have for ages used the ethical elaborations about their actions created by philosophical experts. Such a dialog was present in all other activities. But everybody took for granted that business ethics needed a different approach. Even if it was hard to believe this original and uprooted endeavor would have great results. But only with «Ethics without the sermon» would managers pay heed. Ethics, if it was to enter the boardroom, had to gain a business-like appearance. This meant avoiding references to conscience, character, spirituality, purposes and aims, by stressing the mechanisms and structures inside the firm. Managers wouldn't have it any other way.

Actually, this ethical concern was seen by them as just one among many other demands on companies. As time went, on and societies got more complex, firms had to incorporate many restrictions on their procedures. Labor rights, environment legislations, costumers' protection, quality controls and operational enhancements got to be normal business procedures in the fight for profits. Ethics was just going to be one more element. This integration in the spirit of commercial ventures was what business ethics had to achieve. And did achieve. As long as ethics got to be something like accounting or sanitation, determined by some clear procedures and rules.

In the process, both managers and ethical experts lost sight of the specificity of ethics. Morality is very different from other social demands, as it pertains to them all. By adopting a functional approach to ethics, in the effort to escape the sermons, it can be said business ethics stopped being a part of ethics. This is explicitly stated in some of the most relevant references of the discipline, although hiding the paradoxical consequences.

For example, the already quoted handbook states as its purpose a strange aim: «The goal of ethics education is not character building; but rather, like all college course work, they attempt to share knowledge, build skills, and develop minds» [6 p. 7]. It is thus obvious that the «the goal of ethics education» is not ethics, but something else.

Some researchers are even more explicit, openly stating the purpose of having ethical companies, while precluding the need for ethical managers: «What such proposals for the reform of corporate governance seek to do is enhance the ethical performance of businesses through organizational mechanisms for controlling the behaviour of managers rather than through making those managers morally better people. By changing those mechanisms, it is hoped that managers will be induced to run businesses in ways that are morally preferable to the ways they would otherwise run them.» [4 p. 266]. Many persons, even if not experts in morality, would consider strange such a goal. Ethics is, above all else, a personal attitude and option. How can it be possible to have ethical behavior with no reference to character and conscience? But inside the complex world of management and corporate studies, where results and efficacy are the absolute rules, it was easy to lose sight of the deeper elements involved.

This is even clearer in the explanation of the process of providing ethics, according to the same paper: «So in the sense of looking to organizational structures rather than managerial attitudes as determinants of morally desirable outcomes, this is a strategy which looks to the business itself rather than its managers... So in looking to such mechanisms, business ethics can not only achieve its ultimate aim of morally enhancing business activity without resort to the very fraught and dubious route of morally improving managers, but it can, in principle at least, achieve that aim more effectively» [4 p. 266].

As dealing with persons is seen as a «fraught and dubious route» for moral improving, the good methods are merely corporate mechanisms, which induce such behaviors without any personal participation. Is this even possible? Were such approaches recommended or even feasible in any other aspects of management? Would any manager consider reasonable the implementation of a structure to ensure profits without the contribution of good business decisions? And are these possible without good businessmen? But in ethics things were thought to be different. In fact, as the conclusion states, «To want business ethics to result in the moral improvement of business activity is correct; to expect it to do so merely through the moral improvement of managers as people is, I submit, profoundly unrealistic» [4 p. 264].

Although a bit extreme, this paper states what was and is a very influential mind-set among business ethics theoreticians and practitioners. The Polish poet Stanisław Lec famously asked «Is it a progress if a cannibal is using knife and fork?» [5 p. 78]. Much of today's business ethics could be reduced to attempts to tend to the table manner of business cannibals without changing their diet.

4   The Failure of of Business Ethics?

The surge of business ethics in the latter part of the 20th century followed methods closer to business than ethics. The result, as would be expected, is practical failure among an institutional profusion of mechanisms. Studies repeatedly demonstrate clearly this outcome: «Many scholars view ethics codes as having minimal impact on ethical behaviour within organizations» [9 p. 219]. «It seems that ethical codes are an inferior document in most organizations; it does not really matter whether they exist or not» [7 p. 208].

When faced with the scandals which, has stated, continue to motivate the discipline, researchers repeatedly find the same problem: «Enron ethics means (still ironically) that business ethics is a question of organizational "deep" culture rather than of cultural artifacts like ethics codes, ethics officers and the like» [10 p. 243]. The probable product of such an approach had to be failure.

The economic journal previously quoted is also the one stating clearly the problem. «IF YOU believe what they say about themselves, big companies have never been better citizens. In the past decade, "corporate social responsibility" (CSR) has become the norm in the boardrooms of companies in rich countries, and increasingly in developing economies too. Most big firms now pledge to follow policies that define best practice in everything from the diversity of their workforces to human rights and the environment.» [13]. The description of the influence of ethical industry in corporate culture is preceded by a poisonous «if you believe what they say about themselves», which immediately ethically discredits everything which will be said afterwards. Managers are always seen as wolves in sheep clothing.

This may be obvious, but is still puzzling. If this is true, then the money and efforts spent on ethics represents one of the largest wastes in the history of business. Never was so much squandered by so many with so little results. But where the results really so meager?

One way to access the evolution of the ethical image of business and corporations is to consider two of the numerous surveys conducted on American attitudes by the Gallup Organization. The ones of interest for our quest are those assessing «Confidence in institutions» and «Honesty of professions». As these are yearly surveys which go back to the early 1970s, they show consistent results on the opinions of American citizens. This exercise does not aim establish any kind of empirical analysis, but mere illustration. As the date covers the whole period of the expansion of business ethics industry, one would expect to perceive some kind of positive impact over the image of companies and managers.

In what regards the first survey, the question is «Please tell me how much confidence you, yourself, have in each one – a great deal, quite a lot, some, or very little?» [2].

The questionnaire covers 16 institutions, which range from the «President» and «Congress» to «The church or organized religion» and the «The Military». The results relevant for us are «D. Banks», «I. Organized labor», «N. Big business» and «O. Small business». Table 1 summarizes the results of the survey from 1973 to 2015, showing the percentage of people answering "great deal" and "quite a lot" to the above question.

Table 1. Galllup poll on confidence in economic institutions.

	1973	1975	1977	1979	1981	1983	1984	1985	1986	1987	1988	1989	1990
D	-	-	-	60	46	51	51	51	49	51	49	42	36
I	30	36	39	36	28	26	30	28	29	26	26	-	27
N	26	34	33	32	20	28	29	31	28	-	25	-	25
O	-	-	-	-	-	-	-	-	-	-	-	-	-
	1991	1992	1993	1994	1995	1996	1997	1998	1999	2000	2001	2002	2003
D	32	30	37	35	43	44	41	40	43	46	44	47	50
I	25	22	26	26	26	25	23	26	28	25	26	26	28
N	26	22	22	26	21	24	28	30	30	29	28	20	22
O	-	-	-	-	-	-	63	57	-	-	-	-	-
	2004	2005	2006	2007	2008	2009	2010	2011	2012	2013	2014	2015	
D	53	49	49	41	32	22	23	23	21	26	26	28	
I	31	24	24	19	20	19	20	21	21	20	22	24	
N	24	22	18	18	20	16	19	19	21	22	21	21	
O	-	-	-	59	60	67	66	64	63	65	62	67	

The results span almost all the possible range of answers. Small Business are at the top of trustable institutions, with percentages above 60%. The rise they witness recently has taken them almost to the level of the Military, the highest ranking of all organizations in the survey. «Organized labor» and «Big Business» are at the other extreme, in the bottom of the rankings bellow 25%. In the middle, Banks follow a very dramatic fluctuating path, obviously due to financial crises. They had drastic falls in the second part of the 1980s and the first decade of the new century, having recovered during the 1990s and less so since 2009. It is rather.

In such a diverse scenario, are there any discernible impacts the business ethics evolution? The lack of a systematic tendency is the message this paper should take from this general picture. It is rather obvious there is no visible effect of the important investments made in business ethics in the last decades. If there is any trend visible since the 1970s, where there were no such mechanisms implemented, it is negative.

The second survey relevant for our quest is about honesty of professions. The particular query is «Please tell me how you would rate the honesty and ethical standards of people in these different fields – very high, high, average, low, or very low?» [3]. Again the results are yearly from the mid-1970s. Among the many occupations analyzed, the five which seem to be more relevant for our search are the ones directly related to managing tasks: «business executives», «advertising practitioners», «car salespeople», «insurance salespeople» and «stockbrokers». Table 2 presents the percentage of persons saying "high" and "very high" to the above question.

Table 2. Galllup poll on honesty of economic professions.

	1976	1977	1981	1983	1985	1988	1990	1991	1992	1993	1994
Business executives	20	19	19	18	23	16	25	-	18	-	-
Bankers		39	39	38	38	26	32	30	27	28	27
Advertising pract.	11	10	9	9	12	7	12	19	11	14	10
Car salespeople	-	8	6	6	5	6	6	-	5	-	-
Insurance salesp.	-	15	11	13	10	10	13	-	9	-	-
Stockbrokers	-	-	21	19	20	13	14	-	13	-	-
	1995	1996	1997	1998	1999	2000	2001	2002	2003	2003	2004
Business executives	19	17	20	21	23	23	25	16	17	18	20
Bankers	27	26	34	30	30	37	34		36	35	36
Advertising pract.	10	11	12	10	9	10	11	14	9	12	10
Car salespeople	8	8	5	8	7	8	-	6	7	9	8
Insurance salesp.	11	11	12	11	10	10	13	-	-	12	-
Stockbrokers	16	15	18	19	16	19	19	-	12	15	-
	2005	2006	2007	2008	2009	2010	2011	2012	2013	2014	2015
Business executives	16	18	14	12	12	15	18	21	22	17	17
Bankers	41	37	35	23	19	23	25	28	27	23	25
Advertising pract.	11	11	6	10	11	11	11	11	14	10	10
Car salespeople	7	5	7	6	7	7	7	8	9	8	8
Insurance salesp.	-	13	-	-	10	-	-	15	-	-	-
Stockbrokers	16	17	-	12	9	-	12	11	-	-	13

Again the results are very revealing. In this survey economic occupations are less favorably assessed than most others professions. None of them ever reached the 50% level. The best, around 40% are insurance salespeople and bankers, this last before the large tumble registered in 2008–2009. All other professions mentioned cluster around the lower 10's, with car salespeople reaping the lowest value of all, frequently even for the whole sample.

Once again, there is no visible effect on the trend from the growing concern with business ethics in the recent decades. A positive drift is nowhere visible. A possible conclusion is that, in what concerns public opinion, firms could have maintained the 1970s level of expenditure on ethics, without missing much on their credibility.

5   How to Be Ethical?

Previous analysis indicates there are some serious problems in the realm of business ethics. It is a field where companies have recently invested a lot of money and time, using very sophisticated methods and mechanisms. At the same time universities and research centers have developed a large literature on the subject. But, in spite of all this, the image of businesses has not improved and the number of scandals was maintained, if not increased. There are few other examples of such big investments in companies with so diminutive effects.

Of course these observations do not amount to anything like a scientific proof of the failure thesis. It is possible to argue business ethics was very successful and influential, as things would have been much worst without it. But, nevertheless, there is the lingering suggestion that something should be done to improve the efficacy of the endeavor in the future.

In this brief revision, some obvious aspects leap to our attention. As said, the efforts in the field of business ethics had, from the start, chose an innovative method to improve ethical behavior, attempting be «without the sermon». Thus, the problem might come from this original sin. This stream of morality has always attempted very different methods from all the others lines of applied ethics. Maybe a more regular and traditional approach would be more proficient. After all, that was the way all others fields of human activity proceeded.

It is important to clarify that, whatever the options of the field, applied business behavior is, first and foremost, a personal decision. In order for it to be ethical, it is thus a question of character, of attitude and virtue. To have good companies we need good persons in companies, like there are no profits without good businesspersons, because mechanisms are not enough.

This is not to say that all the contributions of these decades of efforts in the field of business ethics are useless and should be discarded. The mechanisms implemented, codes, commissions, courses, prizes and other elements, are positive and should be kept. But their limits should also be clarified.

In order to understand the relative importance of the several elements present in human behavior, one should analyze the process of ethical decisions. The crucial master in all moral situations is the conscience of the decision maker. The person responsible for the choice is the relevant ethical agent. That person, committed to be ethical, tries to follow the best path available. It is true that, in order to operate, the conscience must be well formed and informed, must know what is expected, what is the right way to proceed. That is where the mechanisms provided by business ethics are relevant.

All methods and efforts of recent decades in business ethics are important as guides for personal consciences. Ethical decision gains a lot from the knowledge of rules, examples, advices and other similar references. Thus, the various mechanisms implemented in companies to promote ethics are very rich contributions towards a formed and informed conscience of managers. But this, in itself represents merely a preparation for the ethical decision. That only takes place in the intimacy of the manager's conscience. And if the person involved has a flawed character, the huge paraphernalia will be dumb and void to influence the real and specific result.

Again it should be noted that this characteristic is very similar to the process in all others corporate decisions. In finance, investment, operation and marketing there are also a lot of methods and mechanisms created to help managers decide. But the final quality of business depends crucially on the personal abilities of the persons managing and deciding. In ethics the process is similar, even if this was mostly omitted in recent decades.

When recruiting new managers, companies are usually very careful in examining all the professional capabilities of candidates, to make sure the hiring gets the best elements. But, because they are confident the ethical mechanisms implemented are enough, they tend to be somewhat sloppy and in what pertains the ethical postures of candidates. Companies want ambitious recruits, filled with creativity, stamina and leadership, but forget about their honesty, virtue and reliability. The results are visible.

Business ethics registered a very dramatic and relevant evolution in recent decades. Most of the advances registered were very useful and relevant, and much was learned about the moral conduct of managers. The only remaining element to be introduced is the moral attitude of the persons managing the companies. Because ethics is always a personal purpose.

References

1. De George, R.T.: History of business ethics. Paper Delivered at The Accountable Corporation, (Third Biennial Global Business Conference) (2005). http://www.scu.edu/ethics/practicing/focusareas/business/conference/presentations/business-ethics-history.html
2. Gallup, Inc. http://www.gallup.com/poll/1597/confidence-institutions.aspx
3. Gallup, Inc. http://www.gallup.com/poll/1654/honesty-ethics-professions.aspx
4. Kaler, J.: Positioning business ethics in relation to management and political philosophy. J. Bus. Ethics **24**, 257–272 (2000)
5. Lec, S.J.: Unkempt Thoughts. St. Martin's Press, New York (1962)
6. Madsen, P., Shafritz, J. (eds.): Essentials of Business Ethics. Penguim Books, New York (1990)
7. Marnburg, E.: The behavioural effects of corporate ethical codes: empirical findings and discussion. Bus. Ethics A Eur. Rev. **9**(3), 200–210 (2000)
8. Nash, L.L.: Ethics without the sermon. Harv. Bus. Rev. **59**, 79–90 (1981)
9. O'Dwyer, B., Madden, G.: Ethical codes of conduct in Irish companies: a survey of code content and enforcement procedures. J. Bus. Ethics **63**, 217–236 (2006)
10. Sims, R.R., Brinkmann, J.: Enron ethics (or: culture matters more than codes). J. Bus. Ethics **45**(3), 243–256 (2003)
11. The Economist: How to be ethical, and still come top, 5 June 1993
12. The Economist: Doing well by doing good, 22 April 2000
13. The Economist: In search of the good company, 6 September 2007
14. The Economist: Just good business;. a special report on corporate social responsibility, 19 January 2008

Automatic Generation of Poetry Inspired by Twitter Trends

Hugo Gonçalo Oliveira[✉]

CISUC, Department of Informatics Engineering, University of Coimbra,
Coimbra, Portugal
hroliv@dei.uc.pt

Abstract. This paper revisits PoeTryMe, a poetry generation platform, and presents its most recent instantiation for producing poetry
inspired by trends in the Twitter social network. The presented system
searches for tweets that mention a given topic, extracts the most frequent words in those tweets, and uses them as seeds for the generation of
new poems. The set of seeds might still be expanded with semantically-
relevant words. Generation is performed by the classic PoeTryMe system, based on a semantic network and a grammar, with a previously
used generate &test strategy. Illustrative results are presented using different seed expansion settings. They show that the produced poems use
semantically-coherent lines with words that, at the time of generation,
were associated with the topic. Resulting poems are not really about
the topic, but they are a way of expressing, poetically, what the system
knows about the semantic domain set by the topic.

Keywords: Computational creativity · Creative systems · Poetry generation · Social media

1   Introduction

Creative systems are computer programs that exhibit behaviours that would be
deemed as creative by unbiased observers [1]. Such behaviours are often rendered in the form of artefacts that go from visual art [2] to linguistic creativity including, but not limited to verbally-expressed humor [3], narratives [4],
metaphors [5], neologisms [6], slogans [7] or poetry, one of the most popular
tasks in this subfield. Poetry generation is a kind of natural language generation
where the resulting text can be seen as a poem. This can be achieved by the
presence of features, such as a regular metre, rhymes, or a figurative language.

PoeTryMe [8,9] is a poetry generation platform that produces lines with the
help of a grammar and a set of relation instances, and combines them according
to a pre-defined strategy, towards the creation of a poem. PoeTryMe has a
versatile architecture that provides a high level of customisation and can be the
starting point for the development of different poetry generation systems. Several
of its components can be changed: its semantic knowledge, the line templates, the

© Springer International Publishing AG 2016
A. Fred et al. (Eds.): IC3K 2015, CCIS 631, pp. 13–27, 2016.
DOI: 10.1007/978-3-319-52758-1_2

generation strategies and, of course, the poem configuration. The combination of all these components enables the generation of diverse poems, thus contributing to a positive perception of creativity.

This paper presents a new instantiation of PoeTryMe where the generation of a poem is inspired by information circulating in the Twitter[1] social network. The process starts with a given topic, which used to retrieve associated words from Twitter, then used as seeds for poetry generation. Resulting poems are not clearly about the topic, but they are, at least, inspired by it, and an abstract connection is usually present. A bot was developed to publish the creations of the presented system in Twitter. Therefore, we see Twitter's role in this process also as a way of continuously retrieving different seeds, and thus contributing to the generation of more diverse poems every time.

In the remaining of the paper, related work, mostly on poetry generation, is first reviewed. Then, a short description of PoeTryMe is provided. Before concluding and discussing cues for future work, the specificities of this instantiation are presented, together with some examples and a critical view.

2 Related Work

Computational Creativity is a multidisciplinary endeavour at the intersection of the fields of artificial intelligence, cognitive psychology, philosophy, and the arts[2]. It is driven towards modelling, simulating or replicating creativity, using a computer, either to: (i) construct programs capable of human-level creativity; (ii) better understand human creativity and formulate an algorithmic perspective on creative behaviour in humans; (iii) design programs that can enhance human creativity without necessarily being creative themselves.

Poetry generation systems are artificial systems that produce text with poetic features and has thus creative value. The automatic generation of poetry is a complex task, as it involves several levels of language (e.g. phonetics, lexical choice, syntax and semantics) and usually demands a considerable amount of input knowledge. However, not all of those levels have to be strictly addressed. On the one hand, poetic text does not have to be extremely precise [10], as several rules, typically present in the production of natural language, need to (or should) be broken [11]. On the other hand, poetry involves a high occurrence of interdependent linguistic phenomena where rhythm, metre, rhyme and other features like alliteration, sentiment, or figurative language play an important role. For instance, it is sometimes enough to have a less clear message, in a trade-off for a pleasant sound given by a highly regular metre and rhymes.

---

[1] https://twitter.com.

[2] Check the website of the Association for Computational Creativity at http://computationalcreativity.net/.

Several poetry generation systems are based on poem or line templates, but most of them go further and combine the previous with other techniques (e.g. [12,13]). Templates learned from human-created poetry are often sequences of words with gaps to be filled by the system, but they can also be sequences of parts-of-speech [14].

Produced word sequences usually evolve to meet the desired constraints, which often include a stress pattern (metre) and may additionally define the position of rhymes, syntactic rules, semantic constrains, and other features like the presence of alliteration or the use of figurative language. Evolution can be made through a generate-and-test approach [10,15] or it can rely on evolutionary algorithms [11,16]. Other approaches include case-based reasoning [17], probabilistic language models [18], constraint programming [13], or multi-agent systems [19].

Besides exhibiting poetic features, produced text should obey linguistic conventions and convey a conceptual message, meaningful under some interpretation [11]. The handling of linguistic rules is typically achieved with the help of natural language processing tools, such as morphological lexicons or grammars. On the other hand, meaningfulness is more subjective and difficult to achieve. Towards this goal, different systems have handled semantics differently. Some start generation from a textual document [12,20,21] or a set of seed words [22–24] to constrain the space of possible generations, in a way that the poem should use these exact words, or others semantically associated. The choice of relevant words may be achieved either by exploring models of semantic similarity, extracted from corpora [13,22,24], with the help of lexical-semantic knowledge bases [14,23], or both [12].

Poetry generation has been mainly addressed for English, but there are attempts in other languages, including Spanish [10,17], Basque [14], Finnish [13,20], Chinese [24], Indonesian [25], or Bengali [26], among others. Our original effort targeted Portuguese [8], which is also the target language of the present work.

3   PoeTryMe

PoeTryMe [8,9] is a poetry generation platform, on the top of which different strategies for poetry generation can be implemented. It relies on a modular architecture (see Fig. 1), which enables the independent development of each module and provides a high level of customisation, depending on the needs of the system and ideas of the user or developer. Among other parameters, users may define the semantic network to use, the rules of the generation grammar, the transmitted sentiment, the generation strategy and the structure of the poem. Developers may reimplement some of the modules and reuse the others.

A Generation Strategy organises lines, such that they suit, as much as possible, the structure of a poetic form and exhibit certain features. A structure file sets the number of stanzas, lines per stanza and of syllables in each line of the poem. An instantiation of the Generation Strategy does not generate the lines, but exploits the Sentence Generator module to retrieve natural language

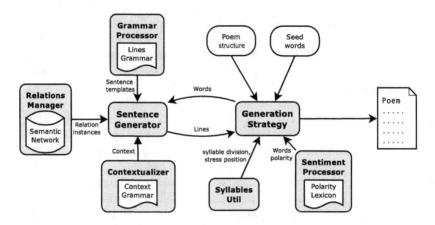

Fig. 1. The architecture of PoeTryMe.

fragments, which might be used as such. Each strategy may differ on the number of fragments requested from the Sentence Generator at any time, and how they are organised into the poem structure, considering for this purpose features like metre, rhyme, coherence between lines or others, depending on the desired goal.

Syllable-related features are evaluated with the help of the Syllable Utils. Given a word, this module may be used to divide it into syllables, to find its stress, or to extract its termination, useful to identify rhymes.

The Sentence Generator is a core module for PoeTryMe. It generates semantically-coherent natural language fragments, with the help of: (i) a semantic network, managed by the Relations Manager, that connects words according to relation predicates; and a generation grammar, processed by the Grammar Processor, with textual renderings for the generation of lines that express semantic relations. The generation of a line is a three-step interaction:

1. A random relation instance, in the form of a $triplet = \{word_1, predicate, word_2\}$, is retrieved from the semantic network. To narrow the space of possible generations, a set of seed words can be provided to the Relations Manager. This set defines the generation domain, represented by a subgraph of the main network that will contain all the triplets involving seed words. A surprise factor, ν, sets the probability of selecting also triplets involving nodes that are two levels far from the seeds.
2. A random rendering for the $triplet$'s predicate is retrieved from the grammar. Grammar rules are natural language renderings of predefined semantic relations. So, there must be a direct mapping between the relation names, in the graph, and the rules' name, in the grammar. Besides terminal tokens, that will be present in the poem without change, rules have placeholders that indicate the position of the relation arguments ($< \mathtt{arg1} >$ and $< \mathtt{arg2} >$). A simple example of a valid rule set, with three hypernymy patterns, is shown in Fig. 2.

3. The resulting fragment is returned after inserting the arguments of the *triplet* in the proper placeholders of the rule. For instance, the rules displayed in Fig. 2 could be used to generate the following fragments: *a tool like a hammer, mango is a delicious fruit, man before animal.*

```
HYPERNYM-OF  ::= a <arg1> like a <arg2>
HYPERNYM-OF  ::= <arg2> is a delicious <arg1>
HYPERNYM-OF  ::= <arg2> before <arg1>
```

Fig. 2. Grammar example rule set.

In addition to the previous modules, the Contextualizer explains why certain words were selected and what is their connection to the seed words, as a list of triplets for each line. It can be used for debugging or evaluation purposes.

Besides the surprise factor, another way of increasing diversity is to expand the set of seeds with semantically-relevant words. For this purpose, before generation, a personalized version of the PageRank [27] algorithm is run in the full semantic network. Initial node weights are randomly distributed across the seeds, while the rest of the nodes have an initial weight of 0. After 30 iterations, nodes will be ranked according to their structural relevance to the seeds. The top-r ranked nodes are added to seed set.

The previous expansion feature can be biased to induce a target sentiment in the poem. For this purpose, the top-r nodes are previously filtered, in order to use only those with a target polarity. The typical polarity of words is obtained from the Sentiment Processor, an interface to a polarity lexicon that lists words and their typical polarities (positive, neutral or negative). For instance, suppose that the top-10 ranked words for the word "blue" are: *grim, blueness, gloomy, sexy, color, dark, dejected, low, dye, down.* When generating a negative poem with, say, the top-3 words, *grim, gloomy* and *dark* would be added to the seed set, together with *blue.* For a positive poem, the word *sexy* would be added, together with the next two positive words in the ranking.

A more detailed description of PoeTryMe's architecture is available elsewhere [9]. Although PoeTryMe was originally developed to produce poetry in Portuguese, its flexible architecture enabled the adaptation to Spanish [28] and English [29]. It has also been used to produce song lyrics for a given melody [29].

4   Poetry Inspired by Current Trends

This section describes a new instantiation of PoeTryMe where seed words are collected from Twitter. Feeding the system with words that, at a certain time, are associated with a topic, enables the generation of different poems every time, with a shallow connection with present events, even if it is not immediately clear.

As the original PoeTryMe, the presented instantiation targets Portuguese although, given our recent adaptations, it could be adapted to Spanish and English with low effort. All the linguistic resources used were the same as those of previous instantiations for Portuguese [9], except for the generation grammar, which is described in the next section. After that, the current approach for producing poetry inspired by Twitter is described; a Twitterbot that publishes poems about trendy topics is introduced; the current setup of this system is detailed; and some illustrative examples are presented, followed by a critical view of our results.

4.1 Generation Grammar

The generation grammar used in this instantiation of PoeTryMe has two main updates: it is more strict and covers different kinds of text. The rules of the grammar are still automatically acquired from human-created poetry, by identifying lines where two words connected in the semantic network co-occur. Yet, current rules were only added to the grammar when the relation arguments matched the desired part-of-speech (POS). Previously, this did not always happen because, depending on the context, the same words might have different POS. For instance, most verbs can also be nouns (e.g. *break, cover*), or many nouns can behave as adjectives (e.g. *red, young*).

Another difference in the grammar is that, in addition to rules learned from human-created poetry, they were also acquired from proverbs and from Wikipedia sentences. As the lines of a poem are already kind of abstract or already involve figurative meanings, when PoeTryMe adds a new level of abstraction, the result can sometimes turn out to be more difficult, if not impossible, to interpret. On the other hand, Wikipedia text is not so creative but easily interpretable. Our intuition is that, combining both kinds of text, the previous issues will be more balanced, and the result may still slightly more clear, even when altered by PoeTryMe.

Being more strict when collecting rules resulted in a much smaller grammar. The current generation grammar, for Portuguese, covers about 1,500 renderings, which is substantially less than the previous 4,100 [9]. Hopefully, low quality grammar rules were left out.

4.2 Approach

The approach of the present system can be divided in three main steps, including the generation of a poem. Before generation, there is a seed acquisition stage, and a seed expansion stage. The seed acquisition stage goes as follows:

1. A topic t, in the form of a word or expression, is given as input.

2. Through the Twitter4J[3] library, m tweets mentioning t are retrieved from Twitter.
3. Each tweet is processed and the top-f most frequent nouns, adjectives or verbs are collected and used as seed words.

If there is one, the main sentiment about the topic can also be estimated by counting the total number of smileys and emojis in the retrieved tweets. For each happy face (positive), a counter c is incremented by 1 ($c = c + 1$), and for each sad or crying face (negative) it is decremented ($c = c - 1$). The estimated polarity depends of the value of c. If $c > \theta$, it is positive, and if $c < -\theta$, it is negative, where θ is a predefined threshold.

Additional seeds can be obtained through the seed expansion procedure described earlier, which can be biased towards the polarity estimated in the previous stage. Alternatively, if there is a Wikipedia article about topic t, open words from its first sentence can also be used as additional seeds. This is an attempt to mix long-term data about the topic, in Wikipedia, with fresh information, from Twitter, and has similarities with what Toivanen et al. [20] do with Wikipedia and recent news.

Generation is the final stage and starts by feeding PoeTryMe with the set of seed words. As most of its previous instantiations (e.g. [28,29]), a generate & test strategy at the line level if followed. This means that, for each line in the target poem structure, text fragments are successively generated and scored against the target metre and presence of rhymes, while the best scoring are kept. The generation of each line stops either after a predefined number of generated candidates (n), or when a candidate line has precisely the target number of syllables and target rhyme, if there is one.

4.3   Twitterbot

Twitter is increasingly becoming a popular tool in the Computational Creativity community, not only as a source of information, but also as a platform for exhibiting the results of creative systems. While Twitterbots are autonomous systems, connected to Twitter that, from time to time, post messages, for creative Twitterbots, messages have a creative value. This includes the production of novel metaphors [30], riddles [31], or Internet memes [32], among others. Poetry has also been produced from the re-organisation of tweets [33].

Following the previous trend, the *@poetartificial*[4] Twitterbot was developed for tweeting poems inspired by the current trends. Every hour, it reads the Twitter trends for Portugal, selects one of the three top trends, and runs the previously described approach for producing a set of poems inspired by the selected trend. The best-scoring poem is selected. Given the size limitations for tweets (140 characters), generated poems are currently blocks of four 10-syllable lines.

---

[3] http://twitter4j.org/.
[4] In Portuguese, *poeta artificial* means artificial poet.

4.4 Setup

PoeTryMe has several customisable parameters, most of them mentioned earlier. For the presented system, including the Twitterbot, the following parameters were set:

- Poem structure: block of four lines, each with 10 syllables;
- Relevant seeds obtained from expansion, $r = 5$;
- Surprise factor, $\nu = 0$;
- Strategy: generate & test
    - Maximum generations / line, $n = 2,500$
    - Increasing factor $\sigma = 0.8$; (to increase the probability of rhymes, $n = n + i * \sigma * n$, where i is the position of the line in the stanza)
- Score:
    - Each syllable missing / out of metre: +1 penalty point
    - Each rhyme: -2 penalty points

It should be highlighted that the surprise factor was set to 0 because we believe that there is already enough richness in the Twitter seeds, especially if they are expanded. Apart from that, there was not a big difference from previous instantiations of the system. The following additional parameters, specific of this instantiation, were set:

- #Retrieved tweets, $m = 200$;
- #Frequent words, $f = 5$;
- Polarity threshold, $\theta = m/20$ (5 % of the retrieved tweets);
- #Wikipedia words, $w = 5$.

The Twitterbot is currently generating 25 poems each hour, but publishing only the one with the best overall generate &test score, out of those that fit in a tweet (140 characters).

4.5 Examples

The new instantiation of PoeTryMe is illustrated by examples displayed in this section. All of them were produced in 19th January, 2016, using, as an example topic, "David Bowie", a well-known musician whose death, 9 days earlier, was still echoed throughout Twitter.

The examples are presented in their original form, in Portuguese, together with a rough English translation. Of course that, due to the vagueness of language and to the inherent abstraction in a poem, the resulting translations were hard to reach, and often resulted in odd constructions. Examples were produced with different sets of seeds words, obtained with different seed expansion settings, enumerated in Table 1, together with the resulting seed words. The examples of Fig. 3 only use the top-5 frequent words in the retrieved tweets, while the others use five additional seeds, obtained by different means. In the poems of Fig. 4, the regular expansion procedure was applied to retrieve relevant words from the semantic network. The poems of Figs. 5 and 6 used the same expansion algorithm but the former is biased towards positive seeds, and the second towards negative seeds. Finally, the poems in Fig. 7 use five additional seeds extracted from the abstract of David Bowie's article in the Portuguese Wikipedia.

The top frequent words in the retrieved tweets include the word 'music', because Bowie was a musician. The word 'homage' is present because several tweets refer planned homages, especially from Lady Gaga, another musician. The verb 'to make' is often used with 'homage' – *to make an homage* – and was thus also frequent. The other two words, 'partnership' and 'good', were used several times to mention that Bowie refused to make a partnership with the band Coldplay, claiming that their music was not good.

Table 1. Seeds collected from Twitter and additional seeds obtained from different expansion settings.

Seeds	*homenagem, msica, fazer, parceria, bom*
	(homage, music, make, partnership, good)
Expansion	*associao, sociedade, lugar, comemorao, promessas*
	(association, society, place, commemoration, promises)
Expansion+	*virtuoso, harmonia, glria, alegria, carola*
	(virtuous, harmony, glory, joy, prayer)
Expansion-	*vo, lbia, treta, vassalas, partida*
	(vain, wordy, bullshit, vassals, departure)
Wikipedia	*musical, artstico, ingls, nome, ator*
	(musical, artistic, english, name, actor)

Besides the original seeds, the poems in Fig. 4 use directly related words, obtained from the semantic network. The exact connection of these words with the domain can be confirmed with the help of the Contextualizer. Related words include synonyms of homage (glory, proof), music (harmony) or of 'to make' (invent, practise, charge), as well as a hypernym of partnership (association). Some of these relations are not held by the same sense of the seed words,

glria da homenagem em cruz
homenagem a prova de luz   glory of homage in the cross
de associaes de parceria   homage is a proof of illumination
sem achar msica, nem harmonia   of partnership associations
                                         without finding music, nor harmony

fazer em frente, ir a inventar   make ahead, go and invent
fazer em frente, ir a praticar   make ahead, go and practice
fazer em frente, ir a facturar   make ahead, go and charge
mais vale fazer que inventar   it is better to do than to invent

Fig. 3. Poems inspired by the topic David Bowie, without seed expansion.

and they are often not the sense we would first thought of. Although the system does not do it intentionally, but because word senses are not handled, we like to see this as an open door to the presence of figurative language.

The poems in Fig. 4 mix the utilisation of the original seeds with other semantically-relevant, together with words directly-related to one of the previous. This includes synonyms of 'to make' (proceed, conclude) or of 'to collect' (gather), a hyponym of recordings (record), also words like 'albums' that may be collected, or 'promises' that should be devoted.

coleccionar sem tocar em lbuns
coligir para coleccionar   collect without touching in albums
melhor fazer que representar   gather to collect
                             it is better to do than to act
um registo de gravaes doiradas   a record of golden recordings

sempre a fazer e a proceder   always making and proceeding
sempre a concluir e a fazer   always completing and making
                             do not want then to give your place
no o queira depois o lugar dar   to have promises without knowing to devote
ter promessas sem saber consagrar

Fig. 4. Poems inspired by the topic David Bowie, with expanded seeds.

The poems in Fig. 5 use the original seeds and other semantically-relevant words with a positive polarity. Words directly related to the previous include synonyms of harmony (communion) and joy (satisfaction), hypernyms of partnership (society) and harmony (art), a hyponym of homage (proof) and an action that causes joy (to rejoice).

alegrar resulta em alegria
de sociedades de parceria   rejoice results in joy
alegrar resulta em alegria   of partnership societies
                             rejoice results in joy
a arte principal a harmonia   the main art is harmony
                             homage is a proof of illumination
homenagem a prova de luz   harmony communion in the cross
comunho da harmonia em cruz   there is not satisfaction without joy
no h satisfao sem alegria   of partnership companies
de sociedades de parceria

Fig. 5. Poem inspired by the topic David Bowie, with expanded positive seeds.

The poems in Fig. 6 use the original seeds and other semantically-relevant words with a negative polarity. Words directly related to the previous include a synonym of vain (lied), a hypernym of bullshit (verbiage), a hyponym of music (cheap music) and an association with tax (efficient), among others.

andar ao palavreado da treta	
com msica santa e musiqueta	wandering to the verbiage of bullshit
eficientes, imposto de mo	with holy music and cheap music
por decreto do cu, mentido e vo	efficient, hand tax by heaven's decree, lied and in vain
tudo comeou a um vo embalde	it all started to one useless vain
tudo comeou a um vo debalde	it all started to one worthless vain
olha os guelfos da nossa partida!	looks the Guelphs of our departure!
de que turma fazer a minha vida	of what class to make my life

Fig. 6. Poem inspired by the topic David Bowie, with expanded negative seeds.

The poems in Fig. 7 use the original seeds and other collected from Wikipedia. As expected, the words collected from Wikipedia are more stable associations with David Bowie: he was an English singer, musical producer, and also an actor. Semantically-related words include synonyms of producer (creator) or name (power), a hypernym of singer (artist) and of musical (movie).

de seu filme, lugar e musical	of your movie, place and music
seu filme de musical palatal	his musical film is palatal
, portanto, um nome de poder	it is therefore a name of power
fazer para chegar a fazer	make to get to make
de criador e de produtor	it is of a creator and a producer
que alma, que produtor, que criador	what a soul, what a producer, what a creator
as responsveis no tlm produtor	the responsible do not have a producer
o artista e depois o cantor	the artist and then the singer

Fig. 7. Poems inspired by the topic David Bowie, with additional seeds from Wikipedia.

Additional real-time examples can be checked in the Twitter feed of the user *@poetartificial*, where the bot operates in real time.

4.6   Critical View

In the displayed examples, features like the regular metre and the frequent presence of rhymes arise. With the current settings, these features are often met. In fact, with the current linguistic resources, meeting them is mostly a matter of increasing the number of generations per line. Grammatical constraints are also frequently satisfied, especially now, with more strict grammar on the POS of the arguments. Though smaller than previous grammars, we can say that the richness of the underlying resources still enables the generation of poems with an interesting degree of variation, which contributes to a positive perception of creativity.

Each line is semantically-coherent, because semantically-related words are basically put in the position of other words that hold the same relation, in what can be seen as a shallow exploitation of analogy. Stranger situations might occur from fixed words in the rule's body, too specific and with a strong connection with the original words, but not so much with the replacements, at least as long as the semantic network is well-structured. At the same time, this is where the poem may become interesting, or when it may fail. But this situation is not that frequent in 10-syllable lines, typically restricted to functional words (e.g. determiners and prepositions), besides the pair of related words.

In the displayed examples, when the previous situations happened, we can say it went quite well in lines such as: *homage is a proof of illumination*, or *harmony communion in the cross*, where the words 'illumination' and 'cross' were fixed. On the other hand, the line *his musical film is palatal*, where the word 'palatal' is fixed, is odder. As most of the more specific words come from the rules learned from Wikipedia, its usage for this purpose should be rethought.

Still on semantics, although they are generated independently, lines end up having some unity, together and with the topic. This happens because they are based on the same semantic domain, set by the topic. Though not frequent, issues might arise from an odd order of the lines. This happens because the system does not have any concern on showing some kind of evolution from the beginning to the end of the poem.

Finally, the connection of the poem with the topic is sometimes too tenuous. Using associated words is not always enough for this purpose. Moreover, there might be strongly-associated words that are not in the semantic network, and will thus never be used. This is why we say that the poem is inspired by the topic, in a sense that it uses related words in semantically-coherent sentences, but we do not claim that it is about the topic.

5   Concluding Remarks

A new instantiation of PoeTryMe, a poetry generation platform, was presented. The singular feature of the presented system is that its poetry is inspired by Twitter trends, more precisely, words that are associated with those.

Presented examples illustrate the potential of this system, but also its limitations. Despite the presence of a regular metre, rhymes, grammatical sentences and semantically-related words, the connection with the topic is not always very clear. If it is not known, it is often hard to identify the original topic. Especially when fresh associations, possibly valid just for a short period of time, are used.

This is why the poems are not about a topic, but inspired by it – given a topic, currently associated words are extracted and fed to PoeTryMe, which expresses "what it knows" about those words, poetically. Writing about the topic would require deeper linguistic processing of the tweets and possibly other sources of knowledge. On this problem, Tobin and Manurung [21] extract predicate-argument structures from an input article and try to keep the same structure during generation. They admit, however, that considering this together with other features results in too much complexity and a long time for producing a poem.

In order to improve the connection with the topic, we are devising an alternative approach for setting the generation domain, but simpler than the previous. In the preprocessing step, whenever a tweet uses two words that, according to our semantic network, are related, this relation will be added to the set of identified relations. Then, instead of using seed words for setting the generation domain, this domain should consist of a graph with the identified relations, which is possible in the PoeTryMe architecture. If the graph is too small, additional relations can be used, based on their distance to the domain, or the current strategy can still be used as a fallback. The issue of only using words from the semantic network remains, though.

Anyway, trends are always changing and people say different things about them. So, independently of writing about the topic or not, the connection to Twitter enables the continuous generation of different poems every time. This can also be seen as a test to the limits of the system, and will certain give hints for further improvements. Moreover, the Twitterbot will hopefully shake a little bit the (still small) Portuguese community of Twitter users.

Acknowledgements. This work was developed in the scope ConCreTe – *Concept Creation Technology*. The project ConCreTe acknowledges the financial support of the Future and Emerging Technologies (FET) programme within the Seventh Framework Programme for Research of the European Commission, under FET grant number 611733.

References

1. Colton, S., Wiggins, G.A.: Computational creativity: the final frontier? In: Proceedings of 20th European Conference on Artificial Intelligence (ECAI 2012). Frontiers in Artificial Intelligence and Applications, Montpellier, France, vol. 242, pp. 21–26. IOS Press (2012)
2. Machado, P., Cardoso, A.: All the truth about NEvAr. Appl. Intell. Spec. Issue Creat. Syst. **16**, 101–119 (2002)
3. Binsted, K., Ritchie, G.: An implemented model of punning riddles. In: Proceedings of 12th National Conference on Artificial Intelligence, AAAI 1994, vol. 1, pp. 633–638. AAAI Press, Menlo Park (1994)
4. Gervás, P., Díaz-Agudo, B., Peinado, F., Hervás, R.: Story plot generation based on CBR. Knowl.-Based Syst. **18**, 235–242 (2005)
5. Veale, T., Hao, Y.: A fluid knowledge representation for understanding and generating creative metaphors. In: Proceedings of 22nd International Conference on Computational Linguistics, COLING 2008, vol. 1, pp. 945–952. ACL Press, Manchester (2008)
6. Smith, M.R., Hintze, R.S., Ventura, D.: Nehovah: a neologism creator nomen ipsum. In: Proceedings of 5th International Conference on Computational Creativity, Ljubljana, Slovenia, ICCC 2014 (2014)
7. Tomašic, P. Znidaršic, M., Papa, G.: Implementation of a slogan generator. In: Proceedings of 5th International Conference on Computational Creativity, Ljubljana, Slovenia. ICCC 2014, pp. 340–343 (2014)

8. Gonçalo Oliveira, H.: PoeTryMe: a versatile platform for poetry generation. In: Proceedings of ECAI 2012 Workshop on Computational Creativity, Concept Invention, and General Intelligence, Montpellier, France, C3GI 2012 (2012)

9. Gonçalo Oliveira, H., Cardoso, A.: Poetry generation with PoeTryMe. In: Besold, T.R., Schorlemmer, M., Smaill, A. (eds.) Computational Creativity Research: Towards Creative Machines. Atlantis Thinking Machines, pp. 243–266. Atlantis/Springer, Lynnwood/Heidelberg (2015)

10. Gervás, P.: WASP: evaluation of different strategies for the automatic generation of Spanish verse. In: Proceedings of AISB 2000 Symposium on Creative and Cultural Aspects and Applications of AI and Cognitive Science, Birmingham, UK, pp. 93–100 (2000)

11. Manurung, H.: An evolutionary algorithm approach to poetry generation. Ph.D. thesis, University of Edinburgh (2003)

12. Colton, S., Goodwin, J., Veale, T.: Full FACE poetry generation. In: Proceedings of 3rd International Conference on Computational Creativity, Dublin, Ireland, ICCC 2012, Dublin, Ireland, pp. 95–102 (2012)

13. Toivanen, J.M., Järvisalo, M., Toivonen, H.: Harnessing constraint programming for poetry composition. In: Proceedings of the 4th International Conference on Computational Creativity, ICCC 2013, pp. 160–167. The University of Sydney, Sydney (2013)

14. Agirrezabal, M., Arrieta, B., Astigarraga, A., Hulden, M.: POS-tag based poetry generation with wordnet. In: Proceedings of the 14th European Workshop on Natural Language Generation, pp. 162–166. ACL Press, Sofia (2013)

15. Manurung, H.: A chart generator for rhythm patterned text. In: Proceedings of 1st International Workshop on Literature in Cognition and Computer (1999)

16. Levy, R.P.: A computational model of poetic creativity with neural network as measure of adaptive fitness. In: Proceedings of the ICCBR-01 Workshop on Creative Systems (2001)

17. Gervás, P.: An expert system for the composition of formal Spanish poetry. J. Knowl.-Based Syst. **14**, 200–201 (2001)

18. Kurzweil, R.: Cybernetic poet. http://www.kurzweilcyberart.com/poetry/rkcp_overview.php

19. Misztal, J., Indurkhya, B.: Poetry generation system with an emotional personality. In: Proceedings of 5th International Conference on Computational Creativity, ICCC 2014, Ljubljana, Slovenia (2014)

20. Toivanen, J.M., Gross, O., Toivonen, H.: The officer is taller than you, who race yourself! Using document specific word associations in poetry generation. In: Proceedings of 5th International Conference on Computational Creativity, Ljubljana, Slovenia, ICCC 2014 (2014)

21. Tobing, B.C.L., Manurung, R.: A chart generation system for topical metrical poetry. In: Proceedings of the 6th International Conference on Computational Creativity, ICCC 2015, Park City, Utah, USA (2015)

22. Wong, M.T., Chun, A.H.W.: Automatic haiku generation using VSM. In: Proceeding of 7th WSEAS International Conference on Applied Computer and Applied Computational Science (ACACOS 2008), Hangzhou, China (2008)

23. Netzer, Y., Gabay, D., Goldberg, Y., Elhadad, M.: Gaiku: generating haiku with word associations norms. In: Proceedings of the NAACL 2009 Workshop on Computational Approaches to Linguistic Creativity, CALC 2009, Boulder, Colorado, pp. 32–39. ACL Press (2009)

24. Yan, R., Jiang, H., Lapata, M., Lin, S.D., Lv, X., Li, X.: I, poet: automatic Chinese poetry composition through a generative summarization framework under constrained optimization. In: Proceedings of the Twenty-Third International Joint Conference on Artificial Intelligence, IJCAI 2013, pp. 2197–2203. AAAI Press (2013)
25. Rashel, F., Manurung, R.: Pemuisi: a constraint satisfaction-based generator of topical Indonesian poetry. In: Proceedings of 5th International Conference on Computational Creativity, ICCC 2014, Ljubljana, Slovenia (2014)
26. Das, A., Gambäck, B.: Poetic machine: computational creativity for automatic poetry generation in Bengali. In: Proceedings of 5th International Conference on Computational Creativity, ICCC 2014, Ljubljana, Slovenia (2014)
27. Brin, S., Page, L.: The anatomy of a large-scale hypertextual web search engine. Comput. Netw. **30**, 107–117 (1998)
28. Gonçalo Oliveira, H., Hervs, R., Daz, A., Gervs, P.: Adapting a generic platform for poetry generation to produce Spanish poems. In: Proceedings of 5th International Conference on Computational Creativity, ICCC 2014, Ljubljana, Slovenia (2014)
29. Gonçalo Oliveira, H.: Tra-la-lyrics 2.0: automatic generation of song lyrics on a semantic domain. J. Artif. Gen. Intell. **6**, 87–110 (2015). Special Issue: Computational Creativity, Concept Invention, and General Intelligence
30. Veale, T., Valitutti, A., Li, G.: Twitter: the best of bot worlds for automated wit. In: Streitz, N., Markopoulos, P. (eds.) DAPI 2015. LNCS, vol. 9189, pp. 689–699. Springer, Heidelberg (2015). doi:10.1007/978-3-319-20804-6_63
31. Guerrero, I., Verhoeven, B., Barbieri, F., Martins, P. Perez y Perez, R.: TheRiddlerBot: a next step on the ladder towards creative Twitter bots. In: Proceedings of the 6th International Conference on Computational Creativity, ICCC 2015, pp. 315–322. Brigham Young University, Park City (2015)
32. Costa, D., Gonçalo Oliveira, H., Pinto, A.: "In reality there are as many religions as there are papers" - first steps towards the generation of Internet memes. In: Proceedings of the 6th International Conference on Computational Creativity, ICCC 2015, Park City, UT, USA, pp. 300–307 (2015)
33. Charnley, J., Colton, S., Llano, M.T.: The FloWr framework: automated flowchart construction, optimisation and alteration for creative systems. In: 5th International Conference on Computational Creativity, ICCC 2014, Ljubljana, Slovenia (2014)

Knowledge Discovery and Information Retrieval

Knowledge Discovery and Colon... Retrieval

Exploiting Guest Preferences with Aspect-Based Sentiment Analysis for Hotel Recommendation

Fumiyo Fukumoto[1]([✉]), Hiroki Sugiyama[2], Yoshimi Suzuki[1], and Suguru Matsuyoshi[1]

[1] Interdisciplinary Graduate School of Medicine and Engineering, University of Yamanashi, Yamanashi 400-8511, Japan
{fukumoto,ysuzuki,sugurum}@yamanashi.ac.jp
[2] Universal Computer Co., Ltd., Osaka 540-6126, Japan

Abstract. This paper presents a collaborative filtering method for hotel recommendation incorporating guest preferences. We used the results of aspect-based sentiment analysis to recommend hotels because whether or not the hotel can be recommended depends on the guest preferences related to the aspects of a hotel. For each aspect of a hotel, we identified the guest preference by using dependency triples extracted from the guest reviews. The triples represent the relationship between aspect and its preference. We calculated transitive association between hotels by using the positive/negative preference on some aspect. Finally, we scored hotels by Markov Random Walk model to explore transitive associations between the hotels. The empirical evaluation showed that aspect-based sentiment analysis improves overall performance. Moreover, we found that it is effective for finding hotels that have never been stayed at but share the same neighborhoods.

Keywords: Collaborative filtering · Markov Random Walk model · Aspect-based sentiment analysis

1 Introduction

Collaborative filtering (CF) identifies the potential preference of a consumer/guest for a new product/hotel by using only the information collected from other consumers/guests with similar products/hotels in the database. It is a simple technique as it is not necessary to apply more complicated content analysis compared to the content-based filtering framework [1]. CF has been very successful in both research and practical systems. It has been widely studied [2–6,11], and many practical systems such as Amazon for book recommendation and Expedia for hotel recommendation have been developed.

Item-based collaborative filtering is one of the major recommendation techniques [6,7]. It assumes that the consumers/guests are likely to prefer product/hotel that are similar to what they have bought/stayed before. Unfortunately, most of them only consider star ratings and leave consumers/guests

© Springer International Publishing AG 2016
A. Fred et al. (Eds.): IC3K 2015, CCIS 631, pp. 31–46, 2016.
DOI: 10.1007/978-3-319-52758-1_3

textual reviews. Several authors focused on the problem, and attempted to improve recommendation results using the techniques on text analysis, *e.g.*, sentiment analysis, opinion mining, or information extraction [8–10]. However, major approaches aim at finding the positive/negative opinions for the product/hotel, and do not take users preferences related to the *aspects* of a product/hotel into account. For instance, one guest is interested in a hotel with nice restaurants for enjoying her/his vacation, while another guest, *e.g.*, a businessman prefers to the hotel which is close to the station. In this case, the aspect of the former is different from the latter.

This paper presents a collaborative filtering method for hotel recommendation incorporating guest preferences. We rank hotels according to scores. The score is obtained using the analysis of different aspects of guest preferences. The method utilizes a large amount of guest reviews which make it possible to solve the item-based filtering problem of data sparseness, *i.e.*, some items were not assigned a label of users preferences. We used the results of aspect-based sentiment analysis to recommend hotels because whether or not the hotel can be recommended depends on the guest preferences related to the aspects of a hotel. For instance, if one guest stays at hotels for her/his vacation, a room with nice views may be an important factor to select hotels, whereas another guest stayed at hotels for business, may select hotels near to the station. We parsed all reviews by using syntactic analyzer, and extracted dependency triples which represent the relationship between aspect and its preference. For each aspect of a hotel, we identified the guest preference related to the aspect to good or not, based on the dependency triples in the guest reviews. The positive/negative opinion on some aspect is used to calculate transitive association between hotels. Finally, we scored hotels by Markov Random Walk (MRW) model, *i.e.*, we used MRW based recommendation technique to explore transitive associations between the hotels. Random Walk based recommendation overcomes the item-based CF problem that the inability to explore transitive associations between the hotels that have never been stayed but share the same neighborhoods [11].

2   Related Work

Sarwar *et al.* mentioned that CF mainly consists of two procedures, *prediction* and *recommendation* [7]. Prediction refers to a numerical value expressing the predicted likeliness of item for user, and recommendation is a list of items that the user will like the most. As the volume of online reviews has drastically increased, sentiment analysis, opinion mining, and information extraction for the process of prediction are a practical problem attracting more and more attention. Several efforts have been made to utilize these techniques to recommend products [9,12]. Cane *et al.* have attempted to elicit user preferences expressed in textual reviews, and map such preferences onto some rating scales that can be understood by existing CF algorithms [8]. They identified sentiment orientations of opinions by using a relative-frequency-based technique that estimates the strength of a word with respect to a certain sentiment class as the relative frequency of its occurrence in the class. The results using movie reviews from the Internet Movie Database (IMDb) for the MovieLens 100k dataset showed the effectiveness of the method. However, the sentiment analysis they used is limited, *i.e.*, they used only adjectives or verbs.

Niklas *et al.* have attempted to improve the accuracy of movie recommendations by using the results of opinion extraction from free-text reviews [9]. They presented three approaches: (i) manual clustering, (ii) semi-automatic clustering by Explicit Semantic Analysis (ESA), and (iii) fully automatic clustering by Latent Dirichlet Allocation (LDA) [13] to extract movie aspects as opinion targets, and used them as features for the collaborative filtering. The results using 100 random users from the IMDb showed that the LDA-based movie aspect extraction yields the best results. Our work is similar to Niklas *et al.* method in the use of LDA. The difference is that our approach applied LDA to the dependency triples, although Niklas applied LDA to single words. Raghavan *et al.* have attempted to improve the performance of collaborative filtering in recommender systems by incorporating quality scores to ratings [10]. They estimated the quality scores of ratings using the review and user data set, and ranked according to the scores. They adapted the Probabilistic Matrix Factorization (PMF) framework. The PMF aims at inferring latent factors of users and items from the available ratings. The experimental evaluation on two product categories of a benchmark data set, *i.e.*, *Book* and *Audio CDs* from Amazon.com showed the efficacy of the method.

In the context of recommendation, several authors have attempted to rank items by using graph-based ranking algorithms [14, 15]. Wijaya and Bressan have attempted to rank items directly from the text of their reviews [16]. They constructed a sentiment graph by using simple contextual relationships such as collocation, negative collocation and coordination by pivot words such as conjunctions and adverbs. They applied PageRank algorithm to the graph to rank items. Li *et al.* proposed a basket-sensitive random walk model for personalized recommendation in the grocery shopping domain [11]. The method extends the basic random walk model by calculating the product similarities through a weighted bi-partite network which allows the current shopping behaviors to influence the product ranking. Empirical results using three real-world data sets, LeShop, TaFeng and an anonymous Belgium retailer showed that a performance improvement of the method over other existing collaborative filtering models, the cosine, conditional probability and the bi-partite network based similarities. However, the transition probability from one product node to another is computed based on a user's purchase frequency of a product with regardless of the users' positive or negative opinions concerning to the product.

There are three novel aspects in our method. Firstly, we propose a method to incorporate different aspect of a hotel into users preferences/criteria to improve quality of recommendation. Secondly, from a ranking perspective, the MRW model we used is calculated based on the polarities of reviews. Finally, from the opinion mining perspective, we propose overcoming with the unknown polarized words by utilizing LDA.

3   Framework of the System

Figure 1 illustrates an overview of the method. It consists of four steps: (1) Aspect analysis, (2) Positive/negative opinion detection based on aspect analysis, (3) Positive/negative review identification, and (4) Scoring hotels by MRW model.

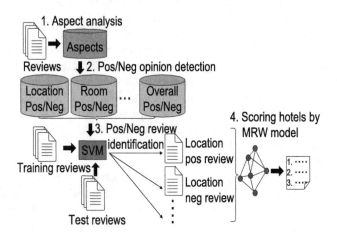

Fig. 1. Overview of the method.

3.1   Aspect Analysis

The first step to recommend hotels based on guest preferences is to extract aspects for each hotel from a guest review corpus. All reviews were parsed by the syntactic analyzer CaboCha [17], and all the dependency triples (rel, x, y) are extracted. Here, x refers to a noun/compound noun word related to the aspect. y shows verb or adjective word related to the preference for the aspect. rel denotes a grammatical relationship between x and y. We classified rel into 9 types of Japanese particle, "*ga(ha)*", "*wo*", "*ni*", "*he*", "*to*", "*de*", "*yori*", "*kara*" and "*made*". For instance, from the sentence "*Cyousyoku (breakfast) ga totemo (very) yokatta (was delicious).*" (The breakfast was very delicious.), we can obtain the dependency triplet, (*ga, cyousyoku, yokatta*). The triplet represents positive opinion, "*yokatta*"(*was delicious*) concerning to the aspect, "*Cyousyoku*"(*breakfast/meal*).

3.2   Positive/Negative Opinion Detection

The second step is to identify positive/negative opinion related to the aspects of a hotel. We classified aspects into seven types: "Location", "Room", "Meal", "Spa", "Service", "Amenity", and "Overall". These types are used in the Rakuten travel data[1] which we used in the experiments. Basically, the identification of positive/negative opinion is done using Japanese sentiment polarity

---

[1] http://rit.rakuten.co.jp/rdr/index.html.

dictionary [18]. More precisely, if y in the triplet (rel, x, y) is classified into positive/negative classes in the dictionary, we regarded the extracted dependency triplet as positive/negative opinion. However, the dictionary makes it nearly impossible to cover all of the words in the review corpus.

For unknown verb or adjective words that were extracted from the review corpus, but did not occur in any of the dictionary classes, we classified them into positive or negative class using a topic model. Topic models such as probabilistic latent semantic indexing [19] and Latent Dirichlet Allocation (LDA) [13] are based on the idea that documents are mixtures of topics, where each topic is captured by a distribution over words. The topic probabilities provide an explicit low-dimensional representation of a document. They have been successfully used in many domains such as text modeling and collaborative filtering [20]. We used LDA and classified unknown words into positive/negative classes. LDA presented by [13] models each document as a mixture of topics, and generates a discrete probability distribution over words for each topic. The generative process of LDA can be described as follows:

1. For each topic $k = 1, \cdots, K$, generate ϕ_k, multinomial distribution of words specific to the topic k from a Dirichlet distribution with parameter β;
2. For each document $d = 1, \cdots, D$, generate θ_d, multinomial distribution of topics specific to the document d from a Dirichlet distribution with parameter α;
3. For each word $n = 1, \cdots, N_d$ in document d;
   (a) Generate a topic z_{dn} of the nth word in the document d from the multinomial distribution θ_d,
   (b) Generate a word w_{dn}, the word associated with the nth word in document d from multinomial ϕ_{zdn}.

Like much previous work on LDA, we used Gibbs sampling to estimate ϕ and θ. The sampling probability for topic z_i in document d is given by:

$$P(z_i \mid z_{\backslash i}, W) = \frac{(n^v_{\backslash i,j} + \beta)(n^d_{\backslash i,j} + \alpha)}{(n_{\backslash i,j} + W\beta)(n^d_{\backslash i,\cdot} + T\alpha)}. \tag{1}$$

$z_{\backslash i}$ refers to a topic set Z, not including the current assignment z_i. $n^v_{\backslash i,j}$ is the count of word v in topic j that does not include the current assignment z_i, and $n_{\backslash i,j}$ indicates a summation over that dimension. W refers to a set of documents, and T denotes the total number of unique topics. After a sufficient number of sampling iterations, the approximated posterior can be used to estimate ϕ and θ by examining the counts of word assignments to topics and topic occurrences in documents. The approximated probability of topic k in the document d, $\hat{\theta}^k_d$, and the assignments word w to topic k, $\hat{\phi}^w_k$ are given by:

$$\hat{\theta}^k_d = \frac{N_{dk} + \alpha}{N_d + \alpha K}. \tag{2}$$

$$\hat{\phi}^w_k = \frac{N_{kw} + \beta}{N_k + \beta V}. \tag{3}$$

For each aspect, we manually collected reviews and created a review set. We applied LDA to each set of reviews consisted of triples. We need to estimate two parameters, *i.e.*, the number of reviews, and the number of topics k for the result obtained by LDA. We note that the result can be regarded as a clustering result: each cluster is positive/negative opinion, and each element of the cluster is positive/negative opinion according to the sentiment polarity dictionary, or unknown words. For each number of reviews, we applied LDA, and as a result, we used Entropy measure which is widely used to evaluate clustering techniques to estimate the number of topics (clusters) k. The Entropy measure is given by:

$$E = -\frac{1}{\log k} \sum_j \frac{N_j}{N} \sum_i P(A_i, C_j) \log P(A_i, C_j). \tag{4}$$

k refers to the number of clusters. $P(A_i, C_j)$ is a probability that the elements of the cluster C_j assigned to the correct class A_i. N denotes the total number of elements and N_j shows the total number of elements assigned to the cluster C_j. The value of E ranges from 0 to 1, and the smaller value of E indicates better result. We chose the parameter k whose value of E is smallest. For each cluster, if the number of positive opinion is larger than those of negative ones, we regarded a triplet including unknown word in the cluster as positive and vice versa.

3.3   Positive/Negative Review Identification

We used the result of positive/negative opinion detection to classify guest reviews into positive or negative related to the aspect. Like much previous work on sentiment analysis based on supervised machine learning techniques [21] or corpus-based statistics, we used Support Vector Machine (SVMs) to annotate automatically [22]. For each aspect, we collected positive/negative opinion (triples) from the results of LDA[2]. Each review in the test data is represented as a vector where each dimension of a vector is positive/negative triplet appeared in the review, and the value of each dimension is a frequency count of the triplet. For each aspect, the classification of each review can be regarded as a two-class problem: positive or negative.

3.4   Scoring Hotels by MRW Model

The final procedure for recommendation is to rank hotels. We used a ranking algorithm, the MRW model that has been successfully used in Web-link analysis, social networks [23], and recommendation [11,14,15]. Given a set of hotels H, $Gr = (H, E)$ is a graph reflecting the relationships between hotels in the set. H is the set of nodes, and each node h_i in H refers to the hotel. E is a set of edges, which is a subset of $H \times H$. Each edge e_{ij} in E is associated with an affinity

---

[2] We used the clusters that the number of positive and negative words is not equal.

weight $f(i \rightarrow j)$ between hotels h_i and $h_j(i \neq j)$. The weight of each edge is a value of transition probability $P(h_j \mid h_i)$ between h_i and h_j, and defined by:

$$P(h_j \mid h_i) = \sum_{k=1}^{|Gr|} \frac{c(g_k, h_j)}{(\sum c(g_k, \cdot))} \cdot \frac{c(g_k, h_i)}{(\sum c(\cdot, h_i))}. \tag{5}$$

Equation (5) shows the preference voting for target hotel h_j from all the guests in Gr stayed at h_i. We note that we classified reviews into positive/negative. We used the results to improve the quality of score. More precisely, we used only the positive review counts to calculate transition probability. $c(g_k, h_j)$ and $c(g_k, h_i)$ in Eq. (5) refer to the lodging count that the guest g_k reviewed the hotel $h_j(h_i)$ as *positive*. $P(h_j \mid h_i)$ in Eq. (5) is the marginal probability distribution over all the guests. The transition probability obtained by Eq. (5) shows a weight assigned to the edge between hotels h_i and h_j.

We used the row-normalized matrix $U_{ij} = (U_{ij})_{|H| \times |H|}$ to describe Gr with each entry corresponding to the transition probability, where $U_{ij} = p(h_j \mid h_i)$. To make U a stochastic matrix, the rows with all zero elements are replaced by a smoothing vector with all elements set to $\frac{1}{|H|}$. The matrix form of the recommendation score $Score(h_i)$ can be formulated in a recursive form as in the MRW model: $\lambda = \mu U^T \lambda + \frac{(1-\mu)}{|H|} e$, where $\lambda = [Score(h_i)]_{|H| \times 1}$ is a vector of saliency scores for the hotels. e is a column vector with all elements equal to 1. μ is a damping factor. We set μ to 0.85, as in the PageRank [24]. The final transition matrix is given by:

$$M = \mu U^T + \frac{(1-\mu)}{|H|} ee^T. \tag{6}$$

Each score is obtained by the principal eigenvector of the new transition matrix M. We applied the algorithm to the graph. The higher score based on transition probability the hotel has, the more suitable the hotel is recommended. For each aspect, we chose the topmost k hotels according to rank score. For each selected hotel, if the negative review is not included in the hotel reviews, we regarded the hotel as a recommendation hotel.

4   Experiments

4.1   Data

We used Rakuten travel data[3]. It consists of 11,468 hotels, 348,564 reviews submitted from 157,729 guests. We used plda[4] to assign positive/negative tag to the aspects. For each aspect, we estimated the number of reviews, and the number of topics (clusters) by searching in steps of 100 from 200 to 1,000. Table 1 shows the minimum entropy value, the number of reviews, and the number of topics for

---

[3] http://rit.rakuten.co.jp/rdr/index.html.
[4] http://code.google.com/p/plda.

Table 1. The minimum entropy value and the # of topics.

Aspect	Entropy	Reviews	Topics
Location	0.209	600	700
Room	0.460	700	600
Meal	0.194	500	700
Spa	0.232	400	500
Service	0.226	500	700
Amenity	0.413	600	600
Overall	0.202	500	700

each aspect. Table 4 shows that the number of reviews ranges from 400 to 700, and the number of topics are from 500 to 700. For each of the seven aspects, we used these numbers of reviews and topics in the experiments. We used linear kernel of SVM-Light [22] and set all parameters to their default values. All reviews were parsed by the syntactic analyzer CaboCha [17], and 633,634 dependency triples are extracted. We used them in the experiments.

We had an experiment to classify reviews into positive or negative. For each aspect, we chose the topmost 300 hotels whose number of reviews are large. We manually annotated these reviews. The evaluation is made by two humans. The classification is determined to be correct if two human judges agree. We obtained 400 reviews consisting 200 positive and 200 negative reviews. 400 reviews are trained by using SVMs for each aspect, and classifiers are obtained. We randomly selected another 100 test reviews from the topmost 300 hotels, and used them as test data. Each of the test data was classified into positive or negative by SVMs classifiers. The process is repeated five times. As a result, the macro-averaged F-score concerning to positive across seven aspects was 0.922, and the F-score for negative was 0.720. For each aspect, we added the reviews classified by SVMs to the original 400 training reviews, and used them as a training data to classify test reviews.

We created the data which is used to test our recommendation method. More precisely, we used the topmost 100 guests staying at a large number of different hotels as recommendation. For each of the 100 guests, we sorted hotels in chronological order. We used these with the latest five hotels as test data. To score hotels by MRW model, we used guest data staying at more than three times. The data is shown in Table 2. "Hotels" and "Different hotels" in Table 2

Table 2. Data used in the experiments.

Hotels	30,358
Different hotels	6,387
Guests	23,042
Reviews	116,033

refer to the total number of hotels, and the number of different hotels that the guests stayed at more than three times, respectively. "Guests" shows the total number of guests who stayed at one of the "Different hotels". "Reviews" shows the number of reviews with these hotels.

We used MAP (Mean-Averaged Precision) as an evaluation measure [25]. For a given set of guests $G = \{g_1, \cdots, g_n\}$, and $H = \{h_1, \cdots, h_{m_j}\}$ be a set of hotels that should be recommended for a guest g_j, the MAP of G is given by:

$$\text{MAP}(G) = \frac{1}{\mid G \mid} \sum_{j=1}^{\mid G \mid} \frac{1}{m_j} \sum_{k=1}^{m_j} Precision(R_{jk}). \tag{7}$$

R_{jk} in Eq. (7) refers to the set of ranked retrieval results from the top result until we get hotel h_k. *Precision* indicates a ratio of correct recommendation hotels by the system divided by the total number of recommendation hotels.

4.2   Basic Results

The results across seven aspects are shown in Table 3. As shown in Table 3, there are no significant difference among seven aspects, and the averaged MAP obtained by our method, aspect-based sentiment analysis (ASA) was 0.392. Table 4 shows sample clusters regarded as positive for three aspects, "location", "room", and "meal" obtained by LDA. Each cluster shows the top 5 triples and content words. We observed that the extracted triples show positive opinion for each aspect. This indicates that aspect extraction contributes to improve over-all performance. In contrast, some words such as *yoi* (be good) and *manzoku* (satisfy) in content word based clusters appear across aspects. Similarly, some words such as *ricchi* (location) and *cyousyoku* (breakfast) which appeared in negative cluster are an obstacle to identify positive/negative reviews in SVMs classification.

Table 3. Basic results.

	Location	Room	Meal	Spa	Service	Amenity	Overall	Avg
MAP	0.391	0.373	0.403	0.392	0.382	0.391	0.414	0.392

We recall that we classified aspects into seven types according to the guest preferences. There are other aspects for the hotels such as hotel types and area. We used three types of the hotels, *i.e.*, "Japanese style inn at a hot spring", "Pension", and "Business hotel". Similarly, we used two area, *i.e.*, "Tokyo" and "Nagano prefecture". We had an experiment to examine how these aspects affect the overall performance of recommendation. The data and the results are shown in Tables 5 and 6. We can see from Table 5 that there are no significant dif-ference among hotel types as the averaged MAP against the hotel types are from 0.384 to 0.394. However, the Map of "Tokyo" related to "Amenity" was 0.376 while that of "Nagano Pref." was only 0.329, and the difference was 0.047.

Table 4. Top 5 triples and content words.

Rank	Aspects		
	Location	Room	Meal
1	(*ni, eki, chikai*) be near to the station	(*ga, heya, yoi*) room was nice	(*ga, shokuji, yoi*) breakfast was nice
2	(*ha, hotel, chikai*) the hotel is close	(*ha, heya, hiroi*) the room is wide	(*ha, shokuji, yoi*) meal was nice
3	(*ni, hotel, chikai*) be near to the hotel	(*ga, heya, kirei*) A room is clean	(*ha, restaurant, good*) a restaurant is good
4	(*ni, parking, chikai*) be near to the parking	(*de, sugoseru, heya*) can spend in the room	(*ha, restaurant, yoi*) restaurant is nice
5	(*ga, konbini, aru*) be near to the convenience store	(*ha, heya, jyuubun*) a room is enough goo	(*ha, buffet, yoi*) Buffet is delicious
Rank	Content words		
	Location	Room	Meal
1	*ricchi* location	*heya* room	*syokuji* meal
2	*eki* station	*hiroi* be wide	**yoi** be good
3	**yoi** be good	*kirei* be clean	*cyousyoku* breakfast
4	*mise* store	**manzoku** satisfy	*oishii* be delicious
5	*subarashii* be great	**yoi** be good	**manzoku** satisfy

Table 5. Data and results (hotel types).

	Data		Results							
	Hotels	Reviews	Location	Room	Meal	Spa	Service	Amenity	Overall	Avg
Hot spring	3,073	52,798	0.393	0.375	0.411	0.394	0.381	0.392	0.411	0.394
Pension	1,845	30,275	0.383	0.364	0.388	0.379	0.376	0.386	0.401	0.384
Business	2,759	38,569	0.394	0.370	0.394	0.386	0.379	0.390	0.408	0.389

Table 6. Data and results (area).

	Data		Results							
	Hotels	Reviews	Location	Room	Meal	Spa	Service	Amenity	Overall	Avg
Tokyo	982	2,902	0.369	0.359	0.387	0.381	0.376	0.382	0.399	0.394
Nagano Pref	897	2,093	0.361	0.353	0.378	0.369	0.329	0.371	0.380	0.367

One reason behind this lies the small number of reviews as the "Amenity" of "Nagano Pref." consisted of only 47 reviews. For future work, we should be extend our method for further efficacy by overcoming the lack of sufficient reviews in data sets.

4.3   Comparative Experiments

We compared the results obtained by our method, ASA with the following four approaches to examine how the results of each method affect the overall performance.

1. Transition probabilities without review (TPWoR)
   The probability $P(h_j \mid h_i)$ used in the method is the preference voting for the target hotel h_j from all the guests in a set G who stayed at h_i, regardless of positive or negative review of G.
2. Content Words (CW)
   The difference between content words method and our method, ASA is that the former applies LDA to the content words.
3. Without reviews classified by SVMs (WoR)
   SVMs used in this method classifies test data by using only the original 400 training reviews.
4. Without negative review filtering (WoNRF)
   The method selected the topmost k hotels according to the MRW model, and the method dose not use negative reviews as a filtering.

Table 7 shows averaged MAP across seven aspects. As we can see from Table 7 that aspect-based sentiment analysis was the best among four baselines, and MAP score attained at 0.392. The result obtained by transition probability without review was worse than any other results. This shows that the use of guest review information is effective for recommendation. Table 7 shows that the result obtained by content words method was worse than the result obtained by aspect-based sentiment analysis, and even worse than the results without reviews classified by SVMs (WoR) and without negative review filtering (WoNRF). Furthermore, we can see from Table 7 that negative review filtering was a small contribution, *i.e.*, the improvement was 0.014 as the result without negative

Table 7. Recommendation results.

Method	MAP
Trans. pro. without review	0.257
Content words	0.304
Without reviews by SVMs	0.356
Without neg review filtering	0.378
Aspect-based SA	0.392

review filtering was 0.378 and aspect-based SA was 0.392. One reason is that the accuracy of negative review identification. The macro-averaged F-score concerning to negative across seven aspects was 0.720, while the F-score for positive was 0.922. Negative review filtering depends on the performance of negative review identification. Therefore, it will be necessary to examine features other than word triples to improve negative review identification.

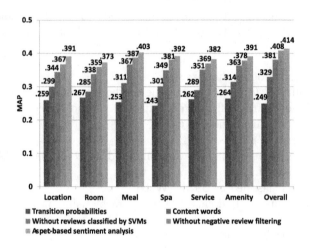

Fig. 2. The results against each aspect.

It is very important to compare the results of our method with four baselines against each aspect. Figure 2 shows MAP against each aspect. The results obtained by aspect-based sentiment analysis were statistically significant compared to other methods except for the aspects "spa" and "overall" in without negative review filtering method.

Table 8 shows a ranked list of the hotels for one guest (guest ID: 2037) obtained by using each method. The aspect is "meal", and each number shows hotel ID. Bold font in Table 8 refers to the correct hotel, *i.e.*, the latest five hotels that the guest stayed at. As can be seen clearly from Table 8, the result obtained by our method includes all of the five correct hotels within the topmost eight hotels, while without negative review filtering (WoNRF) was four. TPWoR, CW, and WoR did not work well as the number of correct hotel was no more than three, and these were ranked seventh and eighth.

It is interesting to note that some recommended hotels are very similar to the correct hotels, while most of the eight hotels did not exactly match these correct hotels except for the result obtained by aspect-based sentiment analysis method. If these hotels were similar to the correct hotels, the method is effective for finding transitive associations between the hotels that have never been stayed but share the same neighborhoods. Therefore, we examined how these hotels are similar to the correct hotels. To this end, we calculated distance between correct hotels

Table 8. Recommendation list for user ID 2037.

Rank	TPWoR	CW	WoR	WoNRF	ASA
1	2349	2203	2614	3022	**449**
2	604	2349	554	604	30142
3	12869	30142	30142	**449**	**18848**
4	90	604	604	30142	531
5	666	12869	3022	**18848**	**769**
6	2149	39502	531	531	**2223**
7	38126	**449**	**449**	**769**	15204
8	**449**	31209	**18848**	**2223**	**20428**

and other hotels within the rank for each method by using seven preferences. The preferences have star rating, *i.e.*, each has been scored from 1 to 5, where 1(bad) is lowest, and 5(good) is the best score. We represented each ranked hotel as a vector where each dimension of a vector is these seven preferences and the value of each dimension is its score value. The distance between correct hotel and other hotels within the rank for each method X is defined as:

$$\text{Dis}(X) = \frac{1}{|G|} \sum_{i=1}^{|G|} \operatorname*{argmin}_{j,k} \ d(R\_h_{ij}, C\_h_{ik}). \tag{8}$$

$| G |$ refers to the number of guests. R_h_{ij} refers to a vector of the j-th ranked hotels except for the correct hotels. Similarly, C_h_{ik} stands for a vector representation of the k-th correct hotel. d refers to Euclidean distance. Equation (8) shows that for each guest, we obtained the minimum value of Euclidean distance between R_h_{ij} and C_h_{ik}. We calculated the averaged summation of the 100 guests. The results are shown in Table 9. The value of "Dis" in Table 9 shows that the smaller value indicates a better result. We can see from Table 9 that the hotels except for the correct hotels obtained by our method are more similar to the correct hotels than those obtained by four baselines. The results show that our method is effective for finding hotels that have never been stayed at but share the same neighborhoods.

Table 9. Distance between correct hotel and another hotel.

Method	Dis
Trans. pro. without review	3.067
Content words	2.859
Without reviews by SVMs	2.721
Without neg review filtering	2.532
Aspect-based SA	2.396

5   Conclusions

We proposed a method for hotel recommendation by incorporating preferences related to the different aspects of a hotel to improve quality of the score. We used the results of aspect-based sentiment analysis to detect guest preferences. We parsed all reviews by the syntactic analyzer, and extracted dependency triples. For each aspect, we identified the guest opinion to positive or negative using dependency triples in the guest review. We calculated transitive association between hotels based on the positive/negative opinion. Finally, we scored hotels by Markov Random Walk model. The comparative results using Rakuten travel data showed that aspect analysis of guest preferences improves overall performance and especially, it is effective for finding hotels that have never been stayed at but share the same neighborhoods.

There are a number of directions for future work. In the aspect-based sentiment analysis for guest preferences, we should be able to obtain further advantages in efficacy by overcoming the lack of sufficient reviews in data sets by incorporating transfer learning approaches [26,27]. We used only surface information of terms (words) and ignore their senses in the aspect-based sentiment analysis. A number of methodologies have been developed for identifying semantic related words in natural language processing research field. This is a rich space for further exploration. We used Rakuten Japanese travel data in the experiments, while the method is applicable to other textual reviews. To evaluate the robustness of the method, experimental evaluation by using other data such as grocery stores: LeShop[5] and movie data: movieLens[6] can be explored in future. Finally, comparison to other recommendation methods, *e.g.*, matrix factorization methods (MF) [28] and combination of MF and the topic modeling [29] will also be considered in the future.

References

1. Balabanovic, M., Shoham, Y.: Fab content-based collaborative recommendation. Commun. ACM **40**, 66–72 (1997)
2. Park, S.T., Pennock, D.M., Madani, O., Good, N., DeCoste, D.: Naive filterbots for robust cold-start recommendations. In: Proceedings of the 12th ACM SIGKDD Conference on Knowledge Discovery and Data Mining, pp. 699–705 (2006)
3. Yildirim, H., Krishnamoorthy, M.S.: A random walk method for alleviating the sparsity problem in collaborative filtering. In: Proceedings of the 3rd ACM Conference on Recommender Systems, pp. 131–138 (2008)
4. Liu, N.N., Yang, Q.: A ranking-oriented approach to collaborative filtering. In: Proceedings of the 31st Annual International ACM SIGIR Conference on Research and Development in Information Retrieval, pp. 83–90 (2008)
5. Lathia, N., Hailes, S., Capra, L., Amatriain, X.: Temporal diversity in recommender systems. In: Proceedings of the 33rd ACM SIGIR Conference on Research and Development in Information Retrieval, pp. 210–217 (2010)

---

[5] www.beshop.ch.
[6] http://www.grouplens.org/node/73.

6. Zhao, X., Zhang, W., Wang, J.: Interactive collaborative filtering. In: Proceedings of the 22nd ACM Conference on Information and Knowledge Management, pp. 1411–1420 (2013)
7. Sarwar, B., Karypis, G., Konstan, J., Reidl, J.: Automatic multimedia cross-model correlation discovery. In: Proceedings of the 10th ACM SIGKDD Conference on Knowledge Discovery and Data Mining, pp. 653–658 (2001)
8. Cane, W.L., Stephen, C.C., Fu-lai, C.: Integrating collaborative filtering and sentiment analysis. In: Proceedings of the ECAI 2006 Workshop on Recommender Systems, pp. 62–66 (2006)
9. Niklas, J., Stefan, H.W., Mark, C.M., Iryna, G.: Beyond the stars: exploiting free-text user reviews to improve the accuracy of movie recommendations. In: Proceedings of the 1st International CIKM Workshop on Topic-Sentiment Analysis for Mass Opinion, pp. 57–64 (2009)
10. Raghavan, S., Gunasekar, S., Ghosh, J.: Review quality aware collaborative filtering. In: Proceedings of the 6th ACM Conference on Recommender Systems, pp. 123–130 (2012)
11. Li, M., Dias, B., Jarman, I.: Grocery shopping recommendations based on basket-sensitive random walk. In: Proceedings of the 15th ACM SIGKDD Conference on Knowledge Discovery and Data Mining, pp. 1215–1223 (2009)
12. Faridani, S.: Using canonical correlation analysis for generalized sentiment analysis product recommendation and search. In: Proceedings of 5th ACM Conference on Recommender Systems, pp. 23–27 (2011)
13. Blei, D.M., Ng, A.Y., Jordan, M.I.: Latent Dirichlet allocation. Mach. Learn. **3**, 993–1022 (2003)
14. Yin, Z., Gupta, M., Weninger, T., Han, J.: A unified framework for link recommendation using random walks. In: Proceedings of the Advances in Social Networks Analysis and Mining, pp. 152–159 (2010)
15. Li, L., Zheng, L., Fan, Y., Li, T.: Modeling and broadening temporal user interest in personalized news recommendation. Expert Syst. Appl. **41**(7), 3163–3177 (2014)
16. Wijaya, D.T., Bressan, S.: A random walk on the red carpet: rating movies with user reviews and pagerank. In: Proceedings of the ACM International Conference on Information and Knowledge Management CIKM 2008, pp. 951–960 (2008)
17. Kudo, T., Matsumoto, Y.: Fast method for Kernel-based text analysis. In: Proceedings of the 41st Annual Meeting of the Association for Computational Linguistics, pp. 24–31 (2003)
18. Kobayashi, N., Inui, K., Matsumoto, Y., Tateishi, K., Fukushima, S.: Collecting evaluative expressions for opinion extraction. J. Nat. Lang. Process. **12**(3), 203–222 (2005)
19. Hofmann, T.: Probabilistic latent semantic indexing. In: Proceedings of the 22nd Annual International ACM SIGIR Conference on Research and Development in Information Retrieval, pp. 50–57 (1999)
20. Li, Y., Yang, M., Zhang, Z.: Scientific articles recommendation. In: Proceedings of the ACM International Conference on Information and Knowledge Management CIKM 2013, pp. 1147–1156 (2013)
21. Turney, P.D.: Thumbs up or thumbs down? Semantic orientation applied to unsupervised classification of reviews. In: Proceedings of the 40th Annual Meeting of the Association for Computational Linguistics, pp. 417–424 (2002)
22. Joachims, T.: SVM Light Support Vector Machine. Department of Computer Science. Cornell University (1998)

23. Xue, G.R., Yang, Q., Zeng, H.J., Yu, Y., Chen, Z.: Exploiting the hierarchical structure for link analysis. In: Proceedings of the 28th ACM SIGIR Conference on Research and Development in Information Retrieval, pp. 186–193 (2005)
24. Brin, S., Page, L.: The anatomy of a large-scale hypertextual web search engine. Comput. Netw. **30**(1–7), 107–117 (1998)
25. Yates, B., Neto, R.: Modern Information Retrieval. Addison Wesley, New York (1999)
26. Blitzer, J., Dredze, M., Pereira, F.: Biographies, bollywood, boom-boxes and blenders: domain adaptation for sentiment classification. In: Proceedings of the 45th Annual Meeting of the Association for Computational Linguistics, pp. 187–295 (2007)
27. Dai, W., Yang, Q., Xue, G., Yu, Y.: Boosting for transfer learning. In: Proceedings of the 24th International Conference on Machine Learning, pp. 193–200 (2007)
28. Koren, Y., Bell, R.M., Volinsky, C.: Matrix factorization techniques for recommender systems. Computer **42**(8), 30–37 (2009)
29. Wang, C., Blei, D.M.: Collaborative topic modeling for recommending scientific articles. In: Proceedings of the 17th ACM SIGKDD Conference on Knowledge Discovery and Data Mining, pp. 448–456 (2011)

Searching the Web by Meaning: A Case Study of Lithuanian News Websites

Tomas Vileiniškis[(⊠)], Algirdas Šukys, and Rita Butkienė

Department of Information Systems, Kaunas University of Technology,
Studentu 50, Kaunas, Lithuania
{tomas.vileiniskis,algirdas.sukys,
rita.butkiene}@ktu.lt

Abstract. The daily growth of unstructured textual information created on the Web raises significant challenges when it comes to serving user information needs. On the other hand, evolving Semantic Web technology has influenced a wide body of research towards meaning-based text processing and information retrieval methods, that go beyond classical keyword-driven approaches. However, most of the work in the field targets English as the primary language of interest. Hence, in this paper we present a very first attempt to process unstructured Lithuanian text at the level of ontological semantics. We introduce an ontology-based semantic search framework capable of answering structured natural Lithuanian language questions, discuss its language-dependent design decisions and draw some observations from the results of a recent case study carried out over domain-specific Lithuanian web news corpus.

Keywords: Semantic search · SBVR · SPARQL · Information retrieval · Ontology · Semantic annotation · Lithuanian language

1 Introduction

In the context of traditional Web search, Information Retrieval (IR) has been known as a task of retrieving documents relevant to user information needs, typically expressed by some form of a query. A general IR model can be characterized by three main building blocks: representation of a user query, document content description and a retrieval function. Early work in IR field highly focused on keyword-based models, such as Boolean and Vector Space Model [1]. The obvious shortcoming of these models is the lack of conceptualization both at the query and document representation end, which eventually results in poor precision and recall rates. Several approaches such as query expansion [2] and word sense disambiguation [3] have been proposed to manage synonymy and polysemy in order to somewhat cope with the limitations of the aforementioned models.

However, the introduction of common Semantic Web standards for semantic data and domain knowledge representation (Resource Description Framework (RDF), RDF Schema (RDFS), Web Ontology Language (OWL2)) followed by a dedicated RDF query language (SPARQL) influenced a wide body of research [4] towards meaning-based IR, which we will refer to as *semantic search* throughout the paper. Advanced Information Extraction (IE) methods (e.g. semantic annotation, ontology population)

© Springer International Publishing AG 2016
A. Fred et al. (Eds.): IC3K 2015, CCIS 631, pp. 47–64, 2016.
DOI: 10.1007/978-3-319-52758-1_4

are employed to complement standard text preprocessing techniques (e.g. tokenization, stemming, stop word removal etc.) behind IR models.

Due to complexity of natural language, IE gets hardly dependent on advances in Natural Language Processing (NLP) techniques. Lithuanian language, once compared to the state-of-the-art of IE oriented NLP research for widely used languages, such as English, is still pretty much an open research field. As a result, the lack of resources for NLP-related tasks behind IE restricts the extent to which Lithuanian (and other less popular languages in general) can be approached practically. In contrast, this is not the case with well-researched languages (e.g. see IBM's Watson project) [5].

The complexity of Lithuanian language raises multiple NLP-specific challenges. First of all, it is highly inflected (7 cases, 2 genders, 2 grammatical numbers, 5 noun declensions) which means that a single word stem may lead to lots of different word forms, each of them belonging to a separate grammatical category. E.g., a nominative singular noun *dokumentas* (document) alone has multiple other grammatical cases reflected by alternating suffixes: *dokumento* (genitive), *dokumentu* (instrumental), *dokumentui* (dative), *dokumente* (locative), *dokumentą* (accusative) etc. Taking into account such declension of nouns and adjectives is crucial when determining grammatical function of a word in a sentence. Lithuanian is also a free word order language, meaning that a single sentence can be expressed in multiple different ways just by switching word positions. This suggests a need for non-standard syntactic parsing strategies that concentrate more on morphological language features [6]. Such challenges generally apply not only for Lithuanian but any other morphologically rich languages (e.g. Slavic) as well.

In this paper we present a combined attempt to semantic content processing and search over Lithuanian web texts. A semantic search framework for the task is proposed. We introduce an ontology population-based IE approach which is tightly coupled with a model-to-model (M2M) transformation-driven IR model. We show how such tight-coupling enables us to serve natural structured language queries over domain-specific data represented in the form of ontology. The applicability of our framework is then evaluated by performing a case study over a crawled Lithuanian news website corpus, focusing on political and economic domains. To the best of our knowledge, this is the first public attempt to Lithuanian text processing at the level of ontological semantics. The rest of the paper is structured as follows. Section 2 gives a brief overview of related work in semantic search area and provides state-of-the-art of NLP research for Lithuanian language. Section 3 presents the architecture of our semantic search framework with emphasis on capturing and maintaining domain-specific semantics throughout the search process. The experimental observations and lessons learned from the case study are presented in Sect. 4. Finally, we draw conclusions and discuss our future research plans in Sect. 5.

2   Related Work

The evolution of Semantic Web technology has made a significant impact on meaning-based IR methods over the last decade. In particular, the introduction of W3C's OWL2, RDFS, RDF and SPARQL to conceptualize, represent and query domain specific knowledge led to an upsurge of research in the field.

[7] Proposed KIM - a framework for semantic annotation and retrieval. The main idea behind KIM is the semantic typing of named entities (NE) by linking them to pre-populated knowledge base entries and/or appropriate domain-ontology classes. [9] Introduced an approach for semantically enhanced IR by adapting the classical vector space model [8]. The IE task used to conceptualize document content is similar to the one proposed by [7]. In addition, [9] use an ontology-based Question Answering (QA) system to interpret the intent behind user queries. This is achieved by deriving linguistic triples from a natural language question and then looking up for answer-bearing ontology concepts by syntactic triple similarity matches [10]. Our approach to capturing user query intents differs substantially: we aim at obtaining a formal SPARQL query model from a structured natural language question (see Sect. 3.2).

Knowledge bases like Freebase or DBpedia have been recently used to tackle the problem of open-domain QA [24, 25]. While their main goal is to retrieve answers to factoid-like questions over structured world's knowledge, our framework is primarily targeted towards mining and searching domain-specific texts in order to satisfy event-centric information needs.

All of the above mentioned approaches target semantic search only from an English language perspective, thus they build upon sophisticated NLP methods that are well known and properly researched. However, this is not the case with Lithuanian NLP research. Perhaps one of the most significant achievements is the early work by [11] who created the first Lithuanian lemmatizer and part-of-speech (POS) tagger called *Lemuoklis*. The syntax of Lithuanian language has been extensively analyzed by [12] [6]. A recent approach to statistical dependency parsing [13] showed the importance of morphological features (especially grammatical case) for the accuracy of results. However, the lack of syntactically annotated data suggests that rule-based parsing is a better choice.

The only publically available case study of NLP-based content processing is presented in [14], where authors apply named entity recognition (NER) among other standard text pre-processing steps to annotate and analyse Lithuanian news media websites.

3   Semantic Search Framework

The architecture of our proposed semantic search framework along its main components is shown in Fig. 1. As was noted in Sect. 1 of the paper, the framework consists of two major tightly coupled parts: information extraction (IE) and information retrieval (IR) modules. For a detailed explanation on how these modules operate together please refer to Sects. 3.1 and 3.2 respectively.

The IE module is dedicated for document text annotation by linguistic components in the NLP pipeline. In other words, IE module is responsible for conceptualizing domain-specific entities and capturing the events they participate in. In order to avoid possible confusion about terminology, a note on the use of the terms "*ontology population*" and "*semantic annotation*" should be given [15]. Our text processing efforts concentrate on *ontology population*, i.e., adding instance data (*A-Box*) to a predefined ontology schema (*T-Box*). In addition, we perform *semantic annotation*, i.e., we link slices of document text to their ontological representation bits (*A-Box*) created in the *ontology population* step. In this aspect, our approach slightly differs from previous

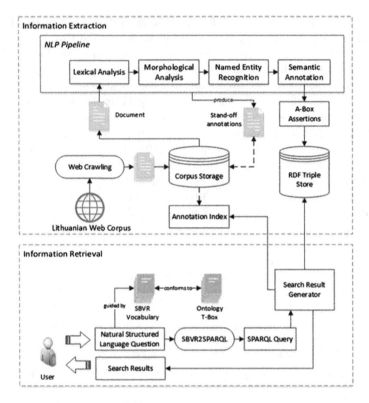

Fig. 1. The architecture of the framework.

works discussed in Sect. 2. The reason for this is two-fold. First, there is no ready-to-use semantic knowledge base that would have sufficient coverage of domain specific entities and relations commonly mentioned in Lithuanian media. The construction of such resource would require a significant amount of manual labor. Although, the existence of multilingual lexical knowledge bases (e.g. BabelNet) [16] is well-known, the entries for entities of a *local importance* (Lithuanian politics, organizations etc.) are rare to be found. Secondly, IR model behind our framework is based on formal SPARQL query execution, thus we expect for all the relevant domain knowledge acquired during text processing to be available in the form of RDF triples at query time.

The IR module behind the framework is highly based on SBVR (Semantics of Business Vocabulary and Business Rules) standard. SBVR is the OMG created metamodel and specification that defines vocabulary and rules for describing business semantics – business concepts, business facts, and business rules using some kind of Controlled Natural Language [17]. SBVR enables to create formal specifications understandable for business people and also interpretable by software tools. This is achieved by the usage of structured natural language for representing meaning as formal logic structures – semantic formulations. SBVR metamodel is based on principle of separating meaning of business concepts and business restrictions from their representation. A number of transformations of SBVR specifications to various software models have been created: Web services [18], BPMN [19], OWL2 [20], etc. We employ specific SBVR metamodel features to capture the meaning behind user's information needs and further to obtain a formal SPARQL query representation by means of model-to-model (M2M) transformation between the two.

3.1 Information Extraction

Information Extraction (IE) module aims to structure natural language document text at the level of ontological semantics, i.e. by analyzing entity mentions and their domain-specific relations we populate a predefined ontology schema with instance data. Formal ontological representation of document content allows taking advantage of implicit knowledge that can be inferred by employing OWL reasoning capabilities.

IE task behind our framework is powered by a pipeline of NLP components for Lithuanian language:

- Lexical analyzer performs stop word removal and standard text tokenization by breaking input text into tokens, sentences and paragraphs.
- Morphological analyzer assigns part-of-speech (POS) tags to each of the word along with its lemma, grammatical number and grammatical case.
- Named Entity recognizer (NER) is based on gazetteer lookups. It detects mentions of entities that belong to three major type categories: organizations, locations and persons.
- Semantic annotator analyses domain-specific relations between entities and produces ontology instance data in the form of RDF triples.

Each of the NLP components produces *stand-off* annotations in a custom data format which gets serialized using JSON. In such way, we keep the documents and their annotations decoupled.

Since the principle behind the three first NLP components in the pipeline is beyond the scope of this paper we will emphasize the fundamental features of our semantic annotator.

Given an ontology schema, semantic annotator attempts to populate it by instantiating classes and their properties with entity and relation mentions found in the analysed text. It follows a fully rule-based approach that looks for specific lexico-semantic patterns, combining information from prior lexical, morphological and named entity annotations.

Our current ruleset targets extraction of political and economic event mentions in their various forms. After collecting the most common reporting verbs (*sakyti* (say), *teigti* (state), *pranešti* (announce) etc.) from the news articles we derived multiple patterns for utterance extraction. Example rules are given below:

Rule I

```
 (c1)              (c2)    (c3)
{„SUBSTANCE",-} <RVERB> <NE> & type(NE) = Person =>
assert(c1:Substance, c2:Talking, c3:Person, talks<c3,c2>,
conveys<c2,c1>, has_talking_type<c2, "Statement">)
```

Rule II

```
(c1)    (c2)                      (c3)
<NE> <RVERB>{,} {kad|jog} {SUBSTANCE} & type(NE) = Person =>
assert(c1:Person, c2:Saying, c3:Substance, talks<c1,c2>, con-
veys<c2,c3>, has_talking_type<c2, "Agreement">)
```

Rule I is based on direct quotation extraction, while Rule II extracts indirect quotations by matching common conjunction *kad, jog* (that) patterns. In both cases the objective is to catch and instantiate the full triple: the agent, the reporting verb and the substance that is being reported. As can be seen from the rules above, specific talking type gets set as well, based on language semantics of different reporting verbs.

Some of the extraction rules are not as straightforward and require paying more attention to language specific morphological features. An example of this is the detection of work performed by persons within organizations (here, PNOUN stands for the position noun like *prezidentas* (president), *ministras* (minister), *teisėjas* (judge) etc.):

Rule III

```
(c1)     (c2)     (c3)
<NE1> <PNOUN> <NE2> & type(NE1) = Organization & type(NE2) =
Person & caseMark(NE1) = genitive =>
assert(c1:Organization, c3:Person, _c:Work, works<c3,_c>,
is performed in< c,c1>)
```

By relying just on lexical term sequence, we could end up with many incorrect extractions. In a sample sentence *Europos Parlamente prezidentė Dalia Grybauskaitė skaitė pranešimą* (President Dalia Grybauskaitė gave a speech at the European Parliament) the locative case of the word *Europos Parlamente* determines its grammatical function - an adverbial modifier of place. Ignoring the case mark, Rule III would result in assertion works<Dalia Grybauskaitė, Work>, is_performed_in<Work, European Parliament> which is not entirely true. Therefore, an additional check for the genitive case must be made in order to avoid incorrect extractions caused by Lithuanian declension.

An example fragment of the output semantic annotator would produce once Rule III gets successfully applied over a sample sentence is given below. Stand-off JSON semantic annotations show the textual boundaries of the asserted entity instances at the token level, while *A-Box* assertions in the form of RDF triples describe the entities and their domain-specific relation at the level of ontological semantics:

[0, 2]   [4, 2]
"*Europos Parlamento prezidentas Martin Schulz skaitė pranešimą.*"
(President of the European Parliament Martin Schulz gave a speech.)

```
"body": "{\"semantics\":
    [
    …
    {\"ref\":[0,2],\"inst_id\":\"
http://semantika.lt/ns/Agents#organization~02a861be33264c6~Age
nts.organization-1\"},
    {\"ref\":[4,2],\"inst_id\":\"http://semantika.lt/ns/Agents#
person~02a861be13326678~Agents.person-1\"},
    ]
```

```
    <rdf:Description
rdf:about="http://semantika.lt/ns/Agents#organization~02a861be
33264c6~Agents.organization-1">
        <rdf:type
rdf:resource="http://semantika.lt/ns/Agents#organization"/>
    ...
        <labels:label_lt xml:lang="lt">Europos
Parlamento</labels:label_lt>
        <semLT:refers_to__document
rdf:resource="http://semantika.lt/ns/SemLT#document~jh79n9kp~S
emLT.document-1"/>
    ...
    </rdf:Description>
```

```
    <rdf:Description
rdf:about="http://semantika.lt/ns/Agents#person~02a861be133266
78~Agents.person-1">
        <rdf:type
rdf:resource="http://semantika.lt/ns/Agents#person"/>
    ...
        <labels:label_lt xml:lang="lt"> Martin Schulz
</labels:label_lt>
        <events:works
rdf:resource="http://semantika.lt/ns/Events#work~bedc0c133~Eve
nts.work-1"/>
        <semLT:refers_to__document
rdf:resource="http://semantika.lt/ns/SemLT#document~jh79n9kp~S
emLT.document-1"/>
    ...
    </rdf:Description>
```

```
    <rdf:Description
rdf:about="http://semantika.lt/ns/Events#work~bedc0c133~Events
.work-1">
        <rdf:type
rdf:resource="http://semantika.lt/ns/Events#work"/>
    ...
        <events:is_performed_in rdf:resource="
http://semantika.lt/ns/Agents#organization~02a861be33264c6~Age
nts.organization-1"/>
        <semLT:refers_to__document
rdf:resource="http://semantika.lt/ns/SemLT#document~jh79n9kp~S
emLT.document-1"/>
    ...
    </rdf:Description>
```

Among the 3 rules presented above, our current ruleset includes over 20 patterns for detecting different kind of events, such as changes of prices, taxes and other abstract objects of interest. Also, any named entity mention always gets instantiated, whether it participates in some event or not.

The final assertions are produced according to the ontology schema that we created for capturing the event-specific knowledge commonly found in Lithuanian news articles. Currently, it has more than 100 classes and 70 relations. A tiny fragment of the ontology relative to the running examples throughout the paper is given in Fig. 2. The link between the document and the recognized objects within the content is established by an object property `<:refers_to_object>` or its inverse form `<:is_referred_in>`. The `Object` class is the top class of all domain entities that we try to detect through the IE process. Thus, the enrichment of ontology with new domain entities is only a matter of sub-classing `Object`.

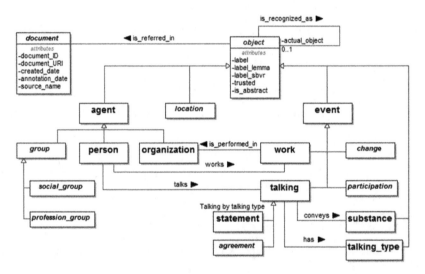

Fig. 2. A fragment of domain-specific event ontology. Ontology classes marked in bold represent directly corresponding SBVR vocabulary concepts.

As every mention of a named entity within the text results in new instance creation, ambiguity issues are unavoidable. The same entities tend to be referred to under different lexical aliases (*Dalia Grybauskaitė, Grybauskaitė, D. Grybauskaitė* etc.) throughout the news articles. Lithuanian declension causes even more suffix alternations (*Daliai Grybauskaitei, Dalios Grybauskaitės, Dalią Grybauskaitę* etc.) in such way having a negative impact on recall with queries including proper names (see Sect. 3.2). Our strategy here is to employ several heuristics in order to disambiguate all the different entity mentions to a single entity we call *trusted*:

– First, we find equal entities by performing common lemma and abbreviation matches.

- Secondly, we determine the main alias behind the *trusted* entity by inflecting its nominative case, and then create a new instance *T*.
- Lastly, we link all the corresponding entities to the *trusted* instance *T* by an object property `<:recognized_as_trusted_object>`.

We iterate the above process for each distinct entity type (organizations, locations, persons) recognizable by NER. In addition, the ontology is pre-populated with a set of well-known *trusted* entities along with their main aliases, which makes the disambiguation process more precise by eliminating the need to perform the second disambiguation step.

As mentioned in Sect. 3, in addition to *A-Box* ontology assertions, semantic annotator produces stand-off *semantic annotations*, i.e. extracted document text fragments get linked to their corresponding ontology entity URIs by token indices. This information is later used in IR phase.

3.2   Information Retrieval

Our semantically enhanced Information Retrieval (IR) model builds upon the framework for querying OWL2 ontologies using structured natural language, as presented in [21]. The operation of this framework depends on SBVR, OWL2 and SPARQL specifications, each of which comes with a formal metamodel, thus making model-to-model transformations possible. SBVR business vocabulary is used to formulate, serialize and transform user's information needs to a SPARQL query which, eventually, retrieves appropriate answers from the ontology. An advantage of such model-driven IR approach is the ability to capture and map domain-specific business restrictions to formal query conditions (triple patterns) in a straightforward way, once the M2M transformation rules are present.

To ensure that the resulting triple patterns of SPARQL query correspond to ontology classes and properties, it is important to keep correspondence between SBVR vocabulary entries (general concepts, verb concepts) and OWL2 ontology entities. The most reliable way to do this is to obtain ontology schema automatically using model transformations from specifications of SBVR business vocabulary and business rules as described in [20] or vice versa [23].

Having business vocabulary and corresponding ontology schema in place, questions can be written using structured natural language, which helps to express the intents more precisely and avoid ambiguities that are common in natural language interpretation. Question formulation using structured natural language under strict grammar rules imposes the need to guide the end user throughout the process. Therefore, we employ EBNF (Extended Backus–Naur Form) to constraint user input to a somewhat relaxed form of possible SBVR formulations present in the vocabulary. As the question is written with the help of contextual suggestions, it gets parsed using EBNF rules and an abstract syntax tree (AST) is created. The latter, which contains recognized statements of the question, is further used to generate SBVR XMI (*XML Metadata Interchange*) model. This model holds the captured meaning of a question that is constructed using a closed projection with restricting logical formulations and projection variables, expressing general concepts that should appear in the answer. SBVR XMI model is further transformed into SPARQL XMI model using ATL model-to-model transformation language. The principles behind transformation process and specific transformation rules are described in more details in [22]. At the final step, SPARQL XMI model is translated to textual representation using a model-to-text generator. The general workflow of this process is shown in Fig. 3.

An illustrative example of transforming question *"Kokie asmenys dirba organizacijose?"* (What persons work in organizations?) to a SPARQL query is given below. For the sake of simplicity, we provide only a small fragment of SBVR vocabulary (Lithuanian and English equivalents), necessary for this particular and following example transformations in the paper:

```
asmuo
organizacija
darbas
asmuo dirba darbą
darbas yra_dirbamas organizacijoje
asmuo dirba organizacijoje
   Definition: asmuo dirba darbą kuris yra_dirbamas
   organizacijoje
turinys
kalbėjimas
kalbėjimo tipas
Teigimo kalbėjimo tipas
   General concept: kalbėjimo tipas
kalbėjimas turi kalbėjimo tipą
asmuo kalba kalbėjimą
kalbėjime kalba turinį
asmuo kalba turinį
   Definition: asmuo kalba kalbėjimą kuriame kalba turinį
turinį kalba asmenys
   See: asmuo kalba turinį
asmuo teigia turinį
   Definition: asmuo kalba kalbėjimą kuriame kalba turinį ir
kalbėjimas turi kalbėjimo tipą Teigimo kalbėjimo tipas
```

The fragment consists of multiple vocabulary entries relevant to the running example question: general concepts asmuo (person), organizacija (organization) and a verb concept asmuo dirba organizacijoje (person works_in organization) denoting the domain specific relation between the prior defined concepts. As this relation is derived through event concept darbas (work), we use SBVR *definition* to describe this derivation. If the verb concept that particular question is based on has definition, that definition is used to transform question to SPARQL query.

```
person
organization
work
person works work
work is_performed_in organization
person works_in organization
   Definition: person works work that is_performed_in
   organization
substance
talking
talking type
Statement type of talking
   General concept: talking type
talking has talking type
person talks talking
talking conveys substance
person talks substance
   Definition: person talks talking which conveys substance
substance is_talked_by person
   See: person talks substance
person states substance
   Definition: person talks talking which conveys substance
and talking has talking type Statement type of talking
```

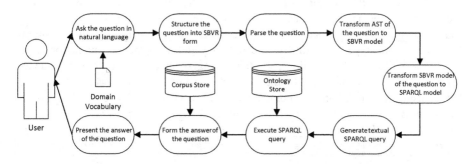

Fig. 3. The general workflow of proposed model-driven IR approach.

The declension of Lithuanian nouns can be clearly seen from the above example, i.e. the word representing general concept organizacija changes its suffix in the verb concept asmuo dirba organizacijoje since the verb dirba governs the locative case. In the English example, the grammatical form of a general concept organization remains the same since its role is determined by the use of the preposition in. We manage such language inflection by referring to the same concepts in different SBVR formulations by their main grammatical form – lemma.

Several heuristics are employed to make the structured question as natural sounding as possible. For example, in certain cases we allow omitting the subject part of the question, which is later derived by performing grammatical case-based matching in the vocabulary entries. As a result, the question in our running example can be expressed in a more user-friendly form "*Kas dirba organizacijose?*" (Who work in organizations?). Similarly, we manage singular and plural noun forms as well. When there's a proper name involved in the question, e.g. "*Kas dirba KTU?*" (Who works in KTU?), we allow omitting the type of individual which is further determined by analysing vocabulary entries and using morphological analyser with NER function in ambiguous cases.

The textual representation of transformed SPARQL query from the question in our running example (English equivalent) is presented below:

```
SELECT ?person_i ?organization_i WHERE {
    ?person_i ?person_works_work ?work_i.
    ?person_works_work :label "person works work".
    ?person_i rdf:type ?person_cl.
    ?person_cl :label "person".
    ?work_i rdf_type ?work_cl.
    ?work_cl :label "work".
    ?work_i ?work_is_performed_in_organization ?organization_i.
    ?work_is_performed_in_organization :label
    "work is performed in organization".
    ?organization_i rdf:type ?organization_cl.
    ?organization_cl :label "organization".
    }
```

Another example question "*Ką kalbėjo asmenys*" (What is_talked_by persons?) finds talks of persons. To transform this question we use verb concept turinį *kalbėjo* asmenys (substance *is_talked_by* person) and its definition. Since we set the type of talking in the ontology during semantic annotation, it is possible to write specific questions, i.e. for finding statements, assertions, announcements, acknowledgments, etc. In the domain vocabulary these specializations are derived using certain definitions. For example, verb concept asmuo *teigia* turinį (person *states* substance) is defined as asmuo *kalbėjo* kalbėjimą kuriame *kalbėjo* turinį ir kalbėjimas *turi* kalbėjimo_tipą Teigimo_kalbėjimo_tipas (person *talks* talking which *conveys* substance and talking *has* talking_type Statement_- type_of_talking). This definition is used as the basis for SPARQL query construction.

After the initial transformation, SELECT clause projects a set of variables V that bind (given the RDF graph data matches) to answer-bearing ontology entity URIs. The basic graph pattern (BGP) consists of multiple triple patterns that reflect the identification of conforming vocabulary and ontology concepts. In particular, we determine the type of each of the projected variables $v \in V$ in two steps:

- A triple pattern *T1* is created that binds a representative literal value of SBVR concept to a non-projected variable n (<?person_cl :label "person">).

- A subsequent triple pattern *T2* is created with *n* in an object position denoting the type of the projected variable *v* (<?person_i rdf:type ?person_cl>).

In a similar way we identify necessary vocabulary-conforming ontology properties.

Once the conformity between ontology classes, properties and SBVR vocabulary entries is established, the query gets further optimized by rewriting redundant triple patterns, i.e. we bind the ontological types of projected variables directly, instead of keeping the aforementioned representative literal values of SBVR concepts:

```
SELECT ?person_i ?organization_i WHERE {
    ?person_i :works ?work_i.
    ?person_i rdf:type :person.
    ?work_i rdf_type :work.
    ?work_i :is_performed_in ?organization_i.
    ?organization_i rdf:type :organization.
}
```

Queries with proper names involved, e.g. *Kas dirba Europos Parlamente*? (Who works in the European Parliament?), are transformed by additionally employing simple heuristics to retrieve disambiguated instances (see Sect. 3.1). In particular, we generate a set of triple patterns that use the <:recognized_as_trusted_object> predicate to bind to all the *non-trusted* instances, thus eventually giving higher recall.

Finally, the query is augmented with triple patterns that require for each of the projected variables to be bound to a single document instance (variable *d*), i.e. for each $v \in V$ we create triple patterns <d :refers_to_object v>. At the last step, we project an additional variable *k* in the SELECT clause that denotes the internal document identifier later on used for snippet generation. Note that $k \notin V$:

```
SELECT ?person_i ?organization_i ?k WHERE {
    ?person_i :works ?work_i.
    ?d :refers_to_object ?person_i.
    ?person_i rdf:type :person.
    ?work_i rdf_type :work.
    ?work_i :is_performed_in ?organization_i.
    ?organization_i rdf:type :organization.
    ?d :refers_to_object ?organization_i.
    ?d :document_ID ?k.
}
```

An ORDER BY clause could be added to sort the results according to document publication date however, the ordering cost proved to be too high on a larger dataset.

At this stage, we have fully-constructed a formal SPARQL query that returns entity URI bindings, essentially performing *data* retrieval. With the original research aim in mind to attempt meaning-based *information* retrieval, our proposed framework includes a component for result snippet generation. The logic behind it is based on the following algorithm:

- For each of the initial SPARQL projection variables $v \in V$ extract their URI bindings $v \rightarrow u$;
- For each *u* retrieve its beginning *b* and ending *e* token indices from the semantic and lexical annotations produced in IE phase;

- Calculate $min(b)$ and $max(e)$ values to determine the range of a text passage;
- If $min(b)$ and $max(e)$ fit within boundaries of a single sentence, extend the range of a text passage to a full sentence;
- Else If $min(b)$ and $max(e)$ overlap to the neighboring sentences, extend the range of a text passage to the boundaries of neighboring sentences;
- Extract the text passage as a final snippet.
- Repeat for every tuple in the binding set.

Given that the lexical and semantic annotations produced in IE phase are correct, the above algorithm results in a snippet containing both the answer-bearing entities, and the original context they were extracted from.

4   Evaluation

An early evaluation of our approach was performed by conducting a case study over a crawled corpus of Lithuanian news texts. We gathered over 500 million domain specific documents from more than 30 news portals. After initial pre-processing steps the documents were annotated producing around 235 million explicit and 273 million implicit RDF triples under OWL-Horst materialization settings in the triple store. A prototype for the search interface was deployed to ease the evaluation of the practical applicability of our approach (see Fig. 4).

Kas dirba Europos Parlamente?

Spauskite *Ctrl+Space* norėdami gauti galimus klausimą sudarančių žodžių pasiūlymus

E. Masiulis: Seimo pasakos, naujas leidimas | Balsas.lt
balsas.lt
2015 sau. 5 d. 19:21 - Kitas atvejis – apie kitą Europos Parlamento narį V. Uspaskichą, kuriam vis dar šmėžuoja, esą jį Lietuvoje kažkas politiškai persekiojo vien dėl to, kad yra ne tos tautybės ir taip toliau.

Minskas lūkuriuoja, o savo pasiuntinių atšaukimą vadina „techniniu" | Balsas.lt
balsas.lt
2015 sau. 11 d. 06:43 - Europos Parlamento pirmininkas Martinas Schulzas dėl tokio Minsko sprendimo pareiškė apgailestavimą.

Fig. 4. A prototype of semantic search interface for the case study. Sample results are shown for a question *Kas dirba Europos Parlamente?* (Who works in the European Parliament?).

In order to evaluate the search results in a quantitative manner, we selected 4 different queries for accuracy calculations: 2 abstract ones and 2 with proper names involved (see Table 1). We then judged the quality of the results on two main criteria: whether the text snippet returned gives a correct answer to the original question and if the answer-related entities are correctly highlighted within the text passage.

Table 1. Query set used for evaluation.

#	Query
Q1	*Ką kalbėjo agentai?* (What did the agents say?)
Q2	*Ką kalbėjo Vladimiras Putinas?* (What did Vladimir Putin say?)
Q3	*Kas dirba organizacijose?* (Who work in organizations?)
Q4	*Kas dirba Europos Parlamente?* (Who works in the European Parliament?)

As a single article could possibly contain multiple distinct answers to the same query, we chose to calculate precision values snippet wise, so the total number of analysed articles differs per query and ranges from 12 to 65. Queries Q1, Q2 and Q3, Q4 were assessed by manually evaluating 61 and 71 snippets respectively. These numbers proved to be enough to observe a general trend in error sources.

Getting correct recall values is not a straightforward task in our current setting since a full set of correct answers to each of the queries is not known in advance. Therefore, we calculated recall only on a working subset of articles, i.e. those that had their snippets evaluated as mentioned above. In particular, we analyzed the content of those articles to collect the number of missed annotations and assertions required to stand as an additional answer (snippet) with respect to the original query.

Table 2 shows the primary results of text snippet evaluation. Here, A_F column stands for the amount of snippets analyzed; A_{FC} – snippets with correct answer, A_{NF} – not found snippets. While most of the queries achieve very high precision rates, Q2 stands out with a bit lower results. We noticed that a common pitfall here is the extraction of indirect quotations, where pure lexico-semantic patterns can't differentiate between the reporting agent and other agents contextually related to the reported substance. Relatively low recall values indicate that our current domain-specific event extraction ruleset is capable of capturing only the most common event expressions.

Table 2. Text snippet accuracy results.

#	A_F	A_{FC}	A_{NF}	Recall	Precision
Q1	61	57	64	0.456	0.934
Q2	61	54	50	0.486	0.885
Q3	71	69	59	0.493	0.971
Q4	71	71	16	0.816	1.000

In addition, we evaluated accuracy of entity highlighting within the correctly returned snippets (see Table 3). Each of the queries from our query set is expected to return an entity tuple, either *agent-substance* (Q1, Q2) or *person-organization* (Q3, Q4), hence we split the results by agent/person and substance/organization columns. A_{FC} column lists the total number of snippets analyzed; A_{AP} – correctly highlighted agent/person entities, A_{SO} – correctly highlighted substance/organization entities. As the results in Table 3 show, Q2, Q3 and Q4 reach near-perfect precision values in both A_{AP} and A_{SO}. This is because these queries mostly return entities that get instantiated by strictly following NE tags. In contrast, Q1 fails significantly on A_{AP}. The *Agent* entity behind Q1 is more general according to our ontology schema (see Fig. 2), therefore not only NE instances are included within the results. In particular, *Agent* subclasses like *Group* (and other specific types of agents) get instantiated by domain list-guided noun lookups during IE. Our efforts here fail short when the looked-up noun only governs the correct entity to be instantiated in a noun phrase and does not stand as an instance by itself. Thus, noun phrase mining should be improved by taking into account more Lithuanian morphological features.

Table 3. Entity highlighting accuracy results.

#	A_{FC}	A_{AP}	A_{SO}	P_{AP}	P_{SO}
Q1	57	36	53	0.632	0.930
Q2	54	54	51	1.000	0.944
Q3	69	69	69	1.000	1.000
Q4	71	71	71	1.000	1.000

The primary experimental evaluation of our approach led to certain observations:

- The precision of search results is mainly affected by the performance of NLP components behind IE task, since the IR phase operates in a Boolean manner, i.e. given a transformed formal SPARQL query the returned bindings hold all the conditions expressed by the set of triple patterns.
- Syntactic parser for Lithuanian is a crucial linguistic component currently missing from the NLP pipeline. Event extraction from complex sentence structures with a free word order is a non-trivial task and can hardly be carried out by solely relying on lexico-semantic patterns.
- Even when NLP-related errors occur at IE stage, our snippet generation approach enables to deliver a correct text passage with a decent level of accuracy.

The evaluation results can be summed up on a qualitative note. As shown in Fig. 4, our strategy for snippet generation attempts to present the user with an answer in a single step, eliminating the need to perform an additional search for the answer-related text passage by opening the whole news article. We see this as one of the main features and advantages of *semantic search* paradigm when compared to classical pure keyword-based approaches.

5 Conclusions and Future Work

This paper introduced an ontology-based semantic search framework with tightly coupled Information Extraction and Information Retrieval modules. In our opinion, such tight coupling between the two once-separate disciplines plays a crucial role in the field of semantic search.

Our ongoing research showed that meaning-based information retrieval methods can be successfully tackled even when the underlying language, like Lithuanian, is highly inflected and has limited resources for natural language processing. Even though this makes the task of information extraction more challenging, paying close detail to language specific linguistic features allows capturing ontological semantics behind the analyzed texts with significant results.

While our efforts were primarily targeted toward Lithuanian language processing, the model-driven approach we took by employing SBVR, OWL2 and SPARQL standards leaves the flexibility to adopt the proposed framework to different languages as well. However, the diverse nature of different languages will always require certain adoptions due to native differences in their grammatical properties.

As the observations gathered from the evaluation results have shown, there is still a lot of room for improvement within our efforts, especially when it comes to information extraction. In order to achieve higher recall rates we plan on extending our event extraction ruleset with more rules and extraction patterns. We will also try to employ Named Entity Linking (NEL) techniques as to expand the typeset of the event-related entities beyond the ones predefined within our ontology. As a final note, the semantic search prototype that we used in our case study is publically available online to enhance the search experience for Lithuanian language users: http://www.semantika.lt/ SemanticSearch/Search/Index.

References

1. Salton, G., Wong, A., Yang, C.S.: A vector space model for automatic indexing. Commun. ACM **18**(11), 613–620 (1975)
2. Carpineto, C., Romano, G.: A survey of automatic query expansion in information retrieval. ACM Comput. Surv. (CSUR) **44**(1), 1 (2012)
3. Stokoe, C., Oakes, M.P., Tait, J.: Word sense disambiguation in information retrieval revisited. In: Proceedings of the 26th Annual International ACM SIGIR Conference on Research and Development in Information Retrieval, pp. 159–166. ACM (2003)
4. Mangold, C.: A survey and classification of semantic search approaches. Int. J. Metadata Semant. Ontol. **2**(1), 23–34 (2007)
5. Ferrucci, D., Brown, E., Chu-Carroll, J., Fan, J., Gondek, D., Kalyanpur, A., Welty, C., et al.: Building Watson: an overview of the DeepQA project. AI Mag. **31**(3), 59–79 (2010)
6. Šveikauskienė, D., Telksnys, L.: Accuracy of the parsing of Lithuanian simple sentences. Inf. Technol. Control **43**(4), 402–413 (2014)
7. Kiryakov, A., Popov, B., Terziev, I., Manov, D., Ognyanoff, D.: Semantic annotation, indexing, and retrieval. Web Semant.: Sci. Serv. Agents World Wide Web **2**(1), 49–79 (2004)
8. Castells, P., Fernandez, M., Vallet, D.: An adaptation of the vector-space model for ontology-based information retrieval. IEEE Trans. Knowl. Data Eng. **19**(2), 261–272 (2007)
9. Fernández, M., Cantador, I., López, V., Vallet, D., Castells, P., Motta, E.: Semantically enhanced information retrieval: an ontology-based approach. Web Semant.: Sci. Serv. Agents World Wide Web **9**(4), 434–452 (2011)

10. Lopez, V., Uren, V., Sabou, M.R., Motta, E.: Cross ontology query answering on the semantic web: an initial evaluation. In: Proceedings of the Fifth International Conference on Knowledge Capture, pp. 17–24. ACM (2009)
11. Zinkevičius, V.: Lemuoklis–morfologinei analizei. Darbai ir dienos **24**, 245–274 (2000)
12. Šveikauskienė, D.: Formal description of the syntax of the Lithuanian language. Inf. Technol. Control **34**(3), 1–12 (2005)
13. Kapociute-Dzikiene, J., Nivre, J., Krupavicius, A.: Lithuanian dependency parsing with rich morphological features. In: Fourth Workshop on Statistical Parsing of Morphologically Rich Languages, p. 12 (2013)
14. Krilavičius, T., Medelis, Ž., Kapočiūtė-Dzikienė, J., Žalandauskas, T.: News media analysis using focused crawl and natural language processing: case of Lithuanian news websites. In: Skersys, T., Butleris, R., Butkiene, R. (eds.) ICIST 2012. CCIS, vol. 319, pp. 48–61. Springer, Heidelberg (2012). doi:10.1007/978-3-642-33308-8_5
15. Amardeilh, F.: Semantic annotation and ontology population. In: Semantic Web Engineering in the Knowledge Society, 424 p. (2008)
16. Navigli, R., Ponzetto, S.P.: BabelNet: the automatic construction, evaluation and application of a wide-coverage multilingual semantic network. Artif. Intell. **193**, 217–250 (2012)
17. OMG. Semantics of Business Vocabulary and Business Rules (SBVR). Version 1.0, December 2008, OMG Document Number: formal/2008-01-02 (2008)
18. Goedertier, S., Vanthienen, J.: A vocabulary and execution model for declarative service orchestration. In: Hofstede, A., Benatallah, B., Paik, H.-Y. (eds.) BPM 2007. LNCS, vol. 4928, pp. 496–501. Springer, Heidelberg (2008). doi:10.1007/978-3-540-78238-4_50
19. Damiani, E., Ceravolo, P., Fugazza, C., Reed, K.: Representing and validating digital business processes. In: Filipe, J., Cordeiro, J. (eds.) WEBIST 2007. LNBIP, vol. 8, pp. 19–32. Springer, Heidelberg (2008). doi:10.1007/978-3-540-68262-2_2
20. Karpovič, J., Kriščiūnienė, G., Ablonskis, L., Nemuraitė, L.: The comprehensive mapping of semantics of business vocabulary and business rules (SBVR) to OWL 2 ontologies. Inf. Technol. Control **43**(3), 289–302 (2014)
21. Sukys, A., Nemuraite, L., Paradauskas, B., Sinkevicius, E.: Transformation framework for SBVR based semantic queries in business information systems. In: The Second International Conference on Business Intelligence and Technology, BUSTECH 2012, pp. 19–24 (2012)
22. Sukys, A., Nemuraite, L., Paradauskas, B.: Representing and transforming SBVR question patterns into SPARQL. In: Skersys, T., Butleris, R., Butkiene, R. (eds.) ICIST 2012. CCIS, vol. 319, pp. 436–451. Springer, Heidelberg (2012). doi:10.1007/978-3-642-33308-8_36
23. Bernotaityte, G., Nemuraite, L., Butkiene, R., Paradauskas, B.: Developing SBVR vocabularies and business rules from OWL2 ontologies. In: Skersys, T., Butleris, R., Butkiene, R. (eds.) ICIST 2013. CCIS, vol. 403, pp. 134–145. Springer, Heidelberg (2013). doi:10.1007/978-3-642-41947-8_13
24. Shekarpour, S., Marx, E., Ngomo, A.C.N., Auer, S.: Sina: semantic interpretation of user queries for question answering on interlinked data. Web Semant.: Sci. Serv. Agents World Wide Web **30**, 39–51 (2015)
25. Yao, X., Van Durme, B.: Information extraction over structured data: question answering with freebase. In: Proceedings of ACL (2014)

Piecewise Factorization for Time Series Classification

Qinglin Cai$^{(\boxtimes)}$, Ling Chen, and Jianling Sun

College of Computer Science and Technology,
Zhejiang University, Hangzhou, China
{qlcai,lingchen,sunjl}@zju.edu.cn

Abstract. In the research field of time series analysis and mining, the nearest neighbor classifier (1NN) based on the dynamic time warping distance (DTW) is well known for its high accuracy. However, the high computational complexity of DTW can lead to the expensive time consumption of the classifier. An effective solution is to compute DTW in the piecewise approximation space (PA-DTW). However, most of the existing piecewise approximation methods must predefine the segment length and focus on the simple statistical features, which would influence the precision of PA-DTW. To address this problem, we propose a novel piecewise factorization model (PCHA) for time series, where an adaptive segment method is proposed and the Chebyshev coefficients of subsequences are extracted as features. Based on PCHA, the corresponding PA-DTW measure named ChebyDTW is proposed for the 1NN classifier, which can capture the fluctuation information of time series for the similarity measure. The comprehensive experimental evaluation shows that ChebyDTW can support both accurate and fast 1NN classification.

Keywords: Time series · Piecewise approximation · Similarity measure

1 Introduction

Time series classification is an important topic in the research field of time series analysis and mining. A plethora of classifiers have been developed for this topic [1, 2], e.g., decision tree, nearest neighbor (1NN), naive Bayes, Bayesian network, random forest, support vector machine, rotation forest, etc. However, the recent empirical evidence [3–5] strongly suggests that, with the merits of robustness, high accuracy, and free parameter, the simple 1NN classifier employing the generic time series similarity measure is exceptionally difficult to beat. Besides, due to the high precision of dynamic time warping distance (DTW), the 1NN classifier based on DTW has been found to outperform an exhaustive list of alternatives [5], including decision trees, multi-scale histograms, multi-layer perception neural networks, order logic rules with boosting, as well as the 1NN classifiers based on many other similarity measures. However, the computational complexity of DTW is quadratic to the time series length, i.e., $O(n^2)$, and the 1NN classifier has to search the entire dataset to classify an object. As a result, the 1NN classifier based on DTW is low efficient for the high-dimensional time series. To address this problem, researchers have proposed to compute DTW in the alternative

© Springer International Publishing AG 2016
A. Fred et al. (Eds.): IC3K 2015, CCIS 631, pp. 65–79, 2016.
DOI: 10.1007/978-3-319-52758-1_5

piecewise approximation space (PA-DTW) [6–9], which transforms the raw data into the feature space based on segmentation, and extracts the discriminatory and low-dimensional features for similarity measure. If the original time series with length n is segmented into $N(N \ll n)$ subsequences, the computational complexity of PA-DTW will reduce to $O(N^2)$.

Many piecewise approximation methods have been proposed so far, e.g., piecewise aggregation approximation (PAA) [6], piecewise linear approximation (PLA) [7, 10], adaptive piecewise constant approximation (APCA) [8], derivative time series segment approximation (DSA) [9], piecewise cloud approximation (PWCA) [11], etc. The most prominent merit of piecewise approximation is the ability of capturing the local characteristics of time series. However, most of the existing piecewise approximation methods need to fix the segment length, which is hard to be predefined for the different kinds of time series, and focus on the simple statistical features, which only capture the aggregation characteristics of time series. For example, PAA and APCA extract the mean values, PLA extracts the linear fitting slopes, and DSA extracts the mean values of the derivative subsequences. If PA-DTW is computed on these methods, its precision would be influenced.

In this paper, we propose a novel piecewise factorization model for time series, named piecewise Chebyshev approximation (PCHA), where a novel code-based segment method is proposed to adaptively segment time series. Rather than focusing on the statistical features, we factorize the subsequences with Chebyshev polynomials, and employ the Chebyshev coefficients as features to approximate the raw data. Besides, the PA-DTW based on PCHA (ChebyDTW) is proposed for the 1NN classification. Since the Chebyshev polynomials with the different degrees represent the fluctuation components of time series, the local fluctuation information can be captured from time series for the ChebyDTW measure. The comprehensive experimental results show that ChebyDTW can support the accurate and fast 1NN classification.

The structure of this paper is as follows: The related work on data representation and similarity measure for time series is reviewed in Sect. 2; Sect. 3 shows the proposed methodology framework; the details of PCHA are presented in Sect. 4; Sect. 5 describes the ChebyDTW measure; Sect. 6 provides the comprehensive experiment results and analysis; Sect. 7 concludes this paper.

2   Related Work

2.1   Data Representation

In many application fields, the high dimensionality of time series has limited the performance of a myriad of algorithms. With this problem, a great number of data representation methods have been proposed to reduce the dimensionality of time series [1, 2]. In these methods, the piecewise approximation methods are prevalent for their simplicity and effectiveness. The first attempt is the PAA representation [6], which segments time series into the equal-length subsequences, and extracts the mean values of the subsequences as features to approximate the raw data. However, the extracted single sort of features only indicates the height of subsequences, which may cause the local information loss. Consecutively, an adaptive version of PAA named piecewise constant approximation (APCA) [8] was proposed, which can segment time series into the subsequences with adaptive lengths and thus can approximate time series with less error. As well, a multi-resolution version of PAA named MPAA [12] was proposed, which can iteratively segment time series into 2^i subsequences. However, both of the variations inherit the poor expressivity of PAA. Another pioneer piecewise representation is the PLA [7, 10], which extracts the linear fitting slopes of the subsequences as features to approximate the raw data. However, the fitting slopes only reflect the movement trends of the subsequences. For the time series fluctuating sharply with high frequency, the effect of PLA on dimension reduction is not prominent. In addition, two novel piecewise approximation methods were proposed recently. One is the DSA representation [9], which takes the mean values of the derivative subsequences of time series as features. However, it is sensitive to the small fluctuation caused by the noise. The other is the PWCA representation [11], which employs the cloud models to fit the data distribution of subsequences. However, the extracted features only reflect the data distribution characteristics and cannot capture the fluctuation information of time series.

2.2 Similarity Measure

DTW [1, 2, 5] is one of the most prevalent similarity measures for time series, which is computed by realigning the indices of time series. It is robust to the time warping and phase-shift, and has high measure precision. However, it is computed by the dynamic programming algorithm, and thus has the expensive $O(n^2)$ computational complexity, which largely limits its application to the high dimensional time series [13]. To overcome this shortcoming, the PA-DTW measures were proposed. The PAA representation based PDTW [14] and the PLA representation based SDTW [10] are the early pioneers, and the DSA representation based DSADTW [9] is the state-of-the-art method. Rather than in the raw data space, they compute DTW in the PAA, PLA, and DSA spaces respectively. Since the segment numbers are much less than the original time series length, the PA-DTW methods can greatly decrease the computational complexity of the original DTW. Nonetheless, the precision of PA-DTWs greatly depends on the used piecewise approximation methods, where both the segment method and the extracted features are crucial factors. As a result, with the weakness of the existing piecewise approximation methods, the PA-DTWs cannot achieve the high precision. In our proposed ChebyDTW, a novel adaptive segment method and the Chebyshev factorization are used, which overcomes the drawback of the fixed segmentation, and can capture the fluctuation information of time series for similarity measure.

3 Methodology Framework

Figure 1 shows the framework of the methods proposed in this paper, which consists of two parts:

(a) Piecewise Chebyshev approximation (PCHA). The time series is first coded into the binary sequence, and then segmented into the subsequences with adaptive

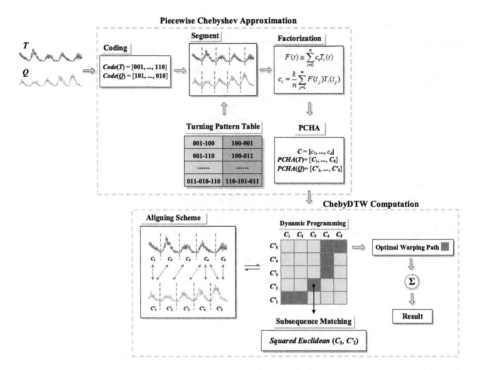

Fig. 1. The framework of the proposed methods.

lengths by matching the turning patterns. After that, the subsequences are factorized with the Chebyshev polynomials and projected into the Chebyshev factorization domain. The Chebyshev coefficients will be extracted as features to approximate the raw data.

(b) ChebyDTW computation. DTW will be computed in the Chebyshev factorization domain. Concretely, in the dynamic programming computation of DTW, the subsequence matching over the Chebyshev features is taken as the subroutine, where the squared Euclidean distance can be employed.

4 Piecewise Factorization

Without loss of generality, the relevant definitions are first given as follows.

Definition 1. (Time Series): The sample sequence of a variable X over n contiguous time moments is called time series, denoted as $T = \{t_1, t_2, \ldots, t_i, \ldots, t_n\}$, where $t_i \in R$ denotes the sample value of X on the i-th moment, and n is the length of T.

Definition 2. (Subsequence): Given a time series $T = \{t_1, t_2, \ldots, t_i, \ldots, t_n\}$, the subset S of T that consists of the continuous samples $\{t_{i+1}, t_{i+2}, \ldots, t_{i+l}\}$, where $0 \leq i \leq n\text{-}l$ and $0 \leq l \leq n$, is called the subsequence of T.

Definition 3. (Piecewise Approximation): Given a time series $T = \{t_1, t_2, ..., t_i, ..., t_n\}$, which is segmented into the subsequence set $S = \{S_1, S_2, ..., S_j, ..., S_N\}$, if $\exists f$: $S_j \rightarrow V_j = [v_1, ..., v_m] \in R^m$, the set $V = \{V_1, V_2, ..., V_j, ..., V_N\}$ is called the piecewise approximation of T.

4.1 Adaptive Segmentation

Inspired by the Marr's theory of vision [15], we regard the turning points, where the trend of time series changes, as a good choice to segment time series. However, the practical time series is mixed with a mass of noise, which results in many trivial turning points with small fluctuation. This problem can be simply solved by the efficient moving average (MA) smoothing method [16].

In order to recognize the significant turning points, we first exhaustively enumerate the location relationships of three adjacent samples t_1–t_3 with their mean μ in time series, as shown in Fig. 2. Six basic cell codes can be defined as Fig. 2(a), which is composed by the binary codes δ_1–δ_3 of t_1–t_3, and denoted as $\Phi(t_1, t_2, t_3) = (\delta_1\delta_2\delta_3)_b$. Six special relationships that one of t_1–t_3 equals to μ are encoded as Fig. 2(b).

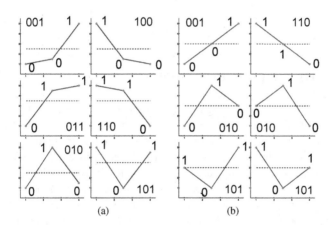

(a)   (b)

Fig. 2. Three adjacent samples with the basic cell codes of (A) basic relationships, and (B) specific relationships.

Based on the cell codes, all the minimum turning patterns (composed with two cell codes) at the turning points can be enumerated as Fig. 3. Note that, the basic cell codes 010 and 101 per se are the turning patterns. Then, we employ a sliding window of length 3 to scan the time series, and encode the samples within each window by Fig. 2. In this process, all the significant turning points can be found by matching Fig. 3, with which time series can be segmented into the subsequences with adaptive lengths.

However, the above segmentation is not perfect. Although the trivial turning points can be removed with the MA, the "singular" turning patterns may exist, i.e., the turning patterns appearing very close. As shown in Fig. 4, a Cricket time series from the UCR time series archive [17] is segmented by the turning patterns (dash line), where the raw data is first smoothed with the smooth degree 10 ($sd = 10$).

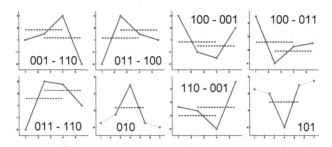

Fig. 3. The minimum turning patterns composed with two cell codes.

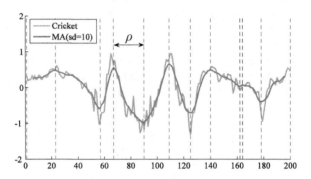

Fig. 4. Segmentation for the Cricket time series ($sd = 10$).

Obviously, the dash lines can significantly segment time series, but the two black dash lines are so close that the segment between them can be ignored. In view of this, we introduce the segment threshold ρ that stipulates the minimum segment length. This parameter can be set as the ratio to the time series length. Since the time series from a specific field exhibit the same fluctuation characteristics, ρ is data-adaptive and can be learned from the labeled dataset. Nevertheless, the segmentation is still primarily established on the recognition of turning patterns, which determines the segment number or lengths adaptively, and is essentially different from the principles of the existing segmentation methods.

4.2   Chebyshev Factorization

At the beginning, it is necessary to z-normalize the obtained subsequences as a pre-processing step. Rather than focusing on the statistical features, PCHA will factorize each subsequence with the first kind of Chebyshev polynomials, and take the Chebyshev coefficients as features. Since the Chebyshev polynomials with different degrees represent the fluctuation components, the local fluctuation information of time series can be captured in PCHA.

The first kind of Chebyshev polynomials are derived from the trigonometric identity $T_n(\cos(\theta)) = \cos(n\theta)$, which can be rewritten as a polynomial of variable t with degree n, as Formula (1).

$$T_n(t) = \begin{cases} \cos(n\cos^{-1}(t)), & t \in [-1, 1] \\ \cosh(n\cosh^{-1}(t)), & t \geq 1 \\ (-1)^n \cosh(n\cosh^{-1}(-t)), & t \leq -1 \end{cases} \tag{1}$$

For the sake of consistent approximation, we only employ the first sub-expression to factorize the subsequences, which is defined over the interval $[-1, 1]$. With the Chebyshev polynomials, a function $F(t)$ can be factorized as Formula (2).

$$F(t) \cong \sum_{i=0}^{n} c_i T_i(t) \tag{2}$$

The approximation is exact if $F(t)$ is a polynomial with the degree of less than or equal to n. The coefficients c_i can be calculated from the Gauss-Chebyshev Formula (3), where k is 1 for c_0 and 2 for the other c_i, and t_j is one of the n roots of $T_n(t)$, which can be get from the formula $t_j = \cos[(j - 0.5)\pi/n]$.

$$c_i = \frac{k}{n} \sum_{j=1}^{n} F(t_j) T_i(t_j) \tag{3}$$

However, the employed Chebyshev polynomials are defined over the interval $[-1, 1]$. If the subsequences are factorized with this "interval function", they must be scaled into the time interval $[-1, 1]$. Besides, the Chebyshev polynomials are defined every-where in the interval, but time series is a discrete function, whose values are defined only at the sample moments. To compute the Chebyshev coefficients, we would process each subsequence with the method proposed in [18], which can extend time series into an interval function. Given a scaled subsequence $S = \{(v_1, t_1), ..., (v_m, t_m)\}$, where $-1 \leq t_1 < ... < t_m \leq 1$, we first divide the interval $[-1, 1]$ into m disjoint subintervals as follows:

$$I_i = \begin{cases} [-1, \frac{t_1+t_2}{2}), & i = 1 \\ [\frac{t_{i-1}+t_i}{2}, \frac{t_i+t_{i+1}}{2}), & 2 \leq i \leq m-1 \\ [\frac{t_{m-1}+t_m}{2}, 1], & i = m \end{cases}$$

Then, the original subsequence can be extended into a step function as Formula (4), where each subinterval $[t_i, t_{i+1}]$ is divided by the mid-point $(t_i + t_{i+1})/2$. The first half takes the value v_i, and the second half takes v_{i+1}.

$$F(t) = v_i, \ t \in I_i, \ 1 \leq i \leq m \tag{4}$$

After the above processing, the Chebyshev coefficients c_i can be computed. For the sake of dimension reduction, we only take the first several coefficients to approximate the raw data, which can reflect the principal fluctuation components of time series.

Figure 5 shows the examples of (a) PAA, (b) APCA, (c) PLA, and (d) PCHA representations for the stock time series of Google Inc. (symbol: GOOG) from The NASDAQ Stock Market, which consists of the close prices at 800 consecutive trading days (2010/10/4-2013/12/5). As shown in Fig. 5(a), PAA extracts the mean values of the subsequences with equal-length as features. In Fig. 5(b), APCA takes the mean values and spans of the subsequences with adaptive-length as features, e.g., [−0.62, 134] for the first subsequence. In Fig. 5(c), PLA takes the linear fitting slopes and spans of the subsequences with adaptive-length as features, e.g., [−0.0035, 96] for the first subsequence. In Fig. 5(d), PCHA factorizes each subsequence and takes the first four Chebyshev coefficients as features, e.g., [−3.8, 0.34, 3, −0.39] for the first subsequence. It is obvious that the approximation of PCHA is different from the others, which can well fit the local fluctuation characteristics of time series.

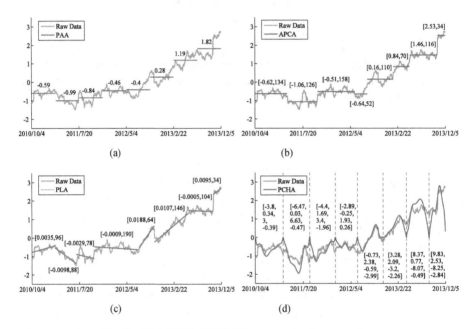

Fig. 5. PAA/APCA/PLA/PCHA representation examples.

In the entire procedure, the time series only needs to be scanned once for the adaptive segmentation and factorization. Thus, the computational complexity of PCHA is $O(kn)$, where k is the extracted Chebyshev coefficient number and much less than the time series length n.

5 Similarity Measure

DTW is one of the most prevalent similarity measures for time series [5]. It exploits the one-to-many aligning scheme to find the optimal alignment between time series, as shown in Fig. 6. Thus, DTW can deal with the intractable basic shape variations, e.g., time warping and phase-shift, etc. Given a sample space F, time series $T = \{t_1, t_2, ..., t_i, ..., t_m\}$ and $Q = \{q_1, q_2, ..., q_j, ..., q_n\}$, $t_i, q_j \in F$, a local distance measure $d: (x, y) \rightarrow \mathbf{R}^+$ should be first set in DTW for measuring two samples. Then, a distance

$$T$$

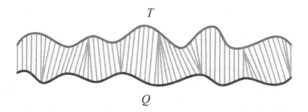

$$Q$$

Fig. 6. One-to-many aligning scheme of DTW.

matrix $C \in R^{m \times n}$ is computed, where each cell records the distance between each pair of samples from T and Q respectively, i.e., $C(i, j) = d(t_i, q_j)$. There is an optimal warping path in C, which has the minimal sum of cells.

Definition 4. (Warping Path): Given the distance matrix $C \in R^{m \times n}$, if the sequence $p = \{c_1, ..., c_l, ..., c_L\}$, where $c_l = (a_l, b_l) \in [1: n] \times [1: m]$ for $l \in [1: L]$, satisfies the conditions that:

(i) $c_1 = (1, 1)$ and $c_L = (m, n)$;
(ii) $c_{l+1} - c_l \in \{(1, 0), (0, 1), (1, 1)\}$ for $l \in [1: L - 1]$;
(iii) $a_1 \leq a_2 \leq ... \leq a_L$ and $b_1 \leq b_2 \leq ... \leq b_L$;

Then, p is called warping path. The sum of cells in p is defined as Formula (5).

$$\Phi_p = C(c_1) + C(c_2) + \cdots + C(c_L) \tag{5}$$

Definition 5. (Dynamic Time Warping Distance): Given the distance matrix $C \in R^{m \times n}$ over time series T and Q, and its warping path set $P = \{p_1, ..., p_i, ..., p_x\}$, $i, x \in R^+$, the minimal sum of cells in the warping paths $\Phi_{min} = \{\Phi_\xi | \Phi_\xi \leq \Phi_\lambda, \xi, \lambda \in P\}$ is defined as the DTW distance between T and Q.

The computation of DTW performs with dynamic programming algorithm, which would lead to the quadratic computational complexity to the time series length, i.e., O (n^2). Figure 7(a) shows the dynamic programming table with the optimal warping path in DTW computation.

Based on PCHA, we propose a novel PA-DTW measure, named ChebyDTW, which contains two layers: subsequence matching and dynamic programming computation. Figure 7(b) shows the dynamic programming table with the optimal-aligned path (red shadow) of ChebyDTW, where each cell records the subsequence matching result over the Chebyshev coefficients. By the intuitive comparison with Fig. 7(a), ChebyDTW would have much lower computational complexity than the original DTW.

With high computational efficiency, the squared Euclidean distance is a proper measure for the subsequence matching. Given d Chebyshev coefficients are employed in PCHA, for the subsequences S_1 and S_2, respectively approximated as $C = [c_1, ..., c_d]$ and $\hat{C} = [\hat{c}_1, ..., \hat{c}_d]$, the squared Euclidean distance between them can be computed as Formula (6).

(a)   **(b)**

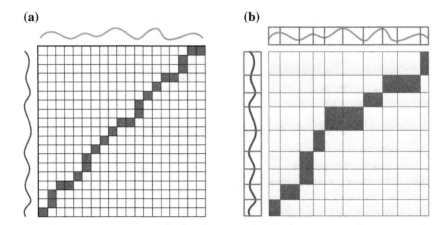

Fig. 7. (a) Dynamic programming table with the optimal-aligned path (red shadow) of DTW, (b) against that of ChebyDTW. (Color figure online)

$$D(\boldsymbol{C}, \hat{\boldsymbol{C}}) = \sum_{i=1}^{d} (c_i - \hat{c}_i)^2 \tag{6}$$

Over the subsequence matching, the dynamic programming computation performs. Given that time series T with length m is segmented into M subsequences, and time series Q with length n is segmented into N subsequences, ChebyDTW can be computed as Formula (7). $\boldsymbol{C}^T$ and $\boldsymbol{C}^Q$ are the PCHA representations of T and Q respectively; C_1^T and C_1^Q are the first coefficient vectors of $\boldsymbol{C}^T$ and $\boldsymbol{C}^Q$ respectively; $rest(\boldsymbol{C}^T)$ means the rest coefficient vectors of $\boldsymbol{C}^T$ except for C_1^T; the same meaning is taken for $rest(\boldsymbol{C}^Q)$.

$$ChebyDTW(T, Q) =$$
$$\begin{cases} 0, & \text{if } m = n = 0 \\ \infty, & \text{if } m = 0 \text{ or } n = 0 \\ D(C_1^T, C_1^Q) + \min \left\{ \begin{array}{l} ChebyDTW[rest(\boldsymbol{C}^T), \boldsymbol{C}^Q], \\ ChebyDTW[\boldsymbol{C}^T, rest(\boldsymbol{C}^Q)], \\ ChebyDTW[rest(\boldsymbol{C}^T), rest(\boldsymbol{C}^Q)] \end{array} \right\} \\ \quad, \text{ otherwise} \end{cases} \tag{7}$$

6   Experiments

We evaluate the 1NN classifier based on ChebyDTW from the aspects of accuracy and efficiency respectively. 12 real-world datasets provided by the UCR time series archive [17] are employed, which come from the various application domains and are characterized by the different series profiles and dimensionality. All the datasets have been

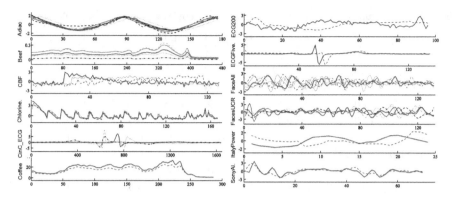

Fig. 8. Sample representative instances from each class of 12 datasets.

z-normalized and partitioned into the training and testing sets by the provider. Figure 8 shows the sample representative instances from each class of the datasets.

All parameters in the measures are learned on the training datasets by the DIRECT global optimization algorithm [19], which is used to seek for the global minimum of multivariate function within a constraint domain. The experiment environment is Intel (R) Core(TM) i5-2400 CPU @ 3.10 GHz; 8G Memory; Windows 7 64-bit OS; MATLAB 8.0_R2012b.

6.1 Classification Accuracy

Firstly, we take four PA-DTWs based on the statistical features as baselines, i.e., PDTW [14], SDTW [10], DTW_{APCA} [8], and DTW_{DSA} [9], which are based on PAA, PLA, APCA, and DSA representations respectively. Secondly, since PA-DTW is computed over the approximate representation, its precision is regarded lower than the measures computed on the raw data. To test this assumption, we also take 4 DTW measures computed on the raw data as baselines, including the original DTW and its variations, i.e., CDTW [3], CIDDTW [20], DDTW [21].

Tables 1 and 2 present the 1NN classification accuracy based on ChebyDTW and the baselines respectively. The best results on each dataset are highlighted in bold. The learned parameters are also presented, which could make each classifier achieve the highest accuracy on each training dataset, including the segment threshold (ρ), the smooth degree (sd), and the extracted Chebyshev coefficient number (θ). For the sake of dimension reduction, we learn the parameter θ in the range of [1, 10] for ChebyDTW.

By the comparison, we find that, (1) the 1NN classifier based on ChebyDTW wins all datasets over that based on the PA-DTW baselines. The superiority mainly derives from the distinctive features extracted in ChebyDTW, which can capture the fluctuation information for similarity measure. Concretely, as shown in Fig. 8, the practical time series in the datasets have the relatively complicated fluctuation that can be transformed into the wide Chebyshev domain, thus the difference between time series can be easily captured by the Chebyshev coefficients. Whereas the statistical features extracted in the baselines only focus on the aggregation characteristics of time series, which would result in much fluctuation information loss.

Table 1. The accuracy of 1NN classifiers based on ChebyDTW and four PA-DTW baselines.

Dataset	ρ	sd	θ	Cheby DTW	PDTW	SDTW	DTW_{APCA}	DTW_{DSA}
Adiac	0.21	22	9	**0.72**	0.61	0.34	0.28	0.38
Beef	0.18	17	5	**0.57**	0.50	0.57	0.57	0.47
CBF	0.98	8	10	**0.98**	**0.98**	0.95	0.91	0.50
Chlorine.	0.73	25	8	**0.65**	0.60	0.55	0.56	0.62
CinC_ECG.	0.29	4	9	**0.81**	0.65	0.63	0.61	0.63
Coffee	0.51	14	9	**0.89**	0.79	0.75	0.82	0.61
ECG200	0.80	7	9	**0.89**	0.80	0.83	0.77	0.81
ECGFive.	0.73	17	9	**0.91**	0.79	0.68	0.68	0.57
FaceAll	0.51	29	10	**0.73**	0.63	0.50	0.63	0.71
FacesUCR	0.51	4	6	**0.80**	0.60	0.57	0.72	0.70
ItalyPower.	0.51	7	5	**0.94**	0.93	0.80	0.90	0.87
SonyAI.	0.95	25	6	**0.80**	0.76	0.73	0.76	0.70

Table 2. The accuracy of 1NN classifiers based on ChebyDTW and four DTW baselines.

Dataset	Cheby DTW	DTW	CDTW	CIDDTW	DDTW
Adiac	**0.72**	0.60	0.61	0.61	0.47
Beef	**0.57**	0.50	0.53	0.50	0.47
CBF	0.98	**1.00**	**1.00**	**1.00**	0.54
Chlorine.	0.65	0.65	0.65	0.64	**0.69**
CinC_ECG.	0.81	0.65	**0.93**	0.70	0.66
Coffee	**0.89**	0.82	0.82	0.82	0.79
ECG200	**0.89**	0.77	0.88	0.81	0.80
ECGFive.	**0.91**	0.77	0.80	0.76	0.65
FaceAll	0.73	0.81	0.81	**0.85**	0.80
FacesUCR	0.80	0.90	**0.91**	0.83	0.76
ItalyPower.	0.94	0.91	0.91	0.88	**0.96**
SonyAI.	0.80	0.95	**0.96**	0.92	0.87

(2) The classifier based on ChebyDTW has higher accuracy on more datasets than the original DTW and its variations. The reason is apparent that, the noise mixed in the time series can be filtered out by the Chebyshev factorization effectively, which is one of the principal factors affecting the precision of similarity measures. Thus, the above assumption that ChebyDTW has lower precision than the measures computed on the raw data is not supported.

6.2 Computational Efficiency

The speedup of computational complexity gained by PA-DTW over the original DTW is $O(n^2/w^2)$, where n is the time series length and w is the segment number. It is positively correlated with the data compression rate (DCR = n/w) of piecewise approximation over the raw data. In Table 3, we present the segment numbers and the DCRs of five PA-DTWs on all datasets. As above, the optimal segment numbers for the 1NN classifiers based on PDTW, SDTW, and DTW_{APCA} are learned on the training datasets, while the average segment numbers on each dataset are computed for ChebyDTW and DTW_{DSA}.

As shown in Table 3, the DCRs of ChebyDTW are not only much larger than the baselines on all datasets, but also robust to the time series length. Thus, it has the highest computational efficiency among the five PA-DTWs. The efficiency superiority of ChebyDTW mainly derives from the precise approximation of PCHA over the raw data, and the data-adaptive segment method, which can segment time series into the less number of subsequences with the adaptive lengths.

Table 3. The DCR results of five PA-DTWs.

Dataset	n	Cheby DTW		PDTW		SDTW		DTW_{APCA}		DTW_{DSA}	
		w	DCR	w	DCR	w	DCR	w	DCR	w	DCR
Adiac	176	3.99	**44.13**	36	4.89	13	13.54	43	4.10	70	2.51
Beef	470	5.18	**90.68**	61	7.70	10	47	61	7.70	192.32	2.44
CBF	128	1	**128.0**	30	4.27	27	4.74	15	8.53	46.19	2.77
Chlorine.	166	2	**83.00**	36	4.61	29	5.72	34	4.88	64.77	2.56
CinC.	1639	4	**409.9**	103	15.91	94	17.44	84	19.51	655.49	2.50
Coffee	286	2	**143.0**	60	4.77	33	8.67	40	7.15	117.34	2.44
ECG200	96	1.93	**49.74**	14	6.86	19	5.05	23	4.17	35.84	2.68
ECGFive.	136	1.61	**84.43**	9	15.11	9	15.11	5	27.20	48.24	2.82
FaceAll	131	2	**65.50**	32	4.09	32	4.09	32	4.09	53.96	2.43
FacesUCR	131	2	**65.50**	24	5.46	32	4.09	31	4.23	54.43	2.41
ItalyPower.	24	1.98	**12.13**	5	4.80	6	4.00	6	4.00	10.61	2.26
SonyAI.	70	1	**70.00**	13	5.38	9	7.78	8	8.75	27.41	2.55

Table 4. The average runtime of 1NN classification based on DTW and ChebyDTW (ms).

Dataset	DTW	ChebyDTW	Ω
Adiac	172.21	5.36	32.11
Beef	105.02	0.54	194.95
CBF	9.34	0.37	25.40
Chlorine.	231.23	5.47	42.24
CinC_ECG	1721.93	0.79	2175.49
Coffee	38.17	0.40	96.05
ECG200	16.15	1.18	13.69
ECGFive.	6.89	0.29	23.60
FaceAll	169.94	8.26	20.57
FacesUCR	61.61	2.37	26.00
ItalyPower.	1.3	0.73	1.78
SonyAI.	2.1	0.25	8.38

In addition, the average runtime of 1NN classification based on DTW and ChebyDTW are presented in Table 4. According to the results, the efficiency speedup (Ω) of ChebyDTW over DTW can achieve as much as 3 orders of magnitude.

7 Conclusions

We proposed a novel piecewise factorization model for time series, i.e., PCHA, where a novel adaptive segment method was proposed, and the subsequences were factorized with the Chebyshev polynomials. We employed the Chebyshev coefficients as features for PA-DTW measure, and thus proposed the ChebyDTW for 1NN classification. The comprehensive experimental results show that ChebyDTW can support the accurate and fast 1NN classification.

Acknowledgements. This work was funded by the Ministry of Industry and Information Technology of China (No. 2010ZX01042-002-003-001), China Knowledge Centre for Engineering Sciences and Technology (No. CKCEST-2014-1-5), and National Natural Science Foundation of China (No. 61332017).

References

1. Esling, P., Agon, C.: Time-series data mining. ACM Comput. Surv. **45**(1), 12 (2012)
2. Fu, T.: A review on time series data mining. Eng. Appl. Artif. Intell. **24**(1), 164–181 (2011)
3. Ding, H., Trajcevski, G., Scheuermann, P., Wang, X., Keogh, E.: Querying and mining of time series data: experimental comparison of representations and distance measures. In: Proceedings of the VLDB Endowment, pp. 1542–1552 (2008)
4. Hills, J., Lines, J., Baranauskas, E., Mapp, J., Bagnall, A.: Classification of time series by shapelet transformation. Data Min. Knowl. Disc. **28**(4), 851–881 (2014)
5. Serra, J., Arcos, J.L.: An empirical evaluation of similarity measures for time series classification. Knowl.-Based Syst. **67**, 305–314 (2014)
6. Keogh, E., Chakrabarti, K., Pazzani, M., Mehrotra, S.: Dimensionality reduction for fast similarity search in large time series databases. Knowl. Inf. Syst. **3**(3), 263–286 (2001)
7. Keogh, E., Chu, S., Hart, D., Pazzani, M.: Segmenting time series: a survey and novel approach. Data Min. Time Ser. Databases **57**, 1–22 (2004)
8. Chakrabarti, K., Keogh, E., Mehrotra, S., Pazzani, M.: Locally adaptive dimensionality reduction for indexing large time series databases. ACM Trans. Database Syst. **27**(2), 188–228 (2002)
9. Gullo, F., Ponti, G., Tagarelli, A., Greco, S.: A time series representation model for accurate and fast similarity detection. Pattern Recognit. **42**(11), 2998–3014 (2009)
10. Keogh, E.J., Pazzani, M.J.: Scaling up dynamic time warping to massive datasets. In: Żytkow, J.M., Rauch, J. (eds.) PKDD 1999. LNCS (LNAI), vol. 1704, pp. 1–11. Springer, Heidelberg (1999). doi:10.1007/978-3-540-48247-5_1
11. Li, H., Guo, C.: Piecewise cloud approximation for time series mining. Knowl.-Based Syst. **24**(4), 492–500 (2011)
12. Lin, J., Vlachos, M., Keogh, E., Gunopulos, D., Liu, J., Yu, S., Le, J.: A MPAA-based iterative clustering algorithm augmented by nearest neighbors search for time-series data streams. In: Ho, T.B., Cheung, D., Liu, H. (eds.) PAKDD 2005. LNCS (LNAI), vol. 3518, pp. 333–342. Springer, Heidelberg (2005). doi:10.1007/11430919_40

13. Rakthanmanon, T., Campana, B., Mueen, A.: Searching and mining trillions of time series subsequences under dynamic time warping. In: Proceedings of the 18th ACM SIGKDD International Conference on Knowledge Discovery and Data Mining, Beijing (2012)
14. Keogh, E.J., Pazzani, M.J.: Scaling up dynamic time warping for data mining applications. In: Proceedings of the 6th ACM SIGKDD International Conference on Knowledge Discovery and Data Mining, pp. 285–289 (2000)
15. Ullman, S., Poggio, T.: Vision: A Computational Investigation into the Human Representation and Processing of Visual Information. MIT Press, Cambridge (1982)
16. Gao, J., Sultan, H., Hu, J., Tung, W.: Denoising nonlinear time series by adaptive filtering and wavelet shrinkage: a comparison. IEEE Sig. Process. Lett. **17**(3), 237–240 (2010)
17. Keogh, E., Zhu, Q., Hu, B., Hao, Y., Xi, X., Wei, L., Ratanamahatana, C. A.: UCR time series classification/clustering (2011). www.cs.ucr.edu/~eamonn/time_series_data/
18. Cai, Y., Ng, R.: Indexing spatio-temporal trajectories with Chebyshev polynomials. In: Proceedings of the 2004 ACM SIGMOD International Conference on Management of Data, pp. 599–610 (2004)
19. Björkman, M., Holmström, K.: Global optimization using the DIRECT algorithm in matlab. Adv. Model. Optim. **1**(2), 17–37 (1999)
20. Batista, G.E., Keogh, E.J., Tataw, O.M., Souza, V.M.: CID: an efficient complexity-invariant distance for time series. Data Min. Knowl. Disc. **28**(3), 634–669 (2014)
21. Keogh, E.J., Pazzani, M.J.: Derivative dynamic time warping. In: Proceedings of SDM, pp. 5–7 (2001)

Experiences in WordNet Visualization
with Labeled Graph Databases

Enrico Giacinto Caldarola[1,2], Antonio Picariello[1], and Antonio M. Rinaldi[1,3(✉)]

[1] Department of Electrical Engineering and Information Technologies,
University of Naples Federico II, Naples, Italy
{enricogiacinto.caldarola,antonio.picariello,
antoniomaria.rinaldi}@unina.it
[2] Institute of Industrial Technologies and Automation,
National Research Council, Bari, Italy
[3] IKNOS-LAB Intelligent and Knowledge Systems,
University of Naples Federico II, LUPT, 80134 Naples, Italy

Abstract. Data and Information Visualization is becoming strategic for
the exploration and explanation of large data sets due to the great impact
that data have from a human perspective. The visualization is the closer
phase to the users within the data life cycle's phases, thus, an effective,
efficient and impressive representation of the analyzed data may result
as important as the analytic process itself. In this paper, we present our
experiences in importing, querying and visualizing graph databases taking one of the most spread lexical database as case study: WordNet. After
having defined a meta-model to translate WordNet entities into nodes
and arcs inside a labeled oriented graph, we try to define some criteria to simplify the large-scale visualization of WordNet graph, providing
some examples and considerations which arise. Eventually, we suggest
a new visualization strategy for WordNet synonyms rings by exploiting
the features and concepts behind tag clouds.

Keywords: Graph database · Big Data · NoSQL · Data visualization ·
WordNet · Neo4J

1 Introduction

We are increasingly confronting with a data deluge in every field and in every
human activity. With data we explore, we understand and get insights from
the whole world. More importantly, with the information coming from data, we
explain and communicate facts or knowledge about the world. Not surprisingly,
with the advent of *Big Data* paradigm, also Data and Information Visualization have become an interesting and wide research field. If the main goal of
Data Visualization is to communicate information clearly and efficiently to users,
involving the creation and study of the visual representation of data – i.e., "information that has been abstracted in some schematic form, including attributes
or variables for the units of information" [1] – the Information Visualization

© Springer International Publishing AG 2016
A. Fred et al. (Eds.): IC3K 2015, CCIS 631, pp. 80–99, 2016.
DOI: 10.1007/978-3-319-52758-1_6

main task is the study of (interactive) visual representations of abstract data to reinforce human cognition. The abstract data may include both numerical and non-numerical data, such as text and geographic information. According to [2], it is possible to distinguish Information Visualization (InfoVis), when the spatial representation is chosen, from Scientific Visualization (SciVis) when the spatial representation is given due to the intrinsic spatial layout of data (e.g., a flow simulation in 3D space). The study presented in this work belongs to the first category being the data we represent abstract and not linked to physical entities, i.e., they do not have pre-defined spatialization, such as, for example, in the case of maps or geographic informative systems. Besides, the challenges that the Big Data imperative [3–5] imposes to data management severely impact on data visualization. The "bigness" of large data sets and their complexity in term of heterogeneity contribute to complicate the representation of data, making the drawing algorithms quite complex: just to make an example, let us consider the popular social network Facebook, in which the nodes represent people and the links represent interpersonal connections; we note that nodes may be accompanied by information such as age, gender, and identity, and links may also have different types, such as colleague relationships, classmate relationships, and family relationships. The effective representation of all the information at the same time is really challenging. A most important solution can be using visual cues, such as color, shape, or transparency to encode different attributes. Also the choice of the graph model is important to fit the nature of the information to be visualized. The Labeled Property Graph model, for example, seems to fit well most of the common scenario from the real life and that why we use this model to convert WordNet synsets in a graph. The availability of large data coming from human activities, exploration and experiments, together with the investigations of new and efficiently ways of visualizing them, open new perspectives from which to view the world we live in and to make business. The *Infographics* become *Infonomic*, a composite term between the term *Information* and *Economics* that wield information as a real asset, a real opportunity to make business and to discover the world. Various techniques have been proposed for graph visualization for the last two decades and they will be presented in the next section. The principled representation methodology we agree on is the Visual Information Seeking Mantra presented by Scheiderman in [6]. It can be summarized as follows: "overview first, zoom and filter, then details-on-demand". The reminder of the paper is organized as follows. After a literature review on the Graph Visualization techniques and methodologies, contained in Sect. 2, a description of the proposed WordNet meta-model is provided in Sect. 3. Afterward, starting from a description of the approach used for the WordNet importing procedure within Neo4j, in Sect. 4, the attempts made in querying and visualizing WordNet are described in Sects. 5 and 6. Section 7 extends the study presented in the previous sections by proposig the adoption of a tag clouds based methodology in order to visualize the WordNet synonims rings in Neo4J. Finally, Sect. 8 draws the conclusion summarizing the major findings and outlining future investigations.

2   Related Works

Since the study conducted in this paper consists in the visual representation of WordNet inside the Neo4j graph DB, this section focuses mainly on a literature review in *Graph Visualization*, referring to other well-known works in the literature for a complete review of the techniques and theories in Information Visualization [7–10]. Graphs are traditional and powerful tools that visually represent sets of data and the relations among them. In the most common sense of the term, a graph is an ordered pair $G = (V, E)$ comprising a set V of vertices or nodes together with a set E of edges or lines, which are 2-element subsets of V (i.e., an edge is related with two vertices, and the relation is represented as an unordered pair of the vertices with respect to the particular edge). Graph visualization usually refers to representation of interconnected nodes arranged in space and navigation through a visual representation to help users understand the global or local original data structures [11]. Graphs are represented visually by drawing a dot or circle for every vertex, and drawing an arc between two vertices if they are connected by an edge. If the graph is directed, the direction is indicated by drawing an arrow. A graph drawing should not be confused with the graph itself (the abstract, non-visual structure) as there are several ways to structure the graph drawing. All that matters is which vertices are connected to which others by how many edges and not the exact layout. In practice it is often difficult to decide if two drawings represent the same graph. Depending on the problem domain some layouts may be better suited and easier to understand than others. The pioneering work of Tutte [12] was very influential in the subject of graph drawing, in particular he introduced the use of linear algebraic methods to obtain graph drawings. The basic graph layout problem is very simple: given a set of nodes with a set of edges, it only needs to calculate the positions of the nodes and draw each edge as curve. Despite the simplicity of the problem, to make graphical layouts understandable and useful is very hard. Basically there are generally accepted aesthetic rules [13,14], which include: distribute nodes and edges evenly, avoid edge crossing, display isomorphic substructures in the same manner, minimize the bends along the edges. However, since it is quite impossible to meet all rules at the same time, some of them conflict with each other or they are very computationally expensive, practical graphical layouts are usually the results of compromise among the aesthetics. Another issue about graph layout is predictability. Due to the task of graph visualization, it is important and necessary to make the results of layout algorithm predictable [15], which means two different results of running the same algorithm with the same or similar data inputs should also look the same or alike.

Below is a brief overview of graph layouts and visualization techniques grouped by categories:

– Node-link layouts.
- • Tree Layout. It uses links between nodes to indicate the parent-child relationships. A very satisfactory solution for node-link layout comes from Reingold and Tilford [16]. Their classical algorithm is simple, fast, predictable, and produces aesthetically pleasing trees on the plane. However,

it makes use of screen space in a very inefficient way. In order to overcome this limitation, some compact tree layout algorithms have been developed to obtain more dense tree, while keeping the classical tree looks [17]. Eades [18] proposes another node-link layout called radial layout that recursively positions children of a sub-tree into a circular wedge shape according to their depths in the tree. Generally, radial views, including its variations [19], share a common characteristic: the focus node is always placed at the center of the layout, and the other nodes radiate outward on separated circles. Balloon layout [20] is similar to radial layout and are formed where siblings of sub-trees are placed in circles around their father node. This can be obtained by projecting cone tree onto the plane.

- Tree Plus Layout. Since large graphs are much more difficult to handle than trees, tree visualization is often used to help users understand graph structures. A straightforward way to visualize graphs is to directly layout spanning trees for them. Munzner [21] finds a particular set of graphs called quasi-hierarchical graphs, which are very suitable to be visualized as minimum spanning trees. However, for most graphs, all links are important. It could be very hard to choose a representative spanning tree. Arbitrary spanning trees can also possibly deliver misleading information.

- Spring Layout. This layout, also known as *Force-Directed* layout, is another popular strategy for general graph layouts. In spring layout, graphs are modeled as physical systems of rings or springs. The attractive idea about spring layout is that the physical analogy can be very naturally extended to include additional aesthetic information by adjusting the forces between nodes. As one of the first few practical algorithms for drawing general graphs, spring layout is proposed by Eades in 1984 [22]. Since then, his method is revisited and improved in different ways [23,24]. Mathematically, Spring layout is based on a cost (energy) function, which maps different layouts of the same graph to different nonnegative numbers. Through approaching the minimum energy, the layout results reaches better and better aesthetically pleasing results. The main differences between different spring approaches are in the choice of energy functions and the methods for their minimization.

– Space Division Layout. In this case, the parent-child relationship is indicated by attaching child node(s) to the parent node. Since the parent-child and sibling relationships are both expressed by adjacency, The layout should have a clear orientation cue to differentiate these two relationships.

– Space Nested Layout. Nested layouts, such as Treemaps [25], draw the hierarchical structure in the nested way. They place child nodes within their parent node.

– 3D Layout. In this case, the extra dimension can give more space and it would be easier to display large structures. Moreover, Due to the general human familiarity with 3D in the real world, there are some attempts to map hierarchical data to 3D objects we are familiar with

– Matrix Layout. Graphs can be presented by their connectivity matrixes. Each row and each column corresponds to a node. The glyph at the interaction of (i, j) encodes the edge from node i to node j. Edge attributes are encoded as visual characteristics of the glyphs, such as color, shape, and size. The major benefit of adjacency matrices is the scalability.

Specifically regarding the visualization of WordNet, there are not many works in the literature. In [26], the authors makes an attempt to visualize the Word-Net structure from the vantage point of a particular word in the database, this in order to overcome the down-side of the large coverage of WordNet, i.e., the difficulty to get a good overview of particular parts of the lexical database. An attempt to apply design paradigms to generate visualizations which maximize the usability and utility of WordNet is made in [27], whereas, in [28] a radial, space-filling layout of hyponymy (IS-A relation) is presented with interactive techniques of zoom, filter, and details-on-demand for the task of document visualization, exploiting the WordNet lexical database. Finally, regarding the comparison between Neo4J Cypher language performances against the traditional SQL-based technologies, an interesting experience has been described in [29] where the authors compare Neo4j back-end different alternatives to each other and to the JPA-based sample back-end running on MySQL.

3   WordNet Case Study

The case study presented in this paper consists in the *reification* of the WordNet database inside the Neo4J GraphDB [30, 31]. WordNet [32, 33] is a large lexical database of English. Nouns, verbs, adjectives and adverbs are grouped into sets of cognitive synonyms (synsets), each expressing a distinct concept. Synsets are interlinked by means of conceptual-semantic and lexical relations.

In this context we have defined and implemented a meta-model for the Word-Net reification using a conceptualization as much as possible close to the way in which the concepts are organized and expressed in human language [34]. We consider concepts and words as nodes in Neo4J, whereas semantic, linguistic and semantic-linguistic relations become Noeo4J links between nodes. For example, the hyponymy property can relate two concept nodes (nouns to nouns or verbs to verbs); on the other hand a semantic property links concept nodes to concepts and a syntactic one relates word nodes to word nodes. Concept and word nodes are considered with *DatatypeProperties*, which relate individuals with a predefined data type. Each word is related to the represented concept by the ObjectProperty *hasConcept* while a concept is related to words that represent it using the ObjectProperty *hasWord*. These are the only properties able to relate words with concepts and vice versa; all the other properties relate words to words and concepts to concepts. Concepts, words and properties are arranged in a class hierarchy, resulting from the syntactic category for concepts and words and from the semantic or lexical type for the properties.

Figures 1(a) and (b) show that the two main classes are: Concept, in which all the objects have defined as individuals and Word which represents all the terms in the ontology.

(a) Concept   (b) Word

Fig. 1. Concept and word.

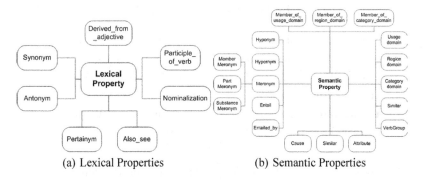

(a) Lexical Properties   (b) Semantic Properties

Fig. 2. Linguistic properties.

The subclasses have been derived from the related categories. There are some union classes useful to define properties domain and co-domain. We define some attributes for Concept and Word respectively: Concept *hasName* that represents the concept name; *Description* that gives a short description of concept. On the other hand Word has Name as attribute that is the word name. All elements have an ID within the WordNet offset number or a user defined ID. The semantic and lexical properties are arranged in a hierarchy (see Fig. 2(a) and (b)). In Table 1 some of the considered properties and their domain and range of definition are shown.

Table 1. Properties.

Property	Domain	Range
hasWord	Concept	Word
hasConcept	Word	Concept
hypernym	NounsAnd	NounsAnd
	VerbsConcept	VerbsConcept
holonym	NounConcept	NounConcept
entailment	VerbWord	VerbWord
similar	AdjectiveConcept	AdjectiveConcept

The use of domain and codomain reduces the property range application. For example, the hyponymy property is defined on the sets of nouns and verbs; if it is applied on the set of nouns, it has the set of nouns as range, otherwise, if it is applied to the set of verbs, it has the set of verbs as range. In Table 2 there are some of defined constraints and we specify on which classes they have been applied w.r.t. the considered properties; the table shows the matching range too.

Table 2. Model constraints.

Constraint	Class	Property	Constraint range
AllValuesFrom	NounConcept	hyponym	NounConcept
AllValuesFrom	AdjectiveConcept	attribute	NounConcept
AllValuesFrom	NounWord	synonym	NounWord
AllValuesFrom	AdverbWord	synonym	AdverbWord
AllValuesFrom	VerbWord	also_see	VerbWord

Sometimes the existence of a property between two or more individuals entails the existence of other properties. For example, being the concept dog a hyponym of animal, we can assert that animal is a hypernymy of dog. We represent this characteristics in OWL, by means of property features shown in Table 3.

Table 3. Property features.

Property	Features
hasWord	*inverse* of hasConcept
hasConcept	*inverse* of hasWord
hyponym	*inverse* of hypernym; *transitivity*
hypernym	*inverse* of hyponym; *transitivity*
cause	*transitivity*
verbGroup	*symmetry* and *transitivity*

4   Importing WordNet into Neo4J

The importing process of WordNet database within Neo4J graphDB [30] has been implemented according to the scheme shown in Fig. 3. The process involves three phases and three components: the *importing from WordNet* module, the *serializer* module and the *importing within Neo4J* module. The first phase has been implemented using a Java-based script that access the WordNet database through JWI (MIT Java Wordnet Interface) API [35, 36] and passes all the information related to synsets, words, semantic relations and lexical relations to the serializer module, producing appropriate serialized data, following a proper schema that will be described in the following.

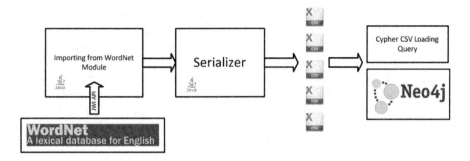

Fig. 3. High-level view of the WordNet importing architecture.

The last component, which is related to the third phase of the process, is responsible for importing the previously serialized information into Neo4J database. The importing from WordNet takes place via five different sub-operations which respectively retrieve: the information related to synsets, the semantic relations among synsets, the words, the lexical relations among words and finally the links between the semantic and the lexical world, i.e., how a word is related to its concepts (or its meaning) and *viceversa*.

The intentional schema of each serialized data is shown as follow:

1. The synset file contains the following fields:
   (a) *Id*: the univoque identifier for the synset;
   (b) *SID*: the Synset ID as reported in the WordNet database;
   (c) *POS*: the synset's part of speech;
   (d) *Gloss*: the synset's gloss which express its meaning.
2. The semantic relations file contains the following fields:
   (a) *Prop*: the semantic relation linking the source and the destination synsets;
   (b) *Src*: the source synset;
   (c) *Dest*: the destination synset;
3. The words file contains the following fields:
   (a) *Id*: the univoque indentifier for the word;
   (b) *WID*: the Word ID as reported in the WordNet database;
   (c) *POS*: the word's part of speech;
   (d) *Lemma*: lexical representation of the word;
   (e) *SID*: the synset Id whose the word is related.
4. The lexical relations file contains the following fields:
   (a) *Prop*: the lexical relation linking the source and the destination words;
   (b) *Src*: the source word;
   (c) *Dest*: the destination word;
5. The lexical-semantic relations file contains the following fields:
   (a) *Word Id*: the word id of the word that is linked to the synset on the right via the *hasConcept* relation;
   (b) *Synset Id*: the synset id of the synset that is linked to the word on the left via the *hasWord* relation;

In order to import all the information contained in the serialized data and translate them into a graph data structure, the meta-model described in the previous section has been used: each synset and word has been converted into a node of the graph with label respectively: *Concept* and *Word*. Each semantic relation has become an edge between two concept nodes with the *type* property expressing the specific semantic relation holding between the concepts. Each lexical relation has been converted into an edge between two word nodes with a type property expressing the specific lexical relation between the word nodes. Finally, the word nodes have been connected to their related concept nodes through the *hasConcept* relation.

The Cypher query code used to import all the serialized information stored into csv lines is shown as follows:

```
USING PERIODIC COMMIT 1000
LOAD CSV WITH HEADERS FROM "PATH_TO_THE_FIRST_FILE"
AS csvLine

CREATE (c: Concept {
id: toInt(csvLine.id),
sid: csvLine.SID, POS:
csvLine.POS,
gloss: csvLine.gloss })

CREATE CONSTRAINT ON (c: Concept)
ASSERT c.id IS UNIQUE

USING PERIODIC COMMIT 1000
LOAD CSV WITH HEADERS FROM "PATH_TO_THE_SECOND_FILE"
AS csvLine
MATCH (src:Concept { id: toInt(csvLine.Src)}),
(dest:Concept { id: toInt(csvLine.Dest)})

CREATE (src)-[:semantic_property
{ type: csvLine.Prop }]->(dest)

USING PERIODIC COMMIT 1000
LOAD CSV WITH HEADERS FROM "PATH_TO_THE_THIRD_FILE"
AS csvLine

CREATE (w: Word {
id: toInt(csvLine.id),
wid: csvLine.WID,
POS: csvLine.POS,
lemma: csvLine.lemma,
sid: toInt(csvLine.SID) })

CREATE CONSTRAINT ON (w: Word)
ASSERT w.id IS UNIQUE

USING PERIODIC COMMIT 1000
LOAD CSV WITH HEADERS FROM "PATH_TO_THE_FOURTH_FILE"
AS csvLine

MATCH (src:Word { id: toInt(csvLine.Src)}),
(dest:Word { id: toInt(csvLine.Dest)})

CREATE (src)-[:lexical_property
{ type: csvLine.Prop }]->(dest)

USING PERIODIC COMMIT 1000
LOAD CSV WITH HEADERS FROM "PATH_TO_THE_FIFTH_FILE"
AS csvLine

MATCH (src:Word { id: toInt(csvLine.Word)}),
(dest:Concept { id: toInt(csvLine.SID)})
CREATE (src)-[:hasConcept]->(dest)
```

Having identified each synset and word with a unique and sequential integer, it has been possible for Neo4J to efficiently create nodes and arcs from csv lines. Furthermore, since the csv file contains a significant number of rows (approaching hundreds of thousands) USING PERIODIC COMMIT can be used to instruct Neo4j to perform a commit after a number of rows. This reduces the memory overhead of the transaction state. Table 4 shows the time performance in importing all the csv file on a laptop computer with an Intel Core i7-4800MQ processor at 2.70 GHz (64-bit) and 8 GB RAM.

Table 4. Neo4J query execution times for each importing query.

Query no	Q1	Q2	Q3	Q4	Q5
[ms]	13015	23779	25787	17358	36907

Listing 1.1. A comparison between Cypher language and standard SQL.

```
// First query comparison

MATCH ( w_src :  Word { lemma :'  politics '}) −[r :  hasConcept]−>(c :  Concept ) −[*..N]−>(d :
    Concept )<−[s :  hasConcept]−(w_dst :  Word )  return  w_src ,  c ,  d ,  w_dst

SELECT  Word . lemma  AS  src_lemma ,  Word . hasConcept  AS  src_concept ,  SemanticRelation . type
    AS  Property_1 ,  SemanticRelation . dst  AS  Intermediate_Concept1 ,  ... ,
    SemanticRelation . type  AS  Property_N ,  SemanticRelation . dst  AS
    Intermediate_ConceptN ,  Word . lemma  AS  dst_lemma
FROM  Word  JOIN  Concept  ON  Word . conceptID  =  Concept . SID  JOIN  SemanticRelation  ON
    SemanticRelation . src  =  Concept . SID  JOIN  SemanticRelation  ON  SemanticRelation . dst
    =  SemanticRelation . src   ...  JOIN  SemanticRelation  ON  SemanticRelation . dst  =
    SemanticRelation . src
WHERE  word . wid  =  (SELECT  Word . wid  FROM  Word  WHERE  Word . lemma  =" politics ")

// Second query comparison

MATCH (n : Concept  { words : '{ politics }'})−[r :  semantic_property *..N { type : 'Hyponym'}]−(m
    :  Concept )  RETURN  COUNT( r )

SELECT  SemanticRelation . src  AS  Source ,  COUNT(*)
FROM  SemanticRelation  Relation_1  JOIN  SemanticRelation  Relation_2  ON  Relation_1 . dst  =
    Relation_2 . src   ....  JOIN  SemanticRelation  Relation_N
ON  Relation_N −1. dst  =  Relation_N . src
WHERE  Source . words  =' politics '  AND  Relation_1 . type  =' Hyponym '   ...  AND  Relation_N . type
    =' Hyponym '

// Third query comparison

MATCH (w : Word  {  lemma :" food "}) ,(v : Word  {  lemma :" lunch "}) ,
p  =  shortestPath ((w) −[*]−(v ))
RETURN  p

(No equivalent )
```

5   Querying the WordNet Graph

The first attempt to visualize the graph-based version of the WordNet database within Neo4j has been carried out using the Neo4j built-in web visualizer. Although it is a power and flexible tool allowing the user to easily customize the view according to her preferences, it suffers a lot when the number of elements to be visualized approaches few hundreds. Figure 4 shows the first result obtained with a simple customization involving the usage of two different colors for the two type of nodes, namely, the green for the *Word* node and the blue for the *Concept* (or *Synset*) node. Semantic relations have been represented with an edge thicker than the one used for the lexical relations or for the *hasConcept* relations.

Each Concept node is labeled with the lexical chain of the synonyms related to such concept. A set edges ends to the synset node and comes from all the words belonging to the synset. These ones also are connected one with each other

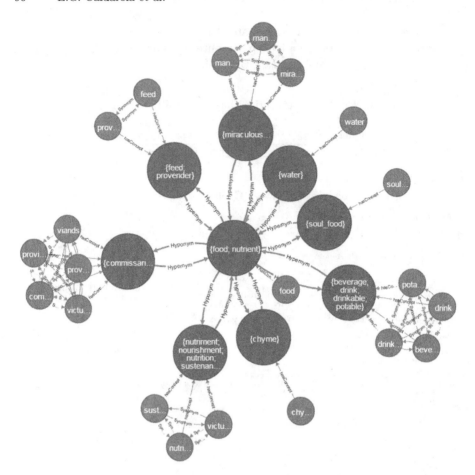

Fig. 4. First view of Neo4j WordNet graph excerpt. (Color figure online)

through the synonymy lexical relation. The concept nodes, in blue, are mainly connected through the *hypernym-hyponym* semantic relations. The greatest value of importing WordNet database into a Neo4j graph, it is not related to the graph visualization capabilities of the web visualizer, but, mainly, to the power of the *Cypher* query language, a declarative graph query language that allows for expressive and efficient querying and updating of the graph store. Since very complicated database queries can easily be expressed through Cypher, this allows the user to focus on the data model domain instead of getting lost in database access. Most of the keywords like WHERE and ORDER BY are inspired by SQL, while pattern matching borrows expression approaches from SPARQL [37].

In the attempt to extract some useful information from the WordNet implementation in Neo4j, we have run few queries and have compared them to an equivalent version expressed in SQL languages. Listing 1.1 reports a comparison of the Cypher-based and SQL-based version of each query. It is not a

Fig. 5. Second view of Neo4j WordNet graph excerpt. (Color figure online)

quantitative comparison but just a qualitative one that clearly shows how complex (or in some circumstances impossible at all) is to translate a query from the graph query language into the relational-based SQL language.

The objective of the first query is to get all concept nodes between the source and the target synset, where the source concept is fixed and has a lemma equal to *politics*, whereas the target node can be any node of the network. The only constraint is that between the first and the last synset there may be N (depth level) relations (*semantic relations* in this case) and $N - 1$ intermediate nodes. The Neo4j web-based tool provides two ways of visualizing results: the table-based and the graph-based. By analysing the query structure in the two columns, it appears quite clear that a graph-based query language is most suitable in order to select sub-graphs, as in this case. In fact, it comes very natural to select a bunch of nodes and relations just by using patterns and pattern-matching, expressed in an intuitive and iconic syntax, to describe the shape of the data you are looking for [29]. On the contrary, the SQL-based version requires more and more intricate combination of JOIN clauses to link synsets from the SemanticRelation and Concept tables. Each JOIN clause involves a cartesian product between the SemanticRelation table, which contains 283.836 rows, and itself with an order that increases with the degree of separation between the source and the target node (N). The second query in table, uses the COUNT aggregation function both in the Cypher and SQL-based version. It gets the number of the *Hyponym* relations holding between the *politics* source concept and any other concept that is at most N hops from the source. According to [38], aggregation operators make worse the performances of the SQL-based query especially when N increases (N < 6) with respect to the equivalent queries in Cypher. Also in this case, the SQL-based query require N JOIN clauses which corresponds to an equal number of cartesian products. Finally, the last query gets the shortest path between the *food* and *lunch* concepts. While the Cypher language offers utility functions like *shortestPath()* or *AllShortestPath()* making very easy to respond to such queries, the SQL language do not has similar ready-to-use functions. Furthermore, a graph-based data structure allows users to use *Traversal API* to specify desired movements through a graph in a programmatic way (Fig. 5).

6   Large-Scale Representation of WordNet Graph

The proposed representation of WordNet, within Neo4j, approaches very closely the Big Data challenges. In particular, the volume dimension must be taken into consideration here: this version of the WordNet graph, in fact, includes near to 2 millions different relations linking more than 3 hundred thousand nodes with each other. With these big numbers, the manipulation, the querying and the visualization of the graph become quite challenging. Before describing the attempts made in this direction, it is important to note that the visualization of the entire structure of WordNet in terms of all synsets, words, semantic and lexical relations in a way that is elegant and human friendly at the same time, is a *chimera*, due to the performance issues of the visualization tools, in particular when sophisticated drawing algorithms are used, and to the strongly connected nature of information to be represented, which often results in a messy and dense structure of nodes and edges. Figure 6 shows a representation of near 15,000 nodes and 30,000 relations of WordNet using the *Cytoscape* v. 3 graph visualization tool [38]. The image has been obtained by limiting to 30,000 relations a simple cypher query that gets some data from the Neo4j implementation of WordNet. The Neo4j running instance has been accessed via the cyNeo4j plugin,

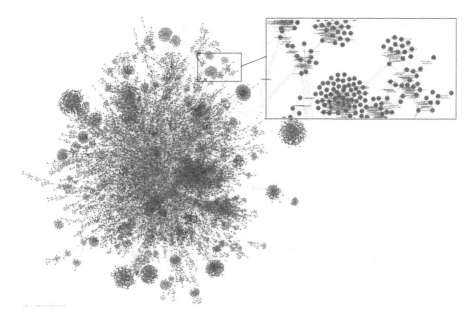

Fig. 6. WordNet graph excerpt with 30000 edges and about 15000 nodes. (Color figure online)

that converts the query results into Cytoscape table format. Afterward, starting from the query tables, a view has been created by defining a custom style and the default layout. This latter is the *Force-directed graph drawing algorithm* that draws graphs in an aesthetically pleasing way by positioning the nodes of a graph in two-dimensional or three-dimensional space, so that all the edges are of more or less equal length and there are as few crossing edges as possible [39]. The resulting figure is more considerable for global analysis than for information that you can retrieve from it. Nevertheless, thanks to the force-directed algorithm, it is possible to observe agglomerates of nodes and edges which correspond to specific semantic categories. The figure also shows a zoomed area selection where it is possible to visualize and read the synset labels belonging to the selected *semantic zone*. Figure 6 shows only synsets and their semantic relations; an attempt to add also the lexical relations and the *Word* nodes results in an even more confused tangle of points and arcs. Thus, it is necessary to simplify the representation of the network by following some functional and esthetic criteria. In this regards, we have selected two simple criteria:

1. the efficiency of the visualization; i.e., avoid the information redundancy and the proliferation of useless signs and graphics as much as possible;
2. the effectiveness of the visualization; i.e., grant that the graphical representation of the network covers the whole informative content of the WordNet graph-based implementation.

3. the clearness of visualization, i.e., use light colors, such as gray, light blue, dark green, etc. with a proper level of brightness and with an appreciable contrast.

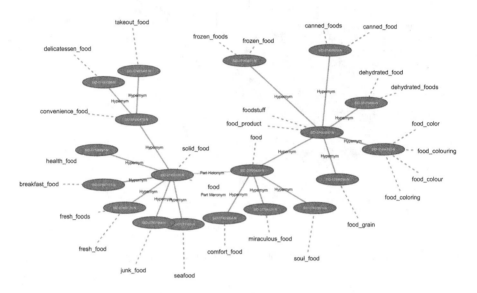

Fig. 7. First layout of WordNet graph excerpt. (Color figure online)

Along the efficiency criteria, we decided to visualize only *Words* labels, avoiding to show again the lexical chain of words representing the corresponding concept into the synset nodes. These ones only show the synset ID as retrieved from the WordNet database inside the stretched blue oval. Furthermore, since for each *Hyponym* relation between synsets corresponds an *Hypernym* relation, we decided to show only one of the two, namely the *Hypernym*, in order to increase the clearness of the representation. The same approach has been adopted for the other pairs of antinomical relations like *Meronym-Holonym*. This approach also satisfies the effectiveness criteria, in fact, even if there is not an explicit representation of the *Hyponym* relations, these ones can be inferred from the corresponding *Hypernym* ones. Each synset is linked to the corresponding lemmas through the *hasConcept* relation which has been represented with a dashed line with a light gray color without an explicit label. This improves the efficiency of the visualization and the effectiveness, since no informative contents is sacrificed for clearness. Figure 7 shows an excerpt of the WordNet representation using the previously described style and layout. It appears evidently clearer than the representation in Fig. 4 (also with respect to the zoomed selected area), also adding the words nodes and the *hasConcept* relations. Figure 7 makes some changes compared to the Fig. 8, mainly regarding the shape of the synset and words nodes, its color and try to distinguish semantic relation by using different colors.

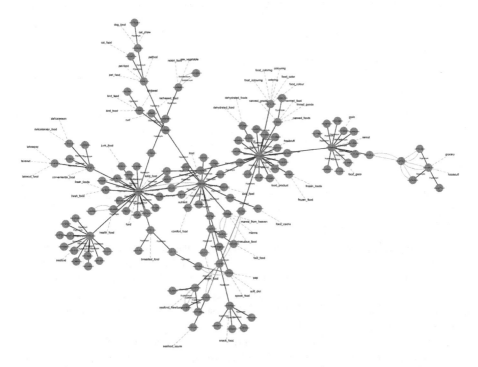

Fig. 8. Second layout of WordNet graph excerpt. (Color figure online)

7   The Tag Cloud-Based Representation of WorNet Synonyms Rings

In addition to the large scale representation previously explained, an advancement in the visualization of WordNet will be proposed in this section as a new exploration path investigated by the authors: the adoption of tag cloud based representation for the WordNet synonyms rings. By using the statistical linguistics measures of *polisemy* and *frequency* of a term as visual cues in drawing the word signs attached to a certain synset, we have improved the visualization style of the WordNet graph in Neo4J. Technically, higher the frequency of the word sense, larger is the font used to represent such word in the corresponding synonym ring, as well as, higher is the polysemy of a word in the whole Word-Net, lighter is the shade of gray used to tag such word. All the word senses are connected to the corresponding synset through a blank gray line and each synset is represented through a short text containing its gloss. Semantic relations between synsets are represented through a transparent green arc showing a label that report the type of the semantic link (e.g., hypernym, hyponim, meronym, etc.). Figure 9 shows the application of the style rules described above to the same cypher query from *home* mentioned in the previous section. For each sense of the term *home*, the figure shows the tag cloud based representation. Some considerations arise from the figure above. The lighter gray used for the term 'home'

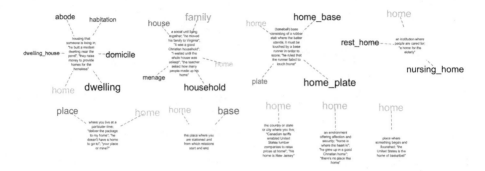

Fig. 9. WordNet synset rings containing the 'home' word with the customized style in Cytoscape.

is due to its high polysemy (9). This color is intentionally weak to demonstrate how vague is the term alone without a context making it meaningful. Things get worse, for example, with terms like *head* or *line* with a polysemy equal to 33 and 29 respectively. On the contrary, the term *home_plate* is large in size and as a strong gray shade because of its low polysemy (1) and high frequency when used in the context of baseball.

Figures 10(a) and (b) show more representations of WordNet excerpts to fully demonstrating the customized style resulting from this work. The figure are obtained through the following Cypher query where 'keyword' is substituted with *book* and *time*:

The figures above also highlights the semantic relations existing between synsets showing a more complete representation of WordNet with the new visualization style described in this work.

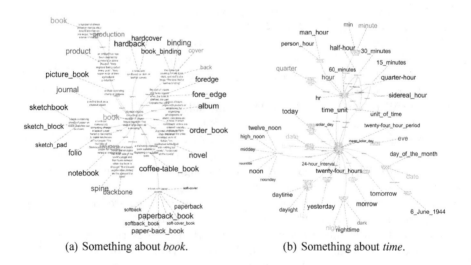

(a) Something about *book*.   (b) Something about *time*.

Fig. 10. WordNet excerpts in Cytoscape with custom style. (Color figure online)

8   Conclusions

In this paper we have described an experience of importing, exploring and visualizing WordNet into a graphDB. To achieve this objective we have faced different issues involving Big Data challenges through all the phases of graph management. The import procedure has been accomplished using he Cypher import functions; queries running has also resulted quite simple by exploiting the potentiality of the Cypher language (such as, its iconicity, the pattern-matching mechanism and the built-in functions to traverse the graph), with respect to the equivalent queries in SQL language. Eventually, the visualization task has resulted more challenging, due to the intricate nature of the WodrNet graph and its "Big" dimensions in terms of nodes and edges, particularly, when it is requested a large-scale visualization. Thus, if we want to simplify the visualization of WordNet, a redefinition of the custom style and also the layout manager is needed. In this regard, we have introduced three criteria to simplify the view, with respect to efficiency, effectiveness and clearness and have adopted them in order to obtain two different representation of WordNet, which results more clear at first glance. In addition, an advancement in the visualization of WordNet has been proposed as a new explanation path by adopting a tag cloud based representation for the WordNet synonyms rings.

Starting from the consideration emerged in this work, at least two different directions could be taken for future investigations or researches. From the one hand, it worth to deepen the comparison between Cypher and SQL languages also through a performance analysis, in order to appreciate the efficiency of the language, in addition to the immediacy of the first language; on the other hand, an improved characterization of the criteria to simplify the view could be investigated, and validated by usability tests in which the user can express a consensus whether the representation is friendly or not, and the information inside WordNet is easily accessible or not. Finally, we think that the adoption of a labeled directed property graph model for the representation of WordNet, as an external and generalistic knowledge base, can be exploited in many contexts, for example, Information Retrieval systems [40] to improve their performance measures or in multimedia knowledge modeling [41, 42] to enhance the expressive power of representations.

References

1. Friendly, M., Denis, D.J.: Milestones in the history of thematic cartography, statistical graphics, and data visualization (2001). http://www.datavis.ca/milestones
2. Munzner, T.: Process and pitfalls in writing information visualization research papers. In: Kerren, A., Stasko, J.T., Fekete, J.-D., North, C. (eds.) Information Visualization. LNCS, vol. 4950, pp. 134–153. Springer, Heidelberg (2008). doi:10.1007/978-3-540-70956-5_6
3. Caldarola, E.G., Picariello, A., Castelluccia, D.: Modern enterprises in the bubble: why big data matters. ACM SIGSOFT Softw. Eng. Notes **40**, 1–4 (2015)

4. Caldarola, E.G., Sacco, M., Terkaj, W.: Big data: the current wave front of the Tsunami. ACS Appl. Comput. Sci. **10**, 7–18 (2014)
5. Caldarola, E.G., Rinaldi, A.M.: Big data: a survey - the new paradigms, methodologies and tools. In: Proceedings of the 4th International Conference on Data Management Technologies and Applications, pp. 362–370 (2015)
6. Bederson, B.B., Shneiderman, B.: The Craft of Information Visualization: Readings and Reflections. Morgan Kaufmann, Burlington (2003)
7. Spence, R.: Information visualization, vol. 1. Springer, Heidelberg (2001)
8. Mazza, R.: Introduction to Information Visualization. Springer Science & Business Media, Heidelberg (2009)
9. Fayyad, U.M., Wierse, A., Grinstein, G.G.: Information Visualization in Data Mining and Knowledge Discovery. Morgan Kaufmann, Burlington (2002)
10. Ware, C.: Information Visualization: Perception for Design. Elsevier, Amsterdam (2012)
11. Cui, W., Qu, H.: A survey on graph visualization. Ph.D. Qualifying Exam (PQE) Report, Computer Science Department, Hong Kong University of Science and Technology, Kowloon, Hong Kong (2007)
12. Tutte, W.T.: How to draw a graph. Proc. Lond. Math. Soc **13**, 743–768 (1963)
13. Purchase, H.: Which aesthetic has the greatest effect on human understanding? In: DiBattista, G. (ed.) GD 1997. LNCS, vol. 1353, pp. 248–261. Springer, Heidelberg (1997). doi:10.1007/3-540-63938-1_67
14. Purchase, H.C., Cohen, R.F., James, M.: Validating graph drawing aesthetics. In: Brandenburg, F.J. (ed.) GD 1995. LNCS, vol. 1027, pp. 435–446. Springer, Heidelberg (1996). doi:10.1007/BFb0021827
15. Herman, I., Delest, M., Melancon, G.: Tree visualisation and navigation clues for information visualisation. Comput. Graph. Forum **17**, 153–165 (1998)
16. Reingold, E.M., Tilford, J.S.: Tidier drawings of trees. IEEE Trans. Softw. Eng. **17**, 223–228 (1981)
17. Beaun, L., Parent, M.A., Vroomen, L.C.: Cheops: a compact explorer for complex hierarchies. In: Proceedings of the Visualization 1996, pp. 87–92. IEEE (1996)
18. Huang, W., Hong, S.H., Eades, P.: Effects of sociogram drawing conventions and edge crossings in social network visualization. J. Graph Algorithms Appl. **11**, 397–429 (2007)
19. Wills, G.J.: NicheWorks — interactive visualization of very large graphs. In: DiBattista, G. (ed.) GD 1997. LNCS, vol. 1353, pp. 403–414. Springer, Heidelberg (1997). doi:10.1007/3-540-63938-1_85
20. Carriere, J., Kazman, R.: Research report. Interacting with huge hierarchies: beyond cone trees. In: Proceedings of the Information Visualization 1995, pp. 74–81. IEEE (1995)
21. Munzner, T.: H3: Laying out large directed graphs in 3D hyperbolic space. In: Proceedings of the IEEE Symposium on Information Visualization 1997, pp. 2–10. IEEE (1997)
22. Eades, P.: A heuristics for graph drawing. Congressus Numerantium **42**, 146–160 (1984)
23. Fruchterman, T.M., Reingold, E.M.: Graph drawing by force-directed placement. Softw. Pract. Exper. **21**, 1129–1164 (1991)
24. Gansner, E.R., North, S.C.: Improved force-directed layouts. In: Whitesides, S.H. (ed.) GD 1998. LNCS, vol. 1547, pp. 364–373. Springer, Heidelberg (1998). doi:10.1007/3-540-37623-2_28

25. Johnson, B., Shneiderman, B.: Tree-maps: a space-filling approach to the visualization of hierarchical information structures. In: Proceedings of the IEEE Conference on Visualization 1991, pp. 284–291. IEEE (1991)
26. Kamps, J., Marx, M.: Visualizing wordnet structure. In: Proceedings of the 1st International Conference on Global WordNet, pp. 182–186 (2002)
27. Collins, C.: Wordnet explorer: applying visualization principles to lexical semantics. Computational Linguistics Group, Department of Computer Science, University of Toronto, Toronto, Ontario, Canada (2006)
28. Collins, C.: DocuBurst: radial space-filling visualization of document content. Technical Report KMDI-TR-2007-1, Knowledge Media Design Institute, University of Toronto (2007)
29. Holzschuher, F., Peinl, R.: Performance of graph query languages: comparison of cypher, gremlin and native access in Neo4j. In: Proceedings of the Joint EDBT/ICDT 2013 Workshops, pp. 195–204. ACM (2013)
30. Webber, J.: A programmatic introduction to Neo4j. In: Proceedings of the 3rd Annual Conference on Systems, Programming, and Applications: Software for Humanity, pp. 217–218. ACM (2012)
31. Robinson, I., Webber, J., Eifrem, E.: Graph Databases. O'Reilly Media Inc., Sebastopol (2013)
32. Fellbaum, C.: Wordnet. The Encyclopedia of Applied Linguistics (1998)
33. Miller, G.A.: WordNet: a lexical database for English. Commun. ACM **38**, 39–41 (1995)
34. Rinaldi, A.M.: A content-based approach for document representation and retrieval. In: Proceedings of the Eighth ACM Symposium on Document Engineering, pp. 106–109. ACM (2008)
35. Finlayson, M.A.: MIT Java Wordnet Interface (JWI) User's Guide, Version 2.2.x (2013)
36. Finlayson, M.A.: Java libraries for accessing the Princeton Wordnet: comparison and evaluation. In: Proceedings of the 7th Global Wordnet Conference, Tartu, Estonia (2014)
37. Van Bruggen, R.: Learning Neo4j. Packt Publishing Ltd., Birmingham (2014)
38. Mathur, A., Dalal, D.: APIARY: A Case for Neo4j? (Equal Experts Labs)
39. Kobourov, S.G.: Spring embedders and force directed graph drawing algorithms. arXiv preprint arXiv:1201.3011 (2012)
40. Albanese, M., Capasso, P., Picariello, A., Rinaldi, A.M.: Information retrieval from the web: an interactive paradigm. In: Candan, K.S., Celentano, A. (eds.) MIS 2005. LNCS, vol. 3665, pp. 17–32. Springer, Heidelberg (2005). doi:10.1007/11551898_5
41. Albanese, M., Maresca, P., Picariello, A., Rinaldi, A.M.: Towards a multimedia ontology system: an approach using TAO_XML. In: DMS, pp. 52–57 (2005)
42. Rinaldi, A.M.: A multimedia ontology model based on linguistic properties and audio-visual features. Inf. Sci. **277**, 234–246 (2014)

Open-Source Search Engines in the Cloud

Khaled Nagi[(✉)]

Faculty of Engineering, Department of Computer and Systems Engineering,
Alexandria University, Alexandria, Egypt
khaled.nagi@alexu.edu.eg

Abstract. The key to the success of the analysis of petabytes of textual data available at our fingertips is to do it in the cloud. Today, several extensions exist that bring Lucene, the open-source de facto standard of textual search engine libraries, to the cloud. These extensions come in three main directions: implementing scalable distribution of the indices over the file system, storing them in NoSQL databases, and porting them to inherently distributed ecosystems. In this work, we evaluate the existing efforts in terms of distribution, high availability, fault tolerance, manageability, and high performance. We are committed to using common open-source technology only. So, we restrict our evaluation to publicly available open-source libraries and eventually fix their bugs. For each system under investigation, we build a benchmarking system by indexing the whole Wikipedia content and submitting hundreds of simultaneous search requests. By measuring the performance of both indexing and searching operations, we report of the most favorable constellation of open-source libraries that can be installed in the cloud.

Keywords: Full-text searching · Indexing · Distributed ecosystems · NoSQL · Cloud

1 Introduction

Since the early 2010s, search engines took over the job of performing intelligent textual analytics of petabytes of data. The most prominent open-source projects in this area are Solr [29] and Elasticsearch [14]. While Solr seems to be more commonly used in information retrieval domains, Elasticsearch is gaining more popularity in the area of data analysis due to its data visualization tool Kibana [9]. Both have Lucene [19] at their heart. Lucene is the de-facto standard search engine library. It is based on research dating back to 1990 [5].

The first attempt to scale Lucene beyond the classical file system is presented in [20]. Lucene storage classes are extended to store the index to and traverse the index from the relational database management systems, such as MySQL. Back then, the storage API was not standardized and undergone many changes. After the wide-spread of NoSQL database management systems and distributed BigData systems, such as Hadoop, the mean-time standardized Lucene storage API is ported to these emerging technologies in several open-source prototypes. A good overview of these efforts are found in [11].

© Springer International Publishing AG 2016
A. Fred et al. (Eds.): IC3K 2015, CCIS 631, pp. 100–117, 2016.
DOI: 10.1007/978-3-319-52758-1_7

In this work, we investigate these open-source implementations as we believe that the key to the success of any large-scale search engine will remain *openness*. We explicitly refrain from adding any customized implementation to the off-the-shelf open-source components. *In the case of the presence of bugs in these publicly available systems, we attempt to solve them.* Other than fixing bugs, we apply only the tweaks supplied by the official performance tuning recommendations from the providers.

Our contribution is the independent evaluation of the existing approaches in terms of support for distribution. The main focus is the evaluation of the performance of both indexing and searching of these systems. Moreover, a robust distributed search engine must support *data partitioning*, *replication* and must be always consistent. So, we investigate the effect of node failures. Furthermore, we take into consideration the ease of management of the cluster.

The rest of the paper is organized as follows. In Sect. 2, we describe the properties of a distributed highly-scalable search engine. We also give a background of the technologies. In Sect. 3, we describe the architecture of each system under investigation. In Sect. 4, we present the performance evaluation based on the benchmarking scenario that we constructed. Section 5 concludes the paper.

2 Background and Related Work

The following properties must be available in any cloud-based large-scale search engine:

- **Partitioning (sharding):** It is splitting the index into several independent sections, usually called *shards*. Each shard is a separate index and is usually indexed independently. Depending on the sharding strategy, a search query is directed to the corresponding shard(s) and each result is merged and returned to the user.
- **Replication:** It means storing various copies of the same data to increase the data availability. On a system built over commodity hardware, such as Amazon EC2 [1] or Microsoft Azure [3], replication protects against the loss of data. Additionally, replication is also used to increase the throughput of index reads.
- **Consistency:** Depending on a relaxed definition of consistency, a newly indexed document is not necessarily made available to the next search request. However, the index data structure must be consistent under whatever storage model used to store it. Consistency between the internal blocks of a single index must be guaranteed all the time, whereas consistency across the independent shards is not a must.
- **Fault-Tolerance:** It means the absence of any Single Point of Failure (SPoF) in the system. In case of the failure of a node, the whole search engine should not be go offline.
- **Manageability:** The administration of the nodes of a cloud-based search engine must be made easily: either through a Command Line Interface (CLI), programmatically embeddable interface, e.g., JMX, or most preferably via web administration consoles.
- **High Performance:** It means that the response time for query processing should be under a couple of seconds under a full load of concurrent search requests. Indexing of new documents should be done in parallel.

2.1   Lucene-Based Search Engines

A full text search index is usually a variation of the inverted index structure [5]. *Indexing* begins with collecting the available set of documents by the crawler. A crawler consists in general of several hundreds of data gathering threads. The parser in these threads converts the collected documents to a stream of plain text. In the analysis phase, the stream of data is tokenized according to predefined delimiters, such as blank, tab, hyphen, etc. Then, all stop words are removed from the tokens. A stem analyzer usually reduces the tokens to their roots to enable phonetic searches. *Searching* begins with parsing the user query using the same parser used in the indexing process. This is a must or else the matching documents will not be exactly the ones needed. The tokens have to be analyzed by the same analyzer used for indexing as well. Then, the index is traversed for possible matches. The fuzzy query processor is responsible for defining the match criteria and the score of the hit according to a calculated distance vector.

Lucene [19] is at the heart of almost every full-text search engine. It provides several useful features, such as ranked searching, fielded searching and sorting. Searching is done through several query types including: phrase queries, wildcard queries, proximity queries, range queries. It allows for simultaneous indexing and searching by implementing a simple pessimistic locking algorithm [17].

An important internal feature of Lucene is that it uses a configurable storage engine. It comes with a codec to store the index on the disc or maintain it in-memory for smaller indices. The internal structure of the index file is public and is platform independent [16]. This ensures its portability. Back in 2007, this concept was used to store the index efficiently into Relational Database Management Systems [20]. The same technique is used today to store the index in other NoSQL databases, such as Cassandra [15] and mongoDB [23].

Apache Solr [29] is built on-top of Lucene. It is a web application that can be deployed in any servlet container. It adds the following functionality to Lucene:

- XML/HTTP/JSON APIs,
- Hit highlighting,
- Faceted search and filtering,
- Range queries,
- Geospatial search,
- Caching,
- Near Real-Time (NRT) searching of newly indexed documents, and
- A web administration interface.

SolrCloud [29] was released in 2012. It allows for both sharding and replication of the Lucene indexes. The management of this distribution is seamlessly integrated into an intuitive web administration console. Figure 1 illustrates the configuration of one of our setups in the web administration console.

Elasticsearch [14] evolved almost in parallel to Solr and SolrCloud. Both bring the same set of features. Both are very performant. Both are open-source and use a different combination of open-source libraries. At their hearts, both have Lucene. In general, Solr seems to be slightly more popular than Elasticsearch in information retrieval domains; whereas Elasticsearch is expanding more in the direction of data analytics.

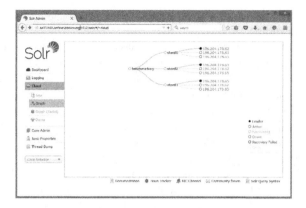

Fig. 1. Screenshot of the web administration console.

2.2 NoSQL Databases

The main strength of NoSQL databases comes from their ability to manage extremely large volumes of data. For this type of applications, ACID transaction properties are too restrictive. More relaxed models emerged such as the CAP theory or eventually consistent emerged [4]. It means that any large-scale distributed DBMS can guarantee for two of three aspects: *Consistency*, *Availability*, and *Partition* tolerance. In order to solve the conflicts of the CAP theory, the BASE consistency model (BAsically, Soft state, Eventually consistent) is defined for modern applications [4]. This principle goes well with information retrieval systems, where intelligent searching is more important than consistent ones.

A good overview of existing NoSQL database management systems can be found in [8]. Mainly, NoSQL database systems fall into four categories:

- graph databases,
- key-value systems,
- column-family systems, and
- document stores.

Graph databases concentrate on providing new algorithms for storing and processing very large and distributed graphs. They are often faster for associative data sets. They can scale more naturally to large data sets as they do not require expensive join operations. Neo4j [21] is a typical example of a graph databases.

Key-value systems use associative arrays (maps) as their fundamental data structure. More complicated data structures are often implemented on top of the maps. Redis [25] is a good example of a basic key-value systems.

The data model of *column-family* systems provides a structured key-value store where columns are added only to specified keys. Different keys can have different number of columns in any given family. A prominent member of the column family stores is Cassandra [15]. Apache Cassandra is a second generation of distributed key value stores; developed at Facebook. It is designed to handle very large amounts of data spread across many commodity servers without a single point of failure. Replication is done even across multiple data centers. Nodes can be added to cluster without downtime.

Document-oriented databases are also a subclass of key-value stores. The difference lies in the way the data is processed. A document-oriented system relies on internal structure in the document order to extract metadata that the data-base engine uses for further optimization. Document databases are schema-less and store all related information together. Documents are addressed in the database via a unique key. Typically, the database constructs an index on the key and all kinds of metadata. mongoDB [23], first developed in 2007, is considered to be the most popular NoSQL nowadays [6]. mongoDB provides high availability with replica sets.

In all attempts to store Lucene index files in NoSQL databases, the contributors take the logical index file as starting point. The set of logical files are broken into logical blocks that are stored in the database. It is therefore clear that plain key-value data stores and graph databases are not suitable for storing a Lucene index. On the other hand, document stores, such as mongoDB, are ideal stores for Lucene indices. One Lucene logical file maps easily to a mongoDB document. Similarly, the Lucene logical directory (files) is mapped to a Cassandra column family (rows), which is captured using an inherited implementation of the abstract Lucene `Directory` class. The files of the directory are broken down into blocks (whose sizes are capped). Each block is stored as the value of a column in the corresponding row.

2.3   Highly Distributed Ecosystems

After the release of [7], Doug Cutting worked on a Java-based MapReduce implementation to solve scalability issues on Nutch [13]; which is an open-source web crawler software project to feed the indexer of the search engine with textual content. This was the base for the Hadoop open source project; which became a top-level Apache Foundation project. Currently, the main Hadoop project includes four modules:

- Hadoop Common: It supports the other Hadoop modules.
- Hadoop Distributed File System (HDFS): A distributed file system.
- Hadoop YARN: A job scheduler and cluster resource management.
- Hadoop MapReduce: A YARN-based system for parallel processing of large data sets.

Each Hadoop task (Map or Reduce) works on the small subset of the data it has been assigned so that the load is spread across the cluster. The map tasks generally load, parse, transform, and filter data. Each reduce task is responsible for handling a subset of the map task output. Intermediate data is then copied from mapper tasks by the reducer tasks in order to group and aggregate the data. *It is definitely appealing to use the MapReduce framework in order to construct the Lucene index using several nodes of a Hadoop cluster.*

The input to a MapReduce job is a set of files that are spread over the Hadoop Distributed File System (HDFS). In the end of the MapReduce operations, the data is written back to HDFS. HDFS is a distributed, scalable, and portable file system. A Hadoop cluster has one *namenode* and a set of *datanodes*. Each datanode serves up blocks of data over the network using a block protocol. HDFS achieves reliability by replicating the data across multiple hosts. Hadoop recommends a replication factor of 3. Since the release of Hadoop 2.0 in 2012, several high-availability capabilities, such as providing automatic failover of the namenode, are implemented. This way, HDFS comes with no single point of failure. HDFS was designed for mostly immutable files [22] and may not be suitable for systems requiring concurrent write-operations. Since the default storage codec for Solr is append-only, it matches HDFS. With the extreme scalability, robustness and widespread of Hadoop clusters, it offers the perfect store for Solr in Cloud-based environments.

Additionally, there are *three* ecosystems that can be used in building distributed search engines: Katta, Blur and Storm.

Katta [12] brings Apache Hadoop and Solr together. It brings search across a completely distributed MapReduce-based cluster. Katta is an open-source project that uses the underlying Hadoop HDFS for storing the indices and providing access to them. Unfortunately, the development of Katta has been stopped. The main reason is the inclusion of several of the Katta features within the SolrCloud project.

Apache *Blur* [2] is a distributed search engine that can work with Apache Hadoop. It is different from the traditional big data systems in that it provides a relational data model-like storage on top of HDFS. Apache Blur does not use Apache Solr; however, it consumes Apache Lucene APIs. Blur provides data indexing using MapReduce and advanced search features; such as a faceted search, fuzzy, pagination, and a wildcard search. Blur shard server is responsible for managing shards. For Synchronization, it uses Apache ZooKeeper [32]. Blur is still in the apache incubator status. The current release version 0.2.3 works with Hadoop 1.x and is not validated using the scalability features coming with Hadoop 2.x.

The third project *Storm* [30] is also in its incubator state at Apache. Storm is a real time distributed computation framework. It processes huge data in real time. Apache Storm processes massive streams of data in a distributed manner. So, it would be a perfect candidate to build Lucene indices over large repositories of documents once it is reaches the release state. Apache Storm uses the concept of Spout and Bolts. Spouts are data inputs; this is where data arrives in the Storm cluster. Bolts process the streams that get piped into it. They can be fed data from spouts or other bolts. The bolts can form a chain of processing, with each bolt performing a unit task in a concept similar to MapReduce.

3 Systems Under Investigation

3.1 Solr on Cassandra

Solandra is an open-source project that uses Cassandra, the column-based NoSQL database, instead of the file system for storing indices in the Lucene index format [16]. The project is very stable. Unfortunately, the last commit dates back to 2010. The current Solandra version available for download uses Apache Solr 3.4 and Cassandra 0.8.6. Solandra does not use SolrCloud since it was not present at the time of the development of the open-source project. The Cassandra-based distributed data storage is implemented behind the Façade `CassandraDirectory`. Solandra uses its own index reader called `SolandraIndexReaderFactory` by overriding the default index reader.

In the Solandra project, Solr and Cassandra run both within the same JVM. However, with change in the configuration and the source code, we run a Cassandra cluster instead of the single database. In a small implementation, the Cassandra cluster spreads over 3 nodes as illustrated in Fig. 2. The larger cluster contains 7 nodes. On Cassandra, each node exchanges information across the cluster every second. This value can be change in its configuration file to match the hardware requirement. A sequentially written commit log on each node captures write activity to ensure data durability. Data is then indexed and written to an in-memory structure. Once the memory structure is full, the data is written to disk in the `SSTable` data file. All writes are automatically partitioned and replicated throughout the cluster. A cluster is arranged as a *ring* of nodes. Clients send read/write requests to any node in the ring; that takes on the role of coordinator node, and forwards the request to the node responsible for servicing it. A *partitioner* decides which nodes store which rows.

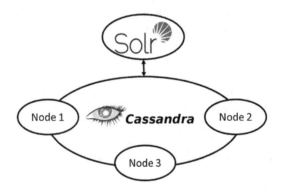

Fig. 2. Our Solandra installation.

Both sharding and replication are automatically made available by the Casandra cluster. Cassandra also guarantees the consistency of the blocks read by its various nodes. Although fault-tolerance is a strong feature of Cassandra, Solr itself is the single point of failure in this implementation, due to the absence of the integration with SolrCloud. Unfortunately, Solandra does not support the administration console of Solr. The only management option is through the Cassandra CLI.

3.2 Lucene on mongoDB

Another open-source NoSQL-based project is LuMongo [18]. LuMongo is a simple JAVA library that implements Lucene APIs and stores the index in mongoDB. All data, including indices and documents, is stored in mongoDB. mongoDB supports sharding and replication out of the box. LuMongo itself operates as another independent cluster. On error, clients can fail to another cluster node. The exchange of the cluster management information is done through an in-memory database called Hazelcast. Nodes in the cluster can be added and removed dynamically through a simple CLI command. Using the CLI, the user can perform the following operations:

- query the health status of cluster,
- list available indices, get their counts,
- submit simple queries, and
- fetch documents.

For example, the following CLI command registers a node in the cluster:

```
bash clusteradmin.sh --command registerNode --mongoConfig
mongo.properties --nodeConfig node.properties
      --address 10.0.0.10
```

The following command starts a node:

```
bash startnode.sh --mongoConfig mongo.properties --
address 10.0.0.10 --hazelcastPort 5702
```

LuMongo indices are broken down into shards called *segments*. Each segment is an independent Lucene standard index. A hash of the document's unique identifier determines which segment a document's indexed fields will be stored into. In our smaller implementation, illustrated in Fig. 3, the segments are stored in a 3 × 3 mongoDB cluster for the small setup and 7 shards and 3 replicas for the larger setup to match the number of LuMongo servers; which is 3 and 7 respectively.

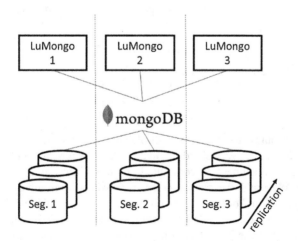

Fig. 3. Our LuMongo implementation.

In this setup, sharding is implemented in both LuMongo and mongoDB. The mongoDB takes care of partitioning seamlessly. mongoDB guarantees the consistency of the index store, while LuMongo guarantees the consistency of the search result. There is *no* single point of failure in mongoDB and LuMongo.

While running our experiments under heavy load of concurrent search requests, *we discover a memory leak within the LuMongo code*. The problem causes the LuMongo node to crash at approx. 60 concurrent search per node. On the positive side, the whole distributed system does not fail. Load is distributed evenly among the available nodes. However, the cost is degrading the performance. We track down the problem to be in fetching the content of the documents after returning the document ids from the search engine. Consequently, we fix the problem and deploy a patched LuMongo to our experiments to eliminate this malefaction.

3.3   Solrcloud

SolrCloud [29] contains a cluster of Solr nodes. Each node runs one or more collections. A collection holds one or more shards. Each shard can be replicated among the nodes. Apache ZooKeeper [32] is responsible for maintaining coordination among various nodes, similar to Hazelcast in the LuMongo project. It provides load-balancing and failover to the Solr cluster. Synchronization of status information of the nodes is done in-memory for speed and is persisted on the disk at fixed checkpoints. Additionally, the ZooKeeper maintains configuration information of the index; such as schema information and Solr configuration parameters. Together, they build a Zookeeper *ensemble*. When the cluster is started, one of the Zookeeper nodes is elected as a *leader*. Although distributed in reality, all Solr nodes retrieve their configuration parameters in a central manner through the Zookeeper *ensemble*. Usually, there are more than one Zookeeper for redundancy. All Zookeeper IPs are stored in a configuration file that is given to each Solr node at startup.

The following commands start ZooKeeper and Solr nodes respectively:

```
./bin/zkServer.sh start conf/zoo.cfg

./bin/solr restart -cloud -z
196.204.178.62:2181,196.204.178.63:2181,196.204.178.65:21
81 -p 9120 -s example/cloud/node2/solr -m 5g
```

SolrCloud distributes search across multiple shards transparently. The request gets executed on all leaders of every shard involved. Search is possible with near-real time (NRT); i.e., after a document is committed. Figure 4 illustrates our small cluster implementation. We build the cluster using a Zookeeper ensemble consisting of 3 nodes. We install 3 SolrCloud instances on three different machines, define 3 shards and replicate them 3 times. In the larger cluster, we extend the Zookeeper ensemble to spread 7 machines. We use 7 SolrCloud instance to master 7 shards while keeping the replication factor at 3.

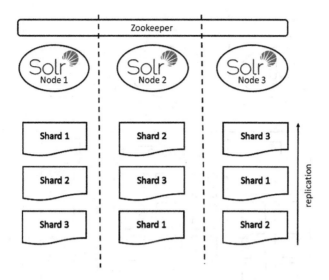

Fig. 4. Our SolrCloud implementation.

3.4 Solrcloud on Hadoop

Building SolrCloud on Hadoop is an extension to the implementation described in Sect. 3.3. The same Zookeeper ensemble and SolrCloud instances are used. Solr is then configured to read and write indices in the HDFS of Hadoop by implementing an `HdfsDirectoryFactory` and implementing a lock type based on HDFS. Both the directory factory and the lock implementation come with the current stable version of Solr [26]. The following command starts SolrCloud with Hadoop as its storage backend.

```
./bin/solr start -cloud -z 196.204.178.62:2181,
196.204.178.63:2181, 196.204.178.65:2181 -p 9152 -m 5g
-Dsolr.directoryFactory=HdfsDirectoryFactory
-Dsolr.lock.type=hdfs
-Dsolr.hdfs.home=hdfs://196.204.178.66:9161/user/nagi/62
```

Figure 5 illustrates our small cluster implementation. We leave replication to the HDFS. We set the replication factor on HDFS to 3 to be consistent with the rest of the setups. For the small cluster, we also use a 3 node Hadoop installation. For the large cluster, we use a 7 node cluster.

Solr provides indexing using MapReduce in two ways. In the first way, the indexing is done at the map side [27]. Each Apache Hadoop mapper transforms the input records into a set of (key, value) pairs, which then get transformed into `SolrInputDocument`. The Mapper task then creates an index from `SolrInputDocument`. The Reducer performs de-duplication of different indices and merges them if needed. In the second way, the indices are generated in the reduce

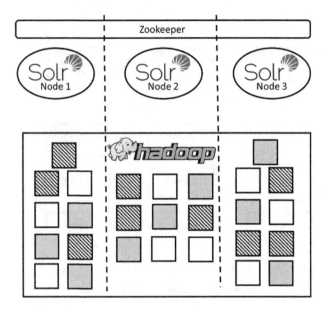

Fig. 5. Our SolrCloud implementation over Hadoop.

phase [28]. Once the indices are created using either ways, they can be loaded by SolrCloud from HDFS and used in searching. We use the first way and employ 20 nodes in the indexing process.

3.5   Functional Comparison

Table 1 summarizes the functional differences between all 4 systems under investigation.

Table 1. Functional comparison of the systems under investigation.

	Solr on Cassandra	Lucene on mongoDB	SolrCloud	SolrCloud on Hadoop
Sharding	Done by Cassandra	Done by mongoDB	Done by Solr	Done by Solr
Replication	Done by Cassandra	Done by mongoDB	Sync. on the level of the file system under to coordination of Zookeeper	Done by HDFS
Consistency	Guaranteed by Cassandra	Guaranteed by LuMungo and mongonDB	Done by Solr and managed by Zookeeper	Guaranteed by HDFS, Solr and Zookeeper
Fault-tolerance	Solr is SPoF	No SPoF	No SPoF	No SPoF
Manageability	CLI	CLI	Web	Web for Solr + web for Hadoop

4   Benchmarking

We build a full text search engine of the English Wikipedia [31] to evaluate the performance of the system described in the previous Section. The index is built over 49 GB of textual content. We develop a benchmarking platform on top of each search engine under investigation as illustrated in Fig. 6.

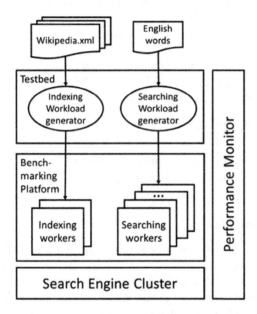

Fig. 6. Components of the benchmarking platform.

The *searching workload generator* composes queries of single terms, which are randomly extracted from a long list of common *English words*. It submits them in parallel to the application. The *indexing workload generator* parses the *Wikipedia dump* and sends the page title, the timestamp, and most important the *content* to the benchmarking platform workers. They pass them to the search engine cluster to be indexed, thus simulating a web crawler. The benchmarking platform manages two connection pools of worker threads. The first pool consists of several hundreds of *searching workers* threads that process the search queries coming from the searching workload generator. The second pool consists of *indexing workers* threads that process the updated content coming from the indexing workload generator. Both worker types submit their requests over http to the search engine cluster under investigation. The performance of the system including that of the search engine cluster is monitored using the *performance monitor* unit.

4.1   Input Parameters and Performance Metrics

We choose the maximum number of fetched hits to be 50. This is a realistic assumption taking into consideration that no more than 25 hits are usually displayed on a web page.

We choose to read the content of these 50 hits and not only the title while fetching the resultset. This exaggerated implementation is intended to artificially stress test the search engines clusters under investigation. The number of search threads is varied from 32 to 320 to match the size of connection pool for the searching worker threads. By relaxing the requirement of prefetching all the pages of the resultset, the number of concurrent searching threads can be increased enormously. In case of high load, the workload generator distributes its searching search threads over 4 physical machines to avoid throttling the requests by the hosting client. Due to locking restrictions inherent in Lucene, we restrict our experiments to maximum one indexing worker per node in the search engine cluster.

In all our experiments, we monitor the response time of the search operations from the moment of submitting the request till receiving the overall result. We also monitor the system throughput in terms of:

- searches per second, and
- index inserts per hour.

Additionally, the performance monitor constantly monitors CPU and memory usages of the machines running the search engine cluster.

4.2   System Configuration

In order to neutralize the effect of using virtualized nodes in globalized data cloud centers, we conduct our experiments in an isolated cluster available at the Internet Archive of the Bibliotheca Alexandrina [10]. The Bibliotheca Alexandrina possesses a huge dedicated computer center for archiving the Internet, digitizing material at Bibliotheca Alexandrina and other digital collections.

The Internet Archive at the Bibliotheca Alexandrina has about 35 racks each rack is comprised of 30 to 40 nodes and a gigabit switch connecting them. The 35 racks are connected also with a gigabit switch. The nodes are based on commodity servers with a total capacity of 7000 TB.

The Bibliotheca Alexandrina dedicated one rack with 20 nodes to our research for approx. one month. The nodes are connected with a gigabit switch and are isolated from the activities of the Internet Archive during the period of our experiments. Each node has an Intel i5 CPU 2.6 GHz, 8 GB RAM, 4 SATA hard disks 3 TB each.

For each search engine cluster, we construct a small version and a larger one as described in Sect. 3. The small cluster consists of three nodes each containing a shard (a portion of the index) while the larger one is built over 7 nodes. In all installations have a replication factor of 3.

4.3   Indexing

Indexing speed varies largely with the number of nodes involved in the index building operation. Lucene; and hence Solr; employs a pessimistic locking mechanism while inserting data into the index. This locking mechanism is being kept for all backend implementations. From our current experiments and from previous ones [20], we conclude that there is no benefit in having more than one indexing thread per Lucene index (or Solr shard).

This means that the increase in number of shards and their dedicated indexing Lucene/Solr yields to a proportional increase in the speed of indexing. The increase is also linear for all systems under investigation. In other words, the indexing speed of a 3 nodes cluster is 3 times that's of a cluster consisting of a single node. Respectively, the indexing speed of a 7 nodes cluster is 2.3 times that's of a cluster consisting of 3 nodes. A clear winner in this contest is SolrCloud on Hadoop that employs MapReduce in indexing. Using all 20 nodes available in the MapReduce operation increases the speed by factor of 18. A minimum overhead is wasted later on in merging the indices into 3 and 7 nodes, respectively.

In order to normalize a comparison between all systems, we plot the throughput of using one indexing thread on a 3 shards, 3 replica cluster in Fig. 7. These numbers are roughly multiplied by the number of nodes involved to get the overall indexing speed.

On the normalized scale, NoSQL backends bring very different results. Casandra has by far the fastest rate of insertion (60% faster than SolrCloud). This experiment confirms the results reported by [24] proving the high throughput of Cassandra as compared to other NoSQL databases. On the other hand, mongoDB-based storage is the slowest. SolrCloud brings very good results on the file system. The overhead of storage on HDFS is about 26% which is very acceptable taking into consideration the advantages of storing data on Hadoop clusters in cloud environments and the huge speed-ups due to the use of MapReduce in indexing.

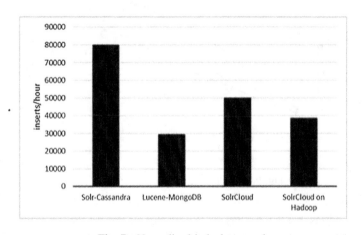

Fig. 7. Normalized indexing speed.

4.4 Searching

Searching is more important than indexing. We repeat the search experiments with the number of search threads varying from 32 to 320. The duration of each experiment is set to 15 min to eliminate any transient effect.

The set of experiments is repeated for both the small cluster and the large cluster. The response time for the small cluster is illustrated in Fig. 8 and the large cluster in Fig. 9. The throughput in terms of number of searches per second versus the number of searching threads is plotted in Fig. 10 for the small cluster and in Fig. 11 for the larger one.

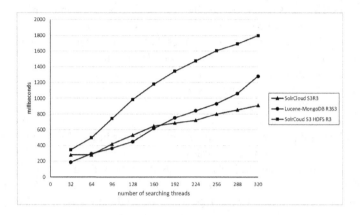

Fig. 8. Search time on the small cluster.

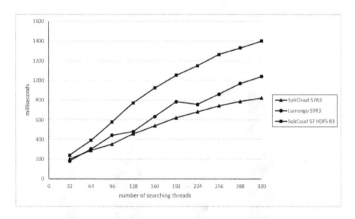

Fig. 9. Search time on the large cluster.

The bad news is that the response time of the single Solr on the Cassandra cluster is far higher than the other systems (>10 s). So, we dropped plotting its values for both clusters. The same applies to the throughput, which was much lower than its counterparts (<50 searches/second). Again this matches the findings in [24], where the high throughput of Cassandra comes at the cost of read latency.

The good news is that the response time for the other systems is very much below the usual 3 s threshold tolerated by a searching user. The maximum search time measured on the small cluster is below 1.8 s and 1.4 s for the larger cluster. The curves

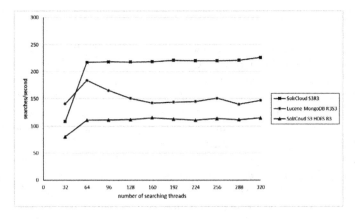

Fig. 10. Throughput of the small cluster.

also show that the response time of the larger cluster is better than the smaller cluster under all settings. This means that the performance of the system is enhanced by the increase of the number of nodes. The system did not achieve its saturation yet.

The figures also illustrate the impact of HDFS on the response time and the overall throughput of the search. Although the search time is increased by almost 40% and the throughput is almost halved, the absolute values remain far below the user threshold of 3 s by retrieving the hits and the contents of each hit for a result-set size of 50 in less than 2 s.

Another important remark is that the performance of all systems degrade gracefully including LuMongo *after fixing the memory leak problem found in the original implementation*.

The throughput curves, Figs. 10 and 11, illustrate that the throughput saturates after a certain number of concurrent search threads. In the small cluster, Fig. 10, the three setups saturate at 64 concurrent threads. On the large cluster, Fig. 11, this number increases to 128.

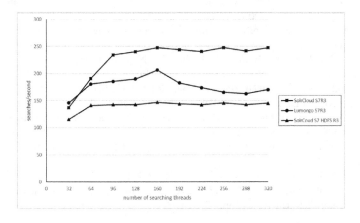

Fig. 11. Throughput of the large cluster.

5 Conclusion

In this paper, we investigate the available options for building large-scale search engines that we deploy in a private Cloud. We restrict ourselves to open-source libraries and do not add extra implementation other that publicly available. Nevertheless, we allow ourselves to fix bugs that we encounter in the available code. We investigate each variation, in terms of scalability through data partitioning, redundancy through replication, consistency, and the ease of management. We build a benchmarking platform on top of the systems under investigation. For each variation, we construct a small and a large cluster. The results of the experiments show that the combination of Solr and Hadoop provide the best tradeoff in terms of scalability, stability and manageability. Search engines based on NoSQL databases offer either a superior indexing speed, or fast searching times, which seem to be mutually exclusive in our settings.

Acknowledgements. We Would like to Thank the Bibliotheca Alexandrina for Providing Us with the Necessary Hardware for Conducting the Benchmarking Experiments.

References

1. Akioka, S., Muraoka, Y.: HPC Benchmarks on Amazon EC2. In: IEEE 24th International Conference on Advanced Information Networking and Applications Workshops (2010)
2. Blur (Incubating) Home. https://incubator.apache.org/blur/. Accessed Jan (2016)
3. Bojanova, I., Samba, A.: Analysis of cloud computing delivery architecture models. In: IEEE Workshops of International Conference on Advanced Information Networking and Applications (2011)
4. Brewer, E.: Towards robust distributed systems. In: ACM Symposium on Principles of Distributed Computing (2000)
5. Cutting, D., Pedersen, J.: Optimizations for dynamic inverted index maintenance. In: SIGIR 1990 (1990)
6. DB-Engines - Knowledge Base of Relational and NoSQL Database Management Systems. http://db-engines.com/en/ranking. Accessed Jan (2016)
7. Dean, J., Ghemawat, S.: MapReduce: simplified data processing on large clusters. Commun. ACM **51**(1), 107–113 (2008)
8. Edlich, S., Friedland, A., Hampe, J., Brauer, B.: NoSQL: Introduction to the World of Non-relational Web 2.0 Databases (In German) NoSQL: Einstieg in die Welt nichtrelationaler Web 2.0 Datenbanken. Hanser Verlag, Munich (2010)
9. Gupta, Y.: Kibana Essentials. Packt Publishing, Birmingham (2015)
10. Internet Archive at Bibliotheca Alexandrina. http://www.bibalex.org/en/project/details?documetid=283. Accessed Jan (2016)
11. Karambelkar, H.V.: Scaling Big Data with Hadoop and Solr, 2nd edn. Packt Publishing, Birmingham (2015)
12. Katta. http://katta.sourceforge.net/. Accessed Jan (2016)
13. Khare, R., et al.: Nutch: a flexible and scalable open-source web search engine. Technical report. Oregon State University, pp. 32–32 (2004)

14. Kuc, R., Rogozinski, M.: Mastering Elasticsearch, 2nd edn. Packt Publishing, Birmingham (2015)
15. Lakshman, A., Malik, P.: Cassandra: a decentralized structured storage system. SIGOPS Oper. Syst. Rev. **44**(2), 35–40 (2010)
16. Lucene - Index File Formats. https://lucene.apache.org/core/3_0_3/fileformats.html. Accessed Jan (2016)
17. Lucene – LockFactory. http://lucene.apache.org/core/4_8_0/core/org/apache/lucene/store/LockFactory.html. Accessed Jan (2016)
18. LuMongo Realtime Time Distributed Search. http://lumongo.org/. Accessed Jan (2016)
19. McCandless, M., Hatcher, E., Gospodnetiæ, O.: Lucene in Action, 2nd edn. Manning, Greenwich (2010)
20. Nagi, K.: Bringing information retrieval back to database management systems. In: International Conference on Information and Knowledge Engineering, IKE 2007 (2007)
21. Neo4j. http://www.neo4j.org. Accessed Jan (2016)
22. Pessach, Y.: Distributed Storage: Concepts, Algorithms, and Implementations. CreateSpace Independent Publishing Platform (2013)
23. Plugge, E., Hawkins, D., Membrey, P.: The Definitive Guide to mongoDB: The NoSQL Database for Cloud and Desktop Computing. Apress, Berkeley (2010)
24. Rabl, T., et al.: Solving big data challenges for enterprise application performance management. VLDB Endow. **5**(12), 1724–1735 (2012)
25. Redis. http://redis.io/. Accessed Jan (2016)
26. Solr - Apache Lucene - The Apache Software Foundation! http://lucene.apache.org/solr/. Accessed Jan (2016)
27. Solr-1045, Build Solr index using Hadoop MapReduce. https://issues.apache.org/jira/browse/SOLR-1045. Accessed Jan (2016)
28. Solr-1301, Add a Solr contrib that allows for building Solr indices via Hadoop's Map-Reduce. https://issues.apache.org/jira/browse/SOLR-1301. Accessed Jan (2016)
29. Smiley, D., Pugh, E., Parisa, K., Mitchell, M.: Apache Solr Enterprise Search Server, 3rd edn. Packt Publishing, Birmingham (2015)
30. Storm - The Apache Software Foundation. https://storm.apache.org/. Accessed Jan (2016)
31. Wikipedia article dump. https://dumps.wikimedia.org/enwiki/. Accessed Jan (2016)
32. Zookeeper. https://zookeeper.apache.org/. Accessed Jan (2016)

Cross-Domain Sentiment Classification via Polarity-Driven State Transitions in a Markov Model

Giacomo Domeniconi, Gianluca Moro, Andrea Pagliarani[✉],
and Roberto Pasolini

DISI, Università Degli Studi di Bologna, Via Venezia 52, Cesena, Italy
{giacomo.domeniconi,gianluca.moro,andrea.pagliarani12,
roberto.pasolini}@unibo.it

Abstract. Nowadays understanding people's opinions is the way to success, whatever the goal. Sentiment classification automates this task, assigning a positive, negative or neutral polarity to free text concerning services, products, TV programs, and so on. Learning accurate models requires a considerable effort from human experts that have to properly label text data. To reduce this burden, cross-domain approaches are advisable in real cases and transfer learning between source and target domains is usually demanded due to language heterogeneity. This paper introduces some variants of our previous work [1], where both transfer learning and sentiment classification are performed by means of a Markov model. While document splitting into sentences does not perform well on common benchmark, using polarity-bearing terms to drive the classification process shows encouraging results, given that our Markov model only considers single terms without further context information.

Keywords: Transfer learning · Sentiment classification · Markov chain · Opinion mining · Polarity-bearing terms · Sentence classification

1 Introduction

When an understanding is required of whether a plain text document has a positive, negative or neutral orientation, *sentiment classification* is involved. This supervised approach learns a model from a training set of documents, labeled with the categories (or classes) we are interested in, then applies it to the test set whose sentiment orientation has to be discovered. Sentiment classification differs from text categorization in the set of classes to be considered: whereas the former usually relies on the same labels (i.e. positive, negative and possibly neutral), in the latter topics are typically what we are interested in and they can range from music to games, to literature, to art, and so on and so forth. However, in both tasks the classification accuracy depends on how much the test documents reproduce the patterns emerged when learning the model on the training set.

Giacomo Domeniconi—This work was partially supported by the project "Toreador".

© Springer International Publishing AG 2016
A. Fred et al. (Eds.): IC3K 2015, CCIS 631, pp. 118–138, 2016.
DOI: 10.1007/978-3-319-52758-1_8

In order to achieve this goal, the most natural approach consists in considering as training set documents belonging to the same domain of those to be classified. The just outlined course of action, referred in literature as *in-domain*, is the most profitable in terms of accuracy, because documents within a unique domain are likely to be lexically and semantically similar. Nevertheless, the implicit assumption in-domain classification relies on in order to build a model, is to always have labeled documents from the same domain of the text set whose sentiment orientation is required to be predicted. This is quite difficult in practice due to some reasons: on the one hand real text sets, like for instance Facebook posts, tweets, discussions in fora, are usually unlabeled. On the other hand, labeling text sets is an onerous activity if manually performed by human experts.

Therefore *cross-domain* approaches are attractive in real cases, because if a model has been built on a certain domain, its reuse in a different domain is definitely desirable due to the aforementioned reasons. For example, let us suppose to have learned a model on a set of reviews about movies and now to be interested in understanding people's thoughts about kitchen appliances. Instead of manually labeling a part of the kitchen appliances reviews and building a different model from scratch, we would exploit the one we already have. However, the test documents might not reflect the regularity of the training set because of language heterogeneity. In fact if a movie review is set to include terms like "amusing" or "boring", a kitchen appliance review is more likely to contain "clean" or "broken". Hence, a *transfer learning* phase is typically demanded to bridge the inter-domain gap.

Many methods have been proposed in literature to transfer knowledge across domains, by either adapting source data to the target domain or representing both in a common space, using approaches such as for example feature expansion and clustering. Remarkable levels of accuracy are reported in experimental evaluations of these methods on topic [2–4] and sentiment classification [5–7]. Given their complexity, these methods often have the drawback of requiring potentially burdensome parameter tuning in order to yield good results in real use cases: at this extent, [8] presents a more straightforward method which obtains comparable experimental results with fixed parameter values.

In our previous work [1], transfer learning was performed along with sentiment classification by folding terms from both source and target domains into the same graph, characterized by a vertex for each term and an edge among terms occurring together in documents. It deserves to be noted that classes were also included into our representation. In this way, since the graph can easily be interpreted as a Markov chain, information flows step by step from source specific terms to target specific ones and reaches categories, allowing both transfer learning and sentiment classification.

In this paper we extend the aforementioned approach with two variants, namely, considering documents as aggregates of sentences rather than as a whole block and using polarity-bearing terms in order to drive state transitions in the Markov chain. The first proposal deals with the idea that not every sentence bears the same sentiment orientation of the document where it appears. Thus, classification is performed sentence by sentence and then results are combined in order to assign the final label to the document. Instead, the second variant aims to classify document sentiment through polarity-bearing terms. Whereas in the basic Markov chain method state transitions are only proportional to the semantic relationships between terms, now they also are function of the terms capability in predicting category. This means that polarity-bearing terms actually are those that contribute the most to the classification process.

Experiments have been performed to compare the underlying variants with the previous version of our Markov chain method. The same benchmark text sets have been chosen to assess performance in 2-classes (i.e. positive and negative) in-domain and cross-domain sentiment classification. On the one hand, results obtained by splitting documents into sentences are unsatisfying; on the other hand, the outcome with polarity-driven state transitions is comparable with both our basic algorithm and other works. Besides preserving the same benefits of our foregoing Markov method, the latter technique improves the way classification is carried out, that is, by taking advantage of polarity-bearing terms. Results are encouraging, especially if considering that the model currently takes only single terms into account, without relying on any kind of context information.

The rest of the paper is organized as follows. Section 2 analyzes the literature about transfer learning, sentiment classification and Markov chains. Section 3 first of all recaps our preceding Markov chain based approach, then explains the variants we introduce in this paper. Section 4 describes the experiments and compares the outcome with other works. Finally, Sect. 5 sums results up and outlines future work.

2   Related Work

Transfer learning generally entails learning knowledge from a *source* domain and using it in a *target* domain. Specifically, *cross-domain* methods are used to handle data of a target domain where labeled instances are only available in a source domain, similar but not equal to the target one. While these methods are used in image matching [9], genomic prediction [10] and many other contexts, classification of text documents by either topic or sentiment is perhaps their most common application. Two major approaches can be distinguished in cross-domain classification [11]: *instance-transfer* directly adjusts source instances to the target domain, while *feature-representation-transfer* maps features of both domains to a different common space. In text categorization by topic, transfer learning has been fulfilled in some ways, for example by clustering together documents and words [2], by extending *probabilistic latent semantic analysis* also to unlabeled instances [3], by extracting latent words and topics, both common and domain specific [4], by iteratively refining target categories representation without a burdensome parameter tuning [8,12].

Apart from the aforementioned, a number of different techniques have been developed solely for sentiment classification. For example, [13] draw on information retrieval methods for feature extraction and to build a scoring function based on words found in positive and negative reviews. In [5,6], a dictionary containing commonly used words in expressing sentiment is employed to label a portion of informative examples from a given domain, in order to reduce the labeling effort and to use the labeled documents as training set for a supervised classifier. Further, lexical information about associations between words and classes can be exploited and refined for specific domains by means of training examples to enhance accuracy [7]. Finally, term weighting could foster sentiment classification as well [14], just like it happens in other mining tasks, from the general information retrieval to specific contexts, such as prediction of gene function annotations in biology [15]. For this purpose, some researchers propose different term weighting schemes: a variant of the well-known *tf-idf* [16], a supervised scheme based on both the importance of a term in a document and the importance of a term in expressing sentiment [17], regularized entropy in combination with singular term cutting and bias term in order to reduce the over-weighting issue [18].

With reference to cross-domain setting, a bunch of methods has been attempted to address the transfer learning issue. Following works are based on some kind of supervision. In [19], some approaches are tried in order to customize a classifier to a new target domain: training on a mixture of labeled data from other domains where such data is available, possibly considering just the features observed in target domain; using multiple classifiers trained on labeled data from diverse domains; including a small amount of labeled data from target. [20] suggest the adoption of a thesaurus containing labeled data from source domain and unlabeled data from both source and target domains. [21] discover a measure of domain similarity contributing to a better domain adaptation. [22] advance a spectral feature alignment algorithm which aims to align words belonging to different domains into same clusters, by means of domain-independent terms. These clusters form a latent space which can be used to improve sentiment classification accuracy of target domain. [23] extend the *joint sentiment-topic* model by adding prior words sentiment, thanks to the modification of the topic-word Dirichlet priors. Feature and document expansion are performed through adding polarity-bearing topics to align domains.

On the other hand, document sentiment classification may be performed by using unsupervised methods as well. In this case, most features are words commonly used in expressing sentiment. For instance, an algorithm is introduced to basically evaluate mutual information between the given sentence and two words taken as reference: "excellent" and "poor" [24]. Furthermore, in another work not only a dictionary of words annotated with both their semantic polarity and their weights is built, but it also includes intensification and negation [25].

Markov chain theory, whose a brief overview can be found in Sect. 3, has been successfully applied in several text mining contexts, such as *information retrieval, sentiment analysis, text classification.*

Markov chains are particularly suitable for modeling hypertexts, which in turn can be seen as graphs, where pages or paragraphs represent states and links represent state transitions. This helps in some information retrieval tasks, because it allows discovering the possible presence of patterns when humans search for information in hypertexts [26], performing link prediction and path analysis [27] or even defining a ranking of Web pages just dealing with their hypertext structure, regardless information about page content [28].

Markov chains, in particular hidden Markov chains, have been also employed to build information retrieval systems where firstly query, document or both are expanded and secondly the most relevant documents with respect to a given query are retrieved [29,30], possibly in a *spoken document retrieval* context [31] or in the *cross-lingual area* [32]. Anyhow, to fulfil these purposes, Markov chains are exploited to model term relationships. Specifically, they are used either in a single-stage or in a multi-stage fashion, the latter just in case indirect word relationships need to be modeled as well [33].

The idea of modeling word dependencies by means of Markov chains is also pursued for *sentiment analysis*. In practice, hidden Markov models (HMMs) aim to find out opinion words (i.e. words expressing sentiment) [34], possibly trying to correlate them with particular topics [35,36]. Typically, transition probabilities and output probabilities between states are estimated by using the Baum-Welch algorithm, whereas the most likely sequence of topics and related sentiment is computed through the Viterbi algorithm. The latter algorithm also helps in *Part-of-speech (POS) tagging*, where Markov chain states not only model terms but also tags [37,38]. In fact, when a tagging for a sequence of words is demanded, the goal is to find the most likely sequence of tags for that sequence of words.

Following works are focused on *text classification*, where the most widespread approach based on Markov models consists in building a HMM for each different category. The idea is, for each given document, to evaluate the probability of being generated by each HMM, finally assigning to that document the class corresponding to the HMM maximizing this probability [39–41]. Beyond directly using HMMs to perform text categorization, they can also be exploited to model inter-cluster associations. For instance, words in documents can be clustered for dimensionality reduction purposes and each cluster can be mapped to a different Markov chain state [42]. Another interesting application is the classification of multi-page documents where, modeling each page as a different bag-of-words, a HMM can be exploited to mine correlation between documents to be classified (i.e. pages) by linking concepts in different pages [43].

3   Method Description

This Section firstly recaps the method based on the Markov chain theory we advanced in [1] to accomplish both in-domain and cross-domain sentiment classification. Then two variants of the underlying approach are proposed, which can also be combined together.

In order for non-expert readers to have a complete understanding, we would like to remind that a Markov chain is a mathematical model that is subject to transitions from one state to another in a states space $\mathcal{S}$. In particular, it is a stochastic process characterized by the so called *Markov property*, namely, future state only depends on current state, whereas it is independent of past states.

3.1   Basic Approach

Before going into details, notice that the entire algorithm can be split into three main stages, namely, the text pre-processing phase, the learning phase and the classification phase. We argue that the learning phase and the classification phase are the most innovative parts of the whole algorithm, because they accomplish both transfer learning and sentiment classification by means of only one abstraction, that is, the Markov chain.

Text Pre-processing Phase. The initial stage of the algorithm is text pre-processing. Starting from a corpus of documents written in natural language, the goal is to transform them in a more manageable, structured format.

Firstly, standard techniques are applied to the plain text, such as word tokenization, punctuation removal, number removal, case folding, stopwords removal and the Porter stemming algorithm [44]. Notice that stemming definitely helps the sentiment classification process, because words having the same morphological root are likely to be semantically similar.

The representation used for documents is the common bag-of-words, that is, a term-document matrix where each document d is seen as a multiset (i.e. bag) of words (or terms). Let $\mathcal{T} = \{t_1, t_2, \ldots, t_k\}$, where k is the cardinality of $\mathcal{T}$, be the dictionary of terms to be considered, which is typically composed of every term appearing in any document in the corpus to be analyzed. In each document d, each word t is associated to a weight w_t^d, usually independent of its position inside d. More precisely, w_t^d only depends on *term frequency* $f(t, d)$, that is, the number of occurrences of t in document d, and in particular, represents *relative frequency* $rf(t, d)$, computed as follows:

$$w_t^d = rf(t, d) = \frac{f(t, d)}{\sum_{\tau \in \mathcal{T}} f(\tau, d)} \tag{1}$$

After having built the bags of words, a feature selection process is performed to fulfil a twofold goal: on the one hand, feature selection allows selecting only the most profitable terms for the classification process. On the other hand, being k higher the more the dataset to be analyzed is large, selecting only a small subset of the whole terms cuts down the computational burden required to perform both the learning phase and the classification phase.

Among the feature selection methods analyzed in [1], the one that performs better was *chi-square* χ^2, defined as in [45], which is a supervised scoring function able to find the most relevant features with respect to its ability to characterize a certain category. The ranking obtained as output is used on the one hand to select the best n features and on the other hand to change term weighting inside documents. In fact, this score $s(t)$ is a global value, stating the relevance of a certain word, whereas relative frequency, introduced by Eq. 1, is a local value only measuring the relevance of a word inside a particular document. Therefore, these values can be combined into a different term weighting to be used for the bag-of-words representation, so that the weight w_t^d comes to be

$$w_t^d = rf(t, d) \cdot s(t) \tag{2}$$

Thus, according to the Eq. 2, both factors (i.e. the global relevance and the local relevance) may be taken into account.

Learning Phase. The learning phase is the second stage of our algorithm. As in any categorization problem, the primary goal is to learn a model from a training set, so that a test set can be accordingly classified. Though, the mechanism should also allow transfer learning in cross-domain setting.

The basic idea consists in modeling term co-occurrences: the more words co-occur in documents the more their connection should be stronger. We could represent this scenario as a graph whose nodes represent words and whose edges represent the strength of the connections between them. Considering a document corpus $\mathcal{D} = \{d_1, d_2, \ldots, d_N\}$ and a dictionary $\mathcal{T} = \{t_1, t_2, \ldots, t_k\}$, $A = \{a_{ij}\}$ is the set of connection weights between the term t_i and the term t_j and each a_{ij} can be computed as follows:

$$a_{ij} = a_{ji} = \sum_{d=1}^{N} w_{t_i}^d \cdot w_{t_j}^d \tag{3}$$

The same strategy could be followed to find the polarity of a certain word, unless having an external knowledge base which states that a word is intrinsically positive, negative or neutral. Co-occurrences between words and classes are modeled for each document whose polarity is given. Again, a graph whose nodes are either terms or classes and whose edges represent the strength of the connections between them is suitable to represent this relationship. In particular, given that $\mathcal{C} = \{c_1, c_2, \ldots, c_M\}$ is the set of categories and $B = \{b_{ij}\}$ is the set of edges between a term t_i and a class c_j, the strength of the relationship between a term t_i and a class c_j is augmented if t_i occurs in documents belonging to the set $D^j = \{d \in \mathcal{D} : c_d = c_j\}$.

$$b_{ij} = \sum_{d \in D^j} w_{t_i}^d \tag{4}$$

Careful readers may have noticed that the graph representing both term co-occurrences and term-class co-occurrences can be easily interpreted as a Markov chain. In fact, graph vertices are simply mapped to Markov chain nodes and graph edges are split into two directed edges (i.e. the edge linking states t_i and t_j is split into one directed edge from t_i to t_j and another directed edge from t_j to t_i). Moreover, for each state a normalization step of all outgoing arcs is enough to satisfy the probability unitarity property. Finally, the Markov property surely holds because each state only depends on directly linked states, since we evaluate co-occurrences considering just two terms (or a term and a class) at a time.

After having explained again the basic idea behind our method, we recap how the learning phase was performed in [1]. Basically, we relied on the assumption that there exist a subset of common terms between source and target domains that can act as a bridge between domain specific terms, allowing and supporting transfer learning. So, these common terms are the key to let information about classes flow from source specific terms to target specific terms, exploiting term co-occurrences, as shown in Fig. 1.

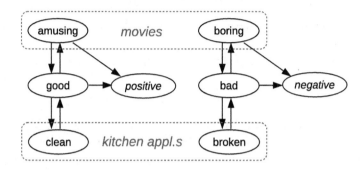

Fig. 1. Transfer learning from book-specific terms to kitchen appliances-specific terms through common terms.

We would like to point out that the just described transfer learning process is not an additional step to be added in cross-domain problems; on the contrary, it is implicit in the Markov chain mechanism and, as such, it is performed in in-domain problems as well. Obviously, if both training set and test set are extracted from the same domain, it is likely that most of the terms in test set documents already have a polarity.

Apart from transfer learning, the Markov chain we propose also fulfils the primary goal of the learning phase, that is, to build a model that can be subsequently used in the classification phase. Markov chain can be represented as a transition matrix (MCTM), composed of four logically distinct submatrices, as shown in Table 1. It is a $(k + M) \times (k + M)$ matrix, having current states as rows and future states as columns. Each entry represents a transition probability, which is computed differently depending on the type of current and future states (term or class), as described below.

Table 1. This table shows the structure of MCTM. It is composed of four submatrices, representing the transition probability that, starting from a current state (i.e. row), a future state (i.e. column) is reached. Both current states and future states can be either terms or classes.

	$t_1, \ldots, t_k$	$c_1, \ldots, c_M$
$t_1, \ldots, t_k$	A'	B'
$c_1, \ldots, c_M$	E	F

Let $\mathcal{D}_{train}$ and $\mathcal{D}_{test}$ be the subsets of document corpus $\mathcal{D}$ chosen as training set and test set respectively. The set A, whose each entry is defined by Eq. 3, is rewritten as

$$a_{ij} = a_{ji} = \begin{cases} 0, & i = j \\ \displaystyle\sum_{d \in \mathcal{D}_{train} \cup \mathcal{D}_{test}} w_{t_i}^d \cdot w_{t_j}^d, & i \neq j \end{cases} \tag{5}$$

and the set B, whose each entry is defined by Eq. 4, is rewritten as

$$b_{ij} = \sum_{d \in D_{train}^j} w_{t_i}^d \tag{6}$$

where $D_{train}^j = \{d \in \mathcal{D}_{train} : c_d = c_j\}$. The submatrices A' and B' are the normalized forms of Eqs. 5 and 6, computed so that each row of the Markov chain satisfies the probability unitarity property. Instead, each entry of the submatrices E and F looks like as follows:

$$e_{ij} = 0 \tag{7}$$

$$f_{ij} = \begin{cases} 1, & i = j \\ 0, & i \neq j \end{cases} \tag{8}$$

Notice that E and F deal with the assumption that classes are absorbing states, which can never be left once reached.

Classification Phase. The last step of the algorithm is the classification phase. The aim is classifying test set documents by using the model learned in the previous step. According to the bag-of-words representation, a document $d_t \in \mathcal{D}_{test}$ to be classified can be expressed as follows:

$$d_t = (w_{t_1}^{d_t}, \ldots, w_{t_k}^{d_t}, c_1, \ldots, c_M) \tag{9}$$

$w_{t_1}^{d_t}, \ldots, w_{t_k}^{d_t}$ is the probability distribution representing the initial state of the Markov chain transition matrix, whereas $c_1, \ldots, c_M$ are trivially set to 0. We initially hypothesize to be in many different states (i.e. every state t_i so that $w_{t_i}^{d_t} > 0$) at the same time. Then, simulating a single step inside the Markov chain transition matrix, we obtain a posterior probability distribution not only over terms, but also over classes. In such a way, estimating the posterior probability that d_t belongs to a certain class c_i, we could assign the most likely label $c_i \in \mathcal{C}$ to d_t. The posterior probability distribution after one step in the transition matrix, starting from document d_t, is:

$$d_t^* = (w_{t_1}^{d_t^*}, \ldots, w_{t_k}^{d_t^*}, c_1^*, \ldots, c_M^*) = d_t \times MCTM \tag{10}$$

where d_t is a column vector having size $(k + M)$ and $MCTM$ is the Markov chain transition matrix, whose size is $(k + M) \times (k + M)$. At this point, the category that will be assigned to d_t is computed as follows:

$$c_{d_t} = \arg \max_{i \in C^*} c_i^* \qquad (11)$$

where $C^* = \{c_1^*, \ldots, c_M^*\}$ is the posterior probability distribution over classes.

Computational Complexity. The computational complexity of our method is the time required to perform both the learning phase and the classification phase. Regarding the learning phase, the computational complexity overlaps the time needed to build the Markov chain transition matrix, say $time(MCTM)$, which is

$$time(MCTM) = time(A) + time(B)$$
$$+time(A^{'} + B^{'}) + time(E) + time(F) \qquad (12)$$

Remember that A and B are the submatrices representing the state transitions having a term as current state. Similarly, E and F are the submatrices representing the state transitions having a class as current state. $time(A^{'} + B^{'})$ is the temporal length of the normalization step, mandatory in order to observe the probability unitarity property. On the other hand, E and F are simply a null and an identity matrix, requiring no computation. Thus, since time complexity depends on these factors, all should be estimated.

The only assumption we can make is that in general $|\mathcal{T}| >> |\mathcal{C}|$. The time needed to compute A is $O(\frac{|\mathcal{T}|^2}{2} \cdot (|\mathcal{D}_{train}| + |\mathcal{D}_{test}|))$, which in turn is equal to $O(|\mathcal{T}|^2 \cdot (|\mathcal{D}_{train}| + |\mathcal{D}_{test}|))$. Regarding transitions from terms to classes, building the submatrix B requires $O(|\mathcal{T}| \cdot |\mathcal{C}| \cdot |\mathcal{D}_{train}|)$ time. In sentiment classification problems we could also assume that $|\mathcal{D}| >> |\mathcal{C}|$ and, as a consequence, the previous time becomes $O(|\mathcal{T}| \cdot |\mathcal{D}_{train}|)$. The normalization step, which has to be computed one time only for both A and B, is $O(|\mathcal{T}| \cdot (|\mathcal{T}| + |\mathcal{C}|) + |\mathcal{T}| + |\mathcal{C}|) = O((|\mathcal{T}| + 1) \cdot (|\mathcal{T}| + |\mathcal{C}|))$, which can be written as $O(|\mathcal{T}|^2)$ given that $|\mathcal{T}| >> |\mathcal{C}|$. Further, building the submatrix E requires $O(|\mathcal{T}|^2)$ time, whereas for submatrix F $O(|\mathcal{T}| \cdot |\mathcal{C}|)$ time is needed, which again can be written as $O(|\mathcal{T}|)$ given that $|\mathcal{T}| >> |\mathcal{C}|$. Therefore, the overall complexity of the learning phase is

$$time(MCTM) \simeq time(A)$$
$$= O(|\mathcal{T}|^2 \cdot (|\mathcal{D}_{train}| + |\mathcal{D}_{test}|)) \qquad (13)$$

In the classification phase, two operations are performed for each document to be categorized: the matrix product in Eq. 10, which requires $time(MatProd)$, and the maximum computation in Eq. 11, which requires $time(Max)$. Hence, as we can see below

$$time(CLASS) = time(MatProd) + time(Max) \qquad (14)$$

the classification phase requires a time that depends on the previous mentioned factors. The matrix product can be computed in $O((|\mathcal{T}| + |\mathcal{C}|)^2 \cdot |\mathcal{D}_{test}|)$ time, which can be written as $O(|\mathcal{T}|^2 \cdot |\mathcal{D}_{test}|)$ given that $|\mathcal{T}| >> |\mathcal{C}|$. On the other hand, the maximum requires $O(|\mathcal{C}| \cdot |\mathcal{D}_{test}|)$ time. Since the assumption that $|\mathcal{T}| >> |\mathcal{C}|$ still holds, the complexity of the classification phase can be approximated by the calculus of the matrix product.

Lastly, the overall complexity of our algorithm, say $time(Algorithm)$, is as follows:

$$time(Algorithm) = time(MCTM) + time(CLASS)$$
$$\simeq time(MCTM) = O(|\mathcal{T}|^2 \cdot (|\mathcal{D}_{train}| + |\mathcal{D}_{test}|)) \tag{15}$$

This complexity is comparable with the best performing state of the art methods.

3.2 Document Splitting into Sentences

The first variant we advance in this work consists in considering the sentence granularity rather than the document granularity. In fact documents can be composed of many sentences, possibly characterized by a different sentiment orientation. It is not seldom to encounter documents expressing positive (resp. negative) opinions about a list of aspects and ending with an overall opposite sentiment summarized in few words. Examples of the underlying behavior are *"My car's steering wheel always vibrates because of [...] The seat is not so comfortable [...] But in general I like my car."* or even *"I read that book. Characters are well described, their psychological profile is meticulously portrayed. However, the plot is definitely boring and I always fall asleep while reading!"*

During the text pre-processing phase, documents are split into sentences using the set of characters $Tok = \{., ;, !, ?\}$ as tokenizers. In this process, we should be careful of some exceptions that can occur and invalidate the splitting. We only handle the most straightforward cases, like for example *Ph.D., Dr.,* websites and emails. Anyway we are aware that, depending on the training set, many nontrivial cases could negatively affect the splitting.

In this context, co-occurrences between terms are no longer to be evaluated inside the same document, but within the same sentence. More formally, considering each document d_z as a set of sentences $d_z = \{p_1^z, p_2^z, \ldots, p_h^z\}$, Eq. 3 can be rewritten as follows:

$$a_{ij} = a_{ji} = \sum_{d \in \mathcal{D}} \sum_{p \in d} w_{t_i}^p \cdot w_{t_j}^p \tag{16}$$

Consequently, Eq. 5 becomes

$$a_{ij} = a_{ji} = \begin{cases} 0, & i = j \\ \displaystyle\sum_{d \in \mathcal{D}_{train} \cup \mathcal{D}_{test}} \sum_{p \in d} w_{t_i}^p \cdot w_{t_j}^p, & i \neq j \end{cases} \tag{17}$$

Similarly, the transition from a term to a class in Eq. 4 can be rewritten as

$$b_{ij} = \sum_{d \in \mathcal{D}} \sum_{p^j \in d} w_{t_i}^{p^j} \tag{18}$$

where $p^j = \{p_t \in d : c_{p_t} = c_j\}$. Likewise, Eq. 6 comes to be

$$b_{ij} = \sum_{d \in \mathcal{D}_{train}} \sum_{p^j \in d} w_{t_i}^{p^j} \tag{19}$$

In the classification phase, we modify Eqs. 10 and 11 to label the single sentences inside a document rather than the document itself:

$$p_t^* = (w_{t_1}^{p_t^*}, \ldots, w_{t_k}^{p_t^*}, c_1^*, \ldots, c_M^*) = p_t \times MCTM \tag{20}$$

$$c_{p_t} = \arg \max_{i \in C^*} c_i^* \tag{21}$$

where p_t is a column vector having size $(k + M)$.

The output labels for each sentence are finally combined by voting in order to obtain the final category for the document to be classified:

$$c_d = \arg \max_{i \in C} |c_i^d| \tag{22}$$

where $|c_i^d| = |\{p_t \in d : c_{p_t} = c_i\}|$. The computational complexity is now function of the number of sentences in the corpus, rather than of the number of documents as in the basic version.

3.3 Polarity-Driven State Transitions

A second alternative, orthogonal to the previous one, has been developed to establish how much a term should be linked with the others and with classes. To this purpose we have to take into account that the probability that the current state has at time t will be redistributed to the other states at time $t + 1$. So far this takes place based on co-occurrences, as stated by Eqs. 5 and 6. Although it might seem reasonable, it is not enough because the capability of both the current state and the future state in discriminating among categories is completely overlooked during state transitions. Some issues related to this behavior could occur. For example, if the current state is not well polarized, not only it will not be able to distinguish among classes, but it will likely be connected to terms having conflicting sentiment. On the other hand, if a term semantically related to the current state is not well polarized, it should not be selected as future state because it will not be useful for classification.

In order to solve the just explained issues, we focus on what it should take place during state transitions. The intuition is that the more the current state is capable of discriminating among categories (i.e. it is polarity-bearing) the more its probability will be given to classes. The remaining part will be distributed to the other terms in a proportional way not only to the semantic relationship between the current state and the future state, but also to the capability of the latter in distinguishing among classes. Everything we need to fulfil this twofold goal already is in our basic version of the Markov chain. In fact, the capability f_i of a term t_i in discriminating among categories can be defined as:

$$f_i = \frac{|b_{i+} - b_{i-}|}{b_{i+} + b_{i-}} \tag{23}$$

where b_{ij} is what has been defined by Eq. 6 and we can notice that $0 \leq f_i \leq 1$. In other words, f_i is the portion of probability that t_i will redistribute to classes in a proportional way to the values computed by Eq. 6. This means that polarity-bearing terms are those that contribute the most to the classification process. The remaining $(1 - f_i)$ will be split among terms according to the following relation:

$$a_{ij} = \begin{cases} 0, & i = j \\ f_j \cdot \displaystyle\sum_{d \in \mathcal{D}_{train} \cup \mathcal{D}_{test}} w_{t_i}^d \cdot w_{t_j}^d, & i \neq j \end{cases} \tag{24}$$

Notice that the transition probability depends not only on the semantic relationship among terms, but also on the capability of the destination term in detecting categories.

We would like to remind that, according to Eq. 9, when classifying a document d_t the initial state of the Markov chain was represented by $w_{t_1}^{d_t}, \ldots, w_{t_k}^{d_t}$. Each weight w_{t_i} has to be multiplied by f_i as well, since only polarity-bearing terms should drive sentiment classification, spreading their probability when performing a step in the Markov chain. The overall computational complexity is aligned with that of the aforementioned basic approach.

4   Experiments

The Markov chain based methods have been implemented in a framework entirely written in Java. Algorithms performance has been evaluated through the comparison with *Spectral feature alignment (SFA)* by Pan et al. [22] and *Joint sentiment-topic model with polarity-bearing topics (PBT)* by He et al. [23], which, to the best of our knowledge, currently are the two best performing approaches in cross-domain sentiment classification.

We used common benchmark datasets to be able to compare results, namely, a collection of Amazon[1] reviews about four domains: Book (B), DVD (D), Electronics (E) and Kitchen appliances (K). Each domain contains 1000 positive and 1000 negative reviews written in English. The text pre-processing phase described in Sect. 3.1 is applied to convert plain text into the bag-of-words representation, possibly splitting documents into sentences when dealing with the variant introduced in Sect. 3.2. Before the learning phase and the classification phase, we perform feature selection by means of χ^2, which turned out to be the best performing technique for this purpose in [1].

Performance of every presented variant is shown below and compared with the state of the art. Differently, the Kitchen domain is ruled out from the analysis, in order for the results to be comparable with those reported in our previous work.

4.1 Setup and Results

From now on we will use MC_S when referring to the Markov chain variant characterized by splitting documents into sentences, MC_P for the polarity-driven transitions one, MC_{SP} for the combination between them, whereas MC_B indicates the basic approach.

In order to compare performance with MC_B, we replicate the same experiment that gave the best outcome in our previous work. Therefore, the training set is composed of 1600 documents and the test set of 400 documents. The best 250 features are selected in the text pre-processing phase by means of χ^2 scoring function. The goodness of results is measured by accuracy, averaged over 10 random source-target splits and evaluated for each particular source-target combination, namely $B \to D$, $D \to B$, $B \to E$, $E \to B$, $D \to E$, $E \to D$, even including in-domain configurations, such as $B \to B$, $D \to D$, $E \to E$.

As we can notice in Fig. 2, both the variants relying on sentence splitting (i.e. MC_S and MC_{SP}) do not perform well. The reason for this outcome is to be found in the learning phase, where the polarity of each sentence should be taken into account, as stated by Eq. 18. However, in the Amazon corpus we are only aware of the whole document polarity and not of the sentiment at sentence level. Consequently we had to make the strong assumption that each sentence

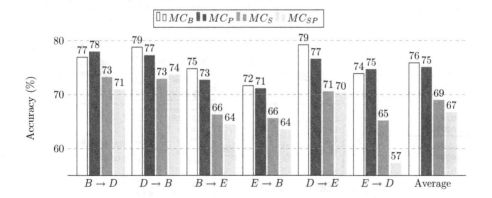

Fig. 2. Cross-domain classification by comparing all the proposed Markov chain based variants. The best 250 features are selected in accordance with the score output by χ^2.

---

[1] www.amazon.com.

in a training set document has the same sentiment of the document itself, even if this is definitely false in general. This could trivially bring to erroneously consider sentiment at term level and, as a consequence, to bad performance.

On the other hand, looking at Fig. 3 we can see a qualitative comparison between MC_P and MC_B. The reported examples show that polarity-driven state transitions could be helpful for the classification process when there are some polarity-bearing terms within the document to be classified. In fact, in such cases MC_P classifier is much more confident with its prediction than MC_B. This could also bring (as in example B) to correctly predict test instances failed by the basic classifier.

Anyway even if MC_P is able to take advantage of polarity-bearing terms, its accuracy does not outperform that of MC_B. This outcome could be explained considering that there is no constraint that forces terms to redistribute their probability to others having the same sentiment. Moreover, a document not con-

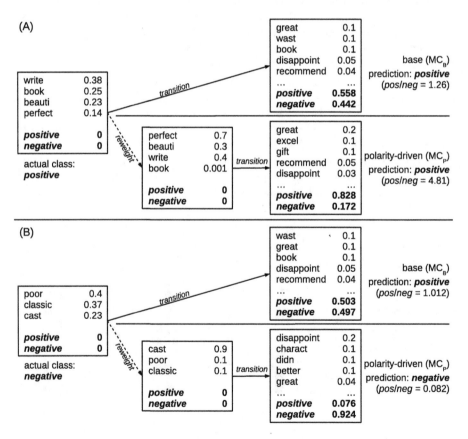

Fig. 3. Two examples of documents (A and B), represented as selected features with associated weights, classified by using either MC_B or MC_P. Each of the rightmost boxes only shows the 5 terms with the highest weights after the transition.

taining most terms bearing its own polarity is likely to be misclassified because, although terms are semantically related, they separately contribute to classification. This happens for example when a positive (resp. negative) document expresses opposite opinions about some aspects before summarizing in few words its overall polarity. Actually the context inside the document to be classified is neglected, because the classification starts assuming to be in many different states at the same time corresponding to the terms occurring in the document itself. Then the step by step evolution of each state is independent of the others; it is only determined by the semantic relationships between each term and the others, learned when training the model.

Table 2 shows a comparison between our Markov chain based methods and other works, namely SFA and PBT. Whereas MC_S and MC_{SP} are far away from the state of the art, MC_B and MC_P achieve comparable results, despite both SFA and PBT perform better on average. On the other side, we would like to emphasize that our algorithms require much fewer features than the others, i.e. 250 with respect to the 2000 needed by PBT and the more than 470000 needed by SFA. Therefore, since the computational complexity quadratically grows with the number of features in all methods, the convergence of both MC_B and MC_P is supposed to be dramatically faster than that of SFA and PBT.

Lastly, we can see that similar considerations can be done in an in-domain setting. Nothing needs to be changed in our methods to perform in-domain sentiment classification, whereas other works use standard classifiers completely bypassing the transfer learning phase.

Table 2. Performance of all the Markov chain based methods in both in-domain and cross-domain sentiment classification, compared with other works. For each dataset, the best accuracy is in bold.

Domain(s)	Other methods		Markov chain method variants			
	SFA	PBT	MC_S	MC_{SP}	MC_B	MC_P
Cross-domain experiments (*source → target*)						
$B \rightarrow D$	**81.50 %**	81.00 %	73.21 %	70.92 %	76.92 %	77.95 %
$D \rightarrow B$	78.00 %	**79.00 %**	72.91 %	73.67 %	78.79 %	77.27 %
$B \rightarrow E$	72.50 %	**78.00 %**	66.24 %	64.45 %	74.80 %	72.68 %
$E \rightarrow B$	**75.00 %**	73.50 %	65.56 %	63.52 %	71.65 %	71.13 %
$D \rightarrow E$	77.00 %	79.00 %	70.54 %	70.28 %	**79.21 %**	76.58 %
$E \rightarrow D$	**77.50 %**	76.00 %	65.15 %	57.32 %	73.91 %	74.68 %
Average	76.92 %	77.75 %	68.94 %	66.69 %	75.88 %	75.05 %
In-domain experiments						
$B \rightarrow B$	**81.40 %**	79.96 %	73.72 %	72.45 %	76.77 %	77.78 %
$D \rightarrow D$	82.55 %	81.32 %	78.34 %	79.60 %	**83.50 %**	82.49 %
$E \rightarrow E$	**84.60 %**	83.61 %	77.78 %	78.54 %	80.90 %	79.55 %
Average	82.85 %	81.63 %	76.61 %	76.86 %	80.39 %	79.94 %

5   Conclusions

In this work we presented some variants of the Markov chain based method already advanced in [1] to accomplish sentiment classification in both in-domain and cross-domain settings.

The first one consists in considering documents at the sentence granularity instead of perceiving them as a whole. Results dealing with this expedient are not good, probably because we made the strong assumption that, when learning the model, sentences have the same polarity of the document including them. A possible walk around is introducing a threshold parameter that establishes the probability that a sentence has the same polarity of the document where it appears. Another viable improvement consists in changing the way sentence labels are folded together to output the final category for the test document. Moreover, when a review is long, the final part usually bears the same sentiment of the entire text, because it contains a summary of the author's thought. On the contrary when it is short, it is likely that the author immediately summarizes his opinion without using terms bearing conflicting sentiment. In both cases, taking only the last few sentences into account could be profitable. A further alternative is using document splitting into sentences just to limit co-occurrences among terms rather than to change the connection between terms and classes.

The second illustrated approach addresses the problem of steering the probability each state has at time t to other states at time $t + 1$ that are capable of discriminating among categories. Although being comparable, this variant does not outperform our basic Markov chain approach. The outcome is somewhat surprising if we think that the classification is driven by polarity-bearing terms. The reason is probably to be found in the fact that there is no constraint that forces terms to redistribute their probability to others having the same sentiment. To overcome this problem we might limit terms spreading their probability to others in a way such that the more the current term is able in discriminating among categories the more it should give its probability to other terms having the same sentiment orientation.

Apart from the mentioned flaws, both the presented variants preserve all the advantages of our basic Markov chain based method. Indeed, they not only act as classifiers, but also allow transfer learning from source domain to target domain in cross-domain problems. The polarity-driven state transitions variant achieves comparable performance in terms of accuracy with the state of the art, whereas the document splitting based ones are outperformed, perhaps due to the strong assumption made. On the other side, all the introduced techniques require lower parameter tuning than previous works. Furthermore, in spite of having a comparable computational complexity, growing quadratically with the number of features, much fewer terms are demanded to obtain good accuracy.

Future work should aim to improve the algorithm effectiveness. In addition to the specific aspects proposed to enhance the particular variants, we believe that the hypotheses we rely on should be better analyzed. For example, the algorithm could suffer from the assumption to be in many different states at the same time, made when a test document is required to be classified. In fact the step by step evolution of each state is independent of the others and consequently context information ends up being overlooked. A possible way to walk could take into account ngram features rather than just unigrams. Another option in order to introduce context information is to consider grammatical relations among terms for the sake of detecting patterns.

After having enhanced the algorithms accuracy, performance in a 3-classes setting (i.e. adding the neutral category) could also be tested. On the other hand, due to their generality, our methods might be applied as is in text categorization problems. Finally, their applicability could be easily extended to other languages, because they only depend on co-occurrences among terms.

References

1. Domeniconi, G., Moro, G., Pagliarani, A., Pasolini, R.: Markov chain based method for in-domain and cross-domain sentiment classification. In: Proceedings of the 7th International Joint Conference on Knowledge Discovery, Knowledge Engineering and Knowledge Management, pp. 127–137 (2015)
2. Dai, W., Xue, G.-R., Yang, Q., Yu, Y.:Co-clustering based classification for out-of-domain documents. In: Proceedings of the 13th ACM SIGKDD International Conference on Knowledge Discovery and Data Mining, pp. 210–219. ACM (2007)
3. Xue, G.-R., Dai, W., Yang, Q., Yu, Y.: Topic-bridged PLSA for cross-domain text classification. In: Proceedings of the 31st Annual International ACM SIGIR Conference on Research and development in Information Retrieval, pp. 627–634. ACM (2008)
4. Li, L., Jin, X., Long, M.: Topic correlation analysis for cross-domain text classification. In: AAAI (2012)
5. Tan, S., Wang, Y., Cheng, X.: Combining learn-based and lexicon-based techniques for sentiment detection without using labeled examples. In: Proceedings of the 31st Annual International ACM SIGIR Conference on Research and Development in Information Retrieval, pp. 743–744. ACM (2008)
6. Qiu, L., Zhang, W., Hu, C., Zhao, K.: Selc: a self-supervised model for sentiment classification. In: Proceedings of the 18th ACM Conference on Information and Knowledge Management, pp. 929–936. ACM (2009)
7. Melville, P., Gryc, W., Lawrence, R.D.: Sentiment analysis of blogs by combining lexical knowledge with text classification. In: Proceedings of the 15th ACM SIGKDD International Conference on Knowledge Discovery and Data Mining, pp. 1275–1284. ACM (2009)
8. Domeniconi, G., Moro, G., Pasolini, R., Sartori, C.: Cross-domain text classification through iterative refining of target categories representations. In: Proceedings of the 6th International Conference on Knowledge Discovery and Information Retrieval (2014)
9. Shrivastava, A., Malisiewicz, T., Gupta, A., Efros, A.A.: Data-driven visual similarity for cross-domain image matching. ACM Trans. Graph. (TOG) **30**, 154 (2011)
10. Domeniconi, G., Masseroli, M., Moro, G., Pinoli, P.: Cross-organism learning method to discover new gene functionalities. Comput. Methods Programs Biomed. **126**, 20–34 (2016)
11. Pan, S.J., Yang, Q.: A survey on transfer learning. IEEE Trans. Knowl. Data Eng. **22**(10), 1345–1359 (2010)

12. Domeniconi, G., Moro, G., Pasolini, R., Sartori, C.: Iterative refining of category profiles for nearest centroid cross-domain text classification. In: Fred, A., Dietz, J.L.G., Aveiro, D., Liu, K., Filipe, J. (eds.) IC3K 2014. CCIS, vol. 553, pp. 50–67. Springer, Heidelberg (2015). doi:10.1007/978-3-319-25840-9_4

13. Dave, K., Lawrence, S., Pennock, D.M.: Mining the peanut gallery: opinion extraction and semantic classification of product reviews. In: Proceedings of the 12th International Conference on World Wide Web, pp. 519–528. ACM (2003)

14. Domeniconi, G., Moro, G., Pasolini, R., Sartori, C.: A comparison of term weighting schemes for text classification and sentiment analysis with a supervised variant of tf.idf. In: Helfert, M., Holzinger, A., Belo, O., Francalanci, C. (eds.) DATA 2015. CCIS, vol. 584, pp. 39–58. Springer, Heidelberg (2016). doi:10.1007/978-3-319-30162-4_4

15. Domeniconi, G., Masseroli, M., Moro, G., Pinoli, P.: Random perturbations of term weighted gene ontology annotations for discovering gene unknown functionalities. In: Fred, A., Dietz, J.L.G., Aveiro, D., Liu, K., Filipe, J. (eds.) IC3K 2014. CCIS, vol. 553, pp. 181–197. Springer, Heidelberg (2015). doi:10.1007/978-3-319-25840-9_12

16. Paltoglou, G., Thelwall, M.: A study of information retrieval weighting schemes for sentiment analysis. In: Proceedings of the 48th Annual Meeting of the Association for Computational Linguistics, pp. 1386–1395. Association for Computational Linguistics (2010)

17. Deng, Z.-H., Luo, K.-H., Yu, H.-L.: A study of supervised term weighting scheme for sentiment analysis. Expert Syst. Appl. **41**(7), 3506–3513 (2014)

18. Wu, H., Gu, X.: Reducing over-weighting in supervised term weighting for sentiment analysis. In: COLING, pp. 1322–1330 (2014)

19. Aue, A., Gamon, M.: Customizing sentiment classifiers to new domains: A case study. In: Proceedings of Recent Advances in Natural Language Processing (RANLP), vol. 1, pp. 2–1 (2005)

20. Bollegala, D., Weir, D., Carroll, J.: Cross-domain sentiment classification using a sentiment sensitive thesaurus. IEEE Trans. Knowl. Data Eng **25**(8), 1719–1731 (2013)

21. Blitzer, J., Dredze, M., Pereira, F., et al.: Biographies, bollywood, boom-boxes and blenders: domain adaptation for sentiment classification. In: ACL, vol. 7, pp. 440–447 (2007)

22. Pan, S.J., Ni, X., Sun, J.-T., Yang, Q., Chen, Z.: Cross-domain sentiment classification via spectral feature alignment. In: Proceedings of the 19th International Conference on World wide web, pp. 751–760. ACM (2010)

23. He, Y., Lin, C., Alani, H.: Automatically extracting polarity-bearing topics for cross-domain sentiment classification. In: Proceedings of the 49th Annual Meeting of the Association for Computational Linguistics: Human Language Technologies, vol. 1, pp. 123–131. Association for Computational Linguistics (2011)

24. Turney, P.D.: Thumbs up or thumbs down?: semantic orientation applied to unsupervised classification of reviews. In: Proceedings of the 40th Annual Meeting on Association for Computational Linguistics, pp. 417–424. Association for Computational Linguistics (2002)

25. Taboada, M., Brooke, J., Tofiloski, M., Voll, K., Stede, M.: Lexicon-based methods for sentiment analysis. Comput. linguist. **37**(2), 267–307 (2011)

26. Qiu, L.: Markov models of search state patterns in a hypertext information retrieval system. J. Am. Soc. Inf. Sci. **44**(7), 413–427 (1993)

27. Sarukkai, R.R.: Link prediction and path analysis using markov chains. Comput. Netw. **33**(1), 377–386 (2000)

28. Page, L., Brin, S., Motwani, R., Winograd, T.: The pagerank citation ranking: bringing order to the web. Technical Report, Stanford University (1999)
29. Mittendorf, E., Schäuble, P.: Document and passage retrieval based on hidden Markov models. In: Croft, B.W., van Rijsbergen, C.J. (eds.) SIGIR 1994, pp. 318–327. Springer, Heidelberg (1994)
30. Miller, D.R., Leek, T., Schwartz, R.M.: A hidden Markov model information retrieval system. In: Proceedings of the 22nd Annual International ACM SIGIR Conference on Research and Development in Information Retrieval, pp. 214–221. ACM (1999)
31. Pan, Y.-C., Lee, H.-Y., Lee, L.-S.: Interactive spoken document retrieval with suggested key terms ranked by a Markov decision process. IEEE Trans. Audio Speech Lang. Process. **20**(2), 632–645 (2012)
32. Xu, J., Weischedel, R.: Cross-lingual information retrieval using hidden Markov models. In: Proceedings of the 2000 Joint SIGDAT Conference on Empirical Methods in Natural Language Processing, Very Large Corpora: held in Conjunction with the 38th Annual Meeting of the Association for Computational Linguistics, vol. 13, pp. 95–103. Association for Computational Linguistics (2000)
33. Cao, G., Nie, J.-Y., Bai, J.: Using Markov chains to exploit word relationships in information retrieval. In: Large Scale Semantic Access to Content (Text, Image, Video, and Sound), pp. 388–402. Le Centre de Hautes Etudes Internationales D'Informatique Documentaire (2007)
34. Li, F., Huang, M., Zhu, X.: Sentiment analysis with global topics and local dependency. In: AAAI, vol. 10, pp. 1371–1376 (2010)
35. Mei, Q., Ling, X., Wondra, M., Su, H., Zhai, C.: Topic sentiment mixture: modeling facets and opinions in weblogs. In: Proceedings of the 16th International Conference on World Wide Web, pp. 171–180. ACM (2007)
36. Jo, Y., Oh, A.H.: Aspect and sentiment unification model for online review analysis. In: Proceedings of the Fourth ACM International Conference on Web Search and Data Mining, pp. 815–824. ACM (2011)
37. Jin, W., Ho, H.H., Srihari, R.K.: Opinionminer: a novel machine learning system for web opinion mining and extraction. In: Proceedings of the 15th ACM SIGKDD International Conference on Knowledge Discovery and Data Mining, pp. 1195–1204. ACM (2009)
38. Nasukawa, T., Yi, J.: Sentiment analysis: capturing favorability using natural language processing. In: Proceedings of the 2nd International Conference on Knowledge Capture, pp. 70–77. ACM (2003)
39. Yi, K., Beheshti, J.: A hidden Markov model-based text classification of medical documents. J. Inf. Sci. **35**, 67–81 (2008)
40. Xu, R., Supekar, K., Huang, Y., Das, A., Garber, A.: Combining text classification and hidden Markov modeling techniques for structuring randomized clinical trial abstracts. In: AMIA Annual Symposium Proceedings 2006, p. 824. American Medical Informatics Association (2006)
41. Yi, K., Beheshti, J.: A text categorization model based on hidden Markov models. In: Proceedings of the Annual Conference of CAIS/Actes du congrès annuel de l'ACSI (2013)
42. Li, F., Dong, T.: Text categorization based on semantic cluster-hidden markov models. In: Tan, Y., Shi, Y., Mo, H. (eds.) ICSI 2013. LNCS, vol. 7929, pp. 200–207. Springer, Heidelberg (2013). doi:10.1007/978-3-642-38715-9_24

43. Frasconi, P., Soda, G., Vullo, A.: Hidden Markov models for text categorization in multi-page documents. J. Intell. Inf. Syst. **18**(2–3), 195–217 (2002)
44. Porter, M.F.: An algorithm for suffix stripping. Program **14**(3), 130–137 (1980)
45. Domeniconi, G., Moro, G., Pasolini, R., Sartori, C.: A study on term weighting for text categorization: a novel supervised variant of tf.idf. In: Proceedings of the 4th International Conference on Data Management Technologies and Applications (2015)

A Word Prediction Methodology Based on Posgrams

Carmelo Spiccia, Agnese Augello$^{(\boxtimes)}$, and Giovanni Pilato

Italian National Research Council (CNR), Istituto di Calcolo e Reti ad Alte
Prestazioni (ICAR), 90100 Palermo, Italy
{spiccia, augello, pilato}@pa.icar.cnr.it

Abstract. This work introduces a two steps methodology for the prediction of missing words in incomplete sentences. In a first step the number of candidate words is restricted to the ones fulfilling the predicted part of speech; to this aim a novel algorithm based on "posgrams" analysis is also proposed. Then, in a second step, a word prediction algorithm is applied on the reduced words set. The work quantifies the advantages in predicting a word part of speech before predicting the word itself, in terms of accuracy and execution time. The methodology can be applied in several tasks, such as Text Autocompletion, Speech Recognition and Optical Text Recognition.

1 Introduction

Word prediction is a well-known challenging task with vast applications in Natural Language Processing. One of the oldest word prediction software described in literature was "The reactive keyboard" [1, 2]: it employed a tree structure to store the frequencies of common ngrams (i.e. sequences of words) typed by the user; suggestions were therefore shown while the user was typing. This was very similar to another software, Profet [3]: for the same task it also used frequency lists of unigrams and bigrams (i.e. ngrams composed by one and two words respectively), but reverted to unigrams when the user started to type a new word. Scientific research on ngrams has focused on improving one of the most important problems of the approach: data sparsity. In fact, even in very large corpora there are legitimate sequences of words that are very rare and do not appear at all. The so-called Laplace smoothing method of assigning a unitary frequency to each ngram whose frequency is zero may lead to an overestimation of those rare ngrams. Therefore new estimation methods have been proposed, in particular: Good-Turing smoothing [4], Kats smoothing [5], Jelinek-Mercer smoothing [6] and Kneser-Ney smoothing [7]. Skipgrams [8] are another approach to data sparsity: one or more words can be "skipped" from the sequence.

In [9] Microsoft released a novel questionnaire to stimulate research on word prediction methods not based on ngrams: the Sentence Completion Challenge questionnaire is composed by 1040 questions; each question consists in a sentence having a missing word and in five candidate words as possible answers. According to the authors, it have been "explicitly designed to be non-solvable using purely N-gram

© Springer International Publishing AG 2016
A. Fred et al. (Eds.): I3K 2015, CCIS 631, pp. 139–154, 2016.
DOI: 10.1007/978-3-319-52758-1_9

based methods". Several alternative methods to ngrams have been developed. In [10] a novel language model is devised that predicts a word on the basis of its estimated syntactic dependency tree probability. Latent Semantic Analysis [11] has been used in [12] and in [13] to create language models by mapping each word in a geometric space. In [14] neural networks are used instead to generate word embeddings for a language model. Noise-contrastive estimation is proposed in [15] as a replacement for importance sampling to reduce neural networks training time requirements. A simplified recurrent neural network called Impulse-Response Language Model (IRLM) is proposed in [16]: overfitting is reduced by employing the random dropout regularization method. However, the best-performing method to date for the Microsoft Sentence Completion Challenge has been proposed in [17]: skipgrams and neural networks are both used.

The questionnaire developed for the Challenge specifically addresses the problem of completing a sentence having a missing word. This task is particularly useful for Text Autocompletion, Speech Recognition and Optical Text Recognition. However, the questionnaire simplifies the task in several ways. First of all, the five possible answers to a given question always have the same part of speech (e.g. adjective). Secondly, some parts of speech are never present in the answers set, stopwords in particular: conjunctions, prepositions, determinants and pronouns. Thirdly, in a real application the entire dictionary should be considered, not just five words. Therefore, real word applications involve processing a larger and more heterogeneous set of candidate words. To handle the general task better, we propose an innovative methodology for predicting a missing word of a sentence. We focused our study on the Italian and English languages, even though the proposed approach is in principle general. The methodology consists in two steps. In the first step the number of candidate words is reduced. In particular, a novel algorithm based on posgrams has been developed for predicting the part of speech of the missing word. Candidate words can therefore be reduced to the ones fulfilling the predicted part of speech. In the second step a word predictor is applied on this reduced words set. This can be accomplished by using any of the word prediction algorithms described in literature. The following sections describe the proposed methodology in more detail, which is also illustrated in Fig. 1.

Section 2 demonstrates why the two steps prediction is advantageous: formulae for the estimation of the success probability of the word prediction and for the estimation of the execution time reduction are derived.

Section 3 quantifies the advantages for the Italian language: a tagged Italian corpus is parsed and statistics about each part of speech of the tagset are collected; the a priori probability of each tag is estimated; the formulae derived in Sect. 2 are then used to estimate the gains in terms of accuracy and execution time.

Section 4 describes how the part of speech prediction step can be accomplished: a novel algorithm based on posgrams is proposed.

Section 5 describes how to use the information achieved in the first step to predict the missing word: a training procedure is described to assert which is the best action to take for a given predicted part of speech.

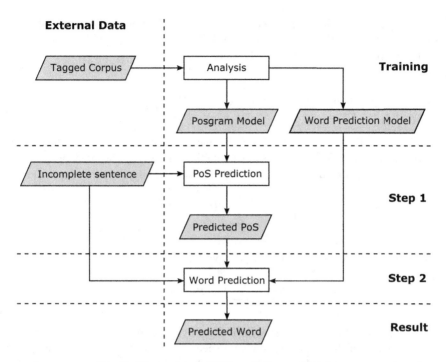

Fig. 1. The two steps word prediction methodology.

Section 6 applies the methodology to the English language: the Brown corpus [18] is parsed and similar considerations to Sects. 3 and 5 are devised.

Section 7 shows the final results: the accuracy of the part of speech prediction algorithm is compared to some baseline methods; the accuracy of the two steps word prediction method is compared to the single step method on two novel questionnaires of incomplete sentences.

2 Missing Word Prediction in Two Steps

Given a sentence having a missing word, predicting the part of speech of the word before predicting the word itself may be convenient. First of all, in the following paragraphs we will consider each word as belonging to one part of speech only: a dictionary which associates each word to its most probable one will be used. While this is a strong assumption, we will show that the methodology works nevertheless, leaving the handling of the more general case for future works. To quantify the advantage, a question is created on the basis of that incomplete sentence, by adding n candidate words as possible answers. Let n_c be the number of candidate words having the part of speech c. Let us suppose that for any part of speech the number of words in the dictionary having that part of speech is far greater than n. When building the answer set, the effect of choosing a word on the probability that the next word will have the same part of speech can be therefore ignored. We will show that this hypothesis is conservative. If the part of speech of the missing word is c, the probability that the answers set contains k words having the same part of speech is:

$$P(n_c = k|c) \cong P(c)^{k-1}(1 - P(c))^{n-k}\binom{n-1}{k-1}. \tag{1}$$

Suppose c is known when answering the question. This gives an advantage to the word predictor. Let S be the success at predicting the word. The probability of the event, by choosing a random word among the answers having the part of speech c, is:

$$P(S|c) = \sum_{k=1}^{n} \frac{P(n_c = k|c)}{k}. \tag{2}$$

The success probability, regardless of c, will be:

$$P(S) = \sum_{c \in C} P(c)P(S|c), \tag{3}$$

where C is the set of the parts of speech. If the previous hypothesis is false, e.g. if the number of words of the dictionary having a specific part of speech is very small or if n is huge, the above formula will give a lower bound estimate of the success probability. This happens because the components $P(n_c = k|c)$ are increasingly overestimated as k grows and they are weighted by the inverse of k.

Knowing the part of speech in advance gives two advantages. First of all, accuracy is improved since:

$$P(S) \geq \frac{1}{k}. \tag{4}$$

The second advantage is the required execution time: by reducing the cardinality of the answers set, less candidate words must be evaluated by the prediction algorithm. In the next section we analyze these advantages in detail for the prediction of the missing word of an Italian incomplete sentence.

3   Prediction of an Italian Word in Two Steps

In order to apply the two steps prediction model to the Italian language, we consider the WaCky Italian Wikipedia Corpus [19, 20], freely available in the CoNLL format. It uses two tagsets conforming to the EAGLES standard [21, 22]: a coarse-grained one (14 tags) and a fine-grained tagset (69 tags). The first tagset is shown in Table 1. It assigns a letter to each part of speech.

For testing the model, we set to five the number n of candidate words per question. Table 2 shows, for each part of speech c, from left to right: the tag; the a priori probability $P(c)$, computed by analyzing the corpus; the probability $P(n_c = k|c)$ of

Table 1. The coarse-grained tagset.

Tag	Part of speech	Tag	Part of speech
A	Adjective	N	Number
B	Adverb	P	Pronoun
C	Conjunction	R	Article
D	Determinant	S	Noun
E	Preposition	T	Predeterminant
F	Punctuation	V	Verb
I	Interjection	X	Other

Table 2. Probabilities related to the construction and usage of a five-answers question.

c	$P(c)$	$P(n_c = 1\|c)$	$P(n_c = 2\|c)$	$P(n_c = 3\|c)$	$P(n_c = 4\|c)$	$P(n_c = 5\|c)$	$P(S\|c)$	$P(c)P(S\|c)$
S	0.2803	0.2683	0.4180	0.2442	0.0634	0.0062	0.5000	0.1401
E	0.1712	0.4719	0.3898	0.1207	0.0166	0.0009	0.6726	0.1151
F	0.1380	0.5521	0.3536	0.0849	0.0091	0.0004	0.7320	0.1010
V	0.1030	0.6474	0.2974	0.0512	0.0039	0.0001	0.7974	0.0821
A	0.0837	0.7051	0.2575	0.0353	0.0021	0.0000	0.8345	0.0698
R	0.0799	0.7166	0.2490	0.0325	0.0019	0.0000	0.8418	0.0673
B	0.0399	0.8498	0.1412	0.0088	0.0002	0.0000	0.9205	0.0367
C	0.0381	0.8560	0.1357	0.0081	0.0002	0.0000	0.9239	0.0352
P	0.0305	0.8834	0.1112	0.0053	0.0001	0.0000	0.9391	0.0287
N	0.0245	0.9056	0.0910	0.0034	0.0001	0.0000	0.9511	0.0233
D	0.0093	0.9631	0.0364	0.0005	0.0000	0.0000	0.9813	0.0092
T	0.0013	0.9947	0.0053	0.0000	0.0000	0.0000	0.9974	0.0013
X	0.0002	0.9990	0.0010	0.0000	0.0000	0.0000	0.9995	0.0002
I	0.0000	0.9998	0.0002	0.0000	0.0000	0.0000	0.9999	0.0000

obtaining k answers having the same part of speech of the missing word, for $k = 1 \ldots 5$, conditioned to c being that part of speech; the probability of success $P(S|c)$ at predicting the missing word by choosing among the answers having part of speech c, conditioned to c being the part of speech of the missing word; the probability of c being the part of speech of the missing word and to succeed, at the same time, at predicting that word.

The sum of the values of the last column is 0.7102. It represents the probability $P(S)$ of success at predicting the missing word by using the two steps methodology. This means that by solving the problem of predicting the part of speech of a missing word, the problem of predicting the word among five possible answers is solved with at least 71% of accuracy. To provide a comparison, the current state of the art single step algorithm achieves an accuracy of 58.9% at predicting the missing word among five possible choices [17].

Figure 2 and 3 show the obtainable advantage in terms of accuracy at different answers set size n for the Italian language. The graph in Fig. 2 compares the two steps model with the single step model: both models use random choice for word prediction, but the first model performs a part of speech prediction step to reduce the number of answers. Accuracy levels are reported. The graph in Fig. 3 shows the ratio between the accuracy of the first model and of the second model. The two steps model always outperforms the single step model, with up to 10 times or greater accuracy.

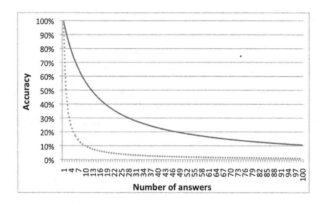

Fig. 2. Accuracy of prediction strategies as the size of the answers set grows; single step (*dotted red*) and two steps (*blue*) random choice word predictors are compared. (Color figure online)

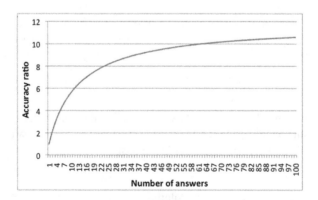

Fig. 3. Accuracy ratio between the two steps and the single step prediction strategies.

Table 3 shows, for each part of speech: the tag; the a priori probability; the number of unique words in a dictionary; the percentage of unique words in the dictionary; the product of the a priori probability with the percentage of unique words. The sum of the values of the last column is 15.11%. It represents the average percentage of the answers set to be processed when the set is a casual sample of the whole dictionary. This can lead to a speedup of about 6.6x (e.g. for LSA five times less scalar products must be computed).

Table 3. Parts of speech a priori probabilities and distribution.

c	P(c)	Words	Words %	P(c) · (Words %)
S	0.2803	32575	30.87%	8.65%
E	0.1712	116	0.11%	0.02%
F	0.1380	18	0.02%	0.00%
V	0.1030	44528	42.20%	4.35%
A	0.0837	25242	23.92%	2.00%
R	0.0799	11	0.01%	0.00%
B	0.0399	1889	1.79%	0.07%
C	0.0381	93	0.09%	0.00%
P	0.0305	128	0.12%	0.00%
N	0.0245	751	0.71%	0.02%
D	0.0093	68	0.06%	0.00%
T	0.0013	5	0.00%	0.00%
X	0.0002	7	0.01%	0.00%
I	0.0000	97	0.09%	0.00%

4 The First Step: Prediction of the Part of Speech

In order to predict the part of speech of the missing word, the first 100,000 words of the corpus have been parsed with a moving window: the frequency of every sequence of one to five parts of speech has been saved to a lookup table. We will refer to these sequences as "posgrams" [23, 24]. The most frequent ones are reported in Tables 4a and 4b.

Given a sentence having a missing word, the part of speech can be predicted by using the posgrams lookup table. The window of five words centered on the missing word is considered: each of the known words is replaced by the corresponding most

Table 4a. Most common posgrams by length and their frequencies on the first 100,000 words of the WaCky Italian Wikipedia Corpus. The length spans from one to three.

1-posgram	Freq.	2-posgram	Freq.	3-posgram	Freq.
S	28039	ES	10072	SES	5201
E	16283	SF	7336	ESE	2657
F	13600	SE	7312	ESF	2587
V	10692	RS	6278	RSE	2260
A	8574	SA	4142	ESA	1678
R	8169	SS	3001	SAF	1663
B	4033	VE	2944	VRS	1606
C	4021	AF	2538	VES	1600
P	3175	SV	2385	ERS	1341
N	2200	AS	2277	SSF	1317

Table 4b. Most common posgrams by length and their frequencies on the first 100,000 words of the WaCky Italian Wikipedia Corpus. The length spans from four to five.

4-posgram	Freq.	5-posgram	Freq.
ESES	1835	SESES	844
RSES	1639	ESESF	529
SESF	1546	VRSES	457
SESE	1191	ESESE	453
SESA	850	RSESF	421
ESAF	733	SESAF	379
VRSE	713	RSESE	372
SAES	641	ERSES	366
ASES	536	VESES	336
SESS	519	RSESA	288

Table 5. Posgram windows, up to the length of five, and their centrality.

Window	Centrality	Window	Centrality	Window	Centrality
X	1	XX_	1	_XXXX	1
_X	1	_XXX	1	X_XXX	2
X_	1	X_XX	2	XX_XX	3
_XX	1	XX_X	2	XXX_X	2
X_X	2	XXX_	1	XXXX_	1

common part of speech; the predicted part of speech of the missing word is the one which maximizes the frequency of the posgram.

In order to improve the prediction accuracy in case of infrequent or absent posgrams in the corpus, an ad-hoc smoothing algorithm has been developed. First of all we define the "centrality" of a window with respect of the part of speech to be predicted as the number of parts of speech, plus one, between the missing element and the nearest extremity of the window. Table 5 shows the centrality of each window up to size five. The part of speech to be predicted is represented by an underscore; the known parts of speech are represented by an "X".

The pseudo-code in Fig. 4 illustrates the smoothing algorithm. It takes in input three parameters: the maximum size p of a posgram; the extended window composed by the concatenation of the $p - 1$ tags on the left of the missing word, an underscore and the $p - 1$ tags on the right of the missing word; a vector $\boldsymbol{\alpha}$ of weights. Subwindows of progressively lower size are processed: the score of each part of speech is incremented for each posgram found in the lookup table. The increment is the product of: the frequency of the posgram; the centrality of the subwindow; the weight α_i, where i is the difference between the size of the first posgram matched and the size of the current subwindow.

```
Function predictPos(maxPosgramSize, window, weights)
   bestScore = 0
   mostProbablePos = "S"
   w = 0
   For size = maxPosgramSize To 1
      For Each pos In tagset
         subwindows = findSubwindows(window, size)
         For Each subwindow In subwindows
            posgram = Replace "_" In window With pos
            scores[pos] += frequency(posgram)
                           * centrality(subwindow)
                           * weights[w]
            If scores[pos] >= bestScore Then
               bestScore = scores[pos]
               mostProabablePos = pos
            End If
         End For Each
      If bestScore > 0 Or w > 0 Then
         w = w + 1
         If w = length(weights) Then Break
      End if
   End For
   Return mostProbablePos
End Function
```

Fig. 4. Pseudocode of the smoothing algorithm employed for part of speech prediction.

5   The Second Step: Prediction of the Missing Word

Given the predicted part of speech, this information can be used to improve the accuracy of the word prediction. First of all, it's convenient to assert the best action to take for each possible part of speech of the tagset. Therefore, the training phase is split into two steps. In the first step, the part of speech predictor and the word predictor are trained. In the second step, a questionnaire is automatically created from the corpus: from each sentence a question is built, by removing a random word; the other candidate words are chosen randomly from the nearby sentences; these words and the removed word constitute the answers set for the question. For each question the part of speech is predicted and the word predictor is invoked on the full answers set. Afterwards, the word predictor is invoked again on the restricted answers set composed by the words having the predicted part of speech. Results statistics are collected and aggregated by the predicted part of speech. After this second step of training, the achieved statistics provide information on which action to take. In particular they tell whether to restrict the answers set to the predicted part of speech, i.e. if the prediction of that part of speech is sufficiently reliable to actually improve word prediction.

6   Application to the English Language

In order to assess the applicability of the approach to the English language, we analyzed the Brown corpus. It consists in 500 English documents taken from different sources covering nine topics: news, religion, hobbies, folklore, literature, government, science, fiction, humor. Its tagset is composed by 226 different tags and it has been mapped to the coarse-grained tagset (14 tags) shown in Table 1.

As for the Italian language, a five-answers question test has been considered. The involved probabilities are shown in Table 6, for each part of speech c, from left to right: the tag; the a priori probability $P(c)$, computed by analyzing the corpus; the probability $P(n_c = k|c)$ of obtaining k answers having the same part of speech of the missing word, for $k = 1 \ldots 5$, conditioned to c being that part of speech; the probability of success $P(S|c)$ at predicting the missing word by choosing among the answers having part of speech c, conditioned to c being the part of speech of the missing word; the probability of c being the part of speech of the missing word and to succeed, at the same time, at predicting that word.

Table 6. Probabilities related to the construction and usage of a five-answers question.

c	$P(c)$	$P(n_c = 1|c)$	$P(n_c = 2|c)$	$P(n_c = 3|c)$	$P(n_c = 4|c)$	$P(n_c = 5|c)$	$P(S|c)$	$P(c)P(S|c)$
S	0.2908	0.2531	0.4150	0.2551	0.0697	0.0071	0.4856	0.1412
E	0.1532	0.5142	0.3721	0.1010	0.0122	0.0006	0.7045	0.1079
V	0.1396	0.5482	0.3556	0.0865	0.0093	0.0004	0.7292	0.1017
F	0.1108	0.6252	0.3116	0.0582	0.0048	0.0002	0.7827	0.0867
R	0.0816	0.7115	0.2528	0.0337	0.0020	0.0000	0.8386	0.0684
A	0.0619	0.7743	0.2045	0.0203	0.0009	0.0000	0.8769	0.0543
P	0.0396	0.8508	0.1403	0.0087	0.0002	0.0000	0.9210	0.0365
B	0.0360	0.8635	0.1291	0.0072	0.0002	0.0000	0.9281	0.0334
C	0.0359	0.8640	0.1287	0.0072	0.0002	0.0000	0.9284	0.0333
D	0.0310	0.8817	0.1128	0.0054	0.0001	0.0000	0.9381	0.0291
N	0.0185	0.9279	0.0701	0.0020	0.0000	0.0000	0.9630	0.0178
X	0.0010	0.9959	0.0041	0.0000	0.0000	0.0000	0.9980	0.0010
I	0.0001	0.9997	0.0003	0.0000	0.0000	0.0000	0.9998	0.0001
T	0.0000	1.0000	0.0000	0.0000	0.0000	0.0000	1.0000	0.0000

This can be compared to Table 2. Surprisingly, the sum of the values of the last column is 0.7115, leading to the same success probability achieved for Italian, i.e. 71%. This probability considers a perfect part of speech predictor for the first step and the worst word predictor (random choice) for the second step. The a priori probabilities of the tags are also very similar among the two languages.

Table 7 shows, for each part of speech: the tag; the a priori probability; the number of unique words in a dictionary; the percentage of unique words in the dictionary; the product of the a priori probability with the percentage of unique words.

The sum of the values of the last column is 21.69%. It represents the average percentage of the answers set to be processed when the set is a casual sample of the whole dictionary. This result is similar to the percentage obtained for the Italian language (15.11%). However while the most common part of speech in the English dictionary is "S" (noun), the most common part of speech in the Italian dictionary is "V" (verb). This happens because in Italian verbs can be inflected in several ways. However in any real usage, e.g. in a corpus, nouns are the most common part of speech in Italian too; in fact, this is reflected into the a priori probabilities.

Table 7. Parts of speech a priori probabilities and distribution.

c	P(c)	Words	Words %	P(c) · (Words %)
S	0.2908	51490	64.36%	18.71%
E	0.1532	191	0.24%	0.04%
V	0.1396	9011	11.26%	1.57%
F	0.1108	16	0.02%	0.00%
R	0.0816	5	0.01%	0.00%
A	0.0620	16022	20.03%	1.24%
P	0.0396	119	0.15%	0.01%
B	0.0360	2135	2.67%	0.10%
C	0.0359	19	0.02%	0.00%
D	0.0310	70	0.09%	0.00%
N	0.0186	654	0.82%	0.02%
X	0.0010	84	0.10%	0.00%
I	0.0000	184	0.23%	0.00%
T	0.0000	0	0.00%	0.00%

As for the WaCky Italian Wikipedia Corpus, the first 100,000 words of the Brown corpus have been parsed with a moving window: the frequency of every sequence of one to five parts of speech has been saved to a lookup table. The most frequent posgrams are reported in Tables 8a and 8b.

The results shown in Tables 8a and 8b can be compared with the results in Tables 4a and 4b for Italian. In particular, while posgrams of size one exhibits similar frequencies, higher order posgrams are more language-sensitive: for example, the most common 5-posgram in Italian ("SESES") is not common in English; similarly, the most common 5-posgram in English ("SSSSS") is rare in Italian. Therefore, a different part of speech predictor should be employed on the first step for each language needed.

Table 8a. Most common posgrams by length and their frequencies on the first 100,000 words of the Brown corpus. The length spans from one to three.

1-posgram	Freq.	2-posgram	Freq.	3-posgram	Freq.
S	29075	SS	7494	ERS	3124
E	15321	SE	7468	SSS	2899
V	13955	SF	6615	RSE	2249
F	11080	RS	5578	SER	2246
R	8158	ER	4464	SES	2237
A	6195	AS	4308	SSF	1858
P	3691	ES	3817	ASE	1501
B	3603	SV	3486	RAS	1431
C	3590	VE	3318	RSS	1353
D	3099	VV	3063	ASF	1189

Table 8b. Most common posgrams by length and their frequencies on the first 100,000 words of the Brown corpus. The length spans from four to five.

4-posgram	Freq.	5-posgram	Freq.
SERS	1613	SSSSS	691
SSSS	1276	SERSS	505
ERSE	1129	RSERS	479
ERSS	839	SERSE	477
ERAS	749	SERAS	363
VERS	682	ERSSS	352
RSER	668	ERSES	347
RSES	662	ASERS	335
SSSF	649	ERSER	324
RASE	644	SERSF	303

7   Results

The proposed part of speech prediction method employed on the first step is not directly comparable with general Part of Speech Tagging (POST) algorithms. In fact those algorithms are concerned with finding the most probable sequence of parts of speech for a complete sentence. While several approaches has been described in literature for handling unknown words, i.e. words not present in the training dataset, no studies have been done, at the best of our knowledge, on handling missing words. POST algorithms address a different task since they assume that no words are missing in the middle of the sentence. For unknown words, they generally take advantage of the word morphology, e.g. prefixes or suffixes, for predicting the part of speech. This cannot be done for missing words. Therefore, Tables 9a and 9b compare the proposed part of speech prediction method with two very baseline methods: random choice and

Table 9a. Accuracy of various part of speech prediction methods for missing words for the Italian language.

Method	Accuracy
Posgrams	43.2%
Always noun	28.0%
Random choice	7.1%

Table 9b. Accuracy of various part of speech prediction methods for missing words for the English language.

Method	Accuracy
Posgrams	45.0%
Always noun	29.1%
Random choice	7.1%

choice of the most probable part of speech, i.e. noun ("S"). The best results are obtained with $p = 5$, $\alpha_0 = 0.5$, $\alpha_1 = 16.7$, $\alpha_2 = 0.2$, achieving an accuracy of 43.2% for the Italian language and of 45.0% for the English language.

State of the art algorithms for word prediction reported in literature have been tested with the Microsoft Sentence Completion Challenge dataset. This is currently the only complete training-test dataset specifically developed for measuring automatic sentence completion algorithms. However, because of its limits exposed in Sect. 1 (uniformity of the part of speech in the answer set of any given question and unrepresented parts of speech among missing words), this dataset could not be employed. Therefore, in order to test the full word prediction methodology two new questionnaires has been built. Their format is the same of the one used for the Microsoft Sentence Completion Challenge: each question is composed by a sentence having a missing word and by five candidate words as answers. However our questionnaires are more general, since they address the aforementioned limits. First of all, each word of the same answers set may belong to a different part of speech. Secondly, the missing word can belong to any part of speech, including: conjunctions, prepositions, determinants and pronouns. The two questionnaires have been built by selecting 368 Italian sentences from the Paisà [25] dataset and 612 English sentences from the book "The adventures of Sherlock Holmes" [26]. From each sentence a question is built, by removing a random word; the other candidate words are chosen randomly from the nearby sentences; these words and the removed word constitute the answers set for the question. We employed three different word prediction methods for the second step: ngrams, Latent Semantic Analysis (LSA) and random choice. Tables 10a and 10b show the results in term of accuracy.

Table 10a. Accuracy of various word prediction methods on the Italian questionnaire, with (2 steps) and without (1 step) employing the proposed part of speech prediction algorithm.

Method	Accuracy
ngrams (2 steps)	51.1%
ngrams (1 step)	50.3%
LSA (2 steps)	30.7%
Random choice (2 steps)	29.3%
LSA (1 step)	25.5%
Random choice (1 step)	20.4%

Table 10b. Accuracy of various word prediction methods on the English questionnaire, with (2 steps) and without (1 step) employing the proposed part of speech prediction algorithm.

Method	Accuracy
ngrams (2 steps)	57.8%
ngrams (1 step)	57.8%
LSA (2 steps)	33.0%
LSA (2 steps)	27.9%
Random choice (1 step)	27.0%
Random choice (1 step)	20.6%

Each two steps method always provides better or equal results than its single step counterpart. Since any part of speech, except punctuation, is admissible in a question answers set, the obtained results are not directly comparable with those reported for the Microsoft Completion Challenge. Furthermore, most stopwords such as conjunctions, prepositions and determiners had to be included during the training phase. This has undermined the quality of the semantic spaces employed by LSA, leading to lower results than those reported in literature for English. Results show that word prediction methods with lower accuracy exhibit greater improvements (up to +8.9% for Random choice) than methods with higher accuracy (up to +0.8% for ngrams). In general, the greater the accuracy of the word prediction method, the greater the part of speech prediction accuracy step is required to be advantageous. While this is not surprising, it should be noted that even a 50.3% accuracy word prediction algorithm (i.e. ngrams for Italian) can be improved by a 43.2% accuracy part of speech predictor: in fact, depending on the predicted part of speech the accuracy of the first step may be greater and therefore still advantageous; when this is not the case the part of speech prediction will be automatically discarded, as described in Sect. 5.

8   Conclusion

In this work we discussed the issues concerning the prediction of missing words in incomplete sentences. As a matter of fact, sentence completion remains one of the more difficult problems within the domain of natural language processing; the best results are achieved when a restricted and well built set of possible completions is given. We therefore proposed an additional step of analysis and prediction with respect to traditional algorithms. The additional preparatory step is aimed at predicting the part of speech of the missing word. This solution allows for an automatic reduction of the number of candidate words and leads to an improvement of the results accuracy and a reduction of the execution time. Experimental results showed that the two steps methodology has always given better or equal results than its single step counterpart. It is important to highlight that small improvements can be obtained even when the average accuracy of the part of speech predictor is relatively low.

In particular the methodology has been tested according to two different experimental settings: one for the Italian language and one for English. To this aim, two different questionnaires have been built and results proved to be consistent among the two languages. They are also consistent with the results of the preliminary analysis conducted on two different corpora, the WaCky Italian Wikipedia corpus and the Brown corpus. From this preliminary analysis resulted that for both languages a priori probabilities for each part of speech were similar; furthermore, the expected theoretical success probability and speedup of the methodology also resulted to be comparable. However, since higher order posgrams showed different statistics among the two languages, two different part of speech predictors has been trained. Experimental results on these first step predictors showed an accuracy of 43.2% for Italian and of 45.0% for English, against a baseline of 7.1% (14 tags). Tests on the full two steps word prediction methodology resulted in accuracy improvements of up to 8.9%.

Future works will focus primarily on improving the first step, i.e. the prediction of the part of speech of the missing word. In particular, we ignored the fact that each word may belong to more parts of speech: while the methodology worked nevertheless, extending the model to handle the general case may improve the results. Secondarily, other word prediction algorithms can be tested for the second step.

References

1. Witten, I.H., Cleary, J.G., Darragh, J.J.: The reactive keyboard: a new technology for text entry (1983)
2. Darragh, J.J., Witten, I.H., James, M.L.: The reactive keyboard: a predictive typing aid. Computer **23**(11), 41–49 (1990)
3. Carlberger, A., Carlberger, J., Magnuson, T., Hunnicutt, S., Palazuelos-Cagigas, S.E., Navarro, S.A.: Profet, a new generation of word prediction: an evaluation study. In: Proceedings, ACL Workshop on Natural Language Processing for Communication Aids, pp. 23–28 (1997)
4. Good, I.J.: The population frequencies of species and the estimation of population parameters. Biometrika **40**(3–4), 237–264 (1953)
5. Katz, S.M.: Estimation of probabilities from sparse data for the language model component of a speech recognizer. IEEE Trans. Acoust. Speech Sig. Process. **35**(3), 400–401 (1987)
6. Jelinek, F., Mercer, R.L.: Interpolated estimation of Markov source parameters from sparse data. In: Proceedings of the Workshop on Pattern Recognition in Practice (1980)
7. Kneser, R., Ney, H.: Improved backing-off for m-gram language modeling. In: 1995 International Conference on Acoustics, Speech, and Signal Processing (ICASSP), vol. 1, pp. 181–184 (1995)
8. Guthrie, D., Allison, B., Liu, W., Guthrie, L., Wilks, Y.: A closer look at skip-gram modelling. In: Proceedings of the 5th International Conference on Language Resources and Evaluation (LREC), pp. 1–4 (2006)
9. Zweig, G., Burges, C.J.C.: The Microsoft Research Sentence Completion Challenge. Microsoft Research Technical report, MSR-TR-2011-129 (2011)
10. Gubbins, J., Vlachos, A.: Dependency language models for sentence completion. In: EMNLP, pp. 1405–1410 (2013)
11. Deerwester, S.C., Dumais, S.T., Landauer, T.K., Furnas, G.W., Harshman, R.A.: Indexing by latent semantic analysis. JASIS **41**(6), 391–407 (1990)
12. Spiccia, C., Augello, A., Pilato, G., Vassallo, G.: A word prediction methodology for automatic sentence completion. In: 2015 IEEE International Conference on Semantic Computing (ICSC), pp. 240–243 (2015)
13. Agostaro, F., Pilato, G., Vassallo, G., Gaglio, S.: A sub-symbolic approach to word modelling for domain specific speech recognition. In: Proceedings, IEEE 7th International Workshop on Computer Architecture for Machine Perception (CAMP), pp. 321–326 (2005)
14. Bengio, Y., Ducharme, R., Vincent, P., Jauvin, C.: A neural probabilistic language model. In: Holmes, D.E., Jain, L.C. (eds.) Innovations in Machine Learning. Studies in Fuzziness and Soft Computing, vol. 194, pp. 137–186. Springer, Heidelberg (2006)
15. Mnih, A., Teh, Y.W.: A fast and simple algorithm for training neural probabilistic language models. arXiv preprint (2012). arXiv:1206.6426
16. Pachitariu, M., Sahani, M.: Regularization and nonlinearities for neural language models: when are they needed? arXiv preprint (2013). arXiv:1301.5650

17. Mikolov, T., Chen, K., Corrado, G., Dean, J.: Efficient estimation of word representations in vector space. arXiv preprint (2013). arXiv:1301.3781
18. Kučera, F., Kučera, H.: A Standard Corpus of Present-Day Edited American English, for use with Digital Computers (Brown). Brown University (1979)
19. Baroni, M., Bernardini, S., Ferraresi, A., Zanchetta, E.: The WaCky wide web: a collection of very large linguistically processed web-crawled corpora. Lang. Resour. Eval. **43**(3), 209–226 (2009)
20. Semantically and Syntactically Annotated Italian Wikipedia. WaCky Corpora. University of Bologna. http://wacky.sslmit.unibo.it/doku.php?id=corpora. Accessed 1 July 2015
21. Calzolari, N., McNaught, J., Zampolli, A.: EAGLES Final Report: EAGLES Editors' Introduction. EAG-EB-EI, Pisa (1996)
22. Tanl POS Tagset, University of Pisa. http://medialab.di.unipi.it/wiki/Tanl_POS_Tagset. Accessed 1 July 2015
23. Stubbs, M.: An example of frequent English phraseology: distributions, structures and functions. Lang. Comput. **62**(1), 89–105 (2007)
24. Lindquist, H.: Corpus Linguistics and the Description of English, pp. 102–103. Edinburg University Press, Edinburgh (2009)
25. Lyding, V., Stemle, E., Borghetti, C., Brunello, M., Castagnoli, S., Dell'Orletta, F., Dittmann, H., Lenci, A., Pirrelli, V.: The PAISA corpus of Italian web texts. In: Proceedings of the 9th Web as Corpus Workshop (WaC-9), 14th Conference of the European Chapter of the Association for Computational Linguistics, pp. 36–43 (2014)
26. Doyle, A.C.: The adventures of Sherlock Holmes. Gutenberg Project, EBook #1661, Edition 12 (2002)

Is Bitcoin's Market Predictable? Analysis of Web Search and Social Media

Martina Matta$^{(\boxtimes)}$, Ilaria Lunesu, and Michele Marchesi

Universita' degli Studi di Cagliari, Piazza d'Armi, 09123 Cagliari, Italy
{martina.matta,ilaria.lunesu,michele}@diee.unica.it

Abstract. In recent years, Internet has completely changed the way real life works. In particular, it has been possible to witness the online emergence of web 2.0 services that have been widely used as communication media. On one hand, services such as blogs, tweets, forums, chats, email have gained wide popularity. On the other hand, due to the huge amount of available information, searching has become dominant in the use of Internet. Millions of users daily interact with search engines, producing valuable sources of interesting data regarding several aspects of the world. Bitcoin, a decentralized electronic currency, represents a radical change in financial systems, attracting a large number of users and a lot of media attention. In this work we studied whether Bitcoin's trading volume is related to the web search and social volumes about Bitcoin. We investigated whether public sentiment, expressed in large-scale collections of daily Twitter posts, can be used to predict the Bitcoin market too. We achieved significant cross correlation outcomes, demonstrating the search and social volumes power to anticipate trading volumes of Bitcoin currency.

Keywords: Bitcoin · Web search media · Social media · Twitter · Sentiment analysis · Google trends · Cross correlation analysis

1 Introduction

The advent of the Internet has completely changed the way real life works. By enabling practically all Internet users to interact at once and to exchange and share information almost cost-free, more efficient decisions on several fields are possible.

The majority of daily activities radically changed, moving towards a "virtual sector", such as web actions, credit card transactions, electronic currencies, navigators, games, and so on. In recent years, web search and social media have emerged online. On one hand, services such as blogs, tweets, forums, chats, email have gained wide popularity. Social media data represent a collective indicator of thoughts and ideas regarding every aspect of the world. It has been possible to assist to deep changes in habits of people in the use of social media and social network [1]. Social media technologies have produced completely new ways of

© Springer International Publishing AG 2016
A. Fred et al. (Eds.): IC3K 2015, CCIS 631, pp. 155–172, 2016.
DOI: 10.1007/978-3-319-52758-1_10

interacting [2], bringing the creation of hundreds of different social media platforms (e.g., social networking, shared photos, podcasts, streaming videos, wikis, blogs).

On the other hand, due to the huge amount of available information, searching has become dominant in the use of Internet. Millions of users daily interact with search engines, producing valuable sources of interesting data regarding several aspects of the world.

Recent studies demonstrated that web search streams could be used to analyze trends about several phenomena [3–5]. In one of the seminal works, Ginsberg et al. proved that search query volume is a sophisticated way to detect regional outbreaks of influenza in USA almost 7 days before CDC surveillance [6].

Kristoufek [7] studied the popularity of the Dow Jones stocks, measured by Google search queries for portfolio diversification. Curme et al. [8] clustered the online searches into groups and showed that mainly politics and business oriented searches are connected to the stock market movements. Preis et al. [9] demonstrated that Google searches, for financial terms, can support profitable trading strategies. Dimpfl et al. found a strong relationship between internet search queries and the leading stock market index. In addition they found a strictly correlation between the Dow Jones' realised volatility and the volume of search queries [10].

There are also studies that report another use of a search engine, namely as a possible predictor of market trends. Bollen et al. showed that search volumes on financial search queries have a predictive power. They compared these volumes with market indexes such as Dow Jones Industrial Average, trading volumes and market volatility, demonstrating the possibility to anticipate financial performances [11]. In this work, Granger causality analysis and a Self Organizing Fuzzy Neural Network are used to investigate the hypothesis that public mood states, as measured by the Opinion Finder and GPOMS mood time series, are predictive of changes in DJIA closing values. Kristoufek proposed the study of Power-law correlations for Google searches queries for Dow Jones Industrial Average (DJIA) component stocks, and their cross-correlations with volatility and traded volume [12].

Bordino et al. proved that search volumes of stocks highly correlate with trading volumes of the corresponding stocks, with peaks of search volume anticipating peaks of trading volume by one day or more [5]. In his work [13], Bulut described that internet search data, via Google Trends, is utilized to nowcast the known variates of two structural exchange rate determinations models. By using internet search data, the author aims to get a timely description of the state of the economy way before the official data are released to the market participants. Kim et al. [14] introduced an analysis system to predict the value fluctuations of virtual currencies used in virtual worlds, and based on user opinion data in selected online communities. In their proposed method, data of user opinions on a predominant community are collected by employing a simple algorithm and guaranteeing a stable prediction of value fluctuations of more than one virtual currency.

Search queries prove to be a useful source of information in financial applications, where the frequency of searches of terms related to the digital currency can be a good measure of interest in the currency [15]. Mondria et al. proved that the number of clicks on search results stemming from a given country correlates with the amount of investment in that country [16]. Further studies showed that changes in query volumes for selected search terms mirror changes in current volumes of stock market transactions [17].

In recent years, social media data assume the role of a collective indicator of thoughts and ideas regarding every aspect of the world. It has been possible to assist to deep changes in habits of people in the use of social media and social network. In particular, we deeply analysed the transmitted sentiment of users regarding a particular topic. Twitter[1], an online social networking website and microblogging service, has become an important tool for businesses and individuals to communicate and share information with a rapid growth and significant adoption. In fact, Java et al. affirmed that it seems to be used to share data and to describe minor daily activities [18].

Twitter and other social media offer a plethora of opportunities to reveal business intuitions, where it remains a challenge to identify the potential social audience. In their work Ling et al. [19] analyzed the Twitter content of an account owner and its list of followers through various text mining methods and machine learning approaches in order to identify a set of users with high-value social audience members. In their paper, Ciulla et al. [20] assessed the usefulness of open source data that come from Twitter for prediction of societal events by analysing in depth the microblogging activity surrounding the voting behaviour on a specific event. Mocanu et al. performed a comprehensive survey of the worldwide linguistic landscape emerging from mining the Twitter microblogging platform [21]. Hick et al. explored the opportunities and challenges in the use of Twitter as platform for playing games, through crawling game that uses Twitter for collaborative creation of game content [22].

Additionally, Twitter has rapidly grown as a mean to share ideas and thoughts on investing decisions. Analyzing in deep the literature related to different uses of social media, and Twitter in particular, we collected information about its use for seeking real world emotions that could predict real financial markets trend [23]. In their paper, Rao and Srivastava studied the complex relationship that exists between tweet board literature (like bullishness, volume, agreement etc.) with the financial market instruments (like volatility, trading volume and stock price) [24].

One of the fascinating phenomena of the Internet era is the emergence of digital currencies. Bitcoin, the most popular among these, has been created in 2008 by Satoshi Nakamoto [25] for the purpose to replace cash, credit cards and bank wire transactions. A digital currency can be defined as an alternative currency which is exclusively electronic and thus has no physical form.

Bitcoin is based on advancements in peer-to-peer networks [26] and cryptographic protocols for security. Due to its properties, Bitcoin is completely

---

[1] https://twitter.com/.

decentralized and is not managed by any government or central bank. More-over, it ensures anonymity. So, it is practically detached from the real economy. Bitcoin is based on a distributed register known as "block-chain" to save trans-actions carried out by users. Like any other currency, a peculiarity of Bitcoin is to facilitate transactions of services and goods with vendors that accept Bitcoins as payment [27], attracting a large number of users and a lot of media attention.

The Bitcoin represents an important new phenomenon in financial markets. Mai et al. examined predictive relationships between social media and Bitcoin returns by considering the relative effect of different social media platforms (Internet forum vs. microblogging) and the dynamics of the resulting relation-ships using auto-regressive vector and error correction vector models [28]. Matta et al. examined the striking similarity between Bitcoin price and the number of queries regarding Bitcoin recovered on Google search engine [29]. In their work, Garcia et al. [30] proved the interdependence between social signals and price in the Bitcoin economy, namely a social feedback cycle based on word-of-mouth effect and a user-driven adoption cycle. They provided evidence that Bitcoins growing popularity causes an increasing search volumes, which in turn result a higher social media activity about Bitcoin. A growing interest inspires the pur-chase of Bitcoins by users, driving the prices up, which eventually feeds back on the search volumes. There are several works that present predictive relationships between social media and Bitcoin volumes where the relative effects of different social media platforms (Internet forum vs. microblogging) and the dynamics of the resulting relationships, are analyzed using cross-correlation [31] or linear regression analysis [32]. Social factors, that are composed of interactions among market actors, may strongly drive the dynamics of Bitcoins economy [30].

In this work we decided to investigate whether social media activity or infor-mation collected by web search media could be profitable and used by investment professionals. We also evaluated the possibility to find a relationship between Bitcoins trading volumes and volumes of exchanges tweets.

We first studied the relationship that exists between trading volumes of Bit-coin currency and the volumes of search engine, then we analyzed a corpus of 2,353,109 tweets in order to discover if the chatter of the community can be used to make qualitative predictions about Bitcoin market, attempting to estab-lish whether there is any correlation between tweets sentiment and the Bitcoins trading volume. The frequency of searches of terms about Bitcoin could have a good explanatory power, so we decided to examine Google, one of the most important search engine. We studied whether web search media activity could be helpful and used by investment professionals, analyzing the search volumes power of anticipate trading volumes of the Bitcoin currency.

We compared USD trade volumes about Bitcoin with search volumes using Google Trends. This is a feature of Google search engine that illustrates how frequently a fixed search term was looked for. Following this kind of approach, we evaluated how much "Bitcoin" term, for the specific time interval, is looked for using Googles search engine.

Simultaneously, we decided to apply automated Sentiment Analysis on shared short messages of users on Twitter in order to automatically analyze peoples opinions, sentiments, evaluations and attitudes. We wondered whether public sentiment, as expressed in large-scale collections of daily Twitter posts, can be used to predict the Bitcoin market. The results of our previous analysis suggest that a significant relationship with future Bitcoins price and volume of tweets exists on a daily level. We found a striking correlation between Bitcoins price spread and changes in query volumes for the "Bitcoin" search term [29].

The body of this paper is organized in five major sections. Section 2, describes the methodology applied in our study, Sect. 3 summarizes and discusses our results and, finally, Sect. 4 presents conclusions and suggestions for future work.

2 Methodology

2.1 Google Trends

Google Trends[2] is a feature of Google Search engine that illustrates how frequently a fixed term is looked for. Through this, you can compare up to five topics at one time to view their relative popularity, allowing you to gain an understanding of the hottest search trends of the moment, along with those developing in popularity over time. This system provides a time series index of the volume of queries made by the users with Google.

Query index is based on the number of web searches, performed with a specific term, and compared to the total amount of searches done over time. Absolute search volumes are not shown, because the data are normalized on a scale from 0 to 100.

Google classifies search queries into 27 categories at the top level and 241 categories at the second level through an automatic classification engine. Indeed, queries are given out to fixed categories, due to natural language processing methods.

The query index data are available as a CSV file in order to facilitate research activities. Figure 1 depicts an example from Google Trends for the query "Bitcoin".

2.2 Blockchain.info

Blockchain.info[3] is an online system that provides detailed information about Bitcoin market. Launched in August 2011, this system shows data on recent transactions, plots on Bitcoin economy and several statistics. It allows users to analyze different Bitcoin aspects:

- Total number of Bitcoins in circulation
- Number of Transactions

---

[2] http://trends.google.com.
[3] http://www.blockchain.info.

Fig. 1. Example of Google Trends usage for the query "Bitcoin".

– Total output volume
– USD Exchange Trade volume
– Market price (USD)

We decided to study a time series regarding the USD trade volume from top exchanges, analyzing its trends.

2.3   Twitter API

The Twitter space can be explored by means of its Application Programming Interface (API)[4], implementing a simple crawler that allows developers to collect the required data. Twitter servers send back XML or JSON responses, that are parsed and processed by the implemented system. Twitter gives the opportunity to work with its API with three different approaches that are listed below.

– *Original REST API* allows users to analyze the Twitter core, in order to update their statuses or profile information in realtime.
– *Search API* is read-only in the search database, and search queries return up to 1500 tweets from up to seven/eight days before.
– *Streaming API* provides real time data access to tweets in filtered forms that return public statuses that match with one or more filters. This approach doesn't allow developers to find all the historical tweets, but it enables the access to data as it is being tweeted.

While Twitter REST API is available and suitable for different applications, we decided to monitor tweets with Twitter Streaming API, that provides immediate updates. As soon as tweets come in, Twitter notifies the implemented system allowing us to store them into the database without the delay of polling the REST API.

---

[4] https://dev.twitter.com/overview/documentation.

2.4 Opinion Mining

The opinion mining is a particular technique that detects automatically the sentiment and subjectivity transmitted in written texts. The user's tweets could express the opinion regarding different topic, trends or brands [33]. For this reason, we decided to monitor the sentiment expressed, day after day, by users on the matter of Bitcoin.

Since the goal of this research is neither to develop a new sentiment analysis nor to improve an existing one, we used "SentiStrenght", a tool developed by a team of researchers in the UK that demonstrated good outcomes [34]. SentiStrength estimates the strength of positive and negative sentiments in short texts. It is based on a dictionary of sentiment words, each one associated with a weight, which is its sentiment strength. In addition, this method uses some rules for non-standard grammar.

Based on the formal evaluation of this system on a large sample of comments from MySpace.com, the accuracy of predicting positive and negative emotions was something similar to that of other systems (72.8% for negative emotions and 60.6% for positive emotions, based on a scale of 1–5). Compared to other methods, SentiStrenght showed the highest correlation with human coders [35]. The tool is able to assess each message separately and, at the end, it returns one singular value.

- +1 if the system identifies a positive sentiment
- −1 if the system identifies a negative sentiment
- 0 if a neutral opinion is identified

2.5 Data Collection

Search query volumes regarding Bitcoin were collected from *Google Trends* website, capturing all searches, inserted from June 2014 to July 2015, with "Bitcoin" word as keyword.

Trading volume data were gathered from blockchain.info website, in order to evaluate daily trends of Bitcoin currency. We assessed the relationship over time between number of daily queries and trading volume of Bitcoin.

We collected a dataset of tweets regarding Bitcoin in the period between January and April 2015 using *Streaming Twitter API*, achieving almost two millions of statuses. The system has been set to raise in real time the tweets that contain "Bitcoin" word in the sentence. Twitter API provides several fields in JSON format. We decided to hold the following tweet's information, saving it in our database.

- Tweet ID
- Tweet text
- Date of creation
- User who posted the status
- User location

For each collected tweet, we detected the language in order to use the correct sentiment dictionary. "Language-detection" is a particular library, implemented in Java language, able to identify the language of a given sentence using Naive Bayesian filter[5]. The system has been tested, achieving a 99% precision for 53 languages. After this step, the sentiment was calculated using the correct language dictionary by means of SentiStrength.

To better understand whether a search engine can be a good predictor of trading volumes, we analysed the correlation between these data expressed as time series, performing a time-lagged cross-correlation study, and a Granger-causality test. We applied the same analysis to the number of tweets, and to the sentiment expressed by users. These analyses based on Twitter have been applied on a shorter period than the search media analysis.

3   Results

We extracted from Google Trends and Blockchain.info data sources time series composed by daily values in the time intervalranging from June 2014 to July 2015, in order to evaluate their relationship and prediction capability. The same approach has been applied to the Twitter, dataset but in the time interval ranging from January 2015 to April 2015. We run statistical analysis and the computation of correlation, cross-correlation and Granger causality test, obtaining interesting results.

3.1   Tweets Analysis

We collected 2,353,109 tweets covering the period between January and April 2015. Using Sentistrength, we computed the sentiment of each tweet, obtaining 418,949 positive tweets and 270,669 negative ones. The remaining tweets are neutral. There are more positive messages than negative ones. We also found that positive messages are almost 2 times more likely to be forwarded than negative ones. After a careful analysis, it was observed that the number of neutral tweets is very high because people very often write non-expressive comments, the price of Bitcoin or simple links that lead to other web pages.

Figure 2 shows the two time series for the period under consideration, representing the positive tweets with a dotted line and the negative tweets with a solid line. Taking a look to the two time series it's possible to see some negative or positive peaks, corresponding to price variations[6]. For instance, the peak of January 26th is due to the top price of the Bitcoin for the same day, 278\$. An other example can be seen on 12 February when there is a negative peak, corresponding to a price decrease at 221.85 dollars. This Figure clearly shows that, most of the times, positive and negative time series grow up and decrease with the same pace in a given day. This is related to the total amount of tweets of

---

[5] https://code.google.com/p/language-detection/.

[6] https://blockchain.info/it/charts/market-price.

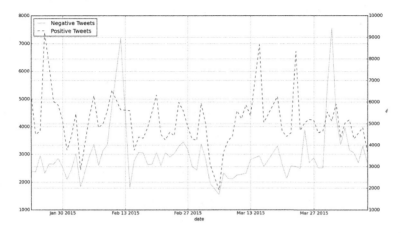

Fig. 2. Representation of the positive tweets with a dotted line and the negative tweets with a solid line for the period between January and April 2015.

the evaluated day. To solve this problem, we developed a simple metric called PT-NT ratio x to predict the trends of the Bitcoin trading volume. We defined the sentiment score x_t of day t as the ratio of positive versus negative messages on the Bitcoin topic. A message is defined as positive if it contains more positive then negative words, and negative in the opposite case.

$$x_t = \frac{count_t(pos.tweet \wedge Bitcoin\ topic)}{count_t(neg.tweet \wedge Bitcoin\ topic)} \tag{1}$$

$$= \frac{p(pos.tweet \mid Bitcoin\ topic, t)}{p(neg.tweet \mid Bitcoin\ topic, t)} \tag{2}$$

With this approach, it's possible to determine the ratio of positive versus negative tweets on a fixed day. The resulting time series were used to study their correlation with the Bitcoin trading volume over a given period of time.

3.2 Pearson Correlation

Pearson's correlation r is a statistical measure that evaluate the strength of a linear association between two time series G and T. Initially, we assumed G as query data and T as trading volumes.

$$r = \frac{\sum_i (G_i - \overline{G})(T_i - \overline{T})}{\sqrt{\sum_i (G_i - \overline{G})^2}\sqrt{\sum_i (T_i - \overline{T})^2}} \tag{3}$$

The Pearson correlation coefficient has values between -1 and $+1$, the bounds denoting maximum anti-correlation or correlation, respectively, whereas 0 indicates no correlation. We calculated the Pearson correlation between queries

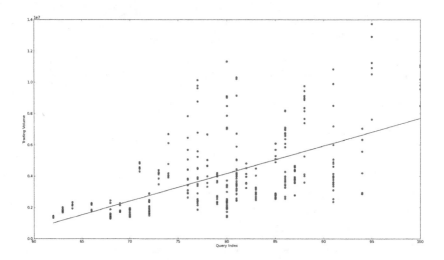

Fig. 3. Correlation between Trading Volume and Queries Volume about Bitcoin.

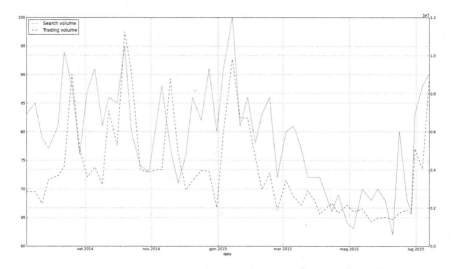

Fig. 4. Bitcoin Trading Volume and Queries Volume about Bitcoin.

search data and trading volume and we found a result equal to 0.60, which is quite high. The correlation is also clearly visible in the Fig. 3.

This result is confirmed by Fig. 4, that shows the two time series. Here peaks in one time series typically occur close to peaks in the other. The solid line, that represents search volumes, very often anticipates the dotted line, that represents trading volumes. The most significant peaks occurred in the interval between August and September 2014, between September and October 2014, between

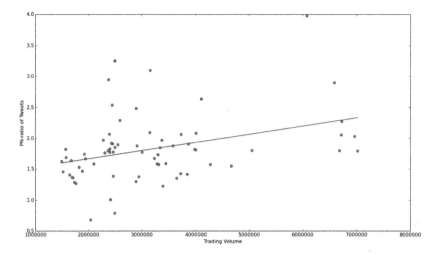

Fig. 5. Pearson Correlation coefficient between Trading Volume and the Volume of Tweets about Bitcoin.

November and December 2014 and between January and February 2015. During other periods, this phenomenon is less evident, but anyway it is present.

Radical changes in peaks are due to several factors. One of the most evident peaks in Fig. 4 corresponds to the interval between end of June and beginning of July. This is the period of the Greek crisis acme, that caused changes also in the Bitcoin market. Indeed, a lot of people started to invest in Bitcoin because people tried to move money out of the country, and Greek government tried to block this process. Bitcoins were seen by many as the only way to move their wealth to other currencies. In fact, Greeks would use Bitcoin to protect the value of their money at home. Ten times more Greek than usual were found at the company *'German Bitcoin.de'*[7] to buy electronic currency. This situation is clearly visible in the right part of Fig. 4, where the curve corresponding to the volumes of queries regarding Bitcoin considerably increases, followed by an increase of the curve of trading volumes after some days. This is confirmed by correlation values.

Comparing the PT-NT ratio of tweets volume with the trading volume of Bitcoins, we found a Pearson correlation equals to 0.37, which is still a remarkable value. The two time series are again quite similar, and this is shown in Fig. 5, that shows a fair correlation, with some points far from the line of best fit. Figure 6 highlights the period between January and February 2015, where we can notice how social peaks correspond to following peaks in the trading volume. This is particularlt evident on January, 25 and on February, 14. The prediction power of the PT-NT ratio on Bitcoin trading volumes, however, is lower than that of web search volumes.

---

[7] https://www.bitcoin.de/.

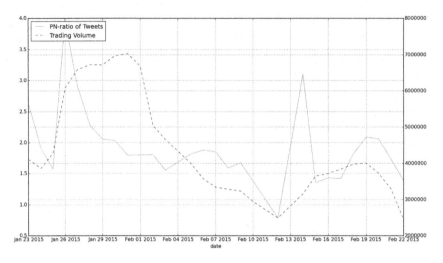

Fig. 6. Bitcoin Trading Volume and PT-NT ratio in the period January, 23 to February, 22, 2015.

3.3   Cross Correlation

We investigated whether query or search volumes can anticipate trading volumes of Bitcoin. We calculated the cross correlation comparing the trading volume T with the query volume G as the time lagged Pearson cross correlation between two time series G and T for all delays d between -5 and 5. We also applied the same analysis, substituting the search data with the social data, according to Eq. 4.

$$r(d) = \frac{\sum_i (G_i - \overline{G})(T_{i-d} - \overline{T})}{\sqrt{\sum_i (G_i - \overline{G})^2} \sqrt{\sum_i (T_{i-d} - \overline{T})^2}} \tag{4}$$

We chose to evaluate a maximum lag of five days and, also in this case, the correlation ranges from -1 to 1.

In Table 1 we report the results obtained from these experiments. Each column shows the cross-correlation corresponding to different time lags. We can observe that cross-correlation for positive lags is always higher than for negative lags. Taking a look to the raw search volume, the results with positive delays take values always higher than 0.64, whereas those with negative delays take values always lower than 0.55. This means that query volumes are able to anticipate trading volumes of 3 days, or even more.

Different outcomes are visible in the bottom raw, corresponding to the PT-NT ratio social volume, where the highest cross correlation result is given with a delay equal to 1. It means, that the social volume is able to anticipate the trading volume in one or two days with 0.37 as the best result. The results with positive delays achieve outcomes always higher than 0.35 and with negative delays report values always lower than 0.22. Although we achieved good outcomes with both search and Twitter functions, the search volume has better predictive power.

Table 1. *Cross-correlation with the trading volume of Bitcoin, compared to the search volume, and to the PT-NT ratio.*

Delay	−5	−4	−3	−2	−1	0	1	2	3	4	5
Search volume	0.363	0.402	0.447	0.502	0.559	0.609	0.649	0.670	**0.682**	0.674	0.645
PT-NT ratio volume	0.058	0.122	0.165	0.169	0.221	0.353	**0.379**	0.374	0.373	0.366	0.343

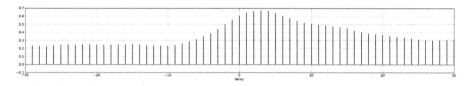

Fig. 7. Cross-Correlation between Trading Volume and Queries Volume about Bitcoin, with a maximum lag of 30 days.

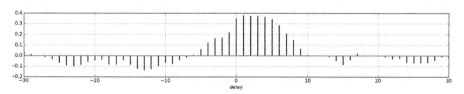

Fig. 8. Cross-Correlation between Trading Volume and PT-NT ratio of social volume about Bitcoin, with a maximum lag of 30 days.

Figures 7 and 8 show the cross correlation results with a maximum lag of 30 days, just to highlight that the media volume (social or search) anticipates the trading volume and that the best result is given by a lag of almost 3 in the Fig. 7 and by a lag of one in the other picture.

3.4   Granger Causality

We performed a Granger causality test in order to verify whether web search queries or the PT-NT ratio of social volume, regarding Bitcoin, are able to cause particular trends in some days. The Granger-causality test is used to determine whether a time series $G(t)$ is a good predictor of another time series $T(t)$ [36]. If G Granger-causes T, then G^{past} should significantly help predicting T^{future} via T^{past} alone.

We compared the two media volumes G, one after the other, with trading volume T with the null hypothesis being that T is not caused by G. An F-test is then used to determine if the null hypothesis can be rejected.

We computed two auto-regression vectors as follows in the Eqs. 5 and 6, where L represents the maximum time lag.

$$T(t) = \sum_{l=1}^{L} a_l T(t-l) + \epsilon_1 \tag{5}$$

$$T(t) = \sum_{l=1}^{L} a_l' T(t-l) + \sum_{l=1}^{L} b_l' G(t-l) + \epsilon_2 \tag{6}$$

We can affirm that G causes T if Eq. 6 is statistically better significant than Eq. 5. We applied the test in both directions, as an instance G → T means that the null hypothesis is "G doesn't Granger-cause T".

Tables 2 and 3 show the results of the Granger causality test, applied to the different media, social and web search. The first column represents the direction of the applied test, the second one the delay, and then the F-test result with its p-value. This parameter represents the probability that statistic test would be at least as extreme as observed, if the null hypothesis were true. So, we reject the null hypothesis if p value is inferior to a certain threshold ($p < 0.05$).

Table 2 demonstrates that trading volumes can be considered Granger-caused by the query volumes. It is clearly shown that time-series G influences T, given by the p value <0.001 for lags ranging from 1 to 5. So, the null hypothesis is strongly rejected. On the other hand, the F-value test applied to the direction T→G reported a p-value always greater than 0.1. Trading volume T doesn't have significant casual relations with changes in queries volumes on Google search engine G. So, null hypothesis cannot be rejected.

Table 3 shows that the Granger causality tests report a p value always higher than 0.25, meaning that the null hypothesis cannot be rejected, so the PT-NT ratio of social volume cannot cause the trading volume. Also in this case, the main problem could be the short temporal period that we took in account, not

Table 2. *Granger-causality tests between trading volume T and web search volume G.*

Direction of causality	Delay	F-value test	P-value
G→T	1	41.8135	$p < 0.001$
	2	15.1435	$p < 0.001$
	3	12.9332	$p < 0.001$
	4	15.1546	$p < 0.001$
	5	12.9279	$p < 0.001$
T→G	1	0.5450	$p = 0.46$
	2	2.3006	$p = 0.10$
	3	1.4878	$p = 0.21$
	4	1.5336	$p = 0.19$
	5	1.2297	$p = 0.29$

Table 3. *Granger-causality tests between PT-NT ratio of social volume G and trading volume T.*

Direction of causality	Delay	F-value test	P-value
G→T	1	2.5175	p = 0.1173
	2	0.1210	p = 0.8863
	3	0.2512	p = 0.8602
	4	1.3807	p = 0.2519
	5	1.1938	p = 0.3243
T→G	1	1.3501	p = 0.24
	2	0.4841	p = 0.61
	3	0.6937	p = 0.55
	4	0.0899	p = 0.98
	5	0.7991	p = 0.55

enough to evaluate the predictive power of anticipation or causation. Nevertheless, we can confirm that search volume is the best predictor, able to cause changes in the trading volume.

4   Conclusions

In this work we studied the relationship that exists between trading volumes of Bitcoin currency and the volumes of search engine. Then we analyzed a corpus of tweets in order to discover if the chatter of the community can be used to make qualitative predictions about Bitcoin market, attempting to establish whether there is any correlation between tweets sentiment and the Bitcoins trading volume.

Initially, we presented an analysis of a collection of queries index about Bitcoin compared to its trading volume. We selected a corpus that lasts the period, between June 2014 and July 2015, using Google Trends media to analyze Bitcoins popularity under the perspective of web search. We examined the Bitcoin trading behaviour comparing its variations with Google Trends data.

We first considered the Pearson correlation analysis and we found that the trading volumes follow the same direction pace of queries volumes. It reveals an obvious correlation due to peaks in one time series that occur close to peaks in the other. Then we continued with cross-correlation and the outcome reports that query volumes is able to anticipate trading volumes in almost 3 days. Finally, we used Granger causality test, confirming that trading volumes can be considered Granger-caused by the query volumes. From results of these analysis, we can affirm that Google Trends is a good predictor, because of its high cross correlation value, where we individuated a value of 0.68.

Then, we analysed a collection of tweets about Bitcoin, considering a total amount of 2,353,109 tweets. The corpus lasts a period of 120 days between

January 2015 and April 2015. We applied automated Sentiment Analysis on these tweets in order to evaluate whether public sentiment could be used to predict Bitcoins market.

We applied the same kind of correlation analysis to the corpus of tweets comparing its variations with the trading volumes of Bitcoin. From results of cross correlation analysis between the time series, we found that the ratio between positive and negative tweets may contribute to predict the movement of Bitcoins trading volume in a few days, achieving 0.37 as outcome. The Granger-causality tests reported that the analyzed PT-NT ratio of social volume cannot cause changes in the trading volume. This can be probably due to the short temporal period that we took in account, not able enough to evaluate the predictive power of anticipation or causation. Anyway, we can confirm that Google Trends can be seen as the best predictor, because of its high cross correlation value with three days of lag.

As future improvements, we are working on the possibility to apply a similar approach to other contexts, in order to better understand the predictive power of web search and social media. We are also working to extend our analysis to evaluate the correlation of Bitcoin Market Volatility.

References

1. Kaplan, A.M., Haenlein, M.: Users of the world, unite! the challenges and opportunities of social media. Bus. Horiz. **53**, 59–68 (2010)
2. Hansen, D., Shneiderman, B., Smith, M.A.: Analyzing Social Media Networks with NodeXL: Insights from a Connected World. Morgan Kaufmann, Burlington (2010)
3. Choi, H., Varian, H.: Predicting the present with google trends. Econ. Rec. **88**, 2–9 (2012)
4. Rose, D.E., Levinson, D.: Understanding user goals in web search. In: Proceedings of the 13th International Conference on World Wide Web, pp. 13–19. ACM (2004)
5. Bordino, I., Battiston, S., Caldarelli, G., Cristelli, M., Ukkonen, A., Weber, I.: Web search queries can predict stock market volumes. PloS One **7**, e40014 (2012)
6. Ginsberg, J., Mohebbi, M.H., Patel, R.S., Brammer, L., Smolinski, M.S., Brilliant, L.: Detecting influenza epidemics using search engine query data. Nature **457**, 1012–1014 (2009)
7. Kristoufek, L.: Can Google trends search queries contribute to risk diversification? Scientific reports 3 (2013)
8. Curme, C., Preis, T., Stanley, H.E., Moat, H.S.: Quantifying the semantics of search behavior before stock market moves. Proc. Nat. Acad. Sci. **111**, 11600–11605 (2014)
9. Preis, T., Moat, H.S., Stanley, H.E.: Quantifying trading behavior in financial markets using Google trends. Scientific reports 3 (2013)
10. Dimpfl, T., Jank, S.: Can internet search queries help to predict stock market volatility? Eur. Finan. Manag. **22**, 171–192 (2015)
11. Bollen, J., Mao, H., Zeng, X.: Twitter mood predicts the stock market. J. Comput. Sci. **2**, 1–8 (2011)
12. Kristoufek, L.: Power-law correlations in finance-related google searches, and their cross-correlations with volatility and traded volume: evidence from the dow jones industrial components. Physica A: Stat. Mech. Appl. **428**, 194–205 (2015)

13. Bulut, L., et al.: Google trends and forecasting performance of exchange rate models. Technical report (2015)
14. Kim, Y.B., Lee, S.H., Kang, S.J., Choi, M.J., Lee, J., Kim, C.H.: Virtual world currency value fluctuation prediction system based on user sentiment analysis. PloS One **10**, e0132944 (2015)
15. Kristoufek, L.: Bitcoin meets Google trends and Wikipedia: quantifying the relationship between phenomena of the internet era. Scientific reports 3 (2013)
16. Mondria, J., Wu, T., Zhang, Y.: The determinants of international investment and attention allocation: using internet search query data. J. Int. Econ. **82**, 85–95 (2010)
17. Preis, T., Reith, D., Stanley, H.E.: Complex dynamics of our economic life on different scales: insights from search engine query data. Philos. Trans. Roy. Soc. Lond. A: Math. Phys. Eng. Sci. **368**, 5707–5719 (2010)
18. Java, A., Song, X., Finin, T., Tseng, B.: Why we Twitter: understanding microblogging usage and communities. In: Proceedings of the 9th WebKDD and 1st SNA-KDD 2007 Workshop on Web Mining and Social Network Analysis, pp. 56–65. ACM (2007)
19. Lo, S.L., Cornforth, D., Chiong, R.: Identifying the high-value social audience from Twitter through text-mining methods. In: Handa, H., Ishibuchi, H., Ong, Y.-S., Tan, K.C. (eds.) Proceedings of the 18th Asia Pacific Symposium on Intelligent and Evolutionary Systems, Volume 1. PALO, vol. 1, pp. 325–339. Springer, Heidelberg (2015). doi:10.1007/978-3-319-13359-1_26
20. Ciulla, F., Mocanu, D., Baronchelli, A., Gonçalves, B., Perra, N., Vespignani, A.: Beating the news using social media: the case study of American idol. EPJ Data Sci. **1**, 1–11 (2012)
21. Mocanu, D., Baronchelli, A., Perra, N., Gonçalves, B., Zhang, Q., Vespignani, A.: The Twitter of babel: mapping world languages through microblogging platforms. PloS One **8**, e61981 (2013)
22. Hicks, K., Gerling, K., Kirman, B., Linehan, C., Dickinson, P.: Exploring Twitter as a game platform; strategies and opportunities for microblogging-based games. In: Proceedings of the 2015 Annual Symposium on Computer-Human Interaction in Play, pp. 151–161. ACM (2015)
23. Kaminski, J., Gloor, P.: Nowcasting the bitcoin market with twitter signals. arXiv preprint arXiv:1406.7577 (2014)
24. Rao, T., Srivastava, S.: Analyzing stock market movements using Twitter sentiment analysis. In: Proceedings of the 2012 International Conference on Advances in Social Networks Analysis and Mining (ASONAM 2012), pp. 119–123. IEEE Computer Society (2012)
25. Nakamoto, S.: Bitcoin: a peer-to-peer electronic cash system. Consulted **1**, 28 (2008)
26. Ron, D., Shamir, A.: Quantitative analysis of the full bitcoin transaction graph. In: Sadeghi, A.-R. (ed.) FC 2013. LNCS, vol. 7859, pp. 6–24. Springer, Heidelberg (2013). doi:10.1007/978-3-642-39884-1_2
27. Grinberg, R.: Bitcoin: an innovative alternative digital currency. Hastings Sci. Tech. LJ **4**, 159 (2012)
28. Mai, F., Bai, Q., Shan, Z., Wang, X.S., Chiang, R.H.: From bitcoin to big coin: the impacts of social media on bitcoin performance (2015)
29. Matta, M., Lunesu, I., Marchesi, M.: Bitcoin spread prediction using social and web search media. In: Proceedings of DeCAT (2015)

30. Garcia, D., Tessone, C.J., Mavrodiev, P., Perony, N.: The digital traces of bubbles: feedback cycles between socio-economic signals in the bitcoin economy. J. Roy. Soc. Interface **11**, 20140623 (2014)
31. Constantinides, E., Romero, C.L., Boria, M.A.G.: Social media: a new frontier for retailers? In: Swoboda, B., Morschett, D., Rudolph, T., Schnedlitz, P., Schramm-Klein, H. (eds.) European Retail Research. European Retail Research, pp. 1–28. Springer, Heidelberg (2009)
32. Mittal, A., Goel, A.: Stock prediction using Twitter sentiment analysis. Standford University, CS229 (2012)
33. Pang, B., Lee, L.: Opinion mining and sentiment analysis. Found. Trends Inf. Retrieval **2**, 1–135 (2008)
34. Thelwall, M., Buckley, K., Paltoglou, G.: Sentiment in Twitter events. J. Am. Soc. Inf. Sci. Technol. **62**, 406–418 (2011)
35. Thelwall, M., Buckley, K., Paltoglou, G., Cai, D., Kappas, A.: Sentiment strength detection in short informal text. J. Am. Soc. Inf. Sci. Technol. **61**, 2544–2558 (2010)
36. Granger, C.W.: Investigating causal relations by econometric models and cross-spectral methods. Econom.: J. Econom. Soc. **37**, 424–438 (1969)

Knowledge Engineering and Ontology Development

A Visual Similarity Metric
for Ontology Alignment

Charalampos Doulaverakis[✉], Stefanos Vrochidis, and Ioannis Kompatsiaris

Information Technologies Institute, Centre for Research and Technology Hellas,
Thessaloniki, Greece
{doulaver,stefanos,ikom}@iti.gr

Abstract. Ontology alignment is the process where two different ontologies that usually describe similar domains are 'aligned', i.e. a set of correspondences between their entities, regarding semantic equivalence, is determined. In order to identify these correspondences several methods have been proposed in literature. The most common features that these methods employ are string-, lexical-, structure- and semantic-based features for which several approaches have been developed. However, what hasn't been investigated is the usage of visual-based features for determining entity similarity. Nowadays the existence of several resources that map lexical concepts onto images allows for exploiting visual features for this purpose. In this paper, a novel method, defining a visual-based similarity metric for ontology matching, is presented. Each ontological entity is associated with sets of images. State of the art visual feature extraction, clustering and indexing for computing the visual-based similarity between entities is employed. An adaptation of a Wordnet-based matching algorithm to exploit the visual similarity is also proposed. The proposed visual similarity approach is compared with standard metrics and demonstrates promising results.

1 Introduction

Semantic Web is providing shared ontologies and vocabularies in different domains that can be openly accessed and used for tasks such as semantic annotation of information, reasoning, querying, etc. The Linked Open Data (LOD) paradigm shows how the different exposed datasets can be linked in order to provide a deeper understanding of information. As each ontology is being engineered to describe a particular domain for usage in specific tasks, it is common for ontologies to express equivalent domains using different terms or structures. These equivalences have to be identified and taken into account in order to enable seamless knowledge integration. Moreover, as an ontology can contain hundreds or thousands of entities, there is a need to automate this process. An example of the above comes from the cultural heritage domain where two ontologies are being used as standards, one is the CIDOC-CRM[1], used for semantically annotating museum content, and the other is the Europeana Data Model[2], which

---

[1] CIDOC-CRM, http://www.cidoc-crm.org.

[2] Europeana Data Model, http://labs.europeana.eu.

© Springer International Publishing AG 2016
A. Fred et al. (Eds.): IC3K 2015, CCIS 631, pp. 175–190, 2016.
DOI: 10.1007/978-3-319-52758-1_11

is used to semantically index and interconnect cultural heritage objects. While these two ontologies have been developed for different purposes, they are used in the cultural heritage domain and correspondences between their entities should exist and be identified.

In ontology alignment the goal is to automatically or semi-automatically discover correspondences between the ontological entities, i.e. their classes, properties or instances. An 'alignment' is a set of mappings that define the similar entities between two ontologies. These mappings can be expressed e.g. using the *owl:equivalentClass* or *owl:equivalentProperty* properties so that a reasoner can automatically access both ontologies during a query. More formally, if $O = (S, A)$ is the definition of an ontology with S being the ontology signature consisting of the lexical terms of concepts and relations, and A being the axioms that restrict the meaning of these terms, then for two ontologies $O_1 = (S_1, A_1)$ and $O_2 = (S_2, A_2)$ a mapping is defined as a morphism $f : S_1 \rightarrow S_2$ such that $A_2 \models f(A_1)$, i.e. all interpretations that satisfy O_1's axioms also satisfy the translated O_2's axioms [13].

While the proposed methodologies in literature have proven quite effective, either alone or combined, in dealing with the alignment of ontologies, there has been little progress in defining new similarity metrics that take advantage of features that haven't been considered so far. In addition existing benchmarks for evaluating the performance of ontology alignments systems, such as the Ontology Alignment Evaluation Initiative[3] (OAEI) have shown that there is still room for improvement in ontology alignment.

In the last 5 years the proliferation of multimedia has generated several annotated resources and datasets that are associated with concepts, such as ImageNet[4] or Flickr[5] thus making their visual representations easily available and retrievable so that they can be further exploited, e.g. for image recognition.

In this paper we propose a novel ontology matching metric that is based on visual similarities between ontological entities. The visual representations of the entities are crafted by different multimedia sources, namely ImageNet and web-based image search, thus assigning each entity to descriptive sets of images. State of the art visual features are extracted from these images and vector representations are generated. The entities are compared in terms of these representations and a similarity value is extracted for each pair of entities, thus the pair with the highest similarity value is considered as valid. The approach is validated in experimental results where it is shown that when it's combined with other known ontology alignment metrics it increases precision and recall of the discovered mappings.

The main contribution of the paper is the introduction of a novel similarity metric for ontology alignment based on visual features. To the best of the authors knowledge this is the first attempt to exploit visual features for ontology

---

[3] OAEI, http://oaei.ontologymatching.org.

[4] ImageNet, http://www.image-net.org/.

[5] Flickr, https://www.flickr.com/.

alignment purposes. We also propose an adaptation of a popular lexical-based matching algorithm where lexical similarity is replaced with visual similarity.

The paper is organized as follows: Sect. 3 describes the methodology in detail, while Sect. 5 presents the experimental results on the popular OAEI conference track dataset. In Sect. 4 an metric that exploits the proposed visual similarity and lexical features is proposed and described. Related work in ontology alignment is documented in Sect. 2. Finally, Sect. 6 concludes the paper and a future work plan is outlined.

2   Related Work

In order to accomplish the automatic discovery of mappings, numerous approaches have been proposed in literature that rely on various features. Ontology matching goes as far back as 2000 where in [18] the Chimaera ontology mapping tool is described. [13] presented a comprehensive ontology alignment methods, approaches and tools. Of the most common are methods that compare the similarity of two strings, e.g. comparing *hasAuthor* with *isAuthoredBy*, are the most used and fastest to compute as they operate on raw strings. Existing string similarity metrics are being used, such as Levenshtein distance, Edit distance, Jaro-Winkler similarity, etc., while string similarity algorithms such as [28] have been developed especially for ontology matching. Other mapping discovery methods rely on lexical processing in order to find synonyms, hypernyms or hyponyms between concepts, e.g. *Author* and *Writer*, where Wordnet is most commonly used. In [16] a survey on methods that use Wordnet [20] for ontology alignment, is carried out. Approaches for exploiting other external knowledge sources have been presented [2,7,24,25]. Other similarity measures rely on the structure of the ontologies by treating ontologies as schemas since schemas can be thought of as ontologies with a reduced relationship type set. Such an approach is the Similarity Flooding [19] algorithm that stems from the relational databases world but has been successfully used for ontology alignment, while others exploit both schema and ontology semantics for mapping discovery. Other works in this field can be found in [17,21]. A comprehensive study of such methods can be found at [26]. Machine learning has also been employed for ontology matching where features as the ones mentioned above are combined in order to train a machine algorithm to decide which pairs of ontology entities will be considered as semantically similar. One can refer to works such as [10] where Support Vector Machines (SVM) are used, [22] where an adaptation of the AdaBoost algorithm which uses Decision Trees as base classifiers is proposed or [4] where a multi-level learning strategy is presented.

In terms of matching systems, there have been proposed numerous approaches that combine matchers or include external resources of the generation of a valid mapping between ontologies. Most available systems have been evaluated in the OAEI benchmarks that are held annually. In [11] the authors use a weighted approach to combine several matchers in order to produce a final matching score between the ontological entities. In [23] the authors go a step further and propose a novel approach to combine elementary matching algorithms using a machine learning approach with decision trees. The system is trained from prior ground truth alignments in order to find the best combination of matchers for each pair of entities. Other systems, such as AML [8,14], make use of external knowledge resources or lexicons to obtain ground truth structure and entity relations. This is especially used when matching ontologies in specialized domains such as in biomedicine. Finally, semantic-based matching is exploited in S-Match [9].

In contrast to the above we have proposed a novel ontology matching framework that corresponds entities with images and makes use of visual features in order to compute similarity between entities. To the authors knowledge, this is the first approach in literature where a visual-based ontology matching algorithm is proposed. Throughout the paper, the term "entity" is used to refer to ontology entities, i.e. classes, object properties, datatype properties, etc.

3   Visual Similarity for Ontology Alignment

The idea for the development of a visual similarity algorithm for ontology alignment originated from the structure of ImageNet where images are assigned to concepts. For example, Fig. 1 shows a subset of images that is found in ImageNet for the words *boat*, *ship* and *motorbike*. Obviously, *boat* and *ship* are more semantically related than *boat* and *motorcycle*. It is also clear from Fig. 1 that the images that correspond to *boat* and *ship* are much more similar in terms of visual appearance than the images of *motorbike*. One can then assume that it is possible to estimate the semantic relatedness of two concepts by comparing their visual representations.

In Fig. 2 the proposed architecture for visual-based ontology alignment is presented. The source and target ontologies are the ontologies to be matched. For every entity in the ontologies, sets of images are assigned through ImageNet by identifying the relevant Wordnet synsets. A synset is a set of words that have the same meaning and these are used to query ImageNet. A single entity might correspond to a number of synsets, e.g. "track" has different meaning in transport and in sports as can be seen in Fig. 3. Thus for each entity a number of image sets are retrieved. For each image in a set, low level visual features are extracted and a numerical vector representation is formed. Therefore for each concept different sets of vectors are generated. Each set of vectors is called a "visual signature". All visual signatures between the source and target ontology are compared in pairs using a modified Jaccard set similarity in order to come up with a list of similarity values assigned to each entity pair. The final list of mappings is generated by employing an assignment optimization algorithm such as the Hungarian method [15].

(a) Images for "boat"   (b) Images for "ship"

(c) Images for "motorbike"

Fig. 1. Images for different synsets. (a) and (b) are semantically more similar than with (c). The visual similarity between (a) and (b) and their difference with (c) is apparent.

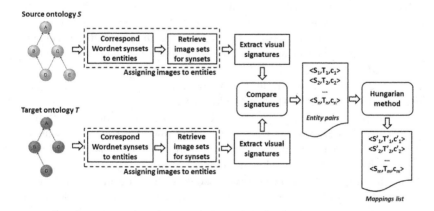

Fig. 2. Architecture of the proposed ontology alignment algorithm.

(a) Images for "track (running)"   (b) Images for "track (train)"

Fig. 3. Images that correspond to different meanings of concept "track". Since we can't be certain of a word meaning (word sense), each concept is associated with all relevant synsets and corresponding image sets from ImageNet.

3.1 Assigning Images to Entities

The main source of images in the proposed work is ImageNet, an image database organized according to the WordNet noun hierarchy in which each node of the hierarchy is associated with a number of images. Users can search the database through a text-search web interface where the user inputs the query words, which are then mapped to Wordnet indexed words and a list of relevant synsets (synonym sets, see [20]) are presented. The user selects the desired synset and the corresponding images are displayed. In addition, ImageNet provides a REST API for retrieving the image list that corresponds to a synset by entering the Wordnet synset id as input and this is the access method we used.

For every entity of the two ontologies to be matched, the following process was followed: A preprocessing procedure is executed where each entity name is first tokenized in order to split it to meaningful words as it is common for names to be in the form of *isAuthorOf* or *is_author_of* thus after tokenization, *isAuthorOf* will be split to the words *is, Author* and *of*. The next step is to filter out stop words, words that do not contain important significance or are very common. In the previous example, the words *is* and *of* are removed, thus after this preprocessing the name that is produced is *Author*.

After the preprocessing step, the next procedure is about identifying the relevant Worndet synset(s) of the entity name and get their ids, which is a rather straightforward procedure. Using these ids, ImageNet is queried in order to retrieve a fixed number of relevant images. However trying to retrieve these images might fail, mainly due to two reasons: either the name does not correspond to a Wordnet synset, e.g. due to misspellings, or the relevant ImageNet synset isn't assigned any images, something which is not uncommon since ImageNet is still under development and is not complete. So, in order not to end up with empty image collections, in the above cases the entity name is used to query Yahoo^TM image search[6] in order to find relevant images. The idea of using web-based search results has been employed in computer vision as in [1] where web image search is used to train an image classifier.

The result of the above-described process is to have each ontological entity C associated with n sets of images I_{iC}, with $i = 1,\ldots,n$, where n is the number of synsets that correspond to entity C.

3.2   Extracting the Visual Signatures of Entities

For allowing a visual-based comparison of the ontological entities, each image set I_{iC} has to be represented using appropriate visual descriptors. For this purpose, a state of the art approach is followed where images are represented as compact numerical vectors. For extracting these vectors the approach which is described in [27] is used as it has been shown to outperform other approaches on standard benchmarks of image retrieval and is quite efficient. In short, SURF (Speeded Up Robust Features) descriptors [?] are extracted for each image in a set. SURF descriptors are numerical representations of important image features and are used to compactly describe image content. These are then represented using the VLAD (Vector of Locally Aggregated Descriptors) representation [12] where four codebooks of size 128 each, were used. The resulting VLAD vectors are PCA-projected to reduce their dimensionality to 100 coefficients, thus ending up with a standard numerical vector representation v_j for each image j in a set. At the end of this process, each image set I_{iC} will be numerically represented by a corresponding vector set. This vector set is termed "visual signature" V_{iC} as it conveniently and descriptively represents the visual content of I_{iC}, thus $V_{iC} = \{v_j\}$, with $j = 1,\ldots,k$ and k being the total number of images in I_{iC}.

The whole processing workflow is depicted in Fig. 4.

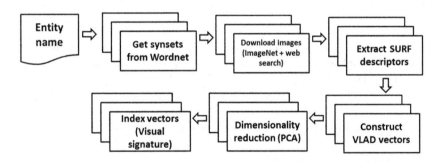

Fig. 4. Block diagram of the process for extracting the visual signatures of an entity.

Algorithm 1 outlines the steps to create visual signatures V_C of entities in an ontology.

---

[6] Yahoo search, https://images.search.yahoo.com.

Algorithm 1. Pseudocode for extracting visual signature V_C of an entity C in ontology O.

Ensure: $V_C = \emptyset$, C is an entity of ontology O
   $C_t \leftarrow removeStopWords(tokenize(C))$
   $W \leftarrow$ find Wordnet synsets of C_t
   for all synsets W_i in W **do**
      $I_{iC} \leftarrow$ download k images from ImageNet
      if $I_{iC} = \emptyset$ **then**
         download k images from web
      end if
      $V_{iC} \leftarrow \emptyset$
      for all images j in I_{iC} **do**
         $v_j \leftarrow extractVisualDescriptors(j)$
         $V_{iC} \leftarrow$ add v_j
      end for
      $V_C \leftarrow$ add V_{iC}
   end for
   return V_C

3.3   Comparing Visual Signatures for Computing Entity Similarity

Having the visual signatures for each entity, the next step is to use an appropriate metric in order to compare these signatures and estimate the similarity between image sets. Several vector similarity and distance metrics exist, such as cosine similarity or euclidean distance, however these are mostly suitable when comparing individual vectors. In the current work, we are interested in establishing the similarity value between vector sets so the Jaccard set similarity measure is more appropriate as it is has been defined exactly for this purpose. It's definition is

$$J_{V_{iCs},V_{jCt}} = \frac{|V_{iCs} \cap V_{jCt}|}{|V_{iCs} \cup V_{jCt}|} \tag{1}$$

where V_{iCs} and V_{jCt} are the i and j different visual signatures of entities C_s and C_t, $|V_{iCs} \cap V_{jCt}|$ is the intersection size of the two sets, i.e. the number of identical images between the sets, and $|V_{iCs} \cap V_{jCt}|$ is the total number of images in both sets. It holds that $0 \leq J_{V_{iCs},V_{jCs}} \leq 1$. For defining if two images A and B are identical, we compute the angular similarity of their vector representations.

$$AngSim_{A,B} = 1 - \frac{\arccos\left(cosineSim(A,B)\right)}{\pi} \tag{2}$$

with $cosineSim(A,B)$ equal to

$$cosineSim(A,B) = \frac{\sum_{k=1}^{n=100} A_k \cdot B_k}{\sqrt{\sum_{k=1}^{n=100} A_k^2} \cdot \sqrt{\sum_{k=1}^{n=100} B_k^2}} \tag{3}$$

For *AngSim*, a value of 0 means that the two images are completely irrelevant and 1 means that they are identical. However, two images might not have $AngSim_{A,B} = 1$ even if they are visually the same but they are acquired from different sources due to e.g. differences in resolution, compression or stored format, thus we risk of having $|V_{iCs} \cap V_{jCt}| = \emptyset$. For this reason instead of aiming to find truly identical images we introduce the concept of "near-identical images" where two images are considered identical if the have a similarity value above a threshold T, thus

$$Identical_{A,B} = \begin{cases} 0 & \text{if } AngSim_{A,B} < T \\ 1 & \text{if } AngSim_{A,B} \geq T \end{cases} \tag{4}$$

T is experimentally defined. Using the above we are able to establish the Jaccard set similarity value of two ontological entities by corresponding each entity to an image set, extracting the visual signature of each set and comparing these signatures. The Jaccard set similarity value J_{V_i,U_j} is computed for every pair i, j of synsets that correspond to the examined entities, V, U. Visual Similarity is defined as

$$VisualSim(C_s, C_t) = \max_{i,j}(J_{V_{iCs}, V_{jCt}}) \tag{5}$$

4   Combining Visual and Lexical Features

The Visual Similarity algorithm can either be exploited as a standalone measure or it can be used as complementary to other ontology matching measures as well. Since in order to construct the visual representation of entities Wordnet is used, one approach is to combine visual with lexical-based features. Lexical-based measures have been used in ontology matching systems in recent OAEI benchmarks, such as in [23] where, among others, the Wu-Palmer [29] Wordnet-based measure has been integrated. The Wu-Palmer similarity value between concepts C_1 and C_2 is defined as

$$WuPalmer_{C_1,C_2} = \frac{2 \cdot N_3}{N_1 + N_2 + 2 \cdot N_3} \tag{6}$$

where C_3 is defined as the least common superconcept (or hypernym) of both C_1 and C_2, N_1 and N_2 are the number of nodes from C_1 and C_2 to C_3, respectively, and N_3 is the number of nodes on the path from C_3 to root. The intuition behind this metric is that since concepts closer to the root have a broader meaning which is made more specific as one moves to the leaves of the hierarchy, if two concepts have a common hypernym closer to them and further from the root, then it's likely that they have a closer semantic relation.

Based on this intuition we have defined a new similarity metric that takes into account the visual features of both concepts and of their least common superconcept. Using the same notation and meaning for C_1, C_2, C_3, the measure we have defined is expressed as

$$LexiVis_{C_1,C_2} = \frac{V_3}{3 - (V_1 + V_2)} \tag{7}$$

Boat	Hydroplane	Craft

Boat-Hydroplane=0.49, Boat-Craft=0.24,
Hydroplane-Craft=0.35

LexiVis (Boat-Hydroplane) = 0.20

Fig. 5. Visual similarity values between the concepts "Boat" and "Hydroplane" which are semantically irrelevant but visually similar. Their common hypernym is "Craft". The LexiVis measure, by taking advantage of lexical features, lowers their similarity value.

where V_3 is the visual similarity value between C_1 and C_2 and V_1,V_2 are the visual similarity values between C_1,C_3 and C_2,C_3 respectively. V_1,V_2 and V_3 are calculated according to Eq. 5. In all cases, $0 \leq LexiVis_{C_1,C_2} \leq 1$. The intuition behind this measure is that semantically related concepts will be each other highly visually similar to each other and also highly similar visually with their closest hypernym. The incorporation of the closest hypernym in the overall similarity estimation of two concepts will allow for corrections in cases where concepts might be visually similar but semantically irrelevant, e.g. "boat" and "hydroplane" pictures depict an object surrounded by a body of water, however when they are visually compared against their common superconcept, in the previous example it is the concept "craft", their pair-wise visual similarity value will be low thus lowering the concepts' similarity. This example is depicted in Fig. 5.

5   Experimental Results

For analyzing the performance of the Visual Similarity ontology matching algorithm we ran it against the Ontology Alignment Evaluation Initiative (OAEI) Conference track of 2014 [5][7]. The OAEI benchmarks are organized annually and have become a standard in ontology alignment tools evaluation. In the conference track, a number of ontologies that are used for the organization of conferences have to be aligned in pairs. The conference track was chosen as, by design, the proposed algorithm requires meaningful entity names that can be visually represented. Other tracks, such as benchmark and anatomy, weren't considered due to this limitation which is further discussed in Sect. 6. Reference alignments are available and these are used for the actual evaluation in an automated manner.

---

[7] OAEI 2014, http://oaei.ontologymatching.org/2014.

The reference alignment that was used is "ra1" since this was readily available for the OAEI 2014 website.

The *VisualSim* and *LexiVis* ontology matching algorithms were integrated in the Alignment API [6] which offers interfaces and sample implementations in order to integrate matching algorithms. The API is recommended from OAEI for participating in the benchmarks. In addition, algorithms to compute standard information retrieval measures, i.e. precision, recall and F-measure, against reference alignments can be found in the API, so these were used for the evaluation of the tests results. In these tests we changed the threshold, i.e. the value under which an entity matching is discarded, and registered the precision, recall and $F1$ measure values.

In order to have a better understanding of the proposed algorithms we compared it against other popular matching algorithms. Ideally the performance of these would be evaluated against other matching algorithms that make use of similar modalities, i.e. visual or other. This wasn't feasible as the proposed algorithms are the first that makes use of visual features, so we compare it with standard algorithms that exploit traditional features such as string-based and Wordnet-based similarity. For this purpose we implemented the ISub string similarity matcher [28] and the Wu-Palmer Wordnet-based matcher which is described in Sect. 4. These matchers have been used in the YAM++ ontology matching system [23] which was one of the top ranked systems in OAEI 2012.

All aforementioned algorithms, ISub, Wu-Palmer, VisualSim and LexiVis, are evaluated using Precision, Recall and $F1$ measure, with

$$F1 = \frac{2 \cdot Precision \cdot Recall}{Precision + Recall} \tag{8}$$

The results of this evaluation are displayed in Fig. 6.

It can be seen from Fig. 6 that *VisualSim* and the *LexiVis* algorithms performs better in all measures than the Wu-Palmer alignment algorithm which confirms with our initial assumption that the semantic similarity between entities can be reflected in their visual representation using imaging modalities. This allows a new range of matching techniques based on modalities that haven't been considered so far to be investigated. However, the string-based ISub matcher displays superior performance, which was expected as string-based matchers are very effective in ontology alignment and matching problems, which points out that the aforementioned new range of matchers should work complementary to the existing and established matchers as these have proven their reliability though time.

An additional performance factor that should be mentioned is the computational complexity and overall execution time for the Visual based algorithm which is much greater than the simpler string-based algorithms. Analyzing Fig. 4, of all the documented steps by far the most time consuming are the image download and visual descriptor extraction. However, ImageNet is already offering visual descriptors which are extracted from the synset images and are freely

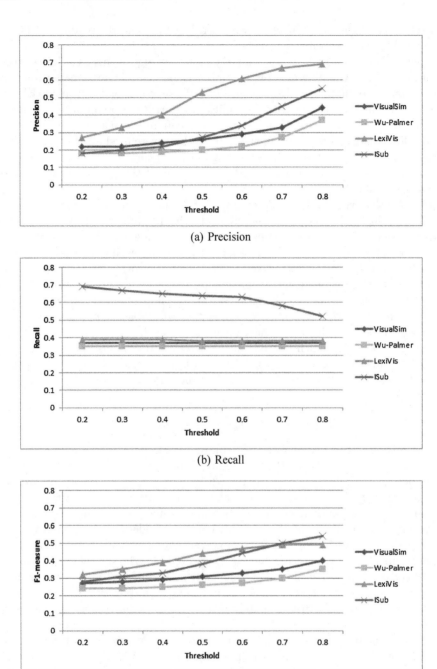

(a) Precision

(b) Recall

(c) F1 measure

Fig. 6. Precision, Recall and F1 diagrams for different threshold values using the conference track ontologies of OAEI 2014.

available to download[8]. The range of images that have been processed is not yet complete but as ImageNet is still in development, the plan is to have the whole image database processed and have the visual descriptors extracted. This availability will make the calculation of the proposed visual-based ontology alignment algorithms faster.

5.1   In Combination with Other Ontology Alignment Algorithms

As a further test, using the Alignment API we integrated the LexiVis matching algorithm and aggregated the matching results with other available matching algorithms in order to have an understanding on how it would perform in a real ontology matching system. We used the LexiVis algorithm as it was shown to perform better than the original Visual Similarity algorithm (Fig. 6). The other algorithms that were used are the ISub and Similarity Flooding matchers in addition to the baseline NameEq matcher. These were used in order to have a combination of matchers that exploit different features, i.e. string, structural and visual. The matchers were combined using an adaptive weighting approach similar to [3]. For this test we again used the conference track benchmark dataset of OAEI 2014. For this dataset, results regarding the performance of the participating matching systems are published in OAEI's website and in [5]. It can be seen from Table 1, in the line denoted with italic font, that the inclusion of the LexiVis ontology matching algorithm in the matching system results in better overall performance than running the system without it. The added value of 0.01 in $F1$ results in an overall $F1$ value of 0.60 which brings our matching system

Table 1. Performance of the LexiVis matching algorithm in combination with other matching algorithms (ISub, Name Equality, Similarity Flooding [19]), and how the performance is compared to matching systems that participated in OAEI 2014 conference track.

System	Precision	Recall	F1-measure
AML	0.85	0.64	0.73
LogMap	0.80	0.59	0.68
LogMap-C	0.82	0.57	0.67
XMap	0.87	0.49	0.63
NameEq + ISub + SimFlood + LexiVis	*0.71*	*0.53*	*0.60*
NameEq + ISub + SimFlood	0.81	0.47	0.59
OMReasoner	0.82	0.46	0.59
Baseline (NameEq)	0.80	0.43	0.56
AOTL	0.77	0.43	0.55
MaasMtch	0.64	0.48	0.55

---

[8] ImageNet visual features download,
http://image-net.org/download-features.

in the top 5 performances. The rather small added value of 0.01 is mainly due to the fact that the benchmark is quite challenging as can be seen from the results of Table 1. For example the XMap system, which is ranked 4th, managed to score 0.07 more in $F1$ than the baseline NameEq matcher which simply compares strings and produces a valid pair if the names are equal. Even this small increase of $F1$ just by including the LexiVis algorithm proves that it can improve results in such a challenging benchmark thus showing its benefit.

6   Conclusions

In this paper a novel ontology matching algorithm which is based on visual features is presented. The algorithm exploits ImageNet's structure which is based on Wordnet in order to correspond image sets to the ontological entities and state of the art visual processing is employed which involves visual feature descriptors extraction, codebook-based feature representation, dimensionality reduction and indexing. The visual-based similarity value is taken by calculating a modified version of the Jaccard set similarity value. A new matcher is also proposed which combines visual and lexical features in order to determine entity similarity. The proposed algorithms have been evaluated using the established OAEI benchmark and has shown to outperform Wordnet-based approaches. In addition it was shown that the combination of the LexiVis algorithm with other standard ontology matchers and metrics in a matching system has led to increase mapping accuracy compared to the same matching system but without including LexiVis. This proves the increased efficiency of our approach.

A limitation of the proposed visual-based matching algorithm is that since it relies of visual depictions of entities, in cases where entity names are not words, e.g. alphanumeric codes, then its performance will be poor as images will be able to be associated with it. A way to tackle this is to extend the approach to include other data, such as *rdfs:label*, which are more descriptive. Another limitations of this approach would be the mapping of concepts that are visually hard to express, e.g. "Idea" or "Freedom", however this is partly leveraged by employing web-based search which likely retrieves relevant images for almost any concept.

Future work will be directed to the optimization of the processing pipeline in order to have visual similarity results in a more timely manner and other approaches such as word sense disambiguation in order to reduce the image sets that correspond to each entity. Finally, as it has already been mentioned, since the current version of the algorithm only uses entity names, we will also consider using additional features such as entity labels and comments.

Acknowledgements. This work was supported by MULTISENSOR (contract no. FP7-610411) and KRISTINA (contract no. H2020-645012) projects, partially funded by the European Commission.

References

1. Chatfield, K., Zisserman, A.: VISOR: towards on-the-fly large-scale object category retrieval. In: Lee, K.M., Matsushita, Y., Rehg, J.M., Hu, Z. (eds.) ACCV 2012. LNCS, vol. 7725, pp. 432–446. Springer, Heidelberg (2013). doi:10.1007/978-3-642-37444-9_34
2. Chen, X., Xia, W., Jiménez-Ruiz, E., Cross, V.: Extending an ontology alignment system with bioportal: a preliminary analysis. In: International Semantic Web Conference (ISWC) (2014)
3. Cruz, I.F., Antonelli, F.P., Stroe, C.: Efficient selection of mappings and automatic quality-driven combination of matching methods. In: ISWC International Workshop on Ontology Matching (OM) CEUR Workshop Proceedings, vol. 551, pp. 49–60. Citeseer (2009)
4. Doan, A., Madhavan, J., Domingos, P., Halevy, A.: Ontology matching: a machine learning approach. In: Staab, S., Studer, R. (eds.) Handbook on Ontologies, pp. 385–403. Springer, Heidelberg (2004)
5. Dragisic, Z., Eckert, K., Euzenat, J., Faria, D., Ferrara, A., Granada, R., Ivanova, V., Jimenez-Ruiz, E., Kempf, A., Lambrix, P., et al.: Results of the ontology alignment evaluation initiative 2014. In: International Workshop on Ontology Matching, pp. 61–104 (2014)
6. Euzenat, J.: An API for ontology alignment. In: McIlraith, S.A., Plexousakis, D., Harmelen, F. (eds.) ISWC 2004. LNCS, vol. 3298, pp. 698–712. Springer, Heidelberg (2004). doi:10.1007/978-3-540-30475-3_48
7. Faria, D., Pesquita, C., Santos, E., Cruz, I.F., Couto, F.M.: Automatic background knowledge selection for matching biomedical ontologies. PLoS ONE 9(11), e111226 (2014)
8. Faria, D., Pesquita, C., Santos, E., Palmonari, M., Cruz, I.F., Couto, F.M.: The agreementmakerlight ontology matching system. In: Meersman, R., Panetto, H., Dillon, T., Eder, J., Bellahsene, Z., Ritter, N., Leenheer, P., Dou, D. (eds.) OTM 2013. LNCS, vol. 8185, pp. 527–541. Springer, Heidelberg (2013). doi:10.1007/978-3-642-41030-7_38
9. Giunchiglia, F., Shvaiko, P., Yatskevich, M.: S-match: an algorithm and an implementation of semantic matching. In: Bussler, C.J., Davies, J., Fensel, D., Studer, R. (eds.) ESWS 2004. LNCS, vol. 3053, pp. 61–75. Springer, Heidelberg (2004). doi:10.1007/978-3-540-25956-5_5
10. Ichise, R.: Machine learning approach for ontology mapping using multiple concept similarity measures. In: Seventh IEEE/ACIS International Conference on Computer and Information Science, ICIS 2008, pp. 340–346. IEEE (2008)
11. Jean-Mary, Y.R., Shironoshita, E.P., Kabuka, M.R.: Ontology matching with semantic verification. Web Semant. Sci. Serv. Agents World Wide Web 7(3), 235–251 (2009)
12. Jégou, H., Douze, M., Schmid, C., Pérez, P.: Aggregating local descriptors into a compact image representation. In: 2010 IEEE Conference on Computer Vision and Pattern Recognition (CVPR), pp. 3304–3311. IEEE (2010)
13. Kalfoglou, Y., Schorlemmer, M.: Ontology mapping: the state of the art. knowl. Eng. Rev. 18(01), 1–31 (2003)
14. Kirsten, T., Gross, A., Hartung, M., Rahm, E.: GOMMA: a component-based infrastructure for managing and analyzing life science ontologies and their evolution. J. Biomed. Semant. 2(6), 1–24 (2011)

15. Kuhn, H.W.: The hungarian method for the assignment problem. Nav. Res. Logistics Q. **2**(1–2), 83–97 (1955)

16. Lin, F., Sandkuhl, K.: A survey of exploiting wordnet in ontology matching. In: Bramer, M. (ed.) IFIP AI 2008. ITIFIP, vol. 276, pp. 341–350. Springer, Heidelberg (2008). doi:10.1007/978-0-387-09695-7_33

17. Madhavan, J., Bernstein, P.A., Rahm, E.: Generic schema matching with cupid. VLDB **1**, 49–58 (2001)

18. McGuinness, D.L., Fikes, R., Rice, J., Wilder, S.: An environment for merging and testing large ontologies. In: KR, pp. 483–493 (2000)

19. Melnik, S., Garcia-Molina, H., Rahm, E.: Similarity flooding: a versatile graph matching algorithm and its application to schema matching. In: 2002 Proceedings of 18th International Conference on Data Engineering, pp. 117–128. IEEE (2002)

20. Miller, G.A.: Wordnet: a lexical database for english. Commun. ACM **38**(11), 39–41 (1995)

21. Milo, T., Zohar, S.: Using schema matching to simplify heterogeneous data translation. In: VLDB 1998, pp. 24–27. Citeseer (1998)

22. Nezhadi, A.H., Shadgar, B., Osareh, A.: Ontology alignment using machine learning techniques. Int. J. Comput. Sci. Inf. Technol. **3**(2), 139 (2011)

23. Ngo, D.H., Bellahsene, Z.: YAM++: a multi-strategy based approach for ontology matching task. In: Teije, A., Völker, J., Handschuh, S., Stuckenschmidt, H., d'Acquin, M., Nikolov, A., Aussenac-Gilles, N., Hernandez, N. (eds.) EKAW 2012. LNCS (LNAI), vol. 7603, pp. 421–425. Springer, Heidelberg (2012). doi:10.1007/978-3-642-33876-2_38

24. Pesquita, C., Faria, D., Santos, E., Neefs, J.-M., Couto, F.M.: Towards visualizing the alignment of large biomedical ontologies. In: Galhardas, H., Rahm, E. (eds.) DILS 2014. LNCS, vol. 8574, pp. 104–111. Springer, Heidelberg (2014). doi:10.1007/978-3-319-08590-6_10

25. Sabou, M., d'Aquin, M., Motta, E.: Using the semantic web as background knowledge for ontology mapping. In: OM 2006 Proceedings of the International Workshop on Ontology Matching (2006)

26. Shvaiko, P., Euzenat, J.: A survey of schema-based matching approaches. In: Spaccapietra, S. (ed.) Journal on Data Semantics IV. LNCS, vol. 3730, pp. 146–171. Springer, Heidelberg (2005). doi:10.1007/11603412_5

27. Spyromitros-Xioufis, E., Papadopoulos, S., Kompatsiaris, I., Tsoumakas, G., Vlahavas, I.: A comprehensive study over VLAD and product quantization in large-scale image retrieval. IEEE Trans. Multimed. **16**(6), 1713–1728 (2014)

28. Stoilos, G., Stamou, G., Kollias, S.: A string metric for ontology alignment. In: Gil, Y., Motta, E., Benjamins, V.R., Musen, M.A. (eds.) ISWC 2005. LNCS, vol. 3729, pp. 624–637. Springer, Heidelberg (2005). doi:10.1007/11574620_45

29. Wu, Z., Palmer, M.: Verbs semantics and lexical selection. In: Proceedings of the 32nd Annual Meeting on Association for Computational Linguistics, pp. 133–138. Association for Computational Linguistics (1994)

Model-Intersection Problems and Their Solution Schema Based on Equivalent Transformation

Kiyoshi Akama[1] and Ekawit Nantajeewarawat[2(✉)]

[1] Information Initiative Center, Hokkaido University, Hokkaido, Japan
akama@iic.hokudai.ac.jp
[2] Computer Science Program, Sirindhorn International Institute of Technology,
Thammasat University, Pathumthani, Thailand
ekawit@siit.tu.ac.th

Abstract. Model-intersection (MI) problems are a very large class of logical problems that includes many useful problem classes, such as proof problems on first-order logic and query-answering (QA) problems in pure Prolog and deductive databases. We propose a general schema for solving MI problems by equivalent transformation (ET), where problems are solved by repeated simplification. The correctness of this solution schema is shown. This general schema is specialized for formalizing solution schemas for QA problems and proof problems. The notion of a target mapping is introduced for generation of ET rules, allowing many possible computation procedures, for instance, computation procedures based on resolution and unfolding. This theory is useful for inventing solutions for many classes of logical problems.

Keywords: Model-intersection problem · Query-answering problem · Equivalent transformation · Problem solving

1 Introduction

This paper introduces a *model-intersection problem* (*MI problem*), which is a pair $\langle Cs, \varphi \rangle$, where Cs is a set of clauses and φ is a mapping, called an *exit mapping*, used for constructing the output answer from the intersection of all models of Cs. More formally, the answer to a MI problem $\langle Cs, \varphi \rangle$ is $\varphi(\bigcap Models(Cs))$, where $Models(Cs)$ is the set of all models of Cs. The set of all MI problems constitutes a very large class of problems and is of great importance.

MI problems includes as an important subclass *query-answering problems* (*QA problems*), each of which is a pair $\langle Cs, a \rangle$, where Cs is a set of clauses and a is a user-defined query atom. The answer to a QA problem $\langle Cs, a \rangle$ is defined as the set of all ground instances of a that are logical consequences of Cs. Characteristically, a QA problem is an "all-answers finding" problem, i.e., all ground instances of a given query atom satisfying the requirement above are to be found. Many logic programming languages, including Datalog, Prolog, and other extensions of Prolog, deal with specific subclasses of QA problems.

© Springer International Publishing AG 2016
A. Fred et al. (Eds.): IC3K 2015, CCIS 631, pp. 191–212, 2016.
DOI: 10.1007/978-3-319-52758-1_12

The class of *proof problems* is also a subclass of MI problems. In contrast to a QA problem, a proof problem, is a "yes/no" problem; it is concerned with checking whether or not one given logical formula is a logical consequence of another given logical formula. Formally, a *proof problem* is a pair $\langle E_1, E_2 \rangle$, where E_1 and E_2 are first-order formulas, and the answer to this problem is defined to be "yes" if E_2 is a logical consequence of E_1, and it is defined to be "no" otherwise.

Historically, among these three kinds of problems, proof problems were first solved, e.g., the tableaux method [8] and the resolution method [12] have been long well known. After that, QA problems on pure Prolog were solved based on the resolution principle. The theory of SLD-resolution was used for the correctness of Prolog computation. So it has been believed that Prolog computation is an inference process. This approach can be called a proof-centered approach. Many solutions proposed so far for some other classes of logical problems are also basically proof-centered.

In this paper, we prove that a QA problem is a MI problem (Theorem 3), and a proof problem is also a MI problem (Theorem 4). Together with the inclusion relation between proof problems and QA problems shown in [2], we have

$$\textsc{Proof} \subset \textsc{QA} \subset \textsc{MI},$$

where PROOF, QA, and MI denote the class of all proof problems, the class of all QA problems, and the class of all MI problems, respectively. The class of all MI problems is larger than the class of all QA problems and that of all proof problems. It is a more natural class to be solved by the method presented in this paper. A general solution method for MI problems can be applied to any arbitrary QA problem and any arbitrary proof problem. This can be called a MI-centered approach.

MI problems are axiomatically constructed on an abstract structure, called a *specialization system*. It consists of abstract atoms and abstract operations (extensions of variable-substitution operations) on atoms, called *specializations*. These abstract components can be any arbitrary mathematical objects as long as they satisfy given axioms. Abstract clauses can be built on abstract atoms. This is a sharp contrast to most of the conventional theories in logic programming, where concrete syntax is usually used. In Prolog, for example, usual first-order atoms and substitutions with concrete syntax are used, and there is no way to give a foundation for other forms of extended atoms and for various specialization operations other than the usual variable-substitution operation.

An axiomatic theory enables us to develop a very general theory. By instantiating a specialization system to a specific domain and by imposing certain restrictions on clauses, our theory can be applied to many subclasses of MI problems.

We propose a general schema for solving MI problems by equivalent transformation (ET), where problems are solved by repeated simplification. We introduce the concept of target mapping and propose three target mappings. Since transformation preserving a target mapping is ET, target mappings provide a strong foundation for inventing many ET rules for solving MI problems on clauses.

An ET-based solution consists of the following steps: (i) formalize an initial MI problem on some specialization system, (ii) prepare ET rules, (iii) construct an ET sequence, and (iv) compute an answer mapping for deriving a solution.

To begin with, Sect. 2 recalls the concept of specialization system and formalizes MI problems on a specialization system. QA problems and proof problems are embedded into the class of all MI problems. Section 3 defines the notions of a target mapping and an answer mapping, and introduces a schema for solving MI problems based on ET preserving answers/target mappings. The correctness of this solution schema is shown. Section 4 introduces three target mappings for clause sets. Among them, MM is a target mapping that associates with each clause set a collection of its specific models computed in a bottom-up manner. Section 5 presents solution schemas for QA problems and proof problems based on the solution schema for MI problems in Sect. 3. Section 6 shows examples of solutions of a QA problem and a proof problem based on the solution schemas in Sect. 5. Section 7 concludes the paper.

The notation that follows holds thereafter. Given a set A, $pow(A)$ denotes the power set of A and $partialMap(A)$ the set of all partial mappings on A (i.e., from A to A). For any partial mapping f from a set A to a set B, $dom(f)$ denotes the domain of f, i.e., $dom(f) = \{a \mid (a \in A) \ \& \ (f(a) \text{ is defined})\}$.

2 Clauses and Model-Intersection Problems

2.1 Specialization Systems

A substitution $\{X/f(a), Y/g(z)\}$ changes an atom $p(X, 5, Y)$ in the term domain into $p(f(a), 5, g(z))$. Generally, a substitution in first-order logic defines a total mapping on the set of all atoms in the term domain. Composition of such mappings is also realized by some substitution. There is a substitution that does not change any atom (i.e., the empty substitution). A ground atom in the term domain is a variable-free atom.

Likewise, in the string domain, substitutions for strings are used. A substitution $\{X/\text{"}aYbc\text{"}, Y/\text{"}xyz\text{"}\}$ changes an atom $p(\text{"}X5Y\text{"})$ into $p(\text{"}aYbc5xyz\text{"})$. Such a substitution for strings defines a total mapping on the set of all atoms that may include string variables. Composition of such mappings is also realized by some string substitution. There is a string substitution that does not change any atom (i.e., the empty substitution). A ground atom in the string domain is a variable-free atom.

A similar operation can be considered in the class-variable domain. Consider, for example, an atom $p(X \colon \textit{animal}, Y \colon \textit{dog}, Z \colon \textit{cat})$ in this domain, where $X \colon \textit{animal}$, $Y \colon \textit{dog}$, and $Z \colon \textit{cat}$ represent an animal object, a dog object, and a cat object, respectively. When we obtain additional information that X is a dog, we can restrict $X \colon \textit{animal}$ into $X \colon \textit{dog}$ and the atom $p(X \colon \textit{animal}, Y \colon \textit{dog}, Z \colon \textit{cat})$ into $p(X \colon \textit{dog}, Y \colon \textit{dog}, Z \colon \textit{cat})$. By contrast, with new information that Z is a dog, we cannot restrict $Z \colon \textit{cat}$ and the above atom since Z cannot be a dog and a cat at the same time. More generally, such a restriction operation may not be applicable to some atoms, i.e., it defines a partial mapping on the set of all atoms. Composition of such partial mappings is also a partial mapping, and we can determine some composition operation corresponding to it. An empty substitution that does not change any atom can be introduced. A ground atom in the class-variable domain is a variable-free atom.

In order to capture the common properties of such operations on atoms, the notion of a specialization system was introduced around 1990.

Definition 1. A *specialization system* Γ is a quadruple $\langle \mathcal{A}, \mathcal{G}, \mathcal{S}, \mu \rangle$ of three sets $\mathcal{A}$, $\mathcal{G}$, and $\mathcal{S}$, and a mapping μ from $\mathcal{S}$ to *partialMap*$(\mathcal{A})$ that satisfies the following conditions:

1. $(\forall s', s'' \in \mathcal{S})(\exists s \in \mathcal{S}) : \mu(s) = \mu(s') \circ \mu(s'')$.
2. $(\exists s \in \mathcal{S})(\forall a \in \mathcal{A}) : \mu(s)(a) = a$.
3. $\mathcal{G} \subseteq \mathcal{A}$.

Elements of $\mathcal{A}$, $\mathcal{G}$, and $\mathcal{S}$ are called *atoms*, *ground atoms*, and *specializations*, respectively. The mapping μ is called the *specialization operator* of Γ. A specialization $s \in \mathcal{S}$ is said to be *applicable* to $a \in \mathcal{A}$ iff $a \in dom(\mu(s))$. $\square$

Assume that a specialization system $\Gamma = \langle \mathcal{A}, \mathcal{G}, \mathcal{S}, \mu \rangle$ is given. A specialization in $\mathcal{S}$ will often be denoted by a Greek letter such as θ. A specialization $\theta \in \mathcal{S}$ will be identified with the partial mapping $\mu(\theta)$ and used as a postfix unary (partial) operator on $\mathcal{A}$ (e.g., $\mu(\theta)(a) = a\theta$), provided that no confusion is caused. Let ϵ denote the identity specialization in $\mathcal{S}$, i.e., $a\epsilon = a$ for any $a \in \mathcal{A}$. For any $\theta, \sigma \in \mathcal{S}$, let $\theta \circ \sigma$ denote a specialization $\rho \in \mathcal{S}$ such that $\mu(\rho) = \mu(\sigma) \circ \mu(\theta)$, i.e., $a(\theta \circ \sigma) = (a\theta)\sigma$ for any $a \in \mathcal{A}$.

2.2   User-Defined Atoms, Constraint Atoms, and Clauses

Let $\Gamma_u = \langle \mathcal{A}_u, \mathcal{G}_u, \mathcal{S}_u, \mu_u \rangle$ and $\Gamma_c = \langle \mathcal{A}_c, \mathcal{G}_c, \mathcal{S}_c, \mu_c \rangle$ be specialization systems such that $\mathcal{S}_u = \mathcal{S}_c$. Elements of $\mathcal{A}_u$ are called *user-defined atoms* and those of $\mathcal{G}_u$ are called *ground user-defined atoms*. Elements of $\mathcal{A}_c$ are called *constraint atoms* and those of $\mathcal{G}_c$ are called *ground constraint atoms*. Hereinafter, assume that $\mathcal{S} = \mathcal{S}_u = \mathcal{S}_c$. Elements of $\mathcal{S}$ are called *specializations*. Let TCON denote the set of all true ground constraint atoms.

A *clause* on $\langle \Gamma_u, \Gamma_c \rangle$ is an expression of the form

$$a_1, \ldots, a_m \leftarrow b_1, \ldots, b_n,$$

where $m \geq 0$, $n \geq 0$, and each of $a_1, \ldots, a_m, b_1, \ldots, b_n$ belongs to $\mathcal{A}_u \cup \mathcal{A}_c$. It is called a *definite clause* iff $m = 1$ and $a_1 \in \mathcal{A}_u$. It is called a *positive unit clause* iff it is a definite clause and each of $b_1, \ldots, b_n$ belongs to $\mathcal{A}_c$. It is a *ground* clause on $\langle \Gamma_u, \Gamma_c \rangle$ iff each of $a_1, \ldots, a_m, b_1, \ldots, b_n$ belongs to $\mathcal{G}_u \cup \mathcal{G}_c$. Let CLS denote the set of all clauses on $\langle \Gamma_u, \Gamma_c \rangle$.

2.3 Interpretations and Models

An *interpretation* is a subset of $\mathcal{G}_u$. Unlike ground user-defined atoms, the truth values of ground constraint atoms are predetermined by TCon (cf. Sect. 2.2) independently of interpretations. A ground constraint atom g is true iff $g \in$ TCon. It is false otherwise.

A ground clause $C = (a_1, \ldots, a_m \leftarrow b_1, \ldots, b_n)$ is *true* with respect to an interpretation $G \subseteq \mathcal{G}_u$ (in other words, G *satisfies* C) iff at least one of the following conditions is satisfied:

1. There exists $i \in \{1, \ldots, m\}$ such that $a_i \in G \cup$ TCon.
2. There exists $j \in \{1, \ldots, n\}$ such that $b_j \notin G \cup$ TCon.

A clause C is *true* with respect to an interpretation $G \subseteq \mathcal{G}_u$ (in other words, G *satisfies* C) iff for any specialization θ such that $C\theta$ is ground, $C\theta$ is true with respect to G. A *model* of a clause set $Cs \subseteq$ Cls is an interpretation that satisfies every clause in Cs.

Note that the standard semantics is taken in this paper, i.e., all models of a formula are considered instead of specific ones, such as those considered in the minimal model semantics [6,11] (i.e., the semantics underlying logic programming) and those considered in stable model semantics [9,10] (i.e., the semantics underlying answer set programming).

2.4 Model-Intersection (MI) Problems

Let *Models* be a mapping that associates with each clause set the set of all of its models, i.e., *Models*(Cs) is the set of all models of Cs for any $Cs \subseteq$ Cls.

Assume that a person A and a person B are interested in knowing which atoms in $\mathcal{G}_u$ are true and which atoms in $\mathcal{G}_u$ are false. They want to know the unknown set G of all true ground atoms. Due to shortage of knowledge, A still cannot determine one unique true subset of $\mathcal{G}_u$. The person A can only limit possible subsets of true atoms by specifying a subset Gs of $pow(\mathcal{G}_u)$. The unknown set G of all true atoms belongs to Gs. One way for A to inform this knowledge to B compactly is to send to B a clause set Cs such that $Gs \subseteq$ *Models*(Cs). Receiving Cs, B knows that *Models*(Cs) includes all possible intended sets of ground atoms, i.e., $G \in$ *Models*(Cs). As such, B can know that each ground atom outside $\bigcup$ *Models*(Cs) is false, i.e., for any $g \in \mathcal{G}_u$, if $g \notin \bigcup$ *Models*(Cs), then $g \notin G$. The person B can also know that each ground atom in $\bigcap$ *Models*(Cs) is true, i.e., for any $g \in \mathcal{G}_u$, if $g \in \bigcap$ *Models*(Cs), then $g \in G$. This shows the importance of calculating $\bigcap$ *Models*(Cs).

A *model-intersection problem* (*MI problem*) is a pair $\langle Cs, \varphi \rangle$, where $Cs \subseteq$ Cls and φ is a mapping from $pow(\mathcal{G}_u)$ to some set W. The mapping φ is called an *exit mapping*. The answer to this problem, denoted by $ans_{MI}(Cs, \varphi)$, is defined by

$$ans_{MI}(Cs, \varphi) = \varphi(\bigcap Models(Cs)),$$

where $\bigcap$ *Models*(Cs) is the intersection of all models of Cs. Note that when *Models*(Cs) is the empty set, $\bigcap$ *Models*(Cs) $= \mathcal{G}_u$.

2.5   Query-Answering (QA) Problems

Let $Cs \subseteq$ CLS. For any $Cs' \subseteq$ CLS, Cs' is a *logical consequence* of Cs, denoted by $Cs \models Cs'$, iff every model of Cs is also a model of Cs'. For any $a \in \mathcal{A}_u$, a is a *logical consequence* of Cs, denoted by $Cs \models a$, iff $Cs \models \{(a \leftarrow)\}$.

A *query-answering problem* (*QA problem*) in this paper is a pair $\langle Cs, a \rangle$, where $Cs \subseteq$ CLS and a is a user-defined atom in $\mathcal{A}_u$. The *answer* to a QA problem $\langle Cs, a \rangle$, denoted by $ans_{QA}(Cs, a)$, is defined by

$$ans_{QA}(Cs, a) = \{a\theta \mid (\theta \in \mathcal{S}) \ \& \ (a\theta \in \mathcal{G}_u) \ \& \ (Cs \models (a\theta \leftarrow))\}.$$

Theorem 1. *For any* $Cs \subseteq$ CLS *and* $a \in \mathcal{A}_u$,

$$ans_{QA}(Cs, a) = rep(a) \cap \left(\bigcap Models(Cs) \right),$$

where $rep(a)$ *denotes the set of all ground instances of* a.

Proof: Let $Cs \subseteq$ CLS and $a \in \mathcal{A}_u$. By the definition of $\models$, for any ground atom $g \in \mathcal{G}_u$, $Cs \models g$ iff $g \in \bigcap Models(Cs)$. Then

$$
\begin{aligned}
ans_{QA}(Cs, a) &= \{a\theta \mid (\theta \in \mathcal{S}) \ \& \ (a\theta \in \mathcal{G}_u) \ \& (Cs \models (a\theta \leftarrow))\} \\
&= \{g \mid (\theta \in \mathcal{S}) \ \& \ (g = a\theta) \& (g \in \mathcal{G}_u) \& (Cs \models (g \leftarrow))\} \\
&= \{g \mid (g \in rep(a)) \ \& \ (Cs \models (g \leftarrow))\} \\
&= \{g \mid (g \in rep(a)) \ \& \ (g \in (\bigcap Models(Cs)))\} \\
&= rep(a) \cap \left(\bigcap Models(Cs) \right).
\end{aligned}
$$

$\square$

Theorem 1 shows the importance of the intersection of all models of a clause set. By this theorem, the answer to a QA problem can be rewritten as follows:

Theorem 2. *Let* $Cs \subseteq$ CLS *and* $a \in \mathcal{A}_u$. *Then* $ans_{QA}(Cs, a) = ans_{MI}(Cs, \varphi_1)$, *where for any* $G \subseteq \mathcal{G}_u$, $\varphi_1(G) = rep(a) \cap G$.

Proof: It follows from Theorem 1 and the definition of φ_1 that $ans_{QA}(Cs, a) = \varphi_1(\bigcap Models(Cs)) = ans_{MI}(Cs, \varphi_1)$. $\square$

This is one way to regard a QA problem as a MI problem, which can be understood as follows: The set $\bigcap Models(Cs)$ often contains too many ground atoms. The set $rep(a)$ specifies a range of interest in the set $\mathcal{G}_u$. The exit mapping φ_1 focuses attention on the part $rep(a)$ by making intersection with it.

Theorem 3 below shows another way to formalize a QA problem as a MI problem.

Theorem 3. *Assume that the specialization system for first-order logic is used. Let* $Cs \subseteq$ CLS *and* $a \in \mathcal{A}_u$. *Then*

$$ans_{QA}(Cs, a) = ans_{MI}(Cs \cup \{(p(x_1, \ldots, x_n) \leftarrow a)\}, \varphi_2),$$

where p is a predicate that appears in neither Cs nor a, the arguments $x_1, \ldots, x_n$ are all the mutually different variables occurring in a, and for any $G \subseteq \mathcal{G}_u$,

$$\varphi_2(G) = \{a\theta \mid (\theta \in \mathcal{S}) \ \& \ (p(x_1, \ldots, x_n)\theta \in G)\}.$$

Proof: By Theorem 1 and the definition of φ_2,

$$ans_{QA}(Cs, a) = rep(a) \cap (\bigcap Models(Cs))$$
$$= \varphi_2(\bigcap Models(Cs \cup \{(p(x_1, \ldots, x_n) \leftarrow a)\}))$$
$$= ans_{MI}(Cs \cup \{(p(x_1, \ldots, x_n) \leftarrow a)\}, \varphi_2).$$

$\square$

In logic programming [11], a problem represented by a pair of a set of definite clauses and a query atom has been intensively discussed. In the description logic (DL) community [4], a class of problems formulated as conjunctions of DL-based axioms and assertions together with query atoms has been discussed [13]. These two problem classes can be formalized as subclasses of QA problems considered in this paper.

2.6 Proof Problems

A *proof problem* is a pair $\langle E_1, E_2 \rangle$, where E_1 and E_2 are first-order formulas, and the answer to this problem, denoted by $ans_{Pr}(E_1, E_2)$, is defined by

$$ans_{Pr}(E_1, E_2) = \begin{cases} \text{``yes'' if } E_2 \text{ is a logical consequence of } E_1, \\ \text{``no'' otherwise.} \end{cases}$$

It is well known that a proof problem $\langle E_1, E_2 \rangle$ can be converted into the problem of determining whether $E_1 \wedge \neg E_2$ is unsatisfiable [5], i.e., whether $E_1 \wedge \neg E_2$ has no model. As a result, $ans_{Pr}(E_1, E_2)$ can be equivalently defined by

$$ans_{Pr}(E_1, E_2) = \begin{cases} \text{``yes'' if } Models(E_1 \wedge \neg E_2) = \emptyset, \\ \text{``no'' otherwise.} \end{cases}$$

Theorem 4. *Let $\langle E_1, E_2 \rangle$ be a proof problem. Let Cs be the set of clauses obtained by transformation of $E_1 \wedge \neg E_2$ preserving satisfiability. Let $\varphi_3 : pow(\mathcal{G}_u) \rightarrow \{\text{``yes''}, \text{``no''}\}$ be defined by: for any $G \subseteq \mathcal{G}_u$,*

$$\varphi_3(G) = \begin{cases} \text{``yes'' if } G = \mathcal{G}_u, \\ \text{``no'' otherwise.} \end{cases}$$

Then $ans_{Pr}(E_1, E_2) = ans_{MI}(Cs, \varphi_3)$.

Proof: Let b be a ground user-defined atom that is not an instance of any user-defined atom occurring in Cs. If m is a model of Cs, then $m - \{b\}$ is also a model of Cs. Obviously, $m - \{b\} \neq \mathcal{G}_u$. Therefore, (i) if $Models(Cs) \neq \emptyset$, then $\bigcap Models(Cs) \neq \mathcal{G}_u$, and (ii) if $Models(Cs) = \emptyset$, then $\bigcap Models(Cs) = \bigcap\{\} = \mathcal{G}_u$. Hence $ans_{Pr}(E_1, E_2) = ans_{MI}(Cs, \varphi_3)$. $\square$

3 Solving MI Problems by Equivalent Transformation

A general schema for solving MI problems based on equivalent transformation is formulated and its correctness is shown (Theorem 9).

3.1 Preservation of Partial Mappings and Equivalent Transformation

Terminologies such as preservation of partial mappings and equivalent transformation are defined in general below. They will be used with a specific class of partial mappings called target mappings, which will be introduced in Sect. 3.2.

Assume that X and Y are sets and f is a partial mapping from X to Y. For any $x, x' \in dom(f)$, transformation of x into x' is said to *preserve* f iff $f(x) = f(x')$. For any $x, x' \in dom(f)$, transformation of x into x' is called *equivalent transformation* (ET) with respect to f iff the transformation preserves f, i.e., $f(x) = f(x')$.

Let $\mathbb{F}$ be a set of partial mappings from a set X to a set Y. Given $x, x' \in X$, transformation of x into x' is called *equivalent transformation* (ET) with respect to $\mathbb{F}$ iff there exists $f \in \mathbb{F}$ such that the transformation preserves f. A sequence $[x_0, x_1, \ldots, x_n]$ of elements in X is called an *equivalent transformation sequence* (ET *sequence*) with respect to $\mathbb{F}$ iff for any $i \in \{0, 1, \ldots, n-1\}$, transformation of x_i into x_{i+1} is ET with respect to $\mathbb{F}$. When emphasis is placed on the initial element x_0 and the final element x_n, this sequence is also referred to as an ET sequence *from x_0 to x_n*.

3.2 Target Mappings and Answer Mappings

Given a MI problem $\langle Cs, \varphi \rangle$, since $ans_{\mathrm{MI}}(Cs, \varphi) = \varphi(\bigcap Models(Cs))$, the answer to this MI problem is determined uniquely by $Models(Cs)$ and φ. As a result, we can equivalently consider a new MI problem with the same answer by switching from Cs to another clause set Cs' if $Models(Cs) = Models(Cs')$. According to the general terminologies defined in Sect. 3.1, on condition that $Models(Cs) = Models(Cs')$, transformation from $x = Cs$ into $x' = Cs'$ preserves $f = Models$ and is called ET with respect to $f = Models$, where (i) $x, x' \in pow(\mathrm{CLS})$ and (ii) $Models(x), Models(x') \in pow(pow(\mathcal{G}_{\mathrm{u}}))$. We can also consider an ET sequence $[Cs_0, Cs_1, \ldots, Cs_n]$ of elements in $pow(\mathrm{CLS})$ with respect to a singleton set $\{Models\}$. MI problems can be transformed into simpler forms by ET preserving $Models$.

In order to use more partial mappings for simplification of MI problems, we extend our consideration from the specific mapping $Models$ to a class of partial mappings, called GSETMAP, defined below.

Definition 2. GSETMAP is the set of all partial mappings from $pow(\mathrm{CLS})$ to $pow(pow(\mathcal{G}_{\mathrm{u}}))$.   □

As defined in Sect. 2.4, *Models*(Cs) is the set of all models of Cs for any $Cs \subseteq \text{CLS}$. Since a model is a subset of $\mathcal{G}_u$, *Models* is regarded as a total mapping from $pow(\text{CLS})$ to $pow(pow(\mathcal{G}_u))$. Since a total mapping is also a partial mapping, the mapping *Models* is a partial mapping from $pow(\text{CLS})$ to $pow(pow(\mathcal{G}_u))$, i.e., it is an element of GSETMAP.

A partial mapping M in GSETMAP is of particular interest if $\bigcap M(Cs) = \bigcap Models(Cs)$ for any $Cs \in dom(M)$. Such a partial mapping is called a *target mapping*.

Definition 3. A partial mapping $M \in$ GSETMAP is a *target mapping* iff for any $Cs \in dom(M)$, $\bigcap M(Cs) = \bigcap Models(Cs)$. $\square$

It is obvious that:

Theorem 5. *The mapping Models is a target mapping.* $\square$

The next theorem provides a sufficient condition for a mapping in GSETMAP to be a target mapping.

Theorem 6. *Let $M \in$ GSETMAP. M is a target mapping if the following conditions are satisfied:*

1. *$M(Cs) \subseteq Models(Cs)$ for any $Cs \in dom(M)$.*
2. *For any $Cs \in dom(M)$ and any $m_2 \in Models(Cs)$, there exists $m_1 \in M(Cs)$ such that $m_1 \subseteq m_2$.*

Proof: Assume that Conditions 1 and 2 above hold. Let $Cs \in dom(M)$. By Condition 1, $\bigcap M(Cs) \supseteq \bigcap Models(Cs)$. We show that $\bigcap M(Cs) \subseteq \bigcap Models(Cs)$ as follows: Assume that $g \in \bigcap M(Cs)$. Let $m_2 \in Models(Cs)$. By Condition 2, there exists $m_1 \in M(Cs)$ such that $m_1 \subseteq m_2$. Since $g \in \bigcap M(Cs)$, g belongs to m_1. So $g \in m_2$. Since m_2 is any arbitrary element of *Models*(Cs), g belongs to $\bigcap Models(Cs)$. It follows that $\bigcap M(Cs) = \bigcap Models(Cs)$. Hence M is a target mapping. $\square$

Definition 4. A partial mapping A from $pow(\text{CLS})$ to a set W is an *answer mapping* with respect to an exit mapping φ iff for any $Cs \in dom(A)$, $ans_{\text{MI}}(Cs, \varphi) = A(Cs)$. $\square$

If M is a target mapping, then M can be used for computing the answers to MI problems. More precisely:

Theorem 7. *Let M be a target mapping and φ an exit mapping. Suppose that A is a partial mapping such that $dom(A) = dom(M)$ and for any $Cs \in dom(M)$, $A(Cs) = \varphi(\bigcap M(Cs))$. Then A is an answer mapping with respect to φ.*

Proof: Let $Cs \in dom(A)$. Since $dom(A) = dom(M)$, Cs belongs to $dom(M)$. Since M is a target mapping, $\bigcap M(Cs) = \bigcap Models(Cs)$. Thus $ans_{\text{MI}}(Cs, \varphi) = \varphi(\bigcap Models(Cs)) = \varphi(\bigcap M(Cs)) = A(Cs)$. So A is an answer mapping with respect to φ. $\square$

3.3   ET Steps and ET Rules

Next, a schema for solving MI problems based on equivalent transformation (ET) preserving answers is formulated. The notions of preservation of answers/target mappings, ET with respect to answers/target mappings, and an ET sequence are obtained by specializing the general definitions in Sect. 3.1.

Let π be a mapping, called *state mapping*, from a given set STATE to the set of all MI problems. Elements of STATE are called *states*.

Definition 5. Let $\langle S, S' \rangle \in$ STATE×STATE. $\langle S, S' \rangle$ is an *ET step* with respect to π iff if $\pi(S) = \langle Cs, \varphi \rangle$ and $\pi(S') = \langle Cs', \varphi' \rangle$, then $ans_{\mathrm{MI}}(Cs, \varphi) = ans_{\mathrm{MI}}(Cs', \varphi')$. ☐

Definition 6. A sequence $[S_0, S_1, \ldots, S_n]$ of elements of STATE is an *ET sequence* with respect to π iff for any $i \in \{0, 1, \ldots, n - 1\}$, $\langle S_i, S_{i+1} \rangle$ is an ET step with respect to π. ☐

The role of ET computation constructing $[S_0, S_1, \ldots, S_n]$ is to start with S_0 and to reach S_n from which the answer to the given problem can be easily computed.

The concept of ET rule on STATE is defined by:

Definition 7. An *ET rule* r on STATE with respect to π is a partial mapping from STATE to STATE such that for any $S \in dom(r)$, $\langle S, r(S) \rangle$ is an ET step with respect to π. ☐

We also define ET rules on $pow(\mathrm{CLS})$ as follows:

Definition 8. An *ET rule* r with respect to a target mapping M is a partial mapping from $pow(\mathrm{CLS})$ to $pow(\mathrm{CLS})$ such that for any $Cs \in dom(r)$, $M(Cs) = M(r(Cs))$. ☐

We can construct an ET rule on STATE based on target mappings.

Theorem 8. *Assume that M is a target mapping and r is a partial mapping from $pow(\mathrm{CLS})$ to $pow(\mathrm{CLS})$. Let $\bar{r}$ be defined as a partial mapping from STATE to STATE such that for any $S \in$ STATE, if $\pi(S) = \langle Cs, \varphi \rangle$, then $\pi(\bar{r}(S)) = \langle r(Cs), \varphi \rangle$. If r is an ET rule with respect to M, then $\bar{r}$ is an ET rule on STATE.*

Proof: Assume that r is an ET rule with respect to M. Let $S \in$ STATE such that $\pi(S) = \langle Cs, \varphi \rangle$. Then

$$ans_{\mathrm{MI}}(Cs, \varphi) = \varphi(\bigcap Models(Cs))$$

$$= (\text{since } M \text{ is a target mapping})$$

$$= \varphi(\bigcap M(Cs))$$

$$= (\text{since } r \text{ is an ET rule with respect to } M)$$

$$= \varphi(\bigcap M(r(Cs)))$$

$$= \text{(since } M \text{ is a target mapping)}$$
$$= \varphi(\bigcap Models(r(Cs)))$$
$$= ans_{\text{MI}}(r(Cs), \varphi).$$

□

3.4 Correct Solutions Based on ET Rules

Assume that

- φ is an exit mapping,
- PROB $= pow(\text{CLS}) \times \{\varphi\}$,
- STATE is a set of states and π is a state mapping for STATE such that for any $S \in$ STATE, the second component of $\pi(S)$ is φ, and
- A is an answer mapping with respect to φ.

The proposed solution schema for MI problems consists of the following steps:

1. Assume that a MI problem $\langle Cs, \varphi \rangle$ in PROB is given as an input problem.
2. Prepare a set R of ET rules on STATE with respect to π.
3. Take S_0 such that $\pi(S_0) = \langle Cs, \varphi \rangle$ to start computation from S_0.
4. Construct an ET sequence $[S_0, \ldots, S_n]$ by applying ET rules in R, i.e., for each $i \in \{0, 1, \ldots, n-1\}$, S_{i+1} is obtained from S_i by selecting and applying $r_i \in R$ such that $S_i \in dom(r_i)$ and $r_i(S_i) = S_{i+1}$.
5. Assume that $\pi(S_n) = \langle Cs_n, \varphi \rangle$. If the computation reaches the domain of A, i.e., $Cs_n \in dom(A)$, then compute the answer by using the answer mapping A, i.e., output $A(Cs_n)$.

Given a set Cs of clauses and an exit mapping φ, the answer to the MI problem $\langle Cs, \varphi \rangle$, i.e., $ans_{\text{MI}}(Cs, \varphi) = \varphi(\bigcap Models(Cs))$, can be directly obtained by the computation shown in the leftmost path in Fig. 1. Instead of taking this computation path, the above solution takes a different one, i.e., the lowest path (from Cs to Cs') followed by the rightmost path (through A) in Fig. 1.

The selection of r_i in R at Step 4 is nondeterministic and there may be many possible computation paths for each MI problem. Every output computed by using any arbitrary computation path is correct. When the set R gives at least one correct computation path for each MI problem in PROB, R can be regarded as an algorithm for PROB.

Theorem 9. *The above procedure gives the correct answer to $\langle Cs, \varphi \rangle$.*

Proof. Since $[S_0, \ldots, S_n]$ is an ET sequence, $ans_{\text{MI}}(Cs, \varphi) = ans_{\text{MI}}(Cs_n, \varphi)$. Since A is an answer mapping with respect to φ, $ans_{\text{MI}}(Cs_n, \varphi) = A(Cs_n)$. Hence $ans_{\text{MI}}(Cs, \varphi) = A(Cs_n)$. □

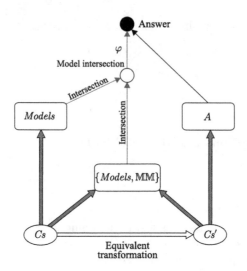

Fig. 1. Target mappings, an answer mapping, and ET computation paths.

3.5 Computation Cost for Solving MI Problems

We next explain the basic structure of computation cost of the solution in Sect. 3.4. For the discussion below, the following notation is assumed:

- For any S, $S' \in$ STATE, let $trans(S, S')$ denote the transformation of S into S', and $time(trans(S, S'))$ denote the computation time required for this transformation step.
- Let π be a state mapping. For any answer mapping A and $S \in$ STATE, let $comp(A, S)$ denote the computation of $A(Cs)$, where $\pi(S) = \langle Cs, \varphi \rangle$, and let $time(comp(A, S))$ denote the amount of time required for this computation.

With this notation, the time of the solution in Sect. 3.4 using an ET sequence $[S_0, S_1, \ldots, S_n]$ and an answer mapping A is evaluated by

$$\mathcal{T}_{\mathrm{MI}} = \Sigma_{i=1}^{n}\, time(trans(S_{i-1}, S_i)) + time(comp(A, S_n)).$$

A basic strategy to obtain an efficient solution is to minimize (i) the time for final computation, i.e., $time(comp(A, S_n))$, (ii) the computation time of each transformation step in the ET sequence from S_0 to S_n, i.e., $time(trans(S_{i-1}, S_i))$ for each $i \in \{1, \ldots, n\}$, and (iii) the total number of computation steps, i.e., n.

4 Target Mappings for Clauses

Next, three target mappings are introduced, i.e., τ_1 for sets of positive unit clauses, τ_2 for sets of definite clauses, and MM for sets of arbitrary clauses.

4.1 A Target Mapping for Sets of Positive Unit Clauses

Let PUCL denote the set of all positive unit clauses. A partial mapping $\tau_1 \in$ GSETMAP is defined as follows:

1. For any $F \subseteq$ PUCL, $\tau_1(F)$ is the singleton set $\{m_F\}$, where

$$m_F = \{a\theta \mid ((a \leftarrow b_1, \ldots, b_n) \in F) \ \& \ (\theta \in \mathcal{S}) \ \& $$
$$(a\theta \in \mathcal{G}_u) \ \& \ (\{b_1\theta, \ldots, b_n\theta\} \subseteq \mathrm{TCON})\}.$$

2. For any $Cs \subseteq$ CLS such that $Cs \not\subseteq$ PUCL, $\tau_1(Cs)$ is undefined.

Theorem 10. τ_1 *is a target mapping.*

Proof: Assume that $F \subseteq$ PUCL. Let $\tau_1(F) = \{m_F\}$. Obviously, m_F is a model of F. Hence $\tau_1(F) = \{m_F\} \subseteq Models(F)$. So the first condition of Theorem 6 is satisfied for τ_1.

Now assume that $F \subseteq$ PUCL and $m \in Models(F)$. Let $\tau_1(F) = \{m_F\}$. Then $m_F \in \tau_1(F)$ and $m_F \subseteq m$. So the second condition of Theorem 6 is also satisfied for τ_1. Thus τ_1 is a target mapping by Theorem 6. □

4.2 A Target Mapping for Sets of Definite Clauses

Let DCL denote the set of all definite clauses. Given a definite clause C, the atom in the left-hand side of C is called the *head* of C, denoted by $head(C)$, and the set of all user-defined atoms and constraint atoms in the right-hand side of C is called the *body* of C, denoted by $body(C)$. Assume that D is a set of definite clauses in DCL. The *meaning* of D, denoted by $\mathcal{M}(D)$, is defined as follows:

1. A mapping T_D on $pow(\mathcal{G})$ is defined by: for any set $G \subseteq \mathcal{G}$, $T_D(G)$ is the set

$$\{head(C\theta) \mid (C \in D) \ \& \ (\theta \in \mathcal{S}) \ \&$$
$$(\text{each user-defined atom in } body(C\theta) \text{ is in } G) \ \&$$
$$(\text{each constraint atom in } body(C\theta) \text{ is true})\}.$$

2. $\mathcal{M}(D)$ is then defined as the set $\bigcup_{n=1}^{\infty} T_D^n(\emptyset)$, where $T_D^1(\emptyset) = T_D(\emptyset)$ and for each $n > 1$, $T_D^n(\emptyset) = T_D(T_D^{n-1}(\emptyset))$.

Then a partial mapping $\tau_2 \in$ GSETMAP is defined below.

1. For any $D \subseteq$ DCL, $\tau_2(D)$ is the singleton set $\{\mathcal{M}(D)\}$.
2. For any $Cs \subseteq$ CLS such that $Cs \not\subseteq$ DCL, $\tau_2(Cs)$ is undefined.

Theorem 11. τ_2 *is a target mapping.*

Proof: Let $D \subseteq$ DCL. Since $\mathcal{M}(D)$ is a model of D, $\tau_2(D) = \{\mathcal{M}(D)\} \subseteq Models(D)$. So the first condition of Theorem 6 is satisfied for τ_2. Let $m \in Models(D)$. Since $\mathcal{M}(D)$ is the least model of D, $\mathcal{M}(D) \subseteq m$. So the second condition of Theorem 6 is satisfied for τ_2. By Theorem 6, τ_2 is a target mapping. □

Theorem 12. *For any* $F \subseteq$ PUCL, $\tau_2(F) = \tau_1(F)$.

Proof: For any $F \subseteq$ PUCL, $\tau_2(F) = \{\mathcal{M}(F)\} = \{m_F\} = \tau_1(F)$. □

4.3 A Target Mapping for All Clause Sets

Given a clause C, the set of all user-defined atoms and constraint atoms in the left-hand side of C is denoted by $lhs(C)$ and the set of all those in the right-hand side of C is denoted by $rhs(C)$. A clause is said to be *positive* if there is at least one user-defined atom in its left-hand side; it is said to be *negative* otherwise.

It is assumed henceforth that (i) for any constraint atom c, $not(c)$ is a constraint atom; (ii) for any constraint atom c and any specialization θ, $not(c)\theta = not(c\theta)$; and (iii) for any ground constraint atom c, c is true iff $not(c)$ is not true.

The following notation is used for defining a target mapping $\mathbb{MM}$ for arbitrary clauses in CLS (Definition 9).

1. Let Cs be a set of clauses possibly with constraint atoms. MVRHS(Cs) is defined as the set $\{$MVRHS$(C) \mid C \in Cs\}$, where for any clause $C \in Cs$, MVRHS(C) is the clause obtained from C as follows: For each constraint atom c in $lhs(C)$, remove c from $lhs(C)$ and add $not(c)$ to $rhs(C)$.
2. Let Cs be a set of clauses with no constraint atom in their left-hand sides. For any $G \subseteq \mathcal{G}$, GINST(Cs, G) is defined as the set

$$\{\text{RMCON}(C\theta) \mid (C \in Cs) \ \& \ (\theta \in \mathcal{S}) \ \&$$
$$\text{(each user-defined atom in } C\theta \text{ is in } G) \ \&$$
$$\text{(each constraint atom in } rhs(C\theta) \text{ is true)}\},$$

   where for any clause C', RMCON(C') is the clause obtained from C' by removing all constraint atoms from it.
3. Let Cs be a set of clauses possibly with constraint atoms. For any $G \subseteq \mathcal{G}$, INST(Cs, G) is defined by

$$\text{INST}(Cs, G) = \text{GINST}(\text{MVRHS}(Cs), G).$$

4. Let Cs be a set of ground clauses with no constraint atom. We can construct a set of definite clauses from Cs as follows: For each clause $C \in Cs$,
   - if $lhs(C) = \emptyset$, then construct a definite clause the head of which is $\perp$ and the body of which is $rhs(C)$, where $\perp$ is a special symbol not occurring in Cs;
   - if $lhs(C) \neq \emptyset$, then (i) select one arbitrary atom a from $lhs(C)$, and (ii) construct a definite clause the head of which is a and the body of which is $rhs(C)$.

   Let DC(Cs) denote the set of all definite-clause sets possibly constructed from Cs in the above way.

Proposition 1. *Let $Cs \subseteq$ CLS. For any $m \subseteq \mathcal{G}$, m is a model of Cs iff m is a model of* INST($Cs, \mathcal{G}$).

Proof: INST($Cs, \mathcal{G}$) is obtained from Cs by (i) moving constraint atoms in the left-hand sides of clauses into their right-hand sides, (ii) instantiation of variables into ground terms, (iii) removal of clauses containing false constraint atoms in their right-hand sides, and (iv) removal of true constraint atoms from the remaining clauses. Each of the operations (i), (ii), (iii), and (iv) preserves models.   □

A mapping MM is defined below.

Definition 9. A mapping $\mathrm{MM} \in \mathrm{GSETMAP}$ is defined by

$$\mathrm{MM}(Cs) = \{\mathcal{M}(D) \mid (D \in \mathrm{DC}(\mathrm{INST}(Cs, \mathcal{G}))) \,\&\, (\bot \notin \mathcal{M}(D))\}$$

for any $Cs \subseteq \mathrm{CLS}$. □

Theorem 13. MM *is a target mapping.*

Proof: First, we show that the first condition of Theorem 6 is satisfied for MM. Let $Cs \subseteq \mathrm{CLS}$. Suppose that $m \in \mathrm{MM}(Cs)$. Let $Cs' = \mathrm{INST}(Cs, \mathcal{G})$. Then there exists D such that $m = \mathcal{M}(D)$, $D \in \mathrm{DC}(Cs')$, and $\bot \notin \mathcal{M}(D)$. We show that m is a model of Cs' as follows:

- Let C_P be a positive clause in Cs'. Since $D \in \mathrm{DC}(Cs')$, there exists $C \in D$ such that $head(C) \in lhs(C_P)$ and $body(C) = rhs(C_P)$. Since m satisfies C, m also satisfies C_P. Hence m satisfies every positive clause in Cs'.
- Let C_N be a negative clause in Cs'. Since $D \in \mathrm{DC}(Cs')$, there exists $C' \in D$ such that $head(C') = \bot$ and $body(C') = rhs(C_N)$. Since $\bot \notin \mathcal{M}(D)$, m does not include $body(C')$. So $rhs(C_N) \not\subseteq m$, whence m satisfies C_N. Hence m satisfies every negative clause in Cs'.

So m is a model of Cs'. By Proposition 1, m is a model of Cs, i.e., $m \in Models(Cs)$.

Next, we show that the second condition of Theorem 6 is satisfied for MM. Let $Cs \subseteq \mathrm{CLS}$. Suppose that $m' \in Models(Cs)$, i.e., m' is a model of Cs. Let $Cs' = \mathrm{INST}(Cs, \mathcal{G})$. By Proposition 1, m' is also a model of Cs'. Let D be a set of definite clauses obtained from Cs' by constructing from each positive clause C in Cs' a definite clause C' as follows:

1. Select an atom a from $lhs(C)$ as follows:
   (a) If $rhs(C) \subseteq m'$, then select an atom $a \in lhs(C) \cap m'$.
   (b) If $rhs(C) \not\subseteq m'$, then select an arbitrary atom $a \in lhs(C)$.
2. Construct C' as a definite clause such that $head(C') = a$ and $body(C') = rhs(C)$.

It is obvious that m' is a model of D. Let $m'' = \mathcal{M}(D)$. Since m'' is the least model of D, $m'' \subseteq m'$. Since m' is a model of Cs', m' satisfies all negative clauses in Cs'. Since $m'' \subseteq m'$, m'' also satisfies all negative clauses in Cs'. It follows that $\bot \notin \mathcal{M}(D)$. Hence $m'' \in \mathrm{MM}(Cs)$.

So MM is a target mapping by Theorem 6. □

Theorem 14. *For any* $D \subseteq \mathrm{DCL}$, $\mathrm{MM}(D) = \tau_2(D)$.

Proof: Let $D \subseteq \mathrm{DCL}$. Then $\mathrm{DC}(\mathrm{INST}(D, \mathcal{G}))$ is the singleton set $\{\mathrm{INST}(D, \mathcal{G})\}$. Obviously, $\mathcal{M}(D) = \mathcal{M}(\mathrm{INST}(D, \mathcal{G}))$ and $\bot \notin \mathcal{M}(\mathrm{INST}(D, \mathcal{G}))$. It follows that

$$\begin{aligned}
\mathrm{MM}(D) &= \{\mathcal{M}(D') \mid (D' \in \mathrm{DC}(\mathrm{INST}(D, \mathcal{G}))) \,\&\, (\bot \notin \mathcal{M}(D'))\} \\
&= \{\mathcal{M}(\mathrm{INST}(D, \mathcal{G}))\} \\
&= \{\mathcal{M}(D)\} \\
&= \tau_2(D).
\end{aligned}$$

□

5   Solutions for QA Problems/Proof Problems

ET solutions for QA problems/proof problems on clauses are given by specializing the solution schema of Sect. 3.

5.1   A Solution for QA Problems

First, we solve a QA problem by using an answer mapping that is determined by the target mapping τ_1 given in Sect. 4.1 (i.e., $\tau_1(F) = \{m_F\}$ for any $F \subseteq \text{PUCL}$).

1. Assume that a QA problem $\langle Cs, a \rangle$ is given.
2. Transform $\langle Cs, a \rangle$ into a MI problem $\langle Cs \cup \{(p(x_1, \ldots, x_n) \leftarrow a)\}, \varphi_2 \rangle$, where p is a predicate that appears in neither Cs nor a, the arguments $x_1, \ldots, x_n$ are all the mutually different variables occurring in a, and φ_2 is the exit mapping defined in Sect. 2.5, i.e., for any $G \subseteq \mathcal{G}_u$, $\varphi_2(G) = \{a\theta \mid (\theta \in \mathcal{S})\ \&\ (p(x_1, \ldots, x_n)\theta \in G)\}$.
3. Prepare a set R of ET rules to transform MI problems preserving their answers. For simplicity, we only use rules that do not change the second argument of a state (i.e., do not change the exit mapping).
4. Transform $\langle Cs \cup \{(p(x_1, \ldots, x_n) \leftarrow a)\}, \varphi_2 \rangle$ equivalently by using ET rules in R.
5. If the transformation sequence reaches a MI problem of the form $\langle F, \varphi_2 \rangle$, where F is a set of positive unit clauses with p-atoms in their heads, then the answer $\varphi_2(m_F)$ is obtained.

5.2   A Solution for Proof Problems

Next, we apply the solution schema of Sect. 3 to proof problems. Since for any clause set Cs, $Models(Cs) = \emptyset$ iff $\bigcap Models(Cs) = \mathcal{G}_u$ by the proof of Theorem 4, we use here an answer mapping that maps obviously unsatisfiable clause sets to "yes" and maps obviously satisfiable ones to "no".

1. Assume that a proof problem $\langle E_1, E_2 \rangle$ is given.
2. Transform $E_1 \wedge \neg E_2$ into a conjunctive normal form Cs in order to obtain a MI problem $\langle Cs, \varphi_3 \rangle$, where φ_3 is the exit mapping defined in Sect. 2.6, i.e., for any $G \subseteq \mathcal{G}_u$, $\varphi_3(G) =$ "yes" if $G = \mathcal{G}_u$, and $\varphi_3(G) =$ "no" otherwise.
3. Prepare a set R of ET rules to transform MI problems preserving their answers. For simplicity, we only use rules that do not change the second argument of a state (i.e., do not change the exit mapping).
4. Transform $\langle Cs, \varphi_3 \rangle$ equivalently by using ET rules in R.
5. If the transformation sequence reaches a MI problem of the form $\langle Cs', \varphi_3 \rangle$, where Cs' is a clause set that contains the empty clause, then the answer is "yes". If the transformation sequence reaches a MI problem of the form $\langle Cs', \varphi_3 \rangle$, where Cs' is a set of positive clauses, then the answer is "no".

For transformation to a conjunctive normal form at Step 2, the usual Skolemization [5] is used. Since resolution and factoring are ET rules [1], the solution for proof problems here includes the usual proof method using resolution. More precisely, for any computation path by the "resolution" proof method using the resolution and factoring inference rules, there is a corresponding MI-based computation path using resolution and factoring ET rules. We may find a better computation path by using other sets of ET rules. For instance, one disadvantage of using inference rules is an increase in the number of clauses, and this disadvantage can be solved by the use of unfolding and definite-clause removal ET rules.

6   Example

Usual first-order atoms are used for illustration below. To apply the proposed theory in this section, a specialization system $\langle \mathcal{A}_u, \mathcal{G}_u, \mathcal{S}, \mu_u \rangle$ corresponding to the usual first-order space is used, where $\mathcal{A}_u$ is the set of all first-order atoms, $\mathcal{G}_u$ is the set of all ground first-order atoms, $\mathcal{S}$ is the set of all substitutions on $\mathcal{A}_u$, and μ_u provides the specialization operation corresponding to the usual application of substitutions in $\mathcal{S}$ to atoms in $\mathcal{A}_u$.

6.1   Problem Description

Let Cs be the set consisting of the clauses C_1–C_{27} in Fig. 2. These clauses are obtained from the *mayDoThesis* problem given in [7] with some modification.[1]

C_1: $FM(x) \leftarrow FP(x)$ C_2: $FP(john) \leftarrow$

C_3: $FP(mary) \leftarrow$ C_4: $teach(john, ai) \leftarrow$

C_5: $St(paul) \leftarrow$ C_6: $AC(ai) \leftarrow$

C_7: $Tp(kr) \leftarrow$ C_8: $Tp(lp) \leftarrow$

C_9: $curr(x, z) \leftarrow exam(x, y), subject(y, z), St(x), Co(y), Tp(z)$

C_{10}: $mayDoThesis(x, y) \leftarrow curr(x, z), expert(y, z), St(x), Tp(z), FP(y), AC(w),$
$\qquad\qquad\qquad\qquad teach(y, w)$

C_{11}: $mayDoThesis(x, y) \leftarrow St(x), NFP(y)$

C_{12}: $exam(paul, ai) \leftarrow$ C_{13}: $subject(ai, kr) \leftarrow$

C_{14}: $subject(ai, lp) \leftarrow$ C_{15}: $expert(john, kr) \leftarrow$

C_{16}: $expert(mary, lp) \leftarrow$ C_{17}: $AC(x) \leftarrow teach(mary, x)$

C_{18}: $\leftarrow AC(x), BC(x)$ C_{19}: $AC(x), BC(x) \leftarrow Co(x)$

C_{20}: $Co(x) \leftarrow AC(x)$ C_{21}: $Co(x) \leftarrow BC(x)$

C_{22}: $FP(x) \leftarrow NFP(x)$ C_{23}: $\leftarrow NFP(x), teach(x, y), Co(y)$

C_{24}: $teach(y, x), NFP(y) \leftarrow FP(y), funcf_0(y, x)$

C_{25}: $Co(x), NFP(y) \leftarrow FP(y), funcf_0(y, x)$

C_{26}: $funcf_0(john, ai) \leftarrow$

C_{27}: $\leftarrow funcf_0(mary, ai)$

Fig. 2. Clauses representing the background knowledge of the modified *mayDoThesis* problem.

---

[1] To represent the original *mayDoThesis* problem in a clausal form, extended clauses with function variables are used. To change atoms with function variables into user-defined atoms, the *funcf_0* predicate is used in the clauses C_{24}–C_{27}.

All atoms appearing in Fig. 2 belong to $\mathcal{A}_u$. The unary predicates *NFP*, *FP*, *FM*, *Co*, *AC*, *BC*, *St*, and *Tp* denote "non-teaching full professor," "full professor," "faculty member," "course," "advanced course," "basic course," "student," and "topic," respectively. The clauses C_9–C_{11} together provide the conditions for a student to do his/her thesis with a professor, where $mayDoThesis(s,p)$, $curr(s,t)$, $expert(p,t)$, $exam(s,c)$, and $subject(c,t)$ are intended to mean "s may do his/her thesis with p," "s studied t in his/her curriculum," "p is an expert in t," "s passed the exam of c," and "c covers t," respectively, for any student s, any professor p, any topic t, and any course c.

Let a be the atom $mayDoThesis(paul, x)$. We consider the QA problem $\langle Cs, a \rangle$, which is to find all students who may do their theses with *paul*. Let φ be defined by: for any $G \subseteq \mathcal{G}_u$,

$$\varphi(G) = \{ mayDoThesis(paul, x) \mid ans(x) \in G \},$$

where *ans* is a unary predicate denoting "answer." The QA problem $\langle Cs, a \rangle$ above can then be transformed into a MI problem $\langle Cs \cup \{C_0\}, \varphi \rangle$, where C_0 is the clause given by:

$$C_0 : ans(x) \leftarrow mayDoThesis(paul, x)$$

Using rules for transformation of clauses given in Sects. 6.2–6.3, how to compute the answer to the MI problem $\langle Cs \cup \{C_0\}, \varphi \rangle$ is illustrated in Sect. 6.4.

6.2   Unfolding Operation

Assume that:

- $Cs \subseteq \text{CLS}$.
- D is a set of definite clauses in CLS.
- occ is an occurrence of an atom b in the right-hand side of a clause C in Cs.

By unfolding Cs using D at occ, Cs is transformed into

$$(Cs - \{C\}) \cup \left(\bigcup \{ resolvent(C, C', b) \mid C' \in D \} \right),$$

where for each $C' \in D$, $resolvent(C, C', b)$ is defined as follows, assuming that ρ is a renaming substitution for usual variables such that C and $C'\rho$ have no usual variable in common:

1. If b and $head(C'\rho)$ are not unifiable, then $resolvent(C, C', b) = \emptyset$.
2. If they are unifiable, then $resolvent(C, C', b) = \{C''\}$, where C'' is the clause obtained from C and $C'\rho$ as follows, assuming that θ is the most general unifier of b and $head(C'\rho)$:
   (a) $lhs(C'') = lhs(C\theta)$
   (b) $rhs(C'') = (rhs(C\theta) - \{b\theta\}) \cup body(C'\rho\theta)$

The resulting clause set is denoted by $\text{UNFOLD}(Cs, D, occ)$.

Theorem 15. *Assume that:*

1. $Cs \subseteq \text{CLS}$.
2. *D is a set of definite clauses in Cs.*
3. *occ is an occurrence of an atom in the right-hand side of a clause in $Cs - D$.*
4. *φ is an exit mapping.*

Then

- $\text{MM}(Cs) = \text{MM}(\text{UNFOLD}(Cs, D, occ))$, *and*
- $ans_{\text{MI}}(Cs, \varphi) = ans_{\text{MI}}(\text{UNFOLD}(Cs, D, occ), \varphi)$.   □

6.3   Other Transformations

Definite-Clause Removal. Assume that (i) D is a set of definite clauses in Cs, (ii) the heads of all definite clauses in D are atoms with the same predicate symbol, say p, (iii) no p-atom appears in any clause in $Cs - D$, and (iv) φ only refers to q-atoms, where $q \neq p$. Then Cs can be transformed into $Cs - D$.

Elimination of Subsumed Clauses and Elimination of Valid Clauses. A clause C_1 is said to *subsume* a clause C_2 iff there exists a substitution θ for usual variables such that $lhs(C_1)\theta \subseteq lhs(C_2)$ and $rhs(C_1)\theta \subseteq rhs(C_2)$. If a clause set Cs contains clauses C_1 and C_2 such that C_1 subsumes C_2, then Cs can be transformed into $Cs - \{C_2\}$.

A clause is *valid* iff all of its ground instances are true. Given a clause C, if some atom in $rhs(C)$ belongs to $lhs(C)$, then C is valid. A valid clause can be removed.

Side-Change Transformation. Assume that p is a predicate occurring in a clause set Cs and p does not appear in a query atom under consideration. The clause set Cs can be transformed by changing the clause sides of p-atoms as follows: First, determine a new predicate *notp* for p. Next, move all p-atoms in each clause to their opposite side in the same clause (i.e., from the left-hand side to the right-hand side and vice versa) with their predicates being changed from p to *notp*.

Side-change transformation is useful for deriving a new set of clauses to which unfolding with respect to p-atoms can be applied, when each clause C in Cs satisfies the following two conditions: (i) there is at most one p-atom in the right-hand side of C, and (ii) if there is exactly one p-atom in the right-hand side of C, then all user-defined atoms in the left-hand side of C are p-atoms.

6.4   ET Computation

The clause set $Cs \cup \{C_0\}$, consisting of C_0–C_{27}, given in Sect. 6.1 is transformed using ET rules provided by Sects. 6.2–6.3 as follows:

C_{28}: $teach(john, ai) \leftarrow$
C_{29}: $AC(ai) \leftarrow$
C_{30}: $AC(x) \leftarrow teach(mary, x)$
C_{31}: $\leftarrow AC(x), BC(x)$
C_{32}: $AC(x), BC(x) \leftarrow Co(x)$
C_{33}: $Co(x) \leftarrow AC(x)$
C_{34}: $Co(x) \leftarrow BC(x)$
C_{35}: $\leftarrow NFP(x), teach(x, y), Co(y)$
C_{36}: $ans(y) \leftarrow NFP(x)$
C_{37}: $ans(john) \leftarrow AC(x), teach(john, x), Co(ai)$
C_{38}: $ans(mary) \leftarrow AC(x), teach(mary, x), Co(ai)$
C_{39}: $ans(john) \leftarrow AC(x), teach(john, x), NFP(john), Co(ai)$
C_{40}: $ans(mary) \leftarrow AC(x), teach(mary, x), NFP(mary), Co(ai)$
C_{41}: $teach(john, ai), NFP(john) \leftarrow$
C_{42}: $Co(ai), NFP(john) \leftarrow$

Fig. 3. Clauses obtained by application of unfolding and application of basic transformation rules

C_{43}: $ans(x), notNFP(x) \leftarrow$
C_{44}: $notNFP(john) \leftarrow$
C_{45}: $ans(john) \leftarrow$
C_{46}: $\leftarrow BC(ai)$

Fig. 4. Clauses obtained by further application of transformation rules

- By (i) unfolding using the definitions of the predicates *mayDoThesis*, *FP*, *Tp*, *curr*, *subject*, *expert*, *St*, *exam*, *funcf$_0$*, and *FM*, (ii) removing these definitions using definite-clause removal, and (iii) removal of valid clauses, the clauses C_0–C_{27} are transformed into the clauses C_{28}–C_{42} in Fig. 3.
- Side-change transformation for *NFP* enables (i) unfolding using the definitions of *teach*, *Co*, and *AC*, (ii) elimination of these definitions using definite-clause removal, (iii) removal of valid clauses, and (iv) elimination of subsumed clauses. By such side-change transformation followed by transformation of these four types, C_{28}–C_{42} are transformed into the clauses C_{43}–C_{46} in Fig. 4.
- Side-change transformation for *notNFP* enables unfolding using the definitions of *BC* and *NFP*. By unfolding and definite-clause removal, C_{43}–C_{46} are transformed into C_{45}, i.e., $(ans(john) \leftarrow)$.

As a result, the MI problem $\langle Cs \cup \{C_0\}, \varphi \rangle$ in Sect. 6.1 is transformed equivalently into the MI problem $\langle \{(ans(john) \leftarrow)\}, \varphi \rangle$. Hence

$$ans_{\mathrm{MI}}(Cs \cup \{C_0\}, \varphi) = ans_{\mathrm{MI}}(\{(ans(john) \leftarrow)\}, \varphi)$$
$$= \varphi(\bigcap Models(\{(ans(john) \leftarrow)\}))$$
$$= \{mayDoThesis(paul, john)\}.$$

6.5   A Proof Problem and Its Solution

By modifying the QA problem in Sect. 6.1, we have a proof problem that determines whether there is a student who may do his/her thesis with *paul*.

This proof problem is formalized as a MI problem $\langle Cs \cup \{C_0\} \cup \{C_{neg}\}, \varphi \rangle$, where C_{neg} is the negative clause $(\leftarrow ans(x))$ and for any $G \subseteq \mathcal{G}_u$, $\varphi(G) = $ "yes" if $G = \mathcal{G}_u$, and $\varphi(G) = $ "no" otherwise.

One way to solve this problem is to transform $Cs \cup \{C_0\} \cup \{C_{neg}\}$ in a way similar to the transformation in Sect. 6.4. Even in the presence of the negative clause C_{neg}, the same transformation rules can be applied and the clause set $Cs \cup \{C_0\} \cup \{C_{neg}\}$ is transformed into $\{(ans(john) \leftarrow)\} \cup \{C_{neg}\}$, from which the clause set

$$\{(ans(john) \leftarrow)\} \cup \{(\leftarrow)\}$$

can be obtained by unfolding. Since this clause set contains the empty clause, $ans_{\mathrm{MI}}(Cs \cup \{C_0\} \cup \{C_{neg}\}, \varphi) = $ "yes".

7   Conclusions

A model-intersection problem (MI problem) is a pair $\langle Cs, \varphi \rangle$, where Cs is a set of clauses and φ is an exit mapping used for constructing the output answer from the intersection of all models of Cs. The class of MI problems considered in this paper has many parameters, such as abstract atoms, specializations, restriction on forms of clauses, etc. By instantiating these parameters, we can obtain theories for subclasses of QA and proof problems corresponding to conventional clause-based theories, such as datalog, Prolog, and many other extensions of Prolog.

The proposed solution schema for MI problems comprises the following steps: (i) formalize a given problem as an MI problem on some specialization system, (ii) prepare ET rules from answers/target mappings, (iii) construct an ET sequence preserving answers/target mappings, and (iv) compute the answer by using some answer mapping (possibly constructed on some target mapping).

We introduced the concept of target mapping and proposed three target mappings, i.e., τ_1 for sets of positive unit clauses, τ_2 for sets of definite clauses, and $\mathbb{MM}$ for arbitrary sets of clauses. These target mappings provide a strong foundation for inventing many ET rules for solving MI problems on clauses. Many kinds of ET rules, including the resolution and factoring ET rules, are realized by transformations that preserve these target mappings. For instance, a proof based on the resolution principle can be regarded as ET computation using the resolution and factoring ET rules. By introducing new ET rules, we can devise a new proof method [2]. By inventing additional ET rules, we have been successful in solving a large class of QA problems [3].

By instantiation, the class of MI problems on specialization systems produces, among others, one of the largest classes of logical problems with first-order atoms and substitutions. The ET solution has been proved to be very general and fundamental since its correctness for such a large class of problems has been shown in this paper. By its generality, the theory developed in this paper makes clear the fundamental and central structure of representation and computation for logical problem solving.

Acknowledgments. This research was partially supported by JSPS KAKENHI Grant Numbers 25280078 and 26540110.

References

1. Akama, K., Nantajeewarawat, E.: Proving theorems based on equivalent transformation using resolution and factoring. In: Proceedings of the Second World Congress on Information and Communication Technologies, WICT 2012, Trivandrum, India, pp. 7–12 (2012)
2. Akama, K., Nantajeewarawat, E.: Embedding proof problems into query-answering problems and problem solving by equivalent transformation. In: Proceedings of the 5th International Conference on Knowledge Engineering and Ontology Development, Vilamoura, Portugal, pp. 253–260 (2013)
3. Akama, K., Nantajeewarawat, E.: Equivalent transformation in an extended space for solving query-answering problems. In: Nguyen, N.T., Attachoo, B., Trawiński, B., Somboonviwat, K. (eds.) ACIIDS 2014. LNCS (LNAI), vol. 8397, pp. 232–241. Springer, Heidelberg (2014). doi:10.1007/978-3-319-05476-6_24
4. Baader, F., Calvanese, D., McGuinness, D.L., Nardi, D., Patel-Schneider, P.F. (eds.): The Description Logic Handbook, 2nd edn. Cambridge University Press, Cambridge (2007)
5. Chang, C.L., Lee, R.C.T.: Symbolic Logic and Mechanical Theorem Proving. Academic Press, Cambridge (1973)
6. Clark, K.L.: Negation as failure. In: Gallaire, H., Minker, J. (eds.) Logic and Data Bases, pp. 293–322. Plenum Press, New York (1978)
7. Donini, F.M., Lenzerini, M., Nardi, D., Schaerf, A.: $\mathcal{AL}$-log: integrating datalog and description logics. J. Intell. Inf. Syst. **16**, 227–252 (1998)
8. Fitting, M.: First-Order Logic and Automated Theorem Proving, 2nd edn. Springer, Berlin (1996)
9. Gelfond, M., Lifschitz, V.: The stable model semantics for logic programming. In: Proceedings of International Logic Programming Conference and Symposium, pp. 1070–1080. MIT Press (1988)
10. Gelfond, M., Lifschitz, V.: Classical negation in logic programs and disjunctive databases. New Gener. Comput. **9**, 365–386 (1991)
11. Lloyd, J.W.: Foundations of Logic Programming, 2nd edn. Springer, Berlin (1987)
12. Robinson, J.A.: A machine-oriented logic based on the resolution principle. J. ACM **12**, 23–41 (1965)
13. Tessaris, S.: Questions and answers: reasoning and querying in description logic. Ph.D. thesis, Department of Computer Science, The University of Manchester, UK (2001)

Representing and Managing Unbalanced Multi-sets

Nouha Chaoued[1,2(✉)], Amel Borgi[1], and Anne Laurent[2]

[1] Université de Tunis El Manar, LIPAH, Tunis, Tunisia
Amel.Borgi@insat.rnu.tn
[2] Université de Montpellier, LIRMM, Montpellier, France
{nouha.chaoued,anne.laurent}@lirmm.fr

Abstract. In Knowledge-Based Systems, experts should model human knowledge as faithful as possible to reality. In this way, it is essential to consider knowledge imperfection. Several approaches have dealt with this kind of data. The most known are fuzzy logic and multi-valued logic. These latter propose a linguistic modeling using linguistic terms that are uniformly distributed on a scale. However, in some cases, we need to assess qualitative aspects by means of variables using linguistic term sets which are not uniformly distributed. We have noticed, in the literature, that in the context of fuzzy logic many researchers have dealt with these term sets. However, it is not the case for multi-valued logic. Thereby, in our work, we aim to establish a methodology to represent and manage this kind of data in the context of multi-valued logic. Two aspects are treated. The first one concerns the representation of terms within an unbalanced multi-set. The second deals with the use of symbolic modifiers within such kind of imperfect knowledge.

Keywords: Imperfect knowledge · Multi-valued logic · Unbalanced terms · Symbolic modifiers

1 Introduction

Knowledge handled by humans is often imperfect. These imperfections may be due to ambiguity, incompleteness, imprecision, uncertainty, inconsistency, etc. Several approaches were suggested in the literature for such knowledge representation and treatment. The most known are fuzzy logic [1] and multi-valued logic [2,3].

Humans are able to perform reasoning without any exact measurements. They mostly use abstract terms of natural language (*young, old, mature,* etc.) and symbolic data rather than numerical values or qualitative ones. Terms can also be composed by using adverbs, such as *little, more or less* and *slightly*. Both fuzzy logic and multi-valued logic propose a linguistic term modeling to allow using words in reasoning process. They use linguistic variables that take values in a set of linguistic terms [4]. These latter express the various nuances of the processed information using words.

© Springer International Publishing AG 2016
A. Fred et al. (Eds.): IC3K 2015, CCIS 631, pp. 213–233, 2016.
DOI: 10.1007/978-3-319-52758-1_13

Generally in Knowledge-Based Systems, experts use linguistic terms that are uniformly and symmetrically distributed on a scale. However, in many cases, linguistic information needs to be defined by unbalanced term sets whose terms are not uniformly and/or not symmetrically distributed. For example, in the evaluation process, we often consider a single negative term, e.g. *Fail*, and many positive terms such as *Medium, Good, Excellent*, etc. The gap between these terms is unequal. Herrera et al. [5,6] deal with this unbalanced linguistic term set in the context of fuzzy logic.

In this paper, we focus on multi-valued logic. It allows to symbolically represent imprecise knowledge using ordered adverbial expressions of natural language [2]. We have noticed that in the context of this logic, few studies have treated unbalanced linguistic term sets [7,8]. The aim of this paper is to introduce a new approach to represent and treat such data in the context of multi-valued logic. It is based on our previous work [8] that expresses unbalanced terms using a uniform multi-set. In the present work, we apply our proposal to an Information Retrieval System (IRS). This latter aims to retrieve a set of documents that satisfies a user query. This system is composed of three units [9]:

1. A **documentary archive** or a database including a set of documents. They are represented by means of index terms describing their subject content.
2. A **query subsystem** presenting user needs by means of weighted queries. It indicates the topics that he/she is asking for.
3. An **evaluation subsystem** allowing the evaluation of documents according to their relevance compared to the user query.

Usually, users are interested in documents whose contents are the most relevant to their queries. This implies the use of more precise labels in the left (positive) interval than in the right (negative) one [9]. Thus, in IRS the use of non-uniformly distributed term sets is recommended. Indeed, it is more appropriate to use such kind of multi-sets to represent the relevance degrees of the documents or to express the weights of index terms in the queries. To achieve this, we will use the unbalanced set $S_{un} = \{None, Low, Medium, High, QuiteHigh, VeryHigh, Total\} = \{N, L, M, H, QH, VH, T\}$ [9] (Fig. 1).

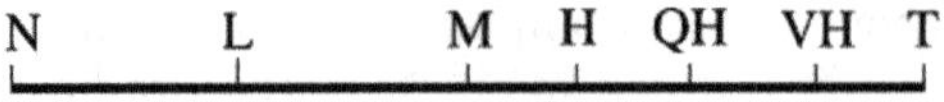

Fig. 1. Unbalanced linguistic term set.

Two aspects are discussed in our work. The first concerns the representation of terms within an unbalanced multi-set. The second deals with the management of such a kind of knowledge. Figure 2 illustrates the different steps of this process. In the first one, we apply the single scale algorithm [8] to express a term belonging to a non uniform multi-set (L in the figure) using a uniform multi-set (L is represented by τ_2'). Afterwards, existing tools designed for balanced sets in

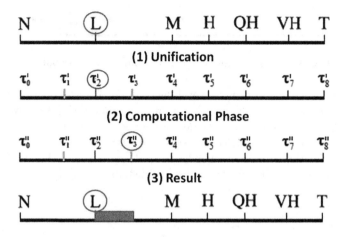

Fig. 2. Management of unbalanced linguistic term set.

the context of multi-valued logic, as aggregation operators or modifiers, can be applied to the obtained term (in this example τ'_2 is modified into τ''_3 using the modifier CR_1). The last step aims to express the result of the computational phase with a term from the initial unbalanced multi-set (*Approximately L* in the figure).

This article is organized as follows. We start, in Sect. 2, with an explanation of what is unbalanced term sets. In Sect. 3, we introduce the basic concepts of multi-valued logic and of symbolic modifiers. Existing works that concern the representation of unbalanced multi-sets are presented in Sect. 4. Then, Sect. 5 introduces our approach to express terms within an unbalanced set (Fig. 2 Step 3). Finally, we propose a new way to use Generalized Symbolic Modifiers (GSM) with unbalanced linguistic terms (Fig. 2 Step 2).

2 Preliminaries

Most Knowledge-Based Systems use linguistic sets with terms that are uniformly and symmetrically distributed. However, there are other cases that need to assess qualitative aspect by means of variables using linguistic terms which are not uniformly and/or symmetrically distributed, named unbalanced linguistic sets. In many real-life situations, these latter are used as in project investment, negotiation process, evaluation process, etc. Asymmetric linguistic information can be a consequence of the nature of the linguistic variables involved in the problem such as personal examination or evaluation system (Fig. 3) [10]. Terms are not equidistant, e.g. the distance between *Poor* and *Average* is greater than between *Average* and *Good*. This difference indicates the expert's interest in having a more precise definition of a part of the domain, that leads to the use of more labels in this interval.

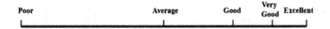

Fig. 3. Set of 5 linguistic terms not uniformly distributed.

Moreover, in many problems, several decision makers are involved. In fact, it is more reliable to obtain a decision based on the opinion of several experts than on a single one. In a multi-experts decision making process, each expert may assess his knowledge with a particular scale having a specific granularity (Fig. 4) [10]. It indicates their different knowledge backgrounds or judging abilities. In each used linguistic set, terms are uniformly distributed.

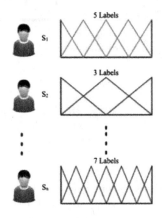

Fig. 4. Fuzzy multi-granular scales used by experts.

Herrera et al. [6] considered that an unbalanced fuzzy set is a set with a minimum term, a maximum term and a central term s_c called also midpoint. The remaining terms are neither uniformly nor symmetrically distributed on both side of central term:

$$S = S_L \cup S_C \cup S_R \tag{1}$$

With:

- S_L: the subset including terms on the left of the central term s_c and $\#(S_L)$ its cardinality;
- S_C: the singleton containing the term s_c;
- S_R: the subset including terms on the right of the central term s_c and $\#(S_R)$ its cardinality.

Hence, S is described by:

$$S = \{(\#(S_L), density_L), 1, (\#(S_R), density_R)\} \tag{2}$$

where 1 is the midpoint. The density corresponds to a high granularity around the central term or the end of fuzzy partition. Therefore, the density value can be *middle* or *extreme*.

For example, the set represented by Fig. 3 will be described by $S = \{(1, extreme), 1, (3, extreme)\}$. In fact, the central term is $\{Average\}$. The left subset includes only the term $\{Poor\}$ which is the minimum term. While the right subset includes $\{Good, Very\ Good, Excellent\}$. These terms are closer to the maximum term *Excellent*. So the density corresponds to *extreme* for the two subsets.

Xu [11] proposed to define the unbalanced set as a set of $(2t - 1)$ labels, with t a positive integer, as follows:

$$S^{(t)} = \{s_\beta^{(t)}|\beta = (1-t), \frac{2}{3}(2-t), \frac{2}{4}(3-t), \ldots, 0, \ldots, \frac{2}{4}(t-3), \frac{2}{3}(t-2), (t-1)\} \quad (3)$$

Thus, the central term has index 0 and other terms have positive or negative indices. The main idea of this definition is that the absolute value of the deviation between the indices of two successive terms increases regularly proceeding from the center term to the end of the scale. Figure 5 [11] illustrates an unbalanced set with $t = 4$. We notice that the terms are symmetrically distributed around the central term.

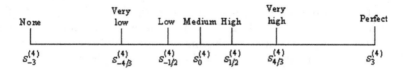

Fig. 5. A set of seven linguistic labels $S^{(4)}$.

In this work, we propose to define an unbalanced linguistic term set S_{un} as a set of degrees that includes a minimum degree, *False*, a maximum degree, *True*, and the remaining ones are neither uniformly nor symmetrically distributed on the scale. The gaps between adjacent terms may be unequal. Set granularity, i.e. its cardinality, can be odd or even. Each term is defined by its position on the scale. They are supplied directly (v_i) or by specifying the gap (d_i) between each successive terms. For example, we consider the set $L_5 = \{A, B, C, D, E\}$. These terms can be specified by they positions (Fig. 6-a):

- $v_0 = v_A = 0$
- $v_1 = v_B = 0.15$
- $v_2 = v_C = 0.25$
- $v_3 = v_D = 0.5$
- $v_4 = v_E = 1$

They can also be defined using the gap (Fig. 6-b) between their positions v_{i+1} and v_i calculated as follows:

$$d_i = v_{i+1} - v_i \tag{4}$$

In our example, the distances are:

- $d_0 = v_1 - v_0 = 0.15$
- $d_1 = v_2 - v_1 = 0.1$
- $d_2 = v_3 - v_2 = 0.25$
- $d_3 = v_4 - v_3 = 0.5$

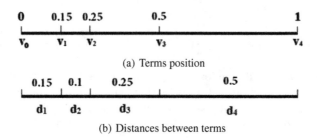

(a) Terms position

(b) Distances between terms

Fig. 6. Terms definitions.

There are two important subjects to treat when dealing with imperfect knowledge: (1) its representation and (2) its management. Many works have dealt with these term sets in the fuzzy logic context [5,9–15]. Researchers have proposed approaches for their representation [5,10,14] and their management, such as operators for unbalanced aggregation [13,15]. In the context of multi-valued logic, only Abchir and we, in [7,8] respectively, proposed algorithms to represent unbalanced term sets within a uniform set. Our present work, developed in the multi-valued context, has two targets: the first one is to propose a new way to represent uniform terms within unbalanced multi-sets; the second intends to define an approach to use symbolic modifiers with such knowledge.

3   Multi-valued Logic and Generalized Symbolic Modifiers

Multi-valued logic is based on De Glas's multi-set theory [2]. In this approach, each linguistic term is represented by a multi-set. Knowledge is expressed thanks to an ordered and finite scale of M symbols denoted by [2,16]:

$$L_M = \{\tau_0, \tau_1, \ldots, \tau_{(M-1)}\}; M \geq 2 \tag{5}$$

τ_i is the membership degree to the multi-set ($i \in [0, M-1]$).

It should be noted that symbolic degrees are connected only by the total order relation $\leq$, defined by [17]:

$$\tau_\alpha \leq \tau_\beta \iff \alpha \leq \beta; \forall \alpha \text{ and } \beta \in [0, M-1] \tag{6}$$

In most existing works on multi-valued logic, these degrees are assumed to be uniformly distributed on the scale. The membership relation in multi-valued logic is partial:

$$x \in_\alpha A \Longleftrightarrow x \text{ belongs to } A \text{ at a degree } \alpha \qquad (7)$$

To express the imprecision of a predicate, a qualifier ϑ_α is associated with each degree:

$$x \text{ is } \vartheta_\alpha A \Longleftrightarrow (x \text{ is } \vartheta_\alpha A) \text{ is true}$$
$$\Longleftrightarrow (x \text{ is } A) \text{ is } \tau_\alpha \text{ true}$$

For example, saying that *the climate is little humid* means that *the climate* satisfies the predicate *humid* with *the degree* τ_2 as shown in the Fig. 7.

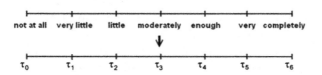

Fig. 7. Representation of a scale of 7 truth-degrees.

The works of Zadeh [18], in fuzzy logic, and more recently those of Akdag et al. [19,20] and Truck [21], in multi-valued logic context, consider that any fuzzy subset or multi-valued symbol can be considered as a modification of another fuzzy subset or multi-valued symbol respectively. A modifier allows modeling knowledge and gradual reasoning to build new terms from the initial ones, or to compare two values by finding the modifier that allows the transformation from one to another.

In multi-valued logic, a membership in a multi-set is characterized by a symbolic membership degree τ_i defined on a scale of ordered degrees L_M. Thus, the data modification is a transformation of a degree and/or the scale of the multi-set. Indeed, some linguistic modifiers preserve the same multi-set, but modify the membership degree. Others transform a multi-set towards another one. Hence, it leads to an expansion or erosion of the original scale.

Symbolic linguistic modifiers have been proposed by Akdag et al. [19,20] and they were generalized and formalized by Truck [21,22]. They were named Generalized Symbolic Modifiers (GSM). According to Truck [21], a GSM (Definition 1) is a triplet of parameters: radius, nature (i.e. dilated, eroded or preserved) and mode (i.e. reinforcing, weakening or central). The radius is denoted by ρ with $\rho \in \mathbb{N}^*$. The higher ρ, the more powerful the modifier.

Definition 1 *[21]. Let τ_i be a symbolic degree, such that $i \in \mathbb{N}$, in a scale L_M of M terms $(M \in \mathbb{N}^* \backslash \{1\})$ with $i < M$. Let m be a GSM with radius ρ denoted m_ρ. The modifier m_ρ is a function that performs a linear transformation of τ_i to a new degree $\tau_{i'} \in L_{M'}$ (where $L_{M'}$ is the linear transformation of L_M) according to a radius $\rho : m_\rho(i) = i'; m_\rho(M) = M'$.*

Table 1. Some examples of generalized symbolic modifiers (GSM) [21].

Mode	Nature	Modifier	Effect
Weakening	Erosion	EW_ρ	$m(i) = \max(0, i - \rho)$ $m(M) = \max(2, M - \rho)$
	Dilatation	DW_ρ	$m(i) = i$ $m(M) = M + \rho$
		DW'_ρ	$m(i) = \max(0, i - \rho)$ $m(M) = M + \rho$
	Conservation	CW_ρ	$m(i) = \max(0, i - \rho)$ $m(M) = M$
Reinforcing	Erosion	ER_ρ	$m(i) = i$ $m(M) = \max(i + 1, M - \rho)$
		ER'_ρ	$m(i) = \min(i + \rho, M - \rho - 1)$ $m(M) = \max(1, M - \rho)$
	Dilatation	DR_ρ	$m(i) = i + \rho$ $m(M) = M - \rho$
	Conservation	CR_ρ	$m(i) = \min(i + \rho, M - 1)$ $m(M) = M$
Central	Erosion	EC_ρ	$m(i) = \max(\lfloor \frac{i}{\rho} \rfloor, 1)$ [a] $m(M) = \max(\lfloor \frac{M}{\rho} \rfloor + 1, 2)$ [a]
	Dilatation	DC_ρ	$m(i) = i\rho$ $m(M) = M\rho - \rho + 1$

[a] $\lfloor . \rfloor$ is the flour function.

For each linguistic degree, Akdag et al. [19] associate a numerical rate, i.e. an intensity level. In fact, an item from the multi-set can be considered as a precision degree of a proposition. This latter, named proportion, is the quotient $prop(\tau_i) = \frac{p(\tau_i)}{M-1}$ associated with each τ_i such that $p(\tau_i)$ is its position in the scale, i.e. i. Thus, $Prop(\tau_i)$ is the weight of the degree τ_i relatively to the linguistic set granularity, i.e. its intensity compared to the truth degree $\tau_{(M-1)}$ (True).

Symbolic modifiers are classified as: weakening when the proportion decreases, i.e. $prop(\tau_{i'}) < prop(\tau_i)$ (the four weakening modifiers defined in [21] are $EW_\rho, DW_\rho, DW'_\rho$ and CW_ρ); reinforcing when it increases, i.e. $prop(\tau_{i'}) > prop(\tau_i)$ (the four reinforcing modifiers proposed in [21] are $ER_\rho, DR_\rho, DR'_\rho$ and CR_ρ) or central if $prop(\tau_i)$ does not change, such modifiers may act as a zoom on the base (the four central modifiers presented in [21] are $EC_\rho, EC'_\rho, DC_\rho$ and DC'_ρ). Some examples of GSM are presented in Table 1.

The Fig. 8 shows the results of applying weakening and reinforcing modifiers to the term τ_3 in L_7 with radius equals to 1.

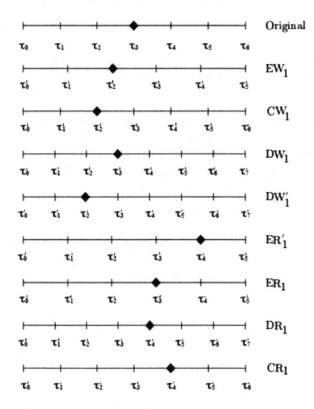

Fig. 8. Results of applying weakening and reinforcing modifiers [21].

4   Representing Unbalanced Multi-sets

In the context of multi-valued logic, few works have treated the non-uniformly dis-tributed knowledge. Abchir proposed in the discussion of his thesis [7] a first approach based on the use of Generalized Symbolic Modifiers (GSM). We also proposed in a previous work [8] a modified version of the Abchir's algorithm to represent unbalanced degrees on a single uniform scale. These two algorithms will be presented in the following subsections.

4.1   Abchir's Algorithm

This approach aims to represent numerical input values $v_1, v_2, \ldots$, representing the position of terms on a scale, as symbolic degrees of a uniform multi-set L_M. The algorithm starts by calculating a γ coefficient multiplier to transform input values (v_i) to integer ones, denoted val_i, if they are not. Then, each obtained integer value is transformed into a couple (τ_i, L_M) according to the following rule:

$$M = max(2, val_{i+1}); \tau_i = \tau_{val_i} \tag{8}$$

The procedure is iterative. It stops when all input values have been treated. The author denotes the first M by M_1. If the next value to treat cannot be integrated in this set, i.e. $val_2 \geq M_1$, a new value M, denoted by M_2, is calculated such that: $M_2 = val_2 + 1$.

Thus, the couple (val_2, M_2) leads to (τ_{val_2}, L_{M_2}). When M changes, previously calculated couples must be represented within the new scale L_{M_2}. The weakening expanding modifier $DW_{(M_2 - M_1)}$ (Table 1) is used to do this.

The process continues for each value, ensuring that the multiplier coefficient γ always generates integer values. If this is not the case, the multiplier γ should be changed in a new γ that allows transforming all considered values into integers, and its old value will be denoted by γ_{old}. Then, all previous values must be recalculated using the central expanding modifier DC_c (Table 1). The multiplier c allows to transform γ_{old} into γ. It is obtained as follows:

$$\gamma = c * \gamma_{old} \tag{9}$$

To illustrate this algorithm, let us consider the linguistic term set $S_{un} = L_7 = \{N, L, M, H, QH, VH, T\}$ [9] (Fig. 1). We aim to represent each term of L_7 using a term from a uniform linguistic set. As input, we consider the positions of terms as:

– Term N: $v_0 = 0$
– Term L: $v_1 = 0.25$
– Term M: $v_2 = 0.5$
– Term H: $v_3 = 0.625$
– Term QH: $v_4 = 0.75$
– Term VH: $v_5 = 0.875$
– Term T: $v_6 = 1$

Let us suppose that the first term to treat is $(M, 0.5)$ then γ is equal to 10. Thus: $val_M = 0.5 * 10 = 5$; $M = max(2, 5 + 1) = 6$. So $(M, 0.5)$ will be represented by (τ_5, L_6).

Considering the couple $(L, 0.25)$ as the second term treated, the corresponding value is: $val_L = 0.25 * 10 = 2.5$. It is not an integer. Therefore, $\gamma_{old} = 10$ and $\gamma = c * 10 = 100$. Then we get $c = 10$ and $val_L = 25$. Consequently, the modifier DC_{10} is applied to the couple (τ_5, L_6). Hence, the term $(M, 0.5)$ is represented by (τ_{50}, L_{51}). While the term $(L, 0.25)$ is represented by (τ_{25}, L_{51}) (Fig. 9).

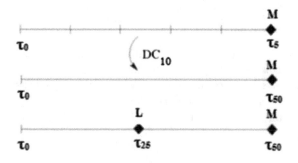

Fig. 9. Example of applying Abchir's approach using GSM.

The corresponding value to the couple $(QH, 0.75)$ is $val_{QH} = 0.75 * 100 = 75 > 51 = M$. The term QH cannot be incorporated into the set L_{51}. Therefore, M is changed from 51 to 76 and the couple $(QH, 0.75)$ is represented by (τ_{75}, L_{76}). It is then necessary to modify the terms L and M already calculated by applying $DW_{(76-51)} = DW_{25}$ (Fig. 10). They become respectively (τ_{25}, L_{76}) and (τ_{50}, L_{76}).

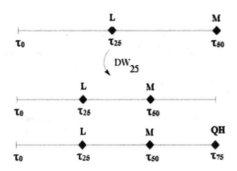

Fig. 10. Second example of applying Abchir's approach using GSM.

The process continues for each value, ensuring that the multiplier coefficient γ always generates integer values and that the value treated could be integrated in the used multi-set. The result of this algorithm, assuming that the values input order is: M, L, QH, H, VH, T, L is described in Fig. 11.

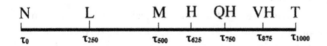

Fig. 11. Result of applying Abchir's approach using GSM.

For this algorithm, neither considered values to treat nor their number is known in advance. Partitioning will be done as the value is specified. The final result remains the same if a change is made in input terms order. The main criticism made to this approach is its iterative part for recalculating pairs representing terms already treated. This is done each time that the used multi-set or the value of the multiplier γ is modified.

4.2 Single Scale Algorithm

The difference between Abchir's algorithm [7] and the single scale approach [8] is in the input data as well as in their treatment. In the single scale approach, input data can be the positions of the values to partition or the distances between them. This last case, not allowed with Abchir's approach, offers more flexibility to the users.

The gap between each pair of successive terms reflects a difference in their meaning. In this proposal, inputs are expressed by numerical values which are supplied directly (d_i) (Fig. 12-a) or specified by the position of each term on the scale (v_i) (Fig. 12-b). In the latter case, the distance between the terms (more precisely their positions) v_{i+1} and v_i are calculated using formula (4).

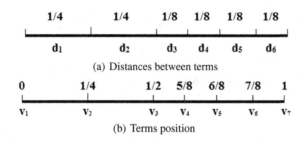

Fig. 12. Algorithm inputs.

This proposal is based on Espinilla et al.'s work [23]. The authors proposed a new definition of linguistic hierarchies: Extended Linguistic Hierarchies (ELH), within fuzzy logic. To build an ELH, they use a finite number of levels $l(t, n(t))$ with:

- t: the number indicating the hierarchy level with t $= 1,\ldots,m$; such as m is the number of experts;
- $n(t)$: the granularity, i.e., the number of terms, of the uniform linguistic term set corresponding to the level t. Each linguistic term has a triangular membership function, uniformly and symmetrically distributed on $[0,1]$. The granularity of each level is always odd.

To express his knowledge, each expert can use a specific level, i.e. a particular granularity [23]. Espinilla et al. proposed to add a new level $l(t^*, n(t^*))$ (Definition 2) to the hierarchy with $t^* = m + 1$. This level retains all the modal points of all the previous levels. The modal points have a membership degree that is equal to one.

Definition 2 *[23]. Let $\{S^{n(1)}, \ldots, S^{n(m)}\}$ be the set of linguistic scales with any odd value of granularity. A new level, $l(t^*, n(t^*))$ with $t^* = m+1$, that keeps the former modal points of the previous m levels can have the following granularity:*

$$n(t^*) = 1 + LCM[(n(1) - 1), (n(2) - 1), \ldots, (n(m + 1) - 1)]$$

With m the number of experts.

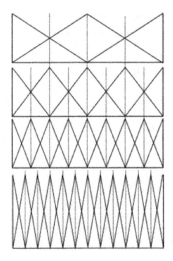

Fig. 13. Extended Linguistic Hierarchy (ELH) with 3, 5, 7 and 13 terms.

Figure 13 [23] shows an ELH with three levels of 3, 5 and 7 terms. The fourth level (t^*) has as granularity the LCM of the previous ones: $1 + LCM(3-1, 5-1, 7-1) = 13$.

In a similar way, we proposed to express unbalanced terms within a new uniform multi-set [8]. In fact, the granularity of the uniform multi-set equals the LCM of the sets representing initial terms granularity. The input data are unbalanced linguistic terms set S_{un} (L_M, M the number of terms) and the gap between terms. These are provided directly (Fig. 12-a) or by specifying the position of each term (Fig. 12-b).

First, the granularity M_k of each uniformly distributed linguistic set L_{M_K} used to represent each input term v_k is calculated. In this set, the distance between each two successive terms is denoted by d_k.

$$M_k = \frac{(1 + d_k)}{d_k}; k = 1, ..., (M - 1) \qquad (10)$$

Once the sets L_{M_K} including all the terms of the initial set are determined, the granularity of the uniform set $L_{M'}$ is deduced as:

$$M' = 1 + LCM_{k=1}^{M-1}(M_k - 1) \qquad (11)$$

Each value will be expressed on the new uniform scale $L_{M'}$ as a couple $(\tau_{\gamma_k}, L_{M'})$ according to the following rule:

$$d_k = \gamma_k * d'; k = 1, ..., (M - 1) \qquad (12)$$

Where d' is the distance between any pair of successive degrees in $L_{M'}$. For a uniform linguistic set, this distance is:

$$d' = \frac{1}{(M' - 1)} \qquad (13)$$

Algorithm 1. Single Scale algorithm.

Input:
M unbalanced linguistic terms
Normalized distances d_k with k=1, ..., (M-1)
begin
    Calculate granularities M_k; $M_k \in \mathbb{N}^* \backslash \{1\}$ with k=1, ..., (M-1)
    Calculate granularity:
    $M' = 1 + LCM_{k=1}^{M-1}(M_k - 1)$
    Deduce distance $d' = \frac{1}{(M'-1)}$
    $s_0 \leftarrow (\tau_0', L_{M'})$
    Calculate $\gamma_k = \frac{d_k}{d'}$ with k=1, ..., (M-1)
    $s_k \leftarrow (\tau_{\gamma_k}', L_{M'})$
Output: $S' = \{(\tau_0', L_{M'}),(\tau_{\gamma_1}', L_{M'}),..., (\tau_{\gamma_{p-1}}', L_{M'})\}$

To express the distances d_k according to the distance d', the multiplier γ_k is calculated using the rule (12).

The minimum term is represented by $(\tau_0', L_{M'})$ and denoted by s_0. The process, presented in Algorithm 1, continues for other values. The obtained couples $(\tau_{\gamma_k}', L_{M'})$ are denoted by s_k. To ensure the succession of terms, the calculated multiplier γ_k is added to the one calculated in the previous iteration named γ_{old}. Thus, the comparison is always done regarding the first term of the set $L_{M'}$, i.e., τ_0'.

This algorithm allows to bring us back to the uniform case which will give us the capacity to use different existing tools as linguistic modifiers [24], aggregation operators, etc.

Considering the complexity, this proposal is less complex ($O(p)$) than Abchir's ($O(p^2)$), with p the number of input terms [8]. Indeed, in this approach the treatment is done after introducing all data. Hence, the couples representing terms are calculated once.

To illustrate this algorithm, let us consider an Information Retrieval System (IRS) and the linguistic term set $S_{un} = L_7 = \{N, L, M, H, QH, VH, T\}$ [9] (Fig. 1). This latter is used to express the relevance of documents in the retrieval process. We aim to represent each term of L_7 using a term from a uniform linguistic set. As input, we consider the normalized distances in L_7 as:

- Distance $N - L : d_1 = \frac{1}{4} = 0.25$
- Distance $L - M : d_2 = \frac{1}{4} = 0.25$
- Distance $M - H : d_3 = \frac{1}{8} = 0.125$
- Distance $H - QH : d_4 = \frac{1}{8} = 0.125$
- Distance $QH - VH : d_5 = \frac{1}{8} = 0.125$
- Distance $VH - T : d_6 = \frac{1}{8} = 0.125$

First, we calculate the values M_k:
$M_1 = \frac{1+d_1}{d_1} = 5 = M_2$;
$M_3 = \frac{1+d_3}{d_3} = 9 = M_4 = M_5 = M_6$

Then, we determine the granularity M' of the uniform set and the distance between any of its successive terms d':

$M' = 1 + LCM(5-1, 9-1) = 9;$

$d' = \frac{1}{M'-1} = 0.125$

We associate the couple (τ'_0, L_9) to the term N, and we treat the other terms:

– For L (k = 1): $\gamma_{old} \leftarrow 0;$
$\gamma_1 \leftarrow \gamma_{old} + \frac{d_1}{d'} = 2;$
$L \leftarrow (\tau_2, L_9)$
– For M (k = 2): $\gamma_{old} \leftarrow 2;$
$\gamma_2 \leftarrow \gamma_{old} + \frac{d_2}{d'} = 4;$
$M \leftarrow (\tau_4, L_9)$
– For H (k = 3): $\gamma_{old} \leftarrow 4;$
$\gamma_3 \leftarrow \gamma_{old} + \frac{d_3}{d'} = 5;$
$H \leftarrow (\tau_5, L_9)$
– For QH (k = 4): $\gamma_{old} \leftarrow 5;$
$\gamma_4 \leftarrow \gamma_{old} + \frac{d_4}{d'} = 6;$
$QH \leftarrow (\tau_6, L_9)$
– For VH (k = 5): $\gamma_{old} \leftarrow 6;$
$\gamma_5 \leftarrow \gamma_{old} + \frac{d_5}{d'} = 7;$
$VH \leftarrow (\tau_7, L_9)$
– For T (k = 6): $\gamma_{old} \leftarrow 7;$
$\gamma_6 \leftarrow \gamma_{old} + \frac{d_6}{d'} = 8;$
$T \leftarrow (\tau_8, L_9)$

The algorithm output is illustrated in Fig. 14.

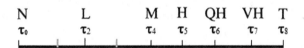

Fig. 14. Result of the single scale algorithm.

5   Representation of Terms Within an Unbalanced Set

In the previous section, we have presented our previous work [8] to represent an unbalanced term set S_{un} within a uniform term set $L_{M'}$. Thereby, existing symbolic modifiers or aggregation operators can be applied to its terms. The obtained values are also expressed using the uniform set $L_{M'}$. However, it is more understandable for the user to represent these results using terms from the initial set S_{un}.

In this section, we focus on the representation of a term τ'_i from the uniform set $L_{M'}$ with a term from S_{un} denoted by L_M. We suppose that the used distances on both sets are normalized. For further precision, we denote the degrees of the uniform set $L_{M'}$ by $\tau'_0, \ldots, \tau'_i \ldots$ and those of S_{un} by $\tau_0, \ldots, \tau_i \ldots$ (knowing that τ_0 and τ'_0 correspond to the same position 0 and τ_M and $\tau'_{M'}$ to 1). We propose a way to determine the nearest position, in S_{un}, to an initial term τ_i', represented by its position v_i in $L_{M'}$. For that purpose, we need to define the proportion of an unbalanced degree from S_{un}.

Definition 3. *Let τ_i be a symbolic degree of an unbalanced multi-set $S_{un}(L_M)$.*

Its proportion is defined as: $Prop(\tau_i) = \sum_{j=1}^{i} d_j$.

With $d_1, d_2, d_j, \ldots d_i$ the normalized distances between each pair of successive degrees in the unbalanced multi-set S_{un}.

We aim to identify the term τ_{pos} in S_{un} which distance between it and the first term (τ_0') of L_M' is the closest to v_i (v_i is the position of τ_i' on L_M'). Thus, we will compare v_i with the sum l_k of the distances separating successive terms in L_M: $l_k = \sum_{j=1}^{k} d_j$.

The process will stop when the value of l_k is higher or equal to v_i. If the values v_i and l_k are equal, the position of τ_i' in $L_{M'}$ is that of τ_{k-1}. If $v_i < l_k$, we check the closest term to v_i between τ_{k-1} and τ_k. In this case, a proportion error, denoted by α, exists and its value is the difference between the position of a term τ_i' and that of the closest term, i.e. τ_{k-1} or τ_k. If these terms are equidistant to τ_i', we choose the term with the smallest index. This approach is described in Algorithm 2.

The inputs for this function are the position v_i of the term τ_i', the M terms of the unbalanced multi-set S_{un} and the normalized distances d_k between its successive terms. As output, we obtain the couple (τ_{pos}, α).

To illustrate our approach, let us continue with the IRS example already described. Let us suppose that a document D is described in the database with a set of 5 index $t_i (i \in \{1, \ldots, 5\})$: $D = 0.7/t_1 + 0.5/t_2 + 0.9/t_3 + 0.6/t_4 + 0.4/t_5$.

In fact, the system indicates for each index term a numeric value corresponding to its importance in describing the subject discussed in the document. If the index term t_i is not linked to the document subject content, its value is equal to 0. However, the index value that is equal to 1 means that t_i is too important in the document subject. We aim to represent each index value, i.e. v_i, by means of a couple (τ_{pos}, α); such τ_{pos} is a membership degree from L_7 (S_{un}).

The counter k is initialized to 1 and the distance l_k to 0. For the first index value v_1:

- k = 1: $l_1 = 0.25 < 0.7 = v_1$
- k = 2: $l_2 = 0.25 + 0.25 = 0.5 < 0.7$
- k = 3: $l_3 = 0.5 + 0.125 = 0.625 < 0.7$
- k = 4: $l_4 = 0.625 + 0.125 = 0.75 > 0.7$

In this case v_1 is lower than l_4 (=0.75). We can say that the index value is between τ_3, *High* (H), and τ_4, *Quite High* (QH). So, we check the closest term between τ_3 and τ_4 by comparing values of $(l_4 - v_1)$ and $(v_1 - (l_4 - d_4))$:

$(0.75 - 0.7) < (0.7 - (0.75 - 0.125))$.

Hence, τ_4 is closer than τ_3 and the proportion error is $\alpha = -0.05$. Thus, v_1 is represented by (QH, -0.05) (Fig. 15). We can say that the degree τ_4 is weakened of 0.05 or that v_1 is *little less* than *QH*. In the present paper we will not treat the use of adverbs to express proportion error but we only consider its numeric value.

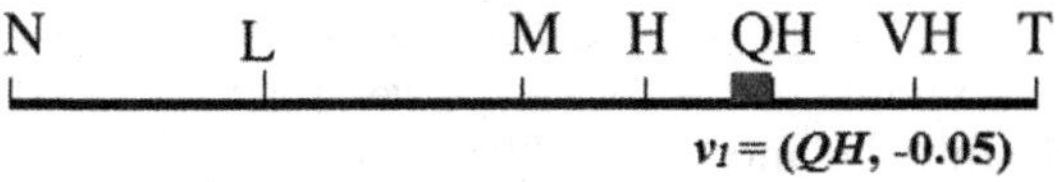

Fig. 15. Representation of the index v_1 within an unbalanced set.

We proceed using the same approach for the other index values. Thus, we obtain:

$$D = (QH, -0.05)/t_1 + (M, 0)/t_2 + (VH, 0.025)/t_3 + (H, -0.025)/t_4 + (M, -0.1)/t_5.$$

Algorithm 2. Representation of terms within unbalanced multi-set.

Input:
The position v_i of the term τ_i', $\tau_i' \in L_{M'}$
M linguistic terms of $L_M(S_{un})$
Normalized distances d_k in L_M; k=1, ..., (M-1)
begin
   $k = 1$
   $l_1 = 0$
   while l_k ¡ v_i **do**
      $l_k = \sum_{j=1}^{k} d_j$
      $k \leftarrow k + 1$
   if $l_k = v_i$ **then**
      pos $\leftarrow$ k-1
      $\alpha \leftarrow 0$
   else
      if $l_k - v_i$ ¡ $v_i - (l_k - d_k)$ **then**
         pos $\leftarrow$ k
         $\alpha \leftarrow - (l_k - v_i)$
      else
         pos $\leftarrow$ k-1
         $\alpha \leftarrow v_i - (l_k - d_k)$
Output: The couple (τ_{pos}, α)

---

6   Linguistic Modifiers with Unbalanced Multi-sets

In multi-valued logic, any symbol can be considered as a modification of another one. Thereby, modifiers allow to represent small variations of imprecise characterizations of a linguistic variable. Symbolic modifiers may transform simultaneously the membership degrees and the scale of the multi-set. We propose in this section to illustrate how to use symbolic modifiers with unbalanced imperfect knowledge. We present an approach to apply Generalized Symbolic Modifiers [21] (Table 1), initially designed for balanced sets, to unbalanced term sets.

To perform this, first, we express the unbalanced term to modify with a term from a balanced multi-set. Afterwards, we apply the GSM modifier on the obtained term within a balanced set. Then, we look for the closest matching term back in the original unbalanced multi-set. Our proposal has as input M linguistic terms of the unbalanced multi-set S_{un} (represented on a scale L_M), a term τ_i from this set, the normalized distances d_k between its successive terms and the GSM m to apply. The treatment will be as follows (Algorithm 3):

1. Express the term τ_i within a uniform multi-set $L_{M'}$ using the single uniform scale approach (Subsect. 4.2). The new term is denoted by τ_i'.
2. Apply the GSM m to the term τ_i'. Thus, we get a new uniform linguistic set L_M'', such that $M'' = m(M')$, and a new term $\tau_i'' = m(\tau_i')$.
3. Represent the term τ_i'' within the initial unbalanced linguistic set L_M (S_{un}) using the approximation function presented in Algorithm 2.

Thus, for this last step, we have to determine the nearest position to τ_i'' in S_{un} (L_M). First, we calculate the position v_i'', i.e. the distance between the term τ_i'' and τ_0''. Afterwards, we determine the term τ_{pos} by comparing v_i'' to the sum l_k of the distances between each successive terms in L_M (d_k).

Algorithm 3. Using GSM with unbalanced multi-set.

Input:
M unbalanced linguistic terms of $L_M(S_{un})$
The term $\tau_i \in L_M$
Normalized distances d_k of L_M; k=1, ..., (M-1)
The GSM m
begin
     $L_{M'} \leftarrow$ Single scale approach (M unbalanced terms, d_k) (Algo. 1)
     Identify the couple ($\tau_{i'}$, $L_{M'}$) corresponding to (τ_i, L_M) in $L_{M'}$
     ($\tau_{i''}$, M'') $\leftarrow$ m ($\tau_{i'}$, M')
     $v_{i''} \leftarrow \dfrac{i''}{(M''-1)}$
     Calculate the closest term τ_{pos} to v_i'' in S_{un} and the proportion error α
     (Algo. 2)
Output: The couple (τ_{pos}, α)

To illustrate our approach, we use the same example of IRS previously presented. We consider the unbalanced set L_7 of relevance degrees (Fig. 16). We apply the reinforcing modifier CR_1 to the relevance degree *Medium*, i.e. τ_2. The process is represented in Fig. 17.

As mentioned before, the distances d_k are: $d_1 = d_2 = 0.25; d_3 = d_4 = d_5 = d_6 = 0.125$.

N	L	M	H	QH	VH	T
τ_0	τ_1	τ_2	τ_3	τ_4	τ_5	τ_6

Fig. 16. Unbalanced set of relevance degrees.

The proportion of the term τ_2 in the unbalanced term set L_7 (Definition 3) is: Prop $(\tau_2) = d_1 + d_2 = 0.25 + 0.25 = 0.5$.

The granularity of the uniform set, previously calculated in Subsect. 4.2, is **9** and the term τ_2 from S_{un} is represented by τ_4' in L_9. We notice that the proportion is preserved $\mathrm{Prop}(\tau_4') = \frac{4}{8} = 0.5$.

For this example, we apply the GSM CR_1 to the couple (τ_4', L_9) (Table 1):
$i'' = m(4) = min(4+1, 9-1) = min(5, 8) = 5$
$M'' = m(9) = 9$.

The obtained couple is (τ_5'', L_9). We remind that the modifier **CR** is a reinforcing one, i.e. the proportion increases:
Prop $(\tau_5'') = \frac{5}{8} > \mathrm{Prop}(\tau_4') = \frac{4}{8}$.

Finally, we must express the term τ_5'' within the initial set L_7. We first calculate the distance: $v_5'' = \frac{i''}{(M''-1)} = \frac{5}{8} = 0.625$.

The counter k is initialized to 1 and the distance l_1 to 0.

– For k = 1; $l_1 \leftarrow 0.25 < v_5''$
– For k = 2; $l_2 \leftarrow 0.25 + 0.25 = 0.5 < v_5''$
– For k = 3; $l_3 \leftarrow 0.5 + 0.125 = 0.625 = v_5''$

In this case, the value of l_3 is equal to v_5''. Thus, τ_3 is the associated term in the unbalanced set L_7. The proportion of this term is Prop $(\tau_3) = 0.625$. It corresponds to the proportion of the term τ_5'' in the balanced multi-set L_9.

We can say that the result of applying CR with radius 1 to the relevance degree *Medium* corresponds to the relevance degree *High*.

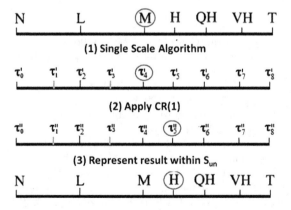

Fig. 17. The process of applying CR_1 to relevance degree M.

7   Conclusions

This work deals with the representation and management of unbalanced multisets. These latter are a particular case of imperfect knowledge. The literature does not propose a formal definition of unbalanced term sets. Besides that, works that use these sets are only concerned with fuzzy logic. Hence, proposing one could be a significant improvement in imperfect knowledge management.

We present in this paper an algorithm for the representation of uniform terms within an unbalanced multi-set. This process uses an approximation function providing the closest unbalanced term to the desired uniform one. Our solution also calculates a proportion error that might exist in some cases. We also propose another algorithm that allows to use symbolic modifiers, initially designed for balanced terms, with unbalanced ones. This proposal is based on our previous work [8] for the representation of unbalanced terms on a single uniform multi-set.

An immediate perspective of this work is to validate our proposals in a real case and compare the obtained results to existing works. For example, we could propose a model to evaluate documents in a retrieval process in the context of multi-valued logic. The evaluation will be done by comparing our results to those of the model proposed by Herrera-Viedma and López-Herrera in [9] in the context of fuzzy logic.

As future work, we will work on a way to treat symbolic cases where the distances between terms are expressed with words instead of numbers. It, also, would be interesting and more understandable by users to express the proportion error by using adverbs like *little, more or less, slightly,...*

Furthermore, another aspect treated in the literature is the approximate reasoning [4]. It is based on the Generalized Modus Ponens (GMP) which aims to deduce from an observation that is approximately equal to the premise rule, a result similar to the conclusion rule. It would be valuable to propose a new GMP rules dealing with unbalanced multi-sets.

References

1. Zadeh, L.A.: Fuzzy sets. Inf. Control **8**, 338–353 (1965)
2. De Glas, M.: Knowledge representation in a fuzzy setting. Rapp. Interne **89**, 48 (1989)
3. Akdag, H., De Glas, M., Pacholczyk, D.: A qualitative theory of uncertainty. Fundamenta Informaticae **17**, 333–362 (1992)
4. Zadeh, L.A.: The concept of a linguistic variable and its application to approximate reasoning - I. Inf. Sci. **8**, 199–249 (1975)
5. Herrera, F., Martínez, L.: A model based on linguistic 2-tuples for dealing with multigranular hierarchical linguistic contexts in multi-expert decision-making. IEEE Trans. Syst. Man Cybern. Part B: Cybern. **31**, 227–234 (2001)
6. Herrera, F., Enrique, H., Martínez, L.: A fuzzy linguistic methodology to deal with unbalanced linguistic term sets. IEEE Trans. Fuzzy Syst. **16**, 354–370 (2008)
7. Abchir, M.A.: Towards fuzzy semantics for geolocation applications. Ph.D. Thesis, University of Paris VIII Vincennes-Saint Denis (2013)

8. Chaoued, N., Borgi, A., Laurent, A.: Representation of unbalanced terms in multi-valued logic. In: 12th IEEE International Multi-conference on Systems, Signals and Devices (2015)
9. Herrera-Viedma, E., López-Herrera, A.G.: A model of an information retrieval system with unbalanced fuzzy linguistic information. Int. J. Intell. Syst. **22**, 1197–1214 (2007)
10. Martínez, L., Herrera, F.: An overview on the 2-tuple linguistic model for computing with words in decision making: extensions, applications and challenges. Inf. Sci. **207**, 1–18 (2012)
11. Xu, Z.: An interactive approach to multiple attribute group decision making with multigranular uncertain linguistic information. Group Decis. Negot. **18**, 119–145 (2009)
12. Wang, B., Liang, J., Qian, Y., Dang, C.: A normalized numerical scaling method for the unbalanced multi-granular linguistic sets. Int. J. Uncertain. Fuzziness Knowl.-Based Syst. **23**, 221–243 (2015)
13. Marin, L., Valls, A., Isern, D., Moreno, A., Merigó, J.M.: Induced unbalanced linguistic ordered weighted average and its application in multiperson decision making. Sci. World J. **2014**, 19 p. (2014). Article ID 642165, doi:10.1155/2014/642165
14. Bartczuk, Ł., Dziwiński, P., Starczewski, J.T.: A new method for dealing with unbalanced linguistic term set. In: Rutkowski, L., Korytkowski, M., Scherer, R., Tadeusiewicz, R., Zadeh, L.A., Zurada, J.M. (eds.) ICAISC 2012. LNCS (LNAI), vol. 7267, pp. 207–212. Springer, Heidelberg (2012). doi:10.1007/978-3-642-29347-4_24
15. Jiang, L., Liu, H., Cai, J.: The power average operator for unbalanced linguistic term sets. Inf. Fusion **22**, 85–94 (2015)
16. Akdag, H., Pacholczyk, D.: Incertitude et logique multivalente, première partie: Etude théorique. BUSEFAL **38**, 122–139 (1989)
17. Adkag, H.: Une approche logique du raisonnement incertain. Ph.D. Thesis, University of Paris, 6 (1992)
18. Zadeh, L.A.: A fuzzy-set-theoretic interpretation of linguistic hedges. J. Cybern. **2**, 4–34 (1972)
19. Akdag, H., Mellouli, N., Borgi, A.: A symbolic approach of linguistic modifiers. In: Information Processing and Management of Uncertainty in Knowledge-Based Systems, Madrid, pp. 1713–1719 (2000)
20. Akdag, H., Truck, I., Borgi, A., Mellouli, N.: Linguistic modifiers in a symbolic framework. Int. J. Uncertain. Fuzziness Knowl.-Based Syst. **9**, 49–61 (2001)
21. Truck, I.: Approches symbolique et floue des modificateurs linguistiques et leur lien avec l'agrégation: application: le logiciel flous. Ph.D. Thesis, University of Reims Champagne-Ardenne (2002)
22. Truck, I., Borgi, A., Akdag, H.: Generalized modifiers as an interval scale: towards adaptive colorimetric alterations. In: Garijo, F.J., Riquelme, J.C., Toro, M. (eds.) IBERAMIA 2002. LNCS (LNAI), vol. 2527, pp. 111–120. Springer, Heidelberg (2002). doi:10.1007/3-540-36131-6_12
23. Espinilla, M., Liu, J., Martínez, L.: An extended hierarchical linguistic model for decision-making problems. Comput. Intell. **27**, 489–512 (2011)
24. Kacem, S.B.H., Borgi, A., Tagina, M.: Extended symbolic approximate reasoning based on linguistic modifiers. Knowl. Inf. Syst. **42**, 633–661 (2015)

Ranking with Ties of OWL Ontology Reasoners Based on Learned Performances

Nourhène Alaya[1,2(✉)], Sadok Ben Yahia[1], and Myriam Lamolle[2]

[1] LIPAH-LR 11ES14, Faculty of Sciences of Tunis, University of Tunis El Manar, 2092 Tunis, Tunisia
sadok.benyahia@fst.rnu.tn

[2] LIASD EA4383, IUT of Montreuil, University of Paris 8, 93526 Saint-Denis, France
{n.alaya,m.lamolle}@iut.univ-paris8.fr

Abstract. Over the last decade, several ontology reasoners have been proposed to overcome the computational complexity of inference tasks on expressive ontology languages such as OWL 2 DL. Nevertheless, it is well-accepted that there is no outstanding reasoner that can outperform in all input ontologies. Thus, deciding the most suitable reasoner for an ontology based application is still a time and effort consuming task. In this paper, we suggest to develop a new system to provide user support when looking for guidance over ontology reasoners. At first, we will be looking at automatically predict a single reasoner empirical performances, in particular its *robustness* and *efficiency*, over any given ontology. Later, we aim at ranking a set of candidate reasoners in a most preferred order by taking into account information regarding their predicted performances. We conducted extensive experiments covering over 2500 well selected real-world ontologies and six state-of-the-art of the most performing reasoners. Our primary prediction and ranking results are encouraging and witnessing the potential benefits of our approach.

Keywords: Ontology · Reasoner · Robustness · Efficiency · Supervised machine learning · Prediction · Ranking

1 Introduction and Related Works

Over the last decade, several semantic reasoning algorithms and engines have been proposed to cope with the computational complexity of inference tasks on expressive ontology languages such as OWL 2 DL [26]. Matentzoglu et al. have recently conducted a detailed survey [24] scrutinizing this considerable research and engineering efforts to design sophisticated reasoners with highly optimized implementation. Nevertheless, it is well-accepted that there is no reasoning engine that can outperform in all input ontologies. It would be quite easy to check this out from the results of the annual reasoner evaluation workshop, ORE [8].

This work is an extension of our previous publication at the 7th International Conference on Knowledge Engineering and Ontology Development KEOD 2015 [2].

© Springer International Publishing AG 2016
A. Fred et al. (Eds.): IC3K 2015, CCIS 631, pp. 234–259, 2016.
DOI: 10.1007/978-3-319-52758-1_14

Generally speaking, the performances of a reasoner is closely depending on the success or the failure of optimizations tricks, set up by reasoner designers to solve hardness features in description logic [4]. Experts pre-established decision rules about the appropriateness of a reasoner for a particular ontology are not efficient in practice, except for trivial cases. In fact, the respective authors of [16,32] have highlighted the lack of knowledge about the factors that are affecting the performances of a specific reasoner on a specific ontology. Empirical results are often surprising the experts as reasoner performances, may remarkably vary on different ontologies standing at the same hardness level. All of these findings gave rise to new challenges in the area of ontology reasoning: we would wonder how to a priori predict a reasoner performances over an input ontology? And how to assist users, with little or no experience about the sophisticated description logic algorithms, in the choice of the most adequate reasoner for the ontology under consideration?

Our first research question was primarily addressed by Kang et al. [20,21], then by Sazonau et al. [31]. Each of the aforementioned authors have deployed machine learning techniques [1] in order to train predictive models of a reasoner runtime. The learning approach consists of using a learning algorithm to model the relation between the ontologies' characteristics, and the relative performance of a specific reasoner. The focus of these authors was about automatically predicting how fast would be a reasoner in achieving a reasoning task over any input ontology. However, they omitted the fact that reasoners can load and process an ontology for some reasoning task, but they can also deliver quite distinct results. Gardiner et al. [12] and more recently Lee et al. [22] have outlined the issue of reasoner disagreement over inferences or query answers, computed over one input ontology. In fact, since ORE 2013 an empirical correctness checking method was established to assess the reasoner derived inferences. The results were quite unforeseen as there were no single reasoner that have correctly processed all the ontologies. Even the fastest reasoner have failed to derive accurate results for some ontologies, while other less performing engines, have succeeded to correctly process them. These findings motivated us to extend the work of reasoner performances prediction by considering the correctness as an additional evaluation criterion. In this chapter, we will describe our methodology of learning reasoner robustness predictive models. By *robustness*, we assign the ability of a reasoner to achieve a reasoning task within a fixed time limit while providing correct results.

Our second research question is linked to the famous problem of algorithm selection, discussed in the first place by John Rice since 1976 [29]. This issue has been at the crossroads of many problem fields, such as machine learning (ML), satisfiability solving (ST), and combinatorial optimization (CP). A common trait among the proposed solutions is that they employ supervised machine learning techniques, regression or ML classification[1], to build an efficient predictor of an

---

[1] It's important to notice that "DL classification" stands for a reasoning task aiming at inferring the subsumption relations between the classes and the properties in given ontology, however the "ML classification" denotes the process of training a predictive model from a labelled dataset.

algorithm's runtime for each problem instance [33]. The final goal was to exploit the complementarity of the algorithms by combining them into an algorithm portfolio [13] in order to obtain improved performances in the average case. The idea of building reasoner portfolio have recently seduced Kang et al. [19]. The latter one have employed his previously proposed reasoner runtime predictive models, to build R_2O_2 a kind of *meta-reasoner* which combines existing reasoners and tries to determine the most efficient one for a given ontology. Their results were remarkable, claiming that R_2O_2 could outperform the most efficient state of art reasoner. However, no rigorous evaluation under no idiosyncratic conditions have been conducted to confirm the Kang's et al. assertions. In fact, the time overheard due to the prediction steps was not discussed. Hence, one would wonder how R_2O_2, which computes 91 ontology metrics, invokes 5 ranking algorithms, compares the predictions and finally, calls the most efficient reasoner in order to process the input ontology, would be faster than straightly and simply calling the reasoner in question?

Given these findings, we chose to address the problem of ontology reasoner selection from a different perspective. We believe that to have reasoners selected as being best suited from an input ontology is a kind of recommendation that would be made for users seeking for guidance. The provided advise would be in its basic form, a ranking of reasoners, in a most preferred order. As previously highlighted by [28], providing a ranking of candidate algorithms is a more informative and flexible solution, than suggesting one single best algorithm. For instance, a user may decide to continue using his favourite reasoner, if its performance is slightly below the topmost one in the ranking. To compute the reasoner ranking we decided to investigate how to take into account information regarding both the robustness and the efficiency of the candidate reasoners. Our proposed ranking method is a supervised one based on learned predictive models of each of the aforementioned reasoner evaluation criteria. The method allows ties between reasoners and follows rules from the bucket order principals. To the best of our knowledge, we are the first to employ more than one criterion to rank the reasoners and we believe it is one important step in the direction of multi-criteria ranking of ontology reasoners. To show the effectiveness of our different proposals, we carried out extensive experiments covering 2500 well selected real-world ontologies and six state-of-the-art of the most performing reasoners. The provided data was trained by known supervised ML algorithms. The obtained assessment results show that we could build highly accuracy reasoner robustness and reasoner efficiency predictive models. Owe to the effectiveness of our training method, accurate ranking of reasoners was obtained witnessing the potential of our approach.

The remainder of this chapter is organized as follows. Section 2 specifies our methodology for building reasoner robustness predictive models. Construction details and evaluation results are given in Sect. 3. We then move to consider ways of extending this prediction system. Section 4 presents our design ideas of a user oriented guidance system for the selection of ontology reasoner. The primarily assessments results are reported in Sect. 5. Finally, Sect. 6 concludes the chapter and gives some of our future directions.

2   Robustness Prediction of the Ontology Reasoners

2.1   Reasoner Empirical Robustness

The notion of *robustness* differs by the field of research. For instance in the software field, Mikolàšek [25] defined the robustness as *the capability of the system to deliver its functions and to avoid service failures under unforeseen conditions*. Recently, Gonçalves et al. [14] bought forward this definition in order to conduct an empirical study about the robustness of DL reasoners. Authors underlined the need to specify the robustness judgement constrains before assessing the reasoners. These constraints would describe some reasoner usage scenario. The authors highlighted that a reasoner may be considered as robust under a certain scenario and non-robust under another.

Given the robustness specification by Gonçalves et al., we started by setting our proper constrains in order to describe an online execution scenario of reasoners where ontologies should be classified as quick as possible. This scenario yields to the following constraints: *(I)* any range of OWL ontologies should be supported; *(II)* functional: the reasoner should be capable to classify the ontology and deliver correct inferences; *(III)* non-functional: the reasoner should complete classification within a fixed cut-off time; and *(IV)* the failure, w.r.t to the 2nd and the 3rd requirements, means either a reasoner crashes when processing the ontology, or exceeds the time limit or delivers unexpected results. To put it a nutshell, we define the robustness of a reasoner given an input ontology as follows:

Definition 1 (Reasoner Robustness). *The robustness of a reasoner stands for its ability to correctly achieve a reasoning task for a given ontology within a fixed cut-off time.*

Respectively, we can define the robustness of a reasoner over an ontology corpus as follows:

Definition 2. *The most robust reasoner over an ontology corpus would be the one satisfying the correctness requirement for the highest number of ontologies while maintaining the shortest computational time.*

One major obstacle we faced when looking to empirically check the robustness of reasoners was about automatically assessing the correctness of a reasoner's results. In fact, only a few attention has been paid to this issue. As previously outlined by Gardiner et al. [12], manually testing the correctness of a reasoner's inferences would be relatively easy for small ontologies, but usually unfeasible for real world ones. The authors claimed that the most straightforward way to automatically determine the correctness is by comparing answers derived by different reasoners for the same ontology. Consistency across multiple answers would imply a high probability of correctness, since the reasoners have been designed and implemented independently and using different algorithms. Luckily, this testing approach was implemented in the ORE evaluation Framework [15]. The reasoner output was checked for correctness by a majority vote, i.e. the result returned by the most reasoners was considered to be correct. Certainly, this is not a faultless method. Improving it would be advantageous for our study, however this issue is out of the scope of this chapter.

2.2   Learning Methodology

The first research question of this paper is to check whether it is feasible to automatically predict the robustness of modern reasoners using the results of their previous running. To accomplish this aim, we choose to work with supervised ML techniques. In any standard machine learning process, every instance in a training dataset is represented using the same set of features. In our case, the instances are ontologies belonging to some corpus $\mathcal{S}(\mathcal{O}) = \{O_1, \ldots, O_N\}$, with N denoting the size of the corpus $n = \mid \mathcal{S}(\mathcal{O}) \mid$. The features are metrics characterizing the ontology content and design. Thus, given a feature space $\mathcal{F}^d$, where d stands for the space dimension, each ontology is represented by a d-dimensional vector $X^{(O_i)} = [x_1^{(i)}, x_2^{(i)}, \ldots, x_d^{(i)}]$ called a *feature vector*, where i refers to the i-th ontology in the dataset, $i \in [1, N]$. The features may be continuous, categorical or binary. The learning is called *supervised* as far as ontologies are given with known labels. In our study, a label would describe a given reasoner robustness when processing the considered ontology for DL classification[2]. Based on the robustness specification given in Sect. 2.1, we designed four labels that would describe the termination state of this reasoning task. The first label describes the state of success and the other ones distinguish three types of failure. These labels are: (1) *Correct* (**C**) standing for an execution achieved within the time limit and delivered expected results; (2) *Unexpected* (**U**) in the case of termination within the time limit but delivered results that are not expected; (3) *Timeout* (**T**) in the case of violating the time limit; and (4) *Halt* (**H**) describing a sudden stop of the reasoning process owing to some error. Thus, the set $\{C, U, T, H\}$ denotes our label space $\mathcal{L}$ admitted for the learning process.

The vector of all ontologies' labels is specific to one reasoner R and denoted by $\mathcal{L}_R = [l(O_1), l(O_2), \ldots, l(O_n)]$, where $l(O_i)$ is the label of the i-th ontology. By gathering the pairs (feature vectors, labels), a dataset is built in and designed by $\mathcal{D}_R = \{(X^{(O_i)}, l_R(O_i))\}, i \in [1, N]$. This dataset illustrates the meta knowledge about the reasoner previous running. Afterwards, the $\mathcal{D}_R$ is provided to a supervised learning algorithm in order to train its data and establish a predictive model about the robustness of the corresponding reasoner. Roughly speaking, a model $\mathcal{M}_R$ is a mapping function from an ontology feature vector $X^{(O)}$ to a specific label describing the reasoner robustness $\hat{l}_R(X^{(O)})$. It would

---

[2] It's important to notice that "DL classification" designs a reasoning task aiming at inferring the class and property subsumption relation in a given ontology, however the "ML classification" denotes the process of training a predictive model from a labeled dataset.

be a mathematical function, a graph or a probability distribution, etc.

$$M_R : X \in \mathcal{F}^d \mapsto \hat{l}_R(X) \in \mathcal{L} \tag{1}$$

When a new ontology, which does not belong to the dataset is introduced, then the task is to compute its features and then predict its exact label, using a reasoner predictive model.

Unluckily, the performances of learning algorithms are highly sensitive to quality of employed features. Uninformative or highly inter-correlated features could deteriorate the accuracy of the learned models. Commonly, feature selection methods [7] are deployed in order to remove irrelevant and redundant features. Plenty of these kind of methods are available in literature. In our study, we will be investigating some of the well known techniques in feature selection. The set of investigated methods is designed by $\mathcal{FS}$. Consequently, different variants of the initial reasoner dataset will be established, called *featured* datasets, each with an eventually reduced subset of features. Worth of cite, we will be also investigate several supervised machine learning algorithms, within a given set denoted by $\mathcal{AL}$. These elaborated investigations would provide us a way to identify the most effective combination, i.e. a learning algorithm associated to a specific feature selection method, which could deliver the most accurate reasoner robustness models. Therefore, the training steps will be repeated as far as the number of algorithms and the number of datasets for a given reasoner. Then, the established models will be evaluated against a bunch of assessment measures. Further, we will introduce a method to compare these models and figure out their "*best*" one. Details about the deployed ML techniques, assessment measures and the selection procedure are given in Sect. 3.2.

Algorithm 1. A Single Reasoner Robustness Learning Procedure.

Input : $\mathcal{S}(O) = \{O_1, \ldots, O_n\}$ the set of the training ontologies, $\mathcal{L}_R = [l(O_1), \ldots, l(O_n)]$ the labels vector of the reasoner R, $\mathcal{AL}$ the set of learning algorithms, $\mathcal{FS}$ the set of feature selection methods.

Output: M_R^{Best} the reasoner R robustness *best* learned model.

1 $[X^{(O_1)}, \ldots, X^{(O_N)}] \leftarrow computeOntologiesFeatureVectors(\mathcal{S}(O))$;

2 $\mathcal{D}_R \leftarrow getTheReasonerDataSet(\mathcal{L}_R, [X^{(O_1)}, \ldots, X^{(O_N)}])$;

3 **foreach** $F \in \mathcal{FS}$ **do**

4 $\quad \mathcal{D}_R^F \leftarrow selectRelevantFeatures(F, \mathcal{D}_R)$;

5 $\quad$ **foreach** $A \in \mathcal{AL}$ **do**

6 $\quad\quad < M^{(F,A)}, \pi^{(F,A)} > \leftarrow trainAndAssessTheModel(A, \mathcal{D}_R^F)$;

7 $\quad\quad \mathcal{M}^F \leftarrow \mathcal{M}^F \cup \{< M^{(F,A)}, \pi^{(F,A)} >\}$;

8 $\quad$ **end**

9 $\quad < M^{(F,best)}, \pi^{(F,best)} > \leftarrow selectBestModelFromOneDataSet(\mathcal{M}^F)$;

10 $\quad \mathcal{M}^{Best} \leftarrow \mathcal{M}^{Best} \cup \{< M^{(F,best)}, \pi^{(F,best)} >\}$;

11 **end**

12 $M_R^{Best} \leftarrow selectBestModelAccrossTheDataSets(\mathcal{M}^{Best})$;

13 **return** M_R^{Best};

Algorithm 1 sum ups the main steps of the aforementioned reasoner robustness learning methodology, where $M^{(F,A)}$ stands for a reasoner robustness predictive model trained by a machine learning algorithm A from a featured dataset by the feature selection method F. The assessment score of this model is denoted by $\pi^{(F,A)}$. Worth noting, the set $\mathcal{FS}$ contains the $\perp$ element, which means no feature selection method to be applied to the initial dataset. This configuration would allow us to keep the initial dataset with its full set of features, denoted by $RAWD$. Later on, models trained from $RAWD$ will be compared to those trained from the featured datasets (Algorithm 1, Line 10).

2.3   Ontology Features

There is no known automatic way of constructing *"good"* ontology feature sets. Instead, distinct domain knowledge should be used to identify properties of ontologies that appear likely to provide useful information about its complexity level. To accomplish our study, we reused some of previously proposed ontology features and defined new ones. Mainly, we discarded those computed based on specific graph translation of the OWL ontology, as there is no agreement of the way an ontology should be translated into a graph [31]. We organized the ontology features into 4 main categories. Some of the latter ones are then refined by splitting them in sub-categories. The categories are as follows:

1. **The Size Description** is a group of metrics characterizing the size of the ontology considering both the amount of its terms and axioms.
2. **The Ontology Expressivity Description** is based on two main features, namely the OWL profile[3] and the DL family name.
3. **Ontology Structural Description** gathers various metrics commonly used by the ontology quality evaluation community [23] to characterize the taxonomic structure of an ontology, i.e. its named class or named properties inheritance hierarchy.
4. **The Ontology Syntactic Description** is a set of metrics proposed in order to quantify some of the general theoretical knowledge about DL complexity sources [4]. Features of the current category are divided in 6 subcategories. Each one of these is specific to a particular OWL syntactic component.

To sum up, we have characterized almost 110 ontology features, described in details in our previous work [3]. Table 1 illustrates the overall organisation of our features and gives the definition of the main ones.

---

[3] For further details about OWL 2 profiles, the reader is kindly referred to http://www.w3.org/TR/owl2-profiles/.

Table 1. OWL ontology features catalogue.

Category	Subcategory	Features	Description
Ontology size	Signature size	SC, SOP, SDP, SI, SDT	Respectively the nbr. of named classes, object properties, data properties, individuals and data types
	Axiom's size	SA, SLA	Axioms count, Logical axioms count
Ontology expressivity		DFN	DL Family name (AL, ALC, SHIF, etc.)
		OPR	OWL Profile (EL, DL)
Ontology structural features	CHierarchy	C(P)_MD	Max. depth of the class (property) hierarchy
	PHierarchy	C(P)_MSB	Max. nbr. of sub-classes (-properties) in the hierarchy
		C(P)_MTangledness	Max. nbr. of super-classes (-properties) in the hierarchy
		C(P)_ASB	Avg. nbr. named sub-classes (-properties)
		C(P)_Tangledness	Avg. nbr. of classes (properties) with multiple direct ancestors
	Cohesion	CCOH, PCOH	Cohesion of respectively class hierarchy, property hierarchy
		OPCOH, OCOH	Cohesion of object property and the cohesion in the ontology
	Richness	RRichness	Ratio of rich relationships (non-hierarchical ones)
		AttrRichness	Ratio of classes having attributes
Ontology syntactic features	Axioms	KBF(set)	Ratio of axioms in *TBox*, *RBox* and *ABox*
		ATF(set)	Frequencies of each OWL axiom types (28 OWL Axiom type)
		AMP, AAP	Respectively the Max. and Avg. parsing depth of axioms
	Constructors	CCF(set)	Frequency of each type of constructors (11 OWL constructors)
		ACCM	Max. nbr. of constructors per axiom
		OCCD	Density of class constructors in the ontology
		CCP(set)	Occurrences of Constructors Coupling Patterns (*UI, EUI, CUI*)
	Classes	CDF(set)	Ratio of class definitions *PCD, NPCD* and *GCI*
		CCYC, CDISJ	Ratio of respectively cyclic and disjoint classes
		CNOM	Ratio of classes defined using nominals
	Properties	OPCF(set)	Avg. frequency of object property hard characteristic. (9 characteristics:*transitivity, symmetry, functional, etc*)
		HVR(set)	The highest value of each nbr. restriction type (min, max, exact)
		AVR	Nbr. restriction Avg. value
	Individuals	NNF	Ratio of nominals in the TBox
		INDISJ, ISAM	Ratio of respectively disjoint and equal individuals

3   Evaluation of Reasoners' Robustness Prediction

Before presenting the learning methods, we start by describing the collected data for the assessment purpose.

3.1   Data Collection

In order to conduct experiments over reasoners in the most reliable way, we have chosen to work with the evaluation Framework of the ORE workshop[4]. The motivation behind our choice is multi-fold: first, the ORE event is widely recognized by the Semantic Web community; the ontology corpus collected by

---

[4] The ORE Framework is available at https://github.com/andreas-steigmiller/ore-20 14-competition-framework/.

ORE community is well diversified and balanced throughout easy and hard cases; and finally the description of the reasoner results is consistent with our definition of the reasoner robustness specified Subsect. 2.1. Using this Framework, we will conduct new DL and EL classification evaluations with 6 reasoners and 2500 ontologies described in the following.

Ontologies. We selected 2500 ontologies from the ORE corpora[5]. The ontologies fall into the OWL 2 DL (1200) and the OWL 2 EL (1300) profiles, binned according to their sizes[6] and cover almost 167 distinct expressivity families. Figure 1 describes the ontologies distribution over size bins w.r.t the OWL profiles, then illustrated the most represented expressivity families in the dataset. We can assert that our ontology sampling is representative of the different ranges of OWL ontologies and it is richly diversified in size and expressivity. Thus, it would reduce the probability for a given reasoner to encounter only problems it is particularly optimized for. We further split up our set of ontologies into $S(O)$ having 2000 ontologies and $T(O)$ gathering the remaining 500 ontologies. The partition was carefully made in a stratified way in order to respect the overall distribution of ontology size and expressivity. Hence, the $S(O)$ set will be used to train and assess the reasoner predictive models, however the $T(O)$'s ontologies will be employed in the evaluation of our second contribution, i.e. the ranking of reasoners.

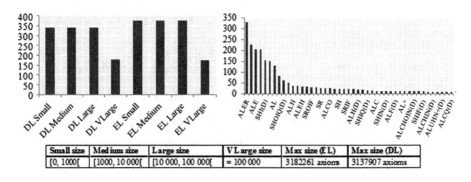

Small size	Medium size	Large size	VLarge size	Max size (EL)	Max size (DL)
[0, 1000[	[1000, 10 000[	[10 000, 100 000[	≈ 100 000	3182261 axioms	3137907 axioms

Fig. 1. Ontologies distribution over size bins and expressivity families.

Reasoners. We investigated the 6 best ranked reasoners[7] in both the DL and EL classification challenges of the ORE'2014 competition. They are namely **Konclude**, **MORe**, **HermiT**, **TrOWL**, **FaCT++** and **JFact**. We excluded **ELK** despite its good results and high rank, as far as it does not support the classification of DL ontologies. We run the ORE Framework in the sequential mode on a machine equipped with an Intel Core I7, CPU running at 3.4 GHz and having 32 GB RAM, where 12 GB were made available for each reasoner. We set the condition of 3 min time limit to classify an ontology by a reasoner. This tight

---

[5] The ORE corpus is available at http://zenodo.org/record/10791.

[6] The size corresponds to the number of the logical axioms.

[7] All ORE'2014 reasoners are available at https://zenodo.org/record/11145/.

schedule would be consistent with the chosen scenario, i.e. the online DL classification of an ontology. Table 2 summarizes the results of the reasoner DL and EL classification challenges[8]. By following Definition 2, reasoners are listed based on their robustness rank over the selected ontology corpus. For each reasoner, Table 2 shows the number of ontologies within each label class (Subsect. 2.1) and the average reasoning runtime for the correctly classified ontologies.

Table 2. Summary of reasoner evaluation results. (Avg. time in milliseconds).

Reasoner	#Correct	#Unexpected	#Timeout	#Halt	Avg. time (Correct)
Konclude	2408	44	2	6	1147.71
MORe	2182	71	241	6	3366.74
HermiT	2167	3	312	18	7918.18
TrOWL	2149	257	74	20	3246.37
FaCT++	1968	18	419	95	4476.26
JFact	1679	41	636	144	8474.55

Building the Reasoner Datasets. Feature vector for each ontology in the study set was computed and recorded. Having the evaluation results and the ontology feature vectors, 6 datasets were established each for a specific reasoner. It's important to notice that our reasoner datasets are **unbalanced** ones. Based on the description made by [18], a dataset is considered as unbalanced, if one of the classes (minority class) contains much smaller number of examples than the remaining classes (majority classes). In our context, the classes are reasoner robustness labels. We take as an example the *Konclude*'s dataset described in Table 2, we can easily notice the huge difference between the number of ontologies labeled as *Correct* #C (the majority) and those labelled as *Unexpected* #U or *Timeout* #T (the minority). However, its obvious that predicting the minor cases is much more of interest for both ontology and reasoner designers, since they highlight a situation of failure. In fact, a user would be interested in knowing whether it would be *safe* to choose *Konclude* to classify its ontology. The learning from unbalanced datasets is considered as one of the most challenging issue in the data mining field [18] and could seriously affect the predictive models' effectiveness. That's why in the remainder, we paid attention to carefully select the most resisting supervised machine learning algoritkhms to the unbalanced data problem.

3.2   Learning Methods

It is worth of mention that all of the learning methods described in the following are available in the machine learning working environment, *Weka* [17]. We have

---

[8] Results of our experiments are available at https://github.com/PhdStudent2015/Classification_Results_2016.

used the Java API of Weka[9] as base tool to implement our prototype of automatic learning of reasoner robustness models.

Feature Selection Methods (FS). In our study, we tried to track down the subset of features, which correlate the most with the robustness of a given reasoner. To achieve this task, we first chose to investigate the utility of employing *supervised discretization* [11], as a feature selection method. Basically, this technique stands for the transformation task of continuous data into discrete bins while considering the distribution of the target labels. When no bins are identified for a given feature, this latter one is considered useless and could be safely removed from the dataset. As an application method, we opted for the well known Fayyad & Irani's *supervised* discretization technique (**MDL**). For the seek of validity, we decided to compare the **MDL**'s results to the ones computed by more common feature selection methods. We chose two well known methods, each representative of a distinct category of feature selection approaches: first, the *Relief* method (**RLF**)[10], which finds relevant features based on a ranking mechanism; then, the *CfsSubset* method (**CFS**), which selects subsets of features that are highly correlated with the target label while having low intercorrelations. In overall, the $\mathcal{FS}$ set is equal to $\{\perp, \text{MDL}, \text{CFS}, \text{RLF}\}$.

Supervised Machine Learning Algorithms (AL). We selected 5 candidate algorithms, each one is representative of a distinct and widely recognized category of machine learning algorithms [1]. All the used implementations were applied with the default parameters of the Weka tool. They are namely: *(1) Random Forest* (**RF**) a combination of C4.5 decision tree classifiers; *(2) Simple Logistic* (**SL**) a linear logistic regression algorithm; *(3) Multilayer Percetron* (**MP**) a back propagation algorithm used to build an Artificial Neural Network model; *(4) SMO* a Support Vector Machine learner with a sequential minimal optimization; and finally *(5) IBk* a K-Nearest-Neighbour algorithm with normalized euclidean distance. To sum it up, the $\mathcal{AL}$ set is equal to $\{\text{RF}, \text{SL}, \text{MP}, \text{SMO}, \text{IBK}\}$.

Prediction Accuracy Measures and Selection Method. In previous works [20], the accuracy standing for the fraction of the correct predicted labels out of the total number of the dataset samples, was adopted as main assessment metric of the predictive models. However, accuracy would be misleading in the case of unbalanced datasets[11]. Thus, we looked for assessment measures known for their appropriateness in the case of unbalanced data. Based on the comparative study conducted by [18], we retained three main measures: *(1) **Kappa coefficient** (κ); (2) **Matthews Correlation Coefficient** (MCC); and (3) **F-Measure** (**FM**). Values of *Kappa* and *MCC* vary between -1 and 1, where 0 represents agreement due to chance. The value 1 represents a complete agreement between

---

[9] The Weka API is available at www.cs.waikato.ac.nz/ml/weka/downloading.html.

[10] *RLF* will be used with a ranking threshold equal to 0.01.

[11] Accuracy places more weight on the majority class(es) than the minority one(s). Consequently, high accuracy rates would be reported, even if the predictive model is not necessarily a good one.

both the real and the predicted labels. *FM* ranges from 0 to 1, the higher is *FM*, the more accurate the model is. However, *Kappa* and *MCC* are known to be more effective than *FM*, whenever the dataset is unbalanced. We computed the *FM* score as it is widely adopted by the machine learning community. In addition, we made use of the standard stratified 10-fold *cross-validation* technique for the evaluation process of the trained models. For the sake of simplicity, we only retained the average values of the computed assessment measures over all the cross-validation steps.

In order to select a reasoner *best* predictive model, we propose to compute a score index per model that would enable us to establish a total order through all the trained predictive models for a particular reasoner. We called it, the *Matthews Kappa Score* (**MKS**). As the acronym would suggest, *MKS* is a weighted aggregation of the *MCC* and the *Kappa* values computed to assess a reasoner model M_R.

$$MKS(M_R) = \frac{\alpha * MCC(M_R) + \beta * Kappa(M_R)}{\alpha + \beta} \qquad (2)$$

such that $(\alpha + \beta) = 1$. Thus, the best predictive model for a given reasoner is the one having the maximal value of the *MKS* score:

$$M_R^{Best} = arg \max_{M_R^i \in \mathcal{M}_R} (MKS(M_R^i)) \qquad (3)$$

where $\mathcal{M}_R$ denotes the set of predictive models learned for a reasoner R. In the case where multiple maximal models are identified, the model with the highest *F-Measure* (FM) is selected. Our aggregation is coherent as both *MCC* and *Kappa* are ranging in $[-1, 1]$.

3.3   Robustness Learning Results and Assessment Discussion

The selection of a reasoner *best* predictive model is achieved through two stages: the best model given a specific reasoner dataset variant (RAWD, MDL, CFS or RLF); then the best model across all the datasets. For the sake of brevity, we only list the results of the second selection stage. In Table 3, the M_{best} column

Table 3. Comparison and selection of a reasoner *best* learned robustness model across the different datasets. #F standing for the number of ontology features used in M_R^{Best}.

	RAWD ($\perp$)			MDL (discrete)			CFS			RLF			M_R^{Best}
	M_{best}	MKS	FM	M_{best}	MKS	FM	M_{best}	MKS	FM	M_{best}	MKS	FM	#F
Konclude	IBK	0.61	0.97	**MP**	**0.63**	**0.97**	RF	0.58	0.97	IBK	0.61	0.97	67
MORe	RF	0.81	0.96	**SMO**	**0.82**	**0.96**	RF	0.79	0.95	RF	0.81	0.96	94
HermiT	RF	0.86	0.96	**MP**	**0.87**	**0.97**	RF	0.85	0.96	RF	0.86	0.96	92
TrOWL	RF	0.77	0.94	**RF**	**0.79**	**0.95**	RF	0.71	0.93	RF	0.75	0.94	95
FaCT++	RF	0.875	0.95	**RF**	**0.879**	**0.95**	RF	0.836	0.94	RF	0.871	0.95	95
JFact	RF	0.900	0.95	RF	0.900	0.95	RF	0.886	0.94	**RF**	**0.903**	**0.95**	58

illustrates the name of the learned algorithm corresponding to the identified reasoner best model given a dataset type. The assessment values of MKS and FM presented in this table are the average across the 10 folds cross-validation. By comparing the MKS score of the different models learned for a given reasoner, it is easy to recognize the final best model M_R^{Best}. This means identifying the most suitable combination (learning algorithm, feature selection method) which have led to the topmost accurate reasoner robustness predictive model. In Table 3, the M_R^{Best} models are highlighted in extra bold face. The number of ontology features deployed within the M_R^{Best} model is reported in the last column of Table 3 (the #F column). In overall, the learned reasoners' best predictive models, i.e. the M_R^{Best} set of models, showed to be highly accurate, achieving in all cases, an over to 0.95 FM value. In addition, they are well resisting to the problem of minor classes, as in all cases, their respective MKS scores were over 0.63. In particular, the MKS of the JFact best predictive model was 0.90, indicating almost a *perfect* predictive model. All of these findings shed light on the high generalizability of the learned model as well as its effectiveness in predicting reasoner robustness for the ontology classification task. A number of important observations can also be made from this table. First, not once the RAWD dataset, with its entire set of almost 110 ontology features, was listed in the final selection of the reasoners' best models. In most cases, the MKS and FM values of reasoner predictive models derived from featured datasets (MDL, CFS or RLF) outperform those ones computed from RAWD models. Therefore, it is quite expectable, that there is a restricted set of key ontology features that are more correlated to the reasoner performances that the other features. It would also be observed that the size of feature vectors of the M_R^{Best} models varies from 58 in the case of JFact going to 95 in the case of TrOWL and FaCT++. This observation pinpoints that the key ontology features, are close to the reasoner under study. Reasoners implement different optimization techniques to overcome particular ontology complexity sources and thus, indicators of their robustness would also vary. In fact, we are not sure which feature selection method would be the most suited in unveiling the ontology key features for all the reasoners. The MDL method has outperformed in major cases, but beaten by the RLF's technique in one reasoner instance, i.e. JFact. On the other hand, SMO, MP and RF have shared the podium of the outstanding ML algorithms. Nevertheless, it can easily be observed that the RF algorithm is the most dominant one except when using the MDL discretization method. Obviously, the discretization improves the predictive quality of the learning algorithms like MP, SMO and SL. Throughout these observation, we would admit that there is no ultimate best combination, (feature selection, ML algorithm), that suits for all the reasoners. Accordingly, we confirm that even if the learning process may be maintained the same, the learning techniques must be diversified to grasp, in a better way, the reasoner-specific empirical behaviours.

4   Ontology Reasoners Ranking with Ties

In the first part of this chapter, we focused on studying individual reasoners seeking to a priori predict their robustness over a given ontology. Now, we put the focus on providing advices to potential users regarding the *best* available reasoners to their ontology based applications. The provided advices are in their basic form, i.e. a ranking of reasoners in a most-preferred order. Our reasoner ranking method allows ties and follows the principals of bucket order. In the remainder, we describe our ranking approach after recalling some basic notions.

4.1   Preliminaries

Following Fagin et al. [10], given a set of data objects, a total order (or a linear order) is a binary relation of objects such that the relation satisfies the three criteria of anti-symmetry, transitivity, and linearity. A bucket order is simply a total order that allows "ties" between objects in a bucket. Formally, let $\mathbf{B}=\{\mathcal{B}_1,\ldots,\mathcal{B}_t\}$ be a set of non empty *buckets* that forms a partition of a set $\mathcal{A}$ of alternatives[12], i.e. $\bigcup_{i=1}^{t}\mathcal{B}_i = \mathcal{A}$ and $\mathcal{B}_i \cap \mathcal{B}_j = \emptyset$ if $i \neq j$, and let $\prec$ be a strict total order on $\mathcal{B}$, i.e. $\mathcal{B}_1 \prec \ldots \prec \mathcal{B}_t$, we say $(\mathcal{B}, \prec)$ forms a *bucket* order on $\mathcal{A}$. $\prec$ is a transitive, strict binary relation such that for any $x, y \in \mathcal{A}$, x precedes y denoted by $x < y$, if and only if there are i, j with $i \neq j$ such that $x \in \mathcal{B}_i$, $y \in \mathcal{B}_j$ and $\mathcal{B}_i \prec \mathcal{B}_j$. If x and y are in the same bucket, then x and y are incomparable and said to be "tied". A *totally ordered set* is a special case of a bucket ordered set in which every bucket is of size 1. A *ranking* σ *with ties* over $\mathcal{A}$ is associated with a *bucket order* $(\mathbf{B}, \prec)$ over $\mathcal{A}$ as $\sigma = \mathcal{B}_1 \prec \ldots \prec \mathcal{B}_t$, where i is the rank of $x \in \mathcal{A}$, i.e. $\sigma(x) = i$, if and only if $x \in \mathcal{B}_i$. $\forall x' \in \mathcal{B}_i \wedge x \neq x'$, then x, x' are equally ranked $\sigma(x) = \sigma(x') = i$ and hence a tie. When x precedes y, i.e. $\sigma(x) < \sigma(y)$, this means a preference for x over y. Following [27], given a bucket order $(\mathbf{B}, \prec)$ over a set of alternatives $\mathcal{A}$, $\mathbf{C_B}$ is a pair ordered matrix capturing the pairwise preferences over the elements of $\mathcal{A}$. $\mathbf{C_B}$ is $|\mathcal{A}| \times |\mathcal{A}|$ matrix such for each $x_i, x_j \in \mathcal{A}$ that $\mathbf{C_B}(i,j) = 1$ if $x_i < x_j$, i.e. $\mathcal{B}_i \prec \mathcal{B}_j$, $\mathbf{C_B}(i,j) = 0.5$ if x_i and x_j belongs to the same bucket and $\mathbf{C_B}(i,j) = 0$ if $x_j < x_i$. By convention, $\mathbf{C_B}(i,i) = 0.5$ and $\mathbf{C_B}(i,j)$ satisfies that $\mathbf{C_B}(i,j) + \mathbf{C_B}(j,i) = 1$. For simplicity, the terms rankings and orders are used interchangeably throughout this chapter.

4.2   Overall Description of the Ranking Approach

We propose a method of ranking with ties over a set of reasoners, which are likely to be good candidates to accomplish a reasoning task over a user input ontology. Our ranking is a supervised one, mainly based on reasoners' robustness models described in Sect. 2. More in details, given a user ontology O_u characterized by a feature vector (Subsect. 2.3) and a set of candidate reasoners $\mathcal{R} = \{R_1, \ldots, R_m\}$,

---

[12] The alternatives are the objects, or elements to be ranked.

we suggest to predict the robustness of each of these reasoners over the input ontology and accordingly compute their ranking through a rigorous application of the bucket order principals. Obviously, by comparing the prediction results of the set of reasoners (i.e. the alternatives), 4 possible reasoner groups could be identified, each corresponds to a robustness predictive label $\{C, U, T, H\}$ (Subsect. 2.2). These groups are disjoint ones, as we would predict only one label per reasoner, their union is equal to the whole set of reasoners, and therefore, they form a partition of $\mathcal{R}$. Hence, we can define the buckets set $\{\mathcal{B}_C, \mathcal{B}_U, \mathcal{B}_T, \mathcal{B}_H\}$, which matches these groups. In addition, a precedence order relation $\prec$ over the described buckets can easily be specified. In our opinion, the most straightforward order is $\mathcal{B}_C \prec \mathcal{B}_U \prec \mathcal{B}_T \prec \mathcal{B}_H$. Intuitively, reasoners belonging to $\mathcal{B}_C$ are the most preferred ones as they are robust over O_u. The $\mathcal{B}_U$'s reasoners could achieve the reasoning task within the time limit but the correctness of their results was not approved by our correctness checking method[13]. Therefore, we think that they are much preferred than reasoners falling in the $\mathcal{B}_T$ bucket, which won't deliver any result to be assessed. Of course, the worse reasoners are in $\mathcal{B}_H$ bucket, as they are not even capable to process the ontology.

Given this configuration, two particular cases need special attention. First when $|\mathcal{B}_C| = 0$, this means no reasoner would be suggested as robust over the input ontology. However, the remaining ordered buckets could still be provided to the user as a good tagging guidance as how reasoners would possibly behave against the O_u ontology. Another interesting case is when $|\mathcal{B}_C| > 1$, here more than one candidate reasoner, may be all of them, would be provided as equally robust reasoners. However, we think that such a ranking would be of little help for the user, since it does not distinguish the most *suited* reasoner over the set of the *robust* ones. Wherefore, we need to break the tie at the top of the ranking list, otherwise further partitioning the $\mathcal{B}_C$ bucket.

For a better discrimination between reasoners falling in the $\mathcal{B}_C$ bucket, we chose to use an additional reasoner evaluation criterion, the efficiency. Applying this method would require training further predictive models for each candidate reasoner, more precisely *regression models*, which map the ontology feature vector $X^{(O_u)}$ to the reasoner's estimated runtime, $\widehat{t}_R \in \mathbb{R}$. Having these models, the $\mathcal{B}_C$ bucket's reasoners will be linearly ordered by a simple decreasing sorting over their predicted computational times. The more efficient a robust reasoner is, the smaller would be its ranking number. Henceforth, each of the *robust* reasoners will have its own corresponding bucket. Nevertheless, the order of reasoners within the remaining buckets, $\mathcal{B}_U \prec \mathcal{B}_T \prec \mathcal{B}_H$, will be maintained the same and their ranking numbers will be updated. For instance, if c reasoners were found to be robust $c = |\mathcal{B}_c|$ and $\mathcal{B}_U \neq \emptyset$, then $\forall R_u \in \mathcal{B}_U, \sigma(R_u) = c + 1$. To meet the aforementioned requirements, we are extending the definition of ranking with ties with particular rules as follows:

---

[13] However, the reasoner results correctness may be approved by a different checking method.

Definition 3 (Reasoners Ranking with Ties). *Given a set of candidate rea-soners $\mathcal{R} = \{R_1, \ldots, R_m\}$ and an ordered bucket partition over $\mathcal{R}$, $\mathbf{B} = \mathcal{B}_C \prec \mathcal{B}_U \prec \mathcal{B}_T \prec \mathcal{B}_H$, a reasoner R_i precedes a reasoner R_j, i.e. $\sigma(R_i) < \sigma(R_j)$ and $\mathbf{C_B}(i, j) = 1$, if and only if, $R_i \in \mathcal{B}_i$ and $R_j \in \mathcal{B}_j \wedge \mathcal{B}_i, \mathcal{B}_j \in \mathbf{B}$ such that $\mathcal{B}_i \prec \mathcal{B}_j$ or $\mathcal{B}_i = \mathcal{B}_j = \mathcal{B}_C$ and $\widehat{t}_{R_i} < \widehat{t}_{R_j}$. Consequently, R_i and R_j are tied, i.e. $\sigma(R_i) = \sigma(R_j)$ and $\mathbf{C_B}(i, j) = 0.5$, if and only if, $\mathcal{B}_i = \mathcal{B}_j \neq \mathcal{B}_C$. The remaining conventions over $\mathbf{C_B}$ matrix won't be changed, this means $\mathbf{C_B}(i, i) = 0.5$ and $\mathbf{C_B}(i, j) + \mathbf{C_B}(j, i) = 1$.*

To sum it up, our reasoner ranking approach operates in two periods: *(i)* an *offline* period for knowledge acquisition about a set of reasoners using state of art supervised learning techniques; and *(ii)* an *online* period for users counselling over the reasoners using our method of ranking with ties. Further details about each period are given in the next subsections.

4.3 Description of the Offline Training Process

The main goal of this stage is to build meta-knowledge about a set of candi-date reasoners $\mathcal{R} = \{R_1, \ldots, R_m\}$, with $m = |\mathcal{R}|$. The meta-knowledge will be acquired using supervised ML techniques and recorded as reasoner profiles. A reasoner profile is defined in the following:

Definition 4 (Reasoner Profile). *Given the performance criteria space of the ontology reasoners $\mathcal{C}^s$, where s stands for the space dimension, a reasoner profile $P(R)$ is represented by a s-dimensional vector of predictive models, $P(R) = [M_R^{C_1}, \ldots, M_R^{C_s}]$, where $M_R^{C_i}$ stands for the predictive model which estimates the quality of the reasoner R according to the performance criterion C_i, $i \in [1, s]$.*

In our case, a couple of reasoner performances criteria are under investigation: the robustness and the efficiency. Thus, two predictive models are needed to be trained for each reasoner: a classification model of the reasoner robustness M_R^B and a regression model of its computational time M_R^E. Given $\mathcal{R}$ the set of candidate reasoners and $\mathcal{S}(O)$ a sampling of n ontologies, reasoners evaluation over these ontologies will be carried on and results will be recorded. Information derived during the evaluation reflects a general empirical behaviour of reasoners, which would serve as meta-knowledge for learning reasoner predictive models.

In fact, by labelling each ontology with the termination state of the reasoning task accomplished by a particular reasoner $R_i \in \mathcal{R}$, we proposed a methodology to learn a predictive model of the overall robustness of this reasoner (details in Subsect. 2.2). The same methodology will also be employed to build the regres-sion models after making some adjustments to the input data and the learn-ing techniques. The first modification to accomplish is about the sampling of ontologies. We recall that a regression model of a reasoner R_i will be invoked only after predicting its robustness over an input ontology. Otherwise, we need to predict how fast would be the reasoner in successfully classifying an ontology. Therefore, we removed from the sampling those ontologies which caused the fail-ure of the reasoner. So, let $\mathcal{S}_c(O) \subset \mathcal{S}(O)$ be the training set of ontologies for efficiency learning task. In a second step, ontologies within $\mathcal{S}_c(O)$ are labelled by the logarithm of the R_i's runtime. The log transformation of runtime was first proposed by Lin et al. [33], whom found it to be of paramount importance to improve the overall accuracy of regression model. The latter one could be deteriorated due to the large variation in runtimes across the ontologies. Thus, the new label vector of the reasoner R_i would be $\mathcal{L}_{R_i} = [log(t(O_1)), \ldots, log(t(O_n))]$, with $t(O_i)$ standing for the processing time spent by R_i to classify O_i. Henceforth, we will learn to approximate, $\hat{t}_i$, the logarithm of R_i' runtime. The $\mathcal{AL}$ set of learning algorithms will be restricted to Random Forest (**RF**), Logistic Regression (**LR**) and the SVM based algorithm (**SMO**), since they are capable to train regression models. On the other hand, the $\mathcal{FS}$ set of feature selection methods will only include the **CFS** and the **RLF** methods in addition to the $\bot$ element. The discretization can't be used for regression, since we are learning continuous values, rather than labels. The predictive accuracy of the trained efficiency models will be adjudged using two well established indices, the root mean squared error (**RMSE**) and the coefficients of determination (R^2) also known as the R-square [1]. Commonly known, a good regression model will have a low RMSE and a high R^2 close to 1. By referring to this rule, we will select the *best* efficiency predictive model for each reasoner. The selection method is similar to the one described in Subsects. 2.2 and 3.2, and having the following selection formula:

$$M_R^{Best} = arg \max_{M_R^i \in \mathcal{M}_R} (R^2(M_R^i)) \wedge arg \min_{M_R^i \in \mathcal{M}_R} (RMSE(M_R^i)) \tag{4}$$

where $\mathcal{M}_R$ denotes the set of predictive models learned for a reasoner R. An aggregation of RMSE and R^2 can not be established since the both metrics have different ranges of values.

As a final result to this knowledge acquisition stage, a set **P** of m reasoner profiles, $\mathbf{P} = [P(R_1), \ldots, P(R_m)]$, is built in and stored in a knowledge base.

4.4   Description of the Online Ranking Process

Once a user introduces its ontology, O_u, to our reasoner recommendation system, features of this ontology are computed, $X^{(O_u)}$, and stored reasoner profiles, $\mathbf{P} = [P(R_1), \ldots, P(R_m)]$, are loaded in order to rank them following the former specified ranking rules (c.f. Definition 3). The ranking procedure is detailed in Algorithm 2. The main purpose of this algorithm is to partition the candidate set into ordered buckets $\mathcal{B}_C \prec \mathcal{B}_U \prec \mathcal{B}_T \prec \mathcal{B}_H$ and accordingly encoding the *reasoner pair order* matrix $\mathbf{C_B}$ ($|\mathbf{P}| \times |\mathbf{P}|$ matrix). Let **B** be a vector of size $|\mathbf{P}|$ recording for each reasoner the bucket it belongs to. For instance, if a reasoner R_i is predicted to be robust over the ontology O_u, then $\mathbf{B}[i] \leftarrow$ 'C'[14]. Initially, the buckets of the reasoners are unknown.

---

[14] We recall that each of the buckets $\mathcal{B}_C \prec \mathcal{B}_U \prec \mathcal{B}_T \prec \mathcal{B}_H$ corresponds to a predictable label $\{C, U, T, H\}$ computed by the robustness classification model.

Algorithm 2. The Reasoner Ranking with ties.

Input : The user input ontology O_u.

Output: The $\mathbf{C_B}$ matrix encoding the pairwise preferences over the candidate reasoners.

1 $X^{(O_u)} \leftarrow$ **computeTheOntologyFeatureVector**(O_u);
2 $\mathbf{P} \leftarrow$ **loadTheReasonerProfiles**() ; //$\mathbf{P} = [P(R_1), \ldots, P(R_m)]$
3 $\mathbf{B} \leftarrow [F, F, \ldots, F]$; //$\mathbf{B}$ *is of size* m
4 $\mathbf{T} \leftarrow [+\infty, +\infty, \ldots, +\infty]$; //$\mathbf{T}$ *is of size* m
5 Initialize the pairs of the $\mathbf{C_B}$ matrix with 0.5 ;
6 **for** $i \leftarrow 1$ *to* $|\mathbf{P}|$ **do**
7 $\quad$ $\widehat{l_i} \leftarrow$ **predictTheReasonerRobustness**$(\mathbf{P}[i], X^{(O_u)})$;
8 $\quad$ $\mathbf{B}[i] \leftarrow \widehat{l_i}$;
9 $\quad$ **if** $\widehat{l_i} = {}'C'$ **then**
10 $\quad\quad$ //*the reasoner is robust*
11 $\quad\quad$ $\mathbf{T}[i] \leftarrow$ **predictTheReasonerRuntime**$(\mathbf{P}[i], X^{(O_u)})$;
12 $\quad$ **end**
13 $\quad$ **for** $j \leftarrow 1$ *to* $|\mathbf{P}|$ **do**
14 $\quad\quad$ **if** $(\mathbf{B}[i] \prec \mathbf{B}[j])$ *or* $(\mathbf{B}[i] = \mathbf{B}[j] = {}'C'$ *and* $\mathbf{T}[i] < \mathbf{T}[j])$ **then**
15 $\quad\quad\quad$ $\mathbf{C_B}[i][j] \leftarrow 1$;
16 $\quad\quad$ **else if** $(\mathbf{B}[i] = \mathbf{B}[j] \neq {}'C')$ **then**
17 $\quad\quad\quad$ $\mathbf{C_B}[i][j] \leftarrow 0.5$;
18 $\quad\quad$ **end**
19 $\quad\quad$ $\mathbf{C_B}[j][i] \leftarrow 1 - \mathbf{C_B}[i][j]$;
20 $\quad$ **end**
21 **end**
22 **return** $\mathbf{C_B}$;

Thus, the $\mathbf{B}$ vector is initialized as a sequence of 'F', standing for *full*, this means the full bucket gathering all the candidates. This particular bucket is placed at the bottom of the buckets, i.e. $\mathcal{B}_C \prec \mathcal{B}_U \prec \mathcal{B}_T \prec \mathcal{B}_H \prec\prec \mathcal{B}_F$. Respectively, we can induce that all the reasoners are initially tied and having the same rank, that's why all the pairs of the $\mathbf{C_B}$ matrix are initialized by 0.5. The partitioning of the reasoners set into disjoint buckets as well as the encoding of $\mathbf{C_B}$ matrix are progressively computed while iterating over the set of reasoners. We recall that reasoners within the $\mathcal{B}_C$ bucket are further ordered according to their predicted runtime. Let $\mathbf{T}$ be the vector of size $|\mathbf{P}|$ recording for each reasoner its predicted runtime. The $\mathbf{T}$ vector is initialized with a sequence of $+\infty$, standing for the maximal possible computing time. This configuration would guarantee that reasoners within any bucket different from $\mathcal{B}_C$, will always be tied at the bottom of the reasoner ranking list.

A step by step example is given in Fig. 2 explaining our algorithm running mode. Given 5 reasoner profiles $\mathbf{P} = [P(R_1), P(R_2), P(R_3), P(R_4), P(R_5)]$ and an ontology O_u, Fig. 2 depicts the different iterations processed in order to encode the $\mathbf{C_B}$ matrix. In each iteration, we show the reasoner R_i predicted label and eventually its predicted runtime, then the corresponding changes over

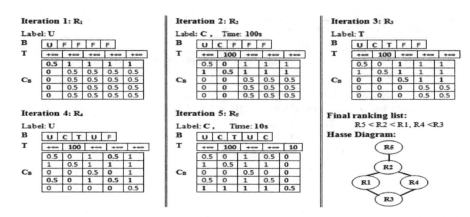

Fig. 2. Reasoner ranking steps with Hasse Diagram representation of the result.

the data structures, **B, T** and **C$_B$**. The computed reasoner pair order matrix could be illustrated using the well known Hasse Diagram [30] or by a simple string representation.

5   Evaluation of the Reasoners Ranking Approach

In this section, we describe the results of the empirical evaluation of our proposed rea-soner ranking approach. First, we report the assessment of the training stage, then the experimental validation of the ranking method.

5.1   Assessment of the Training Stage

As earlier specified, the prime goal of this stage is to build a set of reasoner profiles. Each profile is composed of two reasoner predictive models: the robustness and the efficiency one. In Sect. 3, we described the collected data for the evaluation, then, we reported and discussed the assessment of the trained robustness predictive models of these reasoners. Using the same data and under identical experimental environment, we trained the efficiency predictive models for the each candidate reasoner. Table 4 illustrated the assessment of the best efficiency predictive models trained from a specific reasoner dataset variant. The presented R^2 and $RMSE$ values are the average across the 10 folds cross-validation. The final selection of reasoners' best predictive models across all the datasets, M_R^{Best}, are highlighted in extra bold face. The number of ontology features deployed within the M_R^{Best} is reported in the last column of Table 4 (the #F column).

Three main observations could be drawn from Table 4. First, Random Forest (RF) have beaten Logistic Regression (LR) and SVM (SMOreg) in all dataset variants and for all the reasoners. RF associated to the feature selection method (RLF), have delivered the reasoners' topmost accurate regression models, M_R^{Best}. This is according to both quality measures R^2 and RMSE. Nevertheless, we

Table 4. Comparison and selection of a reasoner *best* learned efficiency models across the different datasets. #F standing for the number of ontology features used in M_R^{Best}.

$\mathcal{FS}$									
Reasoner	RAWD ($\bot$)			CFS			RLF			M_R^{Best}
	M_{best}	R^2	RMSE	M_{best}	R^2	RMSE	M_{best}	R^2	RMSE	#F
Konclude	RF	0.961	0.506	RF	0.961	0.502	**RF**	**0.962**	**0.497**	26
MORe	RF	0.933	0.301	RF	0.921	0.322	**RF**	**0.937**	**0.289**	37
HermiT	RF	0.941	0.451	RF	0.937	0.465	**RF**	**0.942**	**0.450**	36
TrOWL	RF	0.945	0.427	RF	0.943	0.431	**RF**	**0.947**	**0.413**	27
FaCT++	RF	0.921	0.604	RF	0.924	0.587	**RF**	**0.926**	**0.579**	32
JFact	RF	0.947	0.467	RF	0.948	0.457	**RF**	**0.952**	**0.439**	36

would observe that in most cases, the R^2 on the RLF's models differed only in the third decimal place from the R^2 on the RAWD's and on the CFS's models. The same observation also holds with respect to RMSE values. This would indicate that the reduction of the ontology feature space did not remarkably improve the accuracy of the regression models. This may be due to the operating mode of the RF algorithm which by default deploys an internal feature selection mechanism, whatever the initial dataset is featured or not. Nevertheless, the little accuracy boosting obtained thanks to the RLF method is certainly worthwhile. Besides, it provided useful insights about the core ontology features needed to train an accurate regression model. This would be recognized by scrutinizing the number of the final selected features depicted in the last column of Table 4.

Finally, the most important observation that we would make from Table 4 is that our reasoners' best efficiency predictive models are highly accurate. The reported R^2 values are ranging over 0.92 and going to 0.96. We should like to point out that the R^2 values on the regression models of the meta-reasoner R_2O_2 [19] are below our values and vary from 0.73 up to 0.91. This would evidently provide an insight into how well our regression models for the 6 reasoners are fitting the data. This observation would further be confirmed by analysing the RMSE values. We recall the RMSE represents the sample standard deviation of the differences between predicted and observed values. Hence, the smaller the RMSE value, the more accurate the prediction model is. In our case, the RMSE was low on all the M_R^{Best} models predicting the $log_1 0$ of a reasoner runtime. More in details, the lowest reported RMSE is about 0.29 in the case of MORe, which means an average misprediction of a factor of $10^{0.29} < 2$, whereas the largest RMSE is 0.58 in the case of FaCT++, i.e. a factor of $10^{0.58} < 4$. Worth of cite, the RMSE values on the regression models deployed in the portfolio-based algorithm selector SATzilla [33] were ranging in $[0.49, 1.01]$, and this was sufficient to enable SATzilla to win various medals in SAT competitions. Our RMSE values are below those of SATzilla, and this is a valuable proof of the effectiveness of our regression models. Though, it must be noted that currently our goal is not to outperform neither R_2O_2 nor SATZilla, but only to demonstrate that our reasoner runtime regression models could reach sufficiently satisfactory results. The above inspections assert that the trained reasoner efficiency predictive models can be used to accurately predict a given reasoner classification time for a wide spectrum of OWL ontologies.

5.2   Assessment of the Ranking Stage

The empirical evaluation of the reasoner rankings is based on a leave-one-out procedure. We recall that from the beginning of our study we have held out 500 ontologies from the original corpus, for ranking assessment purposes. This is the set of the *unforeseen* ontologies, designed by $T(O)$, that we will use to evaluate our ranking method. The ranking of reasoners recommended for a test ontology is compared to an *ideal* one computed for the same ontology. The *ideal* ranking should represent, as accurately as possible, the correct ordering of the reasoners on the ontology at hand. In our case, it is built by applying rules of Definition 3 on the basis of the actual known reasoner performances acquired during the reasoner evaluation on that ontology. Then, the agreement between both of the rankings is measured using known state-of-the-art metrics. To assess the quality of our recommended ranking, we need to compare it versus a baseline method. This would be a trivial solution of ranking known as the *default ranking*. The latter one is a ranking that remains the same one regardless of the ontology under examination. It is computed on the basis of the reasoner evaluation results described in Subsect. 3.1, this means the overall robustness of a reasoner over the $T(O)$'s ontologies. The default ranking starting from the best reasoner is: Konclude $\prec$ MoRE $\prec$ HermiT $\prec$ TrOWL $\prec$ FaCT++ $\prec$ JFact. A ranking method is interesting, whenever it performs significantly better than this default ranking method.

The agreement between our recommended rankings (resp. the default ones) and the ideal rankings is computed according to five quality measures falling into 2 categories describes in the following:

Correlation Metrics: to assess the agreement on the full computed rankings, we compute the generalized *Kendall Tau* correlation coefficient for ranking with ties, denoted as τ_x, proposed by Emond and Mason [9] and the *Spearman rank correlation coefficient* known as the *Spearman's rho* [6]. In the latter metric, tied reasoners are assigned a rank equal to the average of their positions in the ascending order of the values. For instance, a recommended ranking $r\sigma = (2, 1, 3, 3, 4)$ is transformed into $r\sigma' = (2, 1, 3.5, 3.5, 5)$. According to the both correlation measures, a value of 1 represents perfect agreement, whereas -1 stands for a perfect disagreement. A correlation of 0 means that the rankings are not related.

Information Retrieval Based Metrics [5]: by following the evaluation style of search engines, we will examine how well we are ranking the reasoners at the top of the list. Hence, for each ranking list provided for a test ontology O_t, we measure the *Precision* at position K, denoted as P@K(O_t), which stands for the percentage of correctly predicted ranking of reasoners at that position. Then, we compute the Average of Precision at position K, denoted as AP@K (i.e. $AP@K = \frac{\sum_{i=1}^{K} P@i(O_t)}{K}$)[15]. To obtain an overall idea of precision considering the whole test set, we compute the Mean Average Precision, denoted as MAP@K (i.e. $MAP@K = \frac{\sum_{j=1}^{M} AP@K(O_j)}{M}$, where $M = |T(O)|$). More specifically, our focus is on the 3 first positions since they are most likely to be examined by a user. Thus, we record for each test ontology, the P@1 since it reflects how well we are predicting the most performing reasoner, the AP@3 and finally to sum up the evaluation, we compute MAP@1 and MAP@3.

Table 5 illustrates the mean and standard deviation of the assessment measures computed on our recommended ranking and on the default one. This gives a general idea about the ranking quality across the different ontologies in the test set.

Table 5. Evaluation summary of reasoner ranking methods over $T(O)$'s ontologies.

Method	Kendall Tau X τ_x	Spearman Rho	MAP@1	MAP@3
Recommended $(r\sigma)$	0.67 ± 0.29	0.77 ± 0.26	0.92 ± 0.27	0.76 ± 0.27
Default $(d\sigma)$	0.25 ± 0.26	0.37 ± 0.26	0.92 ± 0.27	0.60 ± 0.17

As it can be seen, the rankings generated by our method, i.e. the recommended, are more correlated to the ideal ranking according to both Spearman's Rho and Kendall's Tau X. The difference with the values of the default ranking is significantly high, witnessing the good quality of our provided rankings. This would further be asserted by looking at the value of MAP@1. The latter one indicates that in 92% of the test cases, our method have succeeded to predict and to correctly rank the topmost performing reasoner. Our value is even better than the MAP@1 achieved by the ranking algorithms deployed in the meta-reasoner R_2O_2 [19]. The authors reported a MAP@1 varying between 88.7% and 90.6%. On the other hand, the value of the mean average precision at position 3, MAP@3, is less good than MAP@1, revealing in some way troubles in ranking reasoners at positions higher than the first place. However, our MAP@3 remains sufficiently higher than the MAP@3 computed for the default rankings.

Even though a good overall ranking accuracy is achieved, we were curious about the behaviour of our method at the individual level of the testing ontologies. To get more insights, we plotted per ontology the assessment values achieved by each ranking method. Figure 3 scrutinizes the comparison established according to the aforementioned assessment measures and having as counterparts our recommended rankings and the default ones. As it can be observed from Fig. 3(a) and (b) in major cases our curve is higher than the one of the default method showing better accuracy. However for some ontologies, with the number varying

---

[15] It is to be noted that when $K = 1$, P@1 is equal to AP@1.

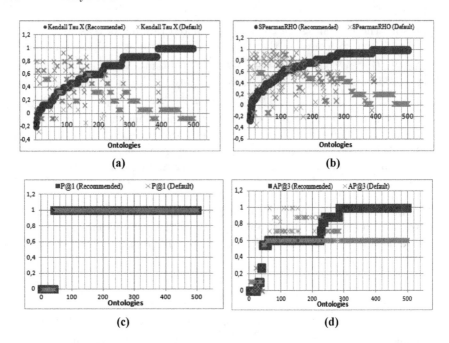

Fig. 3. Per ontology comparison between our recommended rankings (in blue) and the default ones (in red). (Color figure online)

from 50 to 150, we didn't succeed to outperform the default ranking in terms of the correlation with the ideal ranking. Moreover by analysing Fig. 3(c), we notice that our values of P@1 are equal to the ones of the default ranking. We recall, in the latter method, the best reasoner is always Konclude. So, this would indicate that for 50 ontologies over 500 we are still unable to accurately predict the failure of Konclude and most important to predict the alternative reasoner. Worth noting, the Konclude's robustness dataset described in Table 2 is the most biased one towards the *correct* cases that's why the MKS score of its robustness predictive model is the lowest one 0.63. These observations would pinpoint that our ranking method is closely depending on the accuracy of the robustness predictive models of each candidate reasoner. Overpassing the unbalanced nature of the reasoners' datasets is a challenging issue not only in our case but for the whole ML community. The advantage of this kind of analysis is that it enables us to mark off the hard ontologies, which have yield to inaccuracies. Hence, our investigations would be better focussed. On the other hand, our observations would motivate us to inspect different ranking methods like the ones relying on *preferences learning* approaches.

6   Conclusions

In this work, we proposed a new approach looking at providing knowledge for the selection of ontology reasoners. We can point out our contributions to two distinct but complementary fields: *(i)* we applied supervised leaning concepts to the problem of empirically predicting reasoner performances. We proposed a learning methodology relying on a rich set of advanced ontology features. Owing to the benefits of this proposal, we modelled the performance profile of a reasoner, in terms of predictive models of its robustness and efficiency. *(ii)* we provided a novel framework to support non-expert users in reasoner selection task. Our method applies principals of bucket order to establish reasoners' ranking with ties for any given ontology. We conducted a rigorous investigation about 6 well known reasoners and over 2500 ontologies. We put into comparison various supervised learning algorithms and feature selection techniques in order to train best reasoner predictive models. Our trained reasoner predictive models showed to be highly accurate outperforming state-of-art solutions. Moreover, the carried out experiments about our introduced ranking method revealed good results in terms of correlation to the ideal rankings comparing to the default ranking method. However, we point out some difficulties to accurately rank the reasoners for a particular category of ontologies. We designed the latter one as the top hardest ontologies. Predicting the reasoner performances over these ontologies remains a hard task needing more investigation efforts. Therefore, we are intended to establish a method that would enables us to identify these stringent ontologies. We believe that by categorizing the ontologies according to their hardness level, we would be capable to train more targeted and accurate reasoner performances predictive models. Moreover, we are minded to experiment *learning preference* techniques and *learning to rank* methods in order to further improve our approach of ontology reasoner selection. The ultimate purpose of our reasoner learning and ranking methods is to build an efficient ontology reasoner portfolio. Nevertheless, we are convinced that more optimization stages are still needed to be achieved.

References

1. Abbott, D.: Applied Predictive Analytics: Principles and Techniques for the Professional Data Analyst, 1st edn. Wiley, Hoboken (2014)
2. Alaya, N., Ben Yahia, S., Lamolle, M.: Predicting the empirical robustness of the ontology reasoners based on machine learning techniques. In: Proceedings of the 7th International Conference on Knowledge Engineering and Ontology Development KEOD 2015, Lisbon, Portugal, pp. 61–73 (2015)
3. Alaya, N., Lamolle, M., Ben Yahia, S.: Towards unveiling the ontology key features altering reasoner performances. Technical report, IUT of Montreuil, France (2015). http://arxiv.org/abs/1509.08717
4. Baader, F., Calvanese, D., McGuinness, D.L., Nardi, D., Patel-Schneider, P.F.: The Description Logic Handbook: Theory Implementation and Applications, 2nd edn. Cambridge University Press, New York (2010)
5. Baeza-Yates, R.A., Ribeiro-Neto, B.: Modern Information Retrieval. Addison-Wesley Longman Publishing Co., Inc., Boston (1999)
6. Caruso, J.C., Cliff, N.: Empirical size, coverage, and power of confidence intervals for Spearman's rho. J. Educ. Psychol. Meas. **57**, 637–654 (1997)

7. Chandrashekar, G., Sahin, F.: A survey on feature selection methods. Comput. Electr. Eng. **40**(1), 16–28 (2014)
8. Dumontier, M., Glimm, B., Gonçalves, R.S., Horridge, M., Jiménez-Ruiz, E., Matentzoglu, N., Parsia, B., Stamou, G.B., Stoilos, G. (eds.): Informal Proceedings of the 4th International Workshop on OWL Reasoner Evaluation (ORE-2015), Athens, Greece, vol. 1387. CEUR-WS.org (2015)
9. Emond, E.J., Mason, D.: A new rank correlation coefficient with application to consensus ranking problem. J. Multicriteria Decis. Anal. **11**, 17–28 (2002)
10. Fagin, R., Kumar, R., Mahdian, M., Sivakumar, D., Vee, E.: Comparing partial rankings. SIAM J. Discret. Math. **20**, 628–648 (2006)
11. Garcia, S., Luengo, J., Saez, J., Lopez, V., Herrera, F.: A survey of discretization techniques: taxonomy and empirical analysis in supervised learning. IEEE Trans. Knowl. Data Eng. **25**, 734–750 (2013)
12. Gardiner, T., Tsarkov, D., Horrocks, I.: Framework for an automated comparison of description logic reasoners. In: Cruz, I., Decker, S., Allemang, D., Preist, C., Schwabe, D., Mika, P., Uschold, M., Aroyo, L.M. (eds.) ISWC 2006. LNCS, vol. 4273, pp. 654–667. Springer, Heidelberg (2006). doi:10.1007/11926078_47
13. Gomes, C.P., Selman, B.: Algorithm portfolios. Artif. Intell. **126**, 43–62 (2001)
14. Gonçalves, R.S., Matentzoglu, N., Parsia, B., Sattler, U.: The empirical robustness of description logic classification. In: Informal Proceedings of the 26th International Workshop on Description Logics, Ulm, Germany, pp. 197–208 (2013)
15. Gonçalves, R.S., Bail, S., Jiménez-Ruiz, E., Matentzoglu, N., Parsia, B., Glimm, B., Kazakov, Y.: Owl reasoner evaluation (ORE) workshop 2013 results: short report. In: ORE, pp. 1–18 (2013)
16. Gonçalves, R.S., Parsia, B., Sattler, U.: Performance heterogeneity and approximate reasoning in description logic ontologies. In: Cudré-Mauroux, P., et al. (eds.) ISWC 2012. LNCS, vol. 7649, pp. 82–98. Springer, Heidelberg (2012). doi:10.1007/978-3-642-35176-1_6
17. Hall, M., Frank, E., Holmes, G., Pfahringer, B., Reutemann, P.: The weka data mining software: an update. SIGKDD Explor. Newsl. **11**, 10–18 (2009)
18. Hu, B., Dong, W.: A study on cost behaviors of binary classification measures in class-imbalanced problems. CoRR abs/1403.7100 (2014)
19. Kang, Y.-B., Krishnaswamy, S., Li, Y.-F.: R_2O_2: an efficient ranking-based reasoner for OWL ontologies. In: Arenas, M., et al. (eds.) ISWC 2015. LNCS, vol. 9366, pp. 322–338. Springer, Heidelberg (2015). doi:10.1007/978-3-319-25007-6_19
20. Kang, Y.-B., Li, Y.-F., Krishnaswamy, S.: Predicting reasoning performance using ontology metrics. In: Cudré-Mauroux, P., et al. (eds.) ISWC 2012. LNCS, vol. 7649, pp. 198–214. Springer, Heidelberg (2012). doi:10.1007/978-3-642-35176-1_13
21. Kang, Y.B., Li, Y.F., Krishnaswamy, S.: How long will it take? Accurate prediction of ontology reasoning performance. In: Proceedings of the 28th AAAI Conference on Artificial Intelligence, pp. 80–86 (2014)
22. Lee, M., Matentzoglu, N., Sattler, U., Parsia, B.: Verifying reasoner correctness - a justication based method. In: Informal Proceedings of the 4th International Workshop on OWL Reasoner Evaluation (ORE-2015), pp. 46–52 (2015)
23. LePendu, P., Noy, N.F., Jonquet, C., Alexander, P.R., Shah, N.H., Musen, M.A.: Optimize first, buy later: analyzing metrics to ramp-up very large knowledge bases. In: Patel-Schneider, P.F., Pan, Y., Hitzler, P., Mika, P., Zhang, L., Pan, J.Z., Horrocks, I., Glimm, B. (eds.) ISWC 2010. LNCS, vol. 6496, pp. 486–501. Springer, Heidelberg (2010). doi:10.1007/978-3-642-17746-0_31

24. Matentzoglu, N., Leo, J., Hudhra, V., Sattler, U., Parsia, B.: A survey of current, stand-alone OWL reasoners. In: Proceedings of the 4th International Workshop on OWL Reasoner Evaluation, pp. 68–79 (2015)
25. Mikolàšek, V.: Dependability and robustness: state of the art and challenges. In: Proceedings of the First International Workshop on Software Technologies for Future Dependable Distributed Systems (STFSSD), pp. 25–31 (2009)
26. W3C OWL Working Group: OWL 2 Web Ontology Language: Document Overview. W3C Recommendation, October 2009. http://www.w3.org/TR/owl2-overview/
27. Pandit, V., Kenkre, S., Khan, A.: On discovering bucket orders from preference data. In: Proceedings of the Eleventh SIAM International Conference on Data Mining, Arizona, USA, pp. 872–883 (2011)
28. Prudêncio, R.B.C., de Souto, M.C.P., Ludermir, T.B.: Selecting machine learning algorithms using the ranking meta-learning approach. In: Jankowski, N., Duch, W., Grąbczewski, K. (eds.) Meta-Learning in Computational Intelligence. SCI, vol. 358, pp. 225–243. Springer, Heidelberg (2011)
29. Rice, J.R.: The algorithm selection problem. Adv. Comput. **15**, 65–118 (1976)
30. Rosen, K.H.: Discrete Mathematics and Its Applications. McGraw Hill Higher Education, New York (1991)
31. Sazonau, V., Sattler, U., Brown, G.: Predicting performance of OWL reasoners: locally or globally? In: Proceedings of the Fourteenth International Conference on Principles of Knowledge Representation and Reasoning (2014)
32. Weithöner, T., Liebig, T., Luther, M., Böhm, S., Henke, F., Noppens, O.: Real-world reasoning with OWL. In: Franconi, E., Kifer, M., May, W. (eds.) ESWC 2007. LNCS, vol. 4519, pp. 296–310. Springer, Heidelberg (2007). doi:10.1007/978-3-540-72667-8_22
33. Xu, L., Hutter, F., Hoos, H.H., Leyton-Brown, K.: Satzilla: portfolio-based algorithm selection for SAT. J. Artif. Int. Res. **32**, 565–606 (2008)

Refinement by Filtering Translation Candidates and Similarity Based Approach to Expand Emotion Tagged Corpus

Kazuyuki Matsumoto$^{(\boxtimes)}$, Fuji Ren, Minoru Yoshida, and Kenji Kita

Tokushima University, Tokushima, Japan
matumoto@is.tokushima-u.ac.jp

Abstract. Researches on emotion estimation from text mostly use machine learning method. Because machine learning requires a large amount of example corpora, how to acquire high quality training data has been discussed as one of its major problems. The existing language resources include emotion corpora; however, they are not available if the language is different. Constructing bilingual corpus manually is also financially difficult. We propose a method to convert a training data into different language using an existing Japanese-English parallel emotion corpus. With a bilingual dictionary, the translation candidates are extracted against every word of each sentence included in the corpus. Then the extracted translation candidates are narrowed down into a set of words that highly contribute to emotion estimation and we used the set of words as training data. Moreover, when one language's unannotated linguistic resources can be obtained, the words can be expanded based on the word distributed expression. By using this expressions, we can improve accuracy without decreasing information volume of one sentence. Then, we tried the corpus expansion without translating target linguistic resource. As the result of the evaluation experiment using the machine learning algorithm, we could clear the effectiveness of the emotion corpus which expanded based on the original language's unannotated sentences and based on similar sentence. Moreover, when large amount of linguistic resources without annotation can be obtained in one language, their words can be expanded based on distributed expressions of the words. By using distributed expressions, we can improve accuracy without decreasing information volume of one sentence. Then, we attempted to expand corpus without translating target linguistic resource. The result of the evaluation experiment using the machine learning algorithm showed the effectiveness of the expanded emotion corpus based on the original language's unannotated sentences and their similar sentences.

1 Introduction

Recently, there have been a lot of studies on emotion estimation from text such as utterance or weblog articles in the field of sentiment analysis or opinion mining [17,19–23]. Many of them adopted machine learning methods that used words

© Springer International Publishing AG 2016
A. Fred et al. (Eds.): IC3K 2015, CCIS 631, pp. 260–280, 2016.
DOI: 10.1007/978-3-319-52758-1_15

as a feature. When the type of the target sentence for emotion estimation and the type of the sentence prepared as training data are different, as in the case of terminology in the problem of domain adaptation for document classification, the appearance tendency of the emotion words differs. This often causes a problem in fluctuation of accuracy. On the other hand, when a word is used as a feature for emotion estimation, the sentence structure does not have to be considered. As a result, it is easy to apply the method to other languages. Only if we prepare a large number of corpora with annotation of emotion tags on each sentence, emotion would be easily estimated by using the machine learning method. In the machine learning method, because manual definition of a rule is not necessary, we can reduce costs to apply the method to other languages. However, just like the problem in the domain, depending on the types of the languages, sometimes it is difficult to prepare a sufficient amount of tagged corpus. For example, in comparison to English or Chinese, there are not enough tagged corpora in Japanese or Korean, as the people who use such languages are relatively small.

To solve the shortage problem of Chinese emotion corpus, Wan [27] used English emotion corpus as training data that was openly available for free, and attempted to classify Chinese emotions. He proposed a co-training model that combined a method translating the training data and a method translating the test data. Inui et al. [7] mechanically translated Japanese sentences into English sentences and used them as test data or training data to classify review articles into positive or negative by using SVM. They checked whether or not the sentences included evaluation expressions. Then, based on the results, they selected the sentences by judging if the sentence should be added in the training data or in the test data. The experiment obtained approximately 80% accuracy. This accuracy was higher than the accuracy obtained when the untranslated training data was used.

Their method summarizes a document by exclusively using the sentences in a review article that have evaluation expressions. Because the method does not confirm the reliability of the translation results to summarize, it is difficult to deal with the problem of estimation failure caused by low translation accuracy. Because our study does not aim at emotion estimation in document increments, their proposed document summarization technique cannot be applied to our study. We refined the translation candidates of each word in a sentence by narrowing them down under certain condition. Then, because only using narrowing down, it cause lack of versatility, we expanded the training data based on the word similarity. Moreover, we proposed the method to expanding corpus by adding the unannotated sentence similar to sentence which included in the training data.

In this paper, we attempted emotion estimation by our proposed method with machine learning. In the sentences of Japanese-English parallel emotion corpus the translation candidates for each word were obtained in reference to the bilingual dictionary. We used them as training data for machine learning and conducted an emotion estimation experiment. If bilingual dictionaries are used to obtain translation candidates, erroneous translations might be caused as often as or more often than when machine translation is conducted. For that reason, we proposed a refining method that narrowed down the translation candidates according to whether the kind of the sentence's emotion and the word's emotion matched or not. By removing the words that were not likely to contribute to sentence emotion, the translation candidates were refined. The aim of this method is to minimize the effects by translation error. Then, we discussed the effectiveness of the proposed method by evaluating the word expansion method and sentence expansion method. Section 2 describes the related works about emotion estimation based on word feature and emotion estimation based on different languages. To remove noise feature, we propose a method for refining translation candidates extracted from bilingual dictionaries in Sect. 3 and conduct an evaluation experiment in Sect. 4. Then, we discussed the results in Sect. 5. Finally, we summarize this study in Sect. 7.

2   Related Works

The researches on emotion estimation often adopted machine learning method that used words as a feature [10, 18, 28]. Many of these methods do not consider the meanings of the words. Actually, in the task of judging a word's or a phrase's emotion polarity (positive/negative), a certain level of accuracy can be obtained without considering the word's meaning [25, 26].

There are also researches that judge emotion categories of emotional words in a sentence [8]. In the machine learning, the quality or kind of source data used for training data is one of the most important factors that affect the classification accuracy.

To judge the emotion polarity of a sentence belonging to a different domain from the training data, Saiki et al. [24] adapted each domain by using the weighted maximum entropy model to add weight to case. Minato et al. [15] estimated sentence emotion by using appearance frequency weight of word for each emotion category according to Japanese-English parallel emotion corpus. The evaluation result showed that emotion estimation accuracies varied due to small size of the corpus and bias of the number of the sentences in each emotion category.

Balahur et al. [1] treated the problem of sentiment detection in several different languages such as French, German and Spanish. They translated each language resources into English by using the existing machine translation techniques and classified sentence emotion by training the n-gram feature of the translated resources based on Support Vector Machines Sequential Minimal Optimization (SVM SMO).

From the experimental result for the multilingual resources, they concluded that the statistical machine translation (SMT) was mature enough as preprocessing for sentiment classification. However, it is considered that the languages used in their study were easier to translate into English compare to translate Japanese into English. In Japanese language, with only difference of notation or intonation of the word, sense of the word sometimes changes. On the other hand, sentence structure is more complex than English or the other western languages.

Moreover, even if the machine translation system can translate Japanese into English successfully, with a little difference of the translation candidate, the nuance becomes different from the original meaning and the emotion to be conveyed might be changed.

To confirm this, it is necessary to conduct an experiment of emotion estimation by using the translation results based on Japanese-English emotion tagged corpus. Preprocessing was conducted by converting Japanese or English emotion tagged corpora into other language data by machine translation or parallel dictionary. We confirmed whether emotion estimation accuracy could be improved by refining translation candidates or not by the evaluation experiment.

3   Proposed Method

The basic flow of our proposed method is described in Fig. 1. EW Dictionary means a dictionary that stored the words expressing emotion included in the Japanese-English parallel emotion corpus [14]. SW List is a dictionary of stop

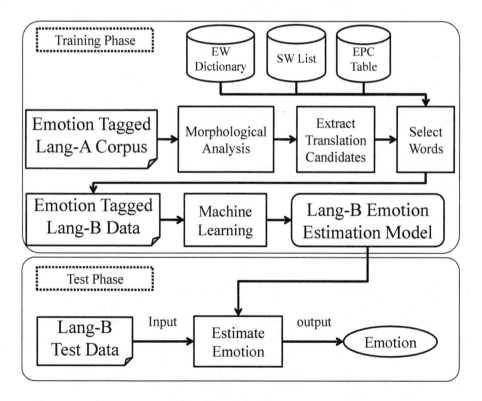

Fig. 1. Emotion estimation in different languages.

word that includes nonsense words (unnecessary words). EPC Table indicates Emotion Polarity Correspondence Table constructed by Takamura [25,26].

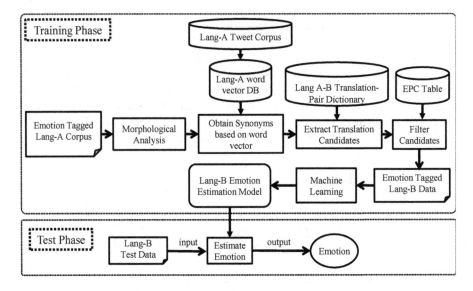

Fig. 2. Synonym based word expansion.

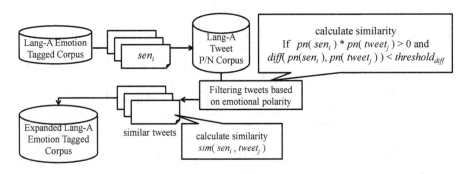

Fig. 3. Sentence similarity based corpus expansion.

In the Training Phase, each sentence in the emotion tagged corpus of language A is morphologically analyzed to obtain the basic forms of the words. Then, for the basic forms of the words the translation candidates are obtained in language B with reference to the bilingual dictionary. The candidate words are refined by our proposed method and the data in language B is created that has annotation of emotion tags in sentence basis. Based on the created data in language B, the emotion estimation model for language B is made by machine learning.

Next, Fig. 2 indicates the method which obtain the synonyms of the word included in sentence translation candidate.

Then, we proposed the method to increase corpus directly by collecting tweets including similar sentence of same language from tweet corpus without translating by sentence or word. The tweet corpus is previously refined into emotion expressed tweets. We expand the corpus based on the emotional polarity of the sentence in original emotion tagged corpus and similarity between sentences. Figure 2 indicates the outline of the method to acquire the translation candidates from the expanded corpus by obtaining the synonyms of the word included in the sentence based on word vector similarity. We also proposed a method to directly increase corpus without translating any sentence or word. In the method the tweets including the similar sentences of the same language were collected from the tweet corpus. The tweet corpus was previously refined into the tweets expressing any emotion. We expanded the corpus based on the emotional polarity of the sentence in the original emotion tagged corpus and the similarity between the sentences (Fig. 3).

3.1   Japanese-English Parallel Emotion Corpus

This section describes the Japanese-English Parallel Emotion Corpus used in this research. The Japanese-English Parallel Emotion Corpus constructed by Minato et al. [14] is a tagged corpus based on a Japanese-English Emotion Expression Dictionary [6].

In the corpus emotion tags are annotated on the Japanese-English bilingual conversation sentences to indicate the speaker's emotion and the emotional expression. In total, 1,190 pairs of the sentences are registered in the corpus.

There are two kinds of tags: sentence emotion tags annotated to sentences and word emotion tags annotated to words or idioms. There are nine kinds of emotion categories: (joy, anger, sorrow, surprise, hate, love, respect, anxiety and neutral).

3.2   Japanese-English Bilingual Dictionary

In this research we used the dictionaries of Eijiro Ver.2.0[1], Edict and Edict2[2] as the Japanese-English bilingual dictionaries. We selected these three dictionaries because these dictionaries included relatively newer words and a larger number of words compared to other dictionaries such as the EDR dictionary.

3.3   Tweet Positive/Negative Corpus

Recently, Twitter is one of the SNS from which can obtain large data. If we can collect the huge amount of tweet corpus for each language, those can be used for expansion of word or expansion of corpus.

In our study, we collected positive tweets and negative tweets of Japanese and English by using the two kinds of symbols such as ':)' and ':('. The number

---

[1] http://www.eijiro.jp/.
[2] http://www.edrdg.org/jmdict/edict.html.

of the collected tweet data is shown in Table 1. However the number of tweets is different by language or positive/negative, these data were obtained at nearly the same time. Recently, Twitter is one of the SNS from which numerous of data are obtainable. If we can collect large scale tweet corpus for each language, they will be able to be used to expand word or corpus.

In our study, we collected positive tweets and negative tweets of Japanese and English by using the two symbols as ':)' and ':('. The number of the collected tweet data is shown in Table 1. Although the numbers of tweets were different depending on languages or positive/negative, these data were obtained at nearly the same time.

Table 1. Positive/Negative tweet corpus.

	Japanese	English
Positive	85,170	144,231
Negative	151,975	179,848

3.4 Refining Translation Candidates

When training data and test data are constructed by extracting translation candidates from the dictionary, the problem is that the translation candidates include unnecessary words that cause estimation error. We considered the following perspectives to refine the translation candidates.

- The stop words included in translations should be removed from training data.
- When the sentence's (writer's) emotion and the emotion of the word included in the sentence are different, the word should be removed from training data.
- The more the word contributes to the emotion expressed by the sentence, the more the weight is added to the word and the word should be included in training data.

We made the training data as shown in Table 2. In Table 2 "Dictionary for Refining" indicates the kind of the dictionaries used for refining the translation candidates. "EW" represents Emotion Word Dictionary, "SW" represents Stop Word List, and "EPC" represents Emotion Polarity Correspondence Table. In these columns, J, E indicates whether the dictionary is Japanese or English. The column of Code represents the abbreviation of each model. In the paper, $J \rightarrow E$ represents extracting English translation candidates from Japanese sentences and $E \rightarrow J$ represents extracting Japanese translation candidates from English sentences. For $M_{E \rightarrow J}, M_{J \rightarrow E}$ in Baseline Model and Machine Translated Model, translation candidates were not refined. The following subsections elaborate the training data used in this paper.

Table 2. Definition of training data.

Model name	Code	Target lang.	Dictionary for refining		
			EW	SW	EPC
Baseline model	$M_{E \to J}$	J	-	-	-
	$M_{J \to E}$	E	-	-	-
	$M'_{E \to J}$	J	J	-	-
	$M'_{J \to E}$	E	E	-	-
Machine translated model	$G_{E \to J}$	J	-	-	-
	$G_{J \to E}$	E	-	-	-
Stop word model	$R_{E \to J}$	J	-	J	-
	$R_{J \to E}$	E	-	E	-
Polarity model	$P_{E \to J}$	J	-	-	J
	$P_{J \to E}$	E	-	-	E
	$P'_{E \to J}$	J	-	-	J
	$P'_{J \to E}$	E	-	-	E

Baseline Method. The Japanese translation candidates corresponding to each word in the English sentences were extracted from the three kinds of bilingual dictionaries. Then, these candidates were used as the feature for creating training data $M_{E \to J}$.

In the same way, $M_{J \to E}$ was created by extracting English translation candidates of each word in Japanese sentence and using these candidates as the feature. Training data $M'_{E \to J}$ and $M'_{J \to E}$ was also created by extracting specific translation candidates of the words that had been marked as emotion expression in the corpus from a bilingual dictionary.

Machine Translated Model. To compare the proposed method and the method using the result of machine translation as training data, the words in Japanese-English sentences were translated by Google translation[3] and used them as feature to create the training data $G_{E \to J}, G_{J \to E}$. Google translation is based on a statistical machine translation method. However, its translation accuracy is not very high. To judge the quality of translation various measures have been proposed.

One of the famous measures is BLEU [16] that uses the word's N-gram precision. IMPACT [5] has a relatively high correlation with evaluation by human in adequacy and fluency. METEOR [2,9] was proposed as an evaluation method without using a word's N-gram. The details of METEOR are described in the paper written by Banerjee et al. [2]. To investigate the accuracy of Google translation in corpus translation, the average score of IMPACT, METEOR and BLEU was calculated. The obtained scores were shown in Table 3.

---

[3] https://translate.google.com.

Table 3. Evaluation of machine translation (Google MT) (METEOR used parameters: $\alpha = 0.9; \beta = 3.0; \gamma = 0.5$)

Evaluation method	$J \to E$	$E \to J$
IMPACT	0.48	0.35
METEOR	0.28	0.34
BLEU	0.19	0.15

The quality of web translation system such as Google translation is highly controversial. However, the average scores obtained were not especially low.

Stop Word Model. Besides the problem of a simple translation error, there is another problem that decreases the accuracy of emotion estimation. The problem is caused by the words that do not contribute to the speaker's (writer's) emotion.

These words should be removed from the training data by refining according to the rule. Therefore, we focused on a method based on stop words. SMART [4] is a famous list of unnecessary words. However, it is an English word list and we cannot apply it to Japanese language. Therefore, we attempted to refine the words by parts of speech for Japanese language. If the part of speech of the word annotated by morphological analysis was not included in Table 4, the word was regarded as an unnecessary word.

Table 4. Part of speech regarded as necessary word.

Part of speech	Sub category
Prefix	Adjective connection, Noun connection
Conjunction	-
Adnominal adjective	-
Noun	General, "Sa"−connection, "Nai"−adjective connection, Adverb-possible, Adjective verb stem
Verb	Independent
Adjective	Independent
Adverb	General, auxiliary word connection
Interjection	-

If the translation candidates of the word w included the unnecessary word sw, the word sw was removed from the training data. In this way, the training data were created and defined as $R_{E \to J}, R_{J \to E}$.

Polarity Model. To refine the translation, we also focused on the emotion polarity of words. If the candidates of the word w include ew, whose emotion polarity is different from the sentence's emotion polarity, the word ew should be removed from the training data. The correspondence between emotion polarity and emotion category is shown in Table 5.

Table 5. Emotion polarity and emotion category.

Positive (+)	Joy, Love, Respect
Negative (-)	Anger, Sorrow, Hate, Anxiety
Neutral (0)	Surprise, Neutral

We used the emotion polarity correspondence table created by Takamura et al. [25] to judge the word's emotion polarity. Because many of the words with extremely small emotion polarity value actually do not express emotion, the threshold value was set for the emotion polarity value. In this paper, the threshold value was set as $th = 0.5$. If the absolute emotion polarity value of a word is smaller than this threshold value, the emotion polarity of the word was set as 0 (neutral). The created training data were $P_{E \to J}, P_{J \to E}$.

However, above method cannot extract features if there are no words that match to the emotion polarity of the sentence. For that reason, we also considered another method to judge whether the word to be included as feature or not according to the degree of contribution to the emotion expression of the sentence.

In this method we used the words that did not match to the emotion polarity of the sentence as feature and we did not to set the threshold for emotion polarity. In this research, the words whose degree of contribution is ranked in the top rc are used as features. How to calculate the contribution degree of the word w_j to the sentence S_i is shown in Eqs. 1–4.

$$\mathrm{EM}(w_j, S_i) = \begin{cases} 1 & if \ \mathrm{EMP}(w_j) = \mathrm{EMP}(S_i) \\ 0 & otherwise \end{cases} \tag{1}$$

$$\mathrm{EMS}(w_j, S_i) = \begin{cases} 1 & if \ \mathrm{EMP}(w_j) = \mathrm{EMP}(S_i) \\ 0.5 & otherwise \end{cases} \tag{2}$$

$$\alpha = \frac{|S_i|}{\sum_{w_k \in S_i} \mathrm{EM}(w_k, S_i)} \tag{3}$$

$$\mathrm{CScore}(w_j \in S_i) = \begin{cases} \alpha \cdot \mathrm{EMS}(w_j, S_i) & if \ \mathrm{EM}(w_j, S_i) = 1 \\ \mathrm{EMS}(w_j, S_i) & otherwise \end{cases} \tag{4}$$

$\mathrm{EMP}(w_j)$, $\mathrm{EMP}(S_i)$ respectively indicate a sign of emotion polarity of the word w_j and sentence S_i. EM shown in Eq. 1 is a function that returns 1 if the signs of emotion polarity of the word and the sentence are the same, otherwise returns 0. EMS shown in Eq. 2 is a function that returns 1 if the signs of emotion polarity of the word and the sentence are the same, otherwise returns 0.5. The Eq. 4 calculates CScore that represents the contribution value of the word to the sentence by using the functions EM and EMS.

$|S_i|$ in the Eq. 3 indicates the total number of the words in the sentence S_i. By multiplying x, the more the weight of the words whose sign of the emotion polarity matched with that of the sentence increases relatively, the less there are the words whose sign of the emotion polarity matched with that of the sentence.

CScore is calculated for each word in each sentence and the words are sorted by descending order of the CScore. By using only the top r_c words as features, training data $P'_{E \to J}, P'_{J \to E}$ are made. We conduct several experiments to calculate the value of r_c and used the value with the highest accuracy as the value of r_c.

With these methods, we thought that we could prevent acquiring unnecessary translation candidates for constructing training data in another language, thereby could create a highly accurate compact estimation model. Moreover, the calculation for training could be reduced.

Classification Method. In the research of emotion estimation from text, the Support Vector Machine (SVM) and the Naive Bayes classifier (NB) are often used for machine learning. In this paper, Naive Bayes classifier was used because it has a simple algorithm and can be easily applied to multiclass classification.

3.5 Word Expansion Based on Word Distributed Representation

This section describes the word expansion method based on word distributed representation.

We used tweet corpus shown in Table 1 to train word distributed representation. As the training algorithm for distributed representation, we used the method proposed by Mikolov et al. [11–13], and used the Word2Vec [4] for training tool. In the training phase, we used the skip-gram model known as accurate method and set the parameters as 200 for the number of dimension of word vector and as 10 for context window size.

The proposed word expansion method is described as follows.

- A word set consisting of the word with similar emotion polarity with the words in the sentence is obtained. We used the cosine similarity between word vectors as similarity.
- Select the words having both the same emotion polarity sign with the original word w_o and the lower absolute value of emotional polarity than a certain threshold value. Equations 5 and 6 indicate the above conditions.
- We obtained the translation candidates by using bilingual dictionary to the refined similar words. In this time, when the word had the same emotion polarity with that of the original sentence, higher weight was added to the word.

$$pn(w_o) \times pn(simw_j) > 0 \qquad (5)$$

$$diff(abs(pn(w_o)), abs(pn(simw_j))) \leq threshold \qquad (6)$$

---

[4] https://code.google.com/p/word2vec/.

3.6 Corpus Expansion Based on Sentence Similarity

This section describes a method to expand corpus based on sentence similarity. In the task of text classification using machine learning, there is a method called as semi-supervised learning method. When the amount of the annotated data is small, the unannotated data is classified based on the annotated data. By adding this annotated data to the training data, the classification models were expanded their range of applications.

Because these methods focused on how improve the general versatility by using unannotated data, they could not avoid decreasing accuracy.

However, the classified data should be increased to solve both sparseness of training data and decreasing of accuracy. In our study, we focused on the similarity between the sentences, the sentences similar with the annotated data were conditionally obtained from unannotated corpus of the same language.

A tweet corpus from Twitter is unannotated data available in large amount. From Tweet P/N corpus used in the previous section, the sentences having both the same polarity with the emotion polarity of the annotated sentences and the higher similarity between the sentences were added to the training data.

Equation 7 indicates the condition for filtering.

$$pn(sen_i) \times pn(tweet_j) > 0 \tag{7}$$

$$diff(pn(sen_i), pn(tweet_j)) < threshold_{diff} \tag{8}$$

4 Experiment

The proposed method was evaluated by experiment. The target data was 1,190 pairs of sentences in Japanese-English parallel emotion corpus and 4,652 pairs of sentences in open Japanese-English parallel corpus with annotation of emotion. The information of the morphemes included in the sentences was used as feature.

Japanese sentences were morphologically analyzed by ChaSen[5]. English sentences were morphologically analyzed by Brill's Tagger [3] then basic forms of the parts of stems were obtained by using the Porter stemming algorithm[6]. We evaluated the results by calculating the accuracy with the Eq. 9.

$$match_i = \begin{cases} 1 \ if \ |T_{o,i} \cap T_{c,i}| \geq 1 \\ 0 \ otherwise \end{cases} \tag{9}$$

$$Accuracy(\%) = \frac{\sum_i^{|S|} match_i}{|S|} \times 100 \tag{10}$$

In the equation, S indicates a set of sentences targeted for the evaluation. $T_{o,i}$, and $T_{c,i}$ respectively indicate a set of tag sets outputted by classifier and a set of correct tags for the sentence $s_i \in S$. $|T_{o,i} \in T_{c,i}|$ indicates the number of the matched tags between the tags outputted by the classifier and the correct

---

[5] http://chasen-legacy.osdn.jp/.
[6] http://tartarus.org/~martin/PorterStemmer/.

tags. $match_i$ means the score of the correct answers for the sentence s_i. Arithmetic average of these scores is calculated as Accuracy. Although the classifier outputs the probability for each emotion category, the category with the highest probability is extracted and evaluated in the experiment.

4.1 Experiment-1

First, to evaluate emotion estimation by using Japanese-English Parallel Emotion Corpus as training data and test data, we created the training data whose feature is the translation candidates of each word in the bilingual sentences.

Then, we conducted Experiment-1 using the test data whose feature is the words in the source language's sentence. Experiment was conducted by using 10-fold cross validation.

4.2 Experiment-2

In experiment-2, we used the corpus improvement method which expand words in sentence and convert those to target language. Then, we evaluated that corpus improvement method.

4.3 Experiment-3

In experiment-3, we used the corpus expansion method which is based on sentence similarity. The accuracy of emotion estimation for open data is evaluated. In experiment-3, we used the corpus expansion method which is based on sentence similarity. The accuracy of emotion estimation for open data is evaluated.

5 Results and Discussions

5.1 Result of Experiment-1

The result of Experiment-1 is shown in Table 6. $P'_{E \to J}, P'_{J \to E}$ were made by setting the threshold rc of contribution value as 2^7.

The result of the training data $M'_{E \to J}, M'_{J \to E}$, which used only emotion expression's translation candidates as the feature, had higher accuracy than using all words' translation candidates as the feature. The result also showed that the conditions of unnecessary words for refining the translation candidates did not greatly increase the accuracy.

However, in the experiment based on Japanese test data, using $R_{E \to J}, P_{E \to J}$ as the training data decreased accuracy only about 2%. Therefore, these refining methods will become more effective when the size of the data increases because the calculation amount decreases.

---

[7] We evaluated the results by setting the threshold values: 1, 2, 3, 4, 5. When the threshold was $r_c = 2$, the accuracy was the highest.

In the experiments with $P'_{E \to J}, P'_{J \to E}$, over 50% accuracies were obtained. When $P'_{E \to J}$ was used as training data and the edict was used as a bilingual dictionary, the best accuracy of 66.7% was obtained.

These results suggested that our method solved the weak points in the Polarity Model, which is considering solely the concordance of emotion polarities refines too much the translation candidates and provides all words the same weights.

- By adding higher weight to the words that contribute to the emotion expression in the sentence, effective features can be emphasized.
- It is able to extract feature even though none of the words in the sentence contribute to the emotion expression in the sentence.

Table 6. Result of experiment-1.

Test	Training	Dictionary			
		-	Edict	Edict2	Eijiro
J	$M_{E \to J}$		50.1	50.3	49.8
	$M'_{E \to J}$		52.3	52.0	51.3
	$G_{E \to J}$	37.5			
	$R_{E \to J}$		48.6	42.4	49.4
	$P_{E \to J}$		47.5	48.2	48.5
	$P'_{E \to J}$		**66.7**	58.1	64.1
E	$M_{J \to E}$		39.4	39.7	35.0
	$M'_{J \to E}$		51.1	51.3	50.0
	$G_{J \to E}$	46.6			
	$R_{J \to E}$		47.5	48.6	49.2
	$P_{J \to E}$		37.5	38.2	48.8
	$P'_{J \to E}$		51.1	50.1	**53.9**

5.2 Result of Experiment-2

Next, we conducted an evaluation experiment by using the synonym based corpus extension model. To expand synonyms, we used 10 for window size and 200 for vector dimension numbers as the parameters of word2vec, and used the skip-gram model as training model. Three types of bilingual dictionaries were used for the experiment. We used logistic regression as machine learning method.

We used a tool of classias[8] as logistic regression, and tried three types of algorithms: "lbfgs.logistic", "pegasos.logistic," and "truncated_gradient.logistic" as logistic regression.

Considering imbalance of the number of the sentences for each emotion in the corpus, we used only four emotions "joy," "anxiety," "hate," and "sorrow"

---

[8] http://www.chokkan.org/software/classias/index.html.ja.

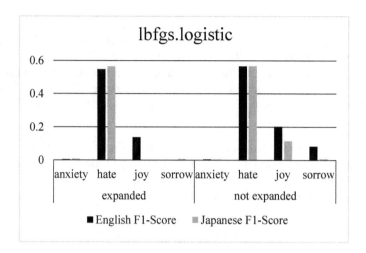

Fig. 4. Result of experiment-2 (lbfgs.logistic).

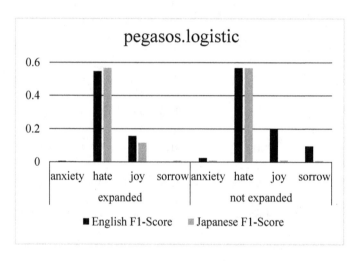

Fig. 5. Result of experiment-2 (pegasos.logistic).

in the evaluation experiment. The method used the corpus without expansion was also used as a comparative method. The experimental result is in Figs. 4, 5, and 6.

As the result, we did not observe effectiveness of word expansion in any emotion. The word expansion based on concept similarity, which we used in the experiment, decreased the quality of emotion estimation.

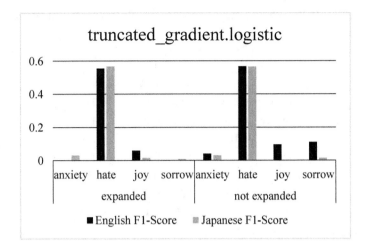

Fig. 6. Result of experiment-2 (truncated_gradient.logistic).

6   Result of Experiment-3

We made an emotion estimation model based on the corpus that was expanded with the positive/negative tweet corpus by using machine learning algorithm, and conducted an emotion estimation experiment. We used logistic regression (lbfgs.logstic, truncated_gradient.logistic, and pegasos.logistic) as machine learning method. The details of the expanded corpus is in Table 7.

Table 7. Statistics of expanded corpus.

Positive/Negative	Emotion	# of tweets	
		J	E
Positive	Love	440	419
	Joy	2177	2041
	Respect	1737	1636
Negative	Anxiety	1200	1134
	Sorrow	300	288
	Anger	780	713
	Hate	4240	4055
Neutral	Surprise	520	456

The experimental result of 10-hold cross validation is in Fig. 7.

To confirm the applicable range of the expanded corpus, we created two emotion estimation models both from the corpus before expansion and after expansion then compared the results of emotion estimation experiments for each model. The results are in Tables 8, 9, and 10. In the tables, "exp." indicates the

Table 8. Result of emotion estimation for open emotion corpus (lbfgs.logistic).

Emotion	F1−Score			
	E		J	
	Not exp.	Exp.	Not exp.	Exp.
Anger	0.023	0.154	0.097	0.112
Anxiety	0.066	0.102	0.228	0.15
Hate	0.191	0.155	0.287	0.256
Joy	0.261	0.254	0.308	0.193
Love	0	0.207	0.047	0.282
Respect	0.234	0.21	0.237	0.201
Sorrow	0.17	0.256	0.173	0.248
Surprise	0.041	0.151	0.112	0.133

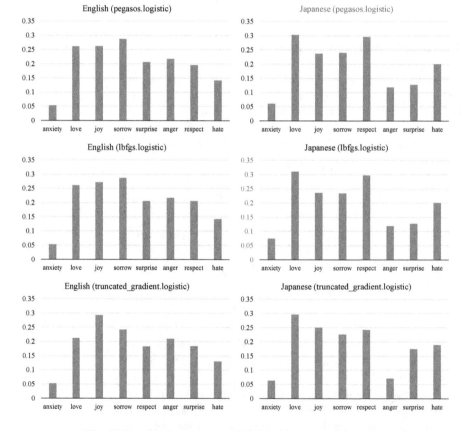

Fig. 7. Result of experiment-3 (10-hold cross validation).

result based on the expanded corpus, and "not exp." indicates the result based on the original corpus.

According to the experimental result, in English corpus, the method based on the expanded corpus obtained higher F1-score than the method based on the original corpus. One of the reasons is thought that the corpus used for expansion was collected from Twitter.

Table 9. Result of emotion estimation for open emotion corpus (pegasos.logistic).

Emotion	F1−Score			
	E		J	
	Not exp.	Exp.	Not exp.	Exp.
Anger	0.023	0.154	0.092	0.112
Anxiety	0.064	0.102	0.244	0.151
Hate	0.195	0.151	0.276	0.254
Joy	0.258	0.254	0.319	0.193
Love	0	0.207	0.047	0.278
Respect	0.231	0.209	0.243	0.196
Sorrow	0.186	0.258	0.185	0.246
Surprise	0.04	0.152	0.157	0.139

Table 10. Result of emotion estimation for open emotion corpus (truncated_gradient.logistic).

Emotion	F1−Score			
	E		J	
	Not exp.	Exp.	Not exp.	Exp.
Anger	0.052	0.176	0.192	0.122
Anxiety	0.15	0.071	0.203	0.123
Hate	0.174	0.142	0.278	0.248
Joy	0.246	0.224	0.306	0.183
Love	0	0.205	0.157	0.259
Respect	0.204	0.199	0.281	0.175
Sorrow	0.214	0.267	0.197	0.246
Surprise	0.109	0.122	0.245	0.218

Japanese language often uses abbreviation expressions and has ambiguity. Therefore, we thought that function words were matched more than content words when obtaining similar sentences.

Because the search function of positive/negative tweet is the original judgement method of Twitter, the method might be more suitable in English than in

Japanese and expansion might be more adequately made in English corpus than in Japanese corpus.

One of the reasons is thought that the corpus used for expansion was collected from Twitter. Japanese language often uses abbreviation expressions and has ambiguity. Therefore, we thought that function words were matched more than content words when obtaining similar sentences.

Because the search function of positive/negative tweet is the original judgement method of Twitter, the method might be more suitable in English than in Japanese and expansion might be more adequately made in English corpus than in Japanese corpus.

7   Conclusions

In this paper, the existing bilingual dictionaries were used to convert the linguistic resources for emotion estimation into another language. To avoid noise feature in the training data caused by converting the resource into other languages, a method to refine the translation candidates was proposed based on emotion polarity or the stop word list.

The evaluation experiment using the basic Japanese-English Bilingual Dictionary obtained approximately 66.7% accuracy in emotion estimation when the translation candidates exclusively corresponding to the emotional expressions were included in the training data. On the other hand, the experimental result using the open data suggested that the process of refining translation candidate worked effectively.

However, the bilingual dictionaries and the emotion polarity correspondent table included unnecessary words for emotion estimation. As the result, noise features could not be removed even though threshold value was set.

We tried to expand applicable range of the corpus by word expansion and confirmed that the word expansion per sentence decreased accuracy. On the other hand, the corpus expansion by obtaining similar sentence from a tweet corpus effectively worked in the English sentences. In future, we would like to improve the method to refine the translation candidates and propose a method to remove unnecessary word from emotion polarity correspondent table and also a method to automatically construct a bilingual dictionary suitable for emotion estimation.

We also would like to propose an effective method for both Japanese and English by combining the translation candidate refining method and the corpus expansion method.

Acknowledgements. This work was supported by JSPS KAKENHI Grant Numbers 15H01712, 15K16077, 15K00425.

References

1. Balahur, A., Turchi, M.: Multilingual sentiment analysis using machine translation? In: The 3rd Workshop on Computational Approaches to Subjectivity and Sentiment Analysis, pp. 52–60 (2012)
2. Banerjee, S.: Meteor: an automatic metric for MT evaluation with improved correlation with human judgments. In: ACL Workshop on Intrinsic and Extrinsic Evaluation Measures for Machine Translation and/or Summarization, pp. 65–72 (2005)
3. Brill, E.: Some advances in transformation-based part of speech tagging. In: National Conference on Artificial Intelligence, pp. 722–727 (1994)
4. Buckley, B., Salton, G., Allan, J., Singhal, A.A.: Automatic query expansion using smart: TREC-3. In: The Third Text REtrieval Conference(TREC-3), pp. 500–238 (1995)
5. Echizen-ya, H., Araki, K.: Automatic evaluation of machine translation based on recursive acquisition of an intuitive common parts continuum. In: The Eleventh Machine Translation Summit (MT SUMMIT XI), pp. 151–158 (2007)
6. Hiejima, I.: Japanese-English Emotion Expression Dictionary. Tokyodo Shuppan (1999). (in Japanese)
7. Inui, T., Yamamoto, M.: Usage of different language translated data on classification of evaluation document. In: The Annual Meeting of Association for Natural Language Processing, pp. 119–122 (2011). (in Japanese)
8. Kang, X., Ren, F., Wu, Y.: Bottom up: exploring word emotions for Chinese sentence chief sentiment classification. In: IEEE International Conference on Natural Language Processing and Knowledge Engineering, pp. 422–426 (2010)
9. Lavie, A., Agarwal, A.: Meteor: an automatic metric for MT evaluation with high levels of correlation with human judgments. In: ACL Second Workshop on Statistical Machine Translation, pp. 228–231 (2007)
10. Matsumoto, K., Ren, F.: Estimation of word emotions based on part of speech and positional information. Comput. Hum. Behav. **2011**(27), 1553–1564 (2011)
11. Mikolov, T., Chen, K., Corrado, G., Dean, J.: Efficient estimation of word representations in vector space. In: ICLR Workshop (2013)
12. Mikolov, T., Sutskever, I., Chen, K., Corrado, G., Dean, J.: Distributed representations of words and phrases and their compositionality. In: NIPS2013 (2013)
13. Mikolov, T., Yih, W.T., Zweig, G.: Linguistic regularities in continuous space word representations. In: NAACL HLT 2013 (2013)
14. Minato, J., Matsumoto, K., Ren, F., Kuroiwa, S.: Corpus-based analysis of Japanese-English of emotional expressions. In: IEEE International Conference on Natural Language Processing and Knowledge Engineering, pp. 413–418 (2007)
15. Minato, J., Matsumoto, K., Ren, F., Tsuchiya, S., Kuroiwa, S.: Evaluation of emotion estimation methods based on statistic features of emotion tagged corpus. Int. J. Innov. Comput. Inf. Control **4**(8), 1931–1941 (2008)
16. Papineni, K., Roukos, S., Ward, T., Zhu, W.: BLEU: a method for automatic evaluation of machine translation. In: The 40th Annual Meeting on Association for Computational Linguistics(ACL 2002), pp. 311–318 (2002)
17. Quan, C., Ren, F.: A blog emotion corpus for emotional expression analysis in chinese. Comput. Speech Lang. **24**(1), 726–749 (2010)
18. Quan, C., Ren, F.: Recognition of word emotion state in sentences. IEEJ Trans. Electr. Electron. Eng. **6**, 34–41 (2011)

19. Quan, C., Ren, F.: Unsupervised product feature extraction for feature-oriented opinion determination. Inf. Sci. **272**(2014), 16–28 (2014)
20. Ren, F.: Affective information processing and recognizing human emotion. Electron. Notes in Theoret. Comput. Sci. **225**, 39–50 (2009)
21. Ren, F., Kang, X., Quan, C.: Examining accumulated emotional traits in suicide blogs with an emotion topic model. IEEE J. Biomed. Health Inform. **20**(5), 1384–1396 (2015)
22. Ren, F., Matsumoto, K.: Semi-automatic creation of youth slang corpus and its application to affective computing. IEEE Trans. Affect. Comput. **7**(2), 176–189 (2015)
23. Ren, F., Wu, Y.: Predicting user-topic opinions in Twitter with social and topical context. IEEE Trans. Affect. Comput. **4**(4), 412–424 (2013)
24. Saiki, Y., Takamura, H., Okumura, M.: Domain adaptaion in sentiment classification by instance weighting. In: IPSJ SIG Notes, pp. 61–67 (2008). (in Japanese)
25. Takamura, H., Inui, T., Okumura, M.: Extracting semantic orientations of words using spin model. In: The 43rd Annual Meeting on Association for Computational Linguistics, pp. 133–140 (2005)
26. Takamura, H., Inui, T., Okumura, M.: Latent variable models for semantic orientations of phrases (in Japanese). Trans. Info. Process. Soc. Jpn. **47**(11), 3021–3031 (2006)
27. Wan, X.: Co-training for cross-lingual sentiment classification. In: the 47th Annual Meeting of the ACL and The 4th IJCNLP of the AFNLP, pp. 235–243 (2009)
28. Wu, Y., Kita, K., Matsumoto, K.: Three predictions are better than one: sentence multi-emotion analysis from different perspectives. IEEJ Trans. Electr. Electron. Eng. (TEEE) **9**(6), 642–649 (2014)

Employing Knowledge Artifacts to Develop Time-Depending Expert Systems

Fabio Sartori$^{(\boxtimes)}$ and Riccardo Melen

Department of Computer Science, Systems and Communication (DISCo),
University of Milan - Bicocca, Viale Sarca 336, 20126 Milan, Italy
{sartori,melen}@disco.unimib.it

Abstract. The integration of heterogeneous information gathered from wearable devices, like watches, bracelets and smartphones, is becoming a very important research trend. The Expert Systems' technology should be able to face this interesting challenge, promoting the development of innovative frameworks runnable on wearables and mobile operating systems. In particular, users and domain experts could be able to interact directly, minimizing the role of knowledge engineer and promoting the real–time updating of knowledge bases when necessary. This paper presents the KAFKA approach to this topic, based on the implementation of the Knowledge Artifact conceptual model supported by Android OS devices.

1 Introduction

Traditional knowledge engineering methodologies, like CommonKads [1] and MIKE [2], considered knowledge acquisition and representation as centralized activities, in which the main point was trying to build up facts' and rules' bases which could model tacit knowledge in order to use and maintain it. Indeed, the explosion of the Internet and related technologies, like Semantic Web, Ontologies, Linked Data ... has radically changed the point of view on knowledge engineering (and knowledge acquisition in particular), highlighting its intrinsically distributed nature.

Finally, the growing diffusion of more and more sophisticated mobile devices, like smartphones and tablets, equipped with higher and higher performing hardware and operating systems allows the distributed implementation on such mobile devices of several key components of the knowledge engineering process. A direct consequence of these considerations is the need for knowledge acquisition and representation frameworks that are able to quickly and effectively understand when knowledge should be integrated, minimizing the role of knowledge engineers and allowing the domain experts to manage knowledge bases in a simpler way.

A relevant work in this field has been recently proposed in [3]: the HeKatE methodology aims at the development of complex rule–based systems focusing on groups of similar rules rather than on single rules. Doing so, efficient inference is assured, since only the rules necessary to reach the goal are fired. Indeed, this

© Springer International Publishing AG 2016
A. Fred et al. (Eds.): IC3K 2015, CCIS 631, pp. 281–301, 2016.
DOI: 10.1007/978-3-319-52758-1_16

characteristic helps the user in understanding how the system works, as well as how to extend it in case of need, minimizing the knowledge engineer role. Our goal is (partially) different: we intend to build a tool for supporting the user in understanding when the knowledge base should be updated, according to the domain conditions he/she detects on the field. For this reason, we don't aim to realize new inference engines, like in the HeKatE project, but a good framework to use existing tools, usable in varying conditions and portable devices.

In this paper, we present *KAFKA* (Knowledge Acquisition Framework based on Knowledge Artifacts), a knowledge engineering framework developed under Android: the most interesting feature of KAFKA is the possibility for the user to design and implement his/her own knowledge–based system without the need for a knowledge engineer. The framework exploits Android as the running OS, since it is the most popular mobile operating system in the world. The final aim of KAFKA is the execution of expert systems written in Jess. Due to the difficulties in importing Jess under Android, KAFKA has been currently developed as a client–server architecture. The most important characteristic of Jess, from the KAFKA point of view, is the possibility to implement facts as Java objects rather than simple (attribute, value) pairs, exploiting the *shadow fact* construct. This point, together with the conceptual model of Knowledge Artifact adopted in designing the expert system, makes possible to understand when new rules and observations should be modeled according to the evolution of the underlying domain.

The rest of the paper is organized as follows: Sect. 2 briefly introduces the research field, focusing on distributed expert systems and Knowledge Artifacts. Section 3 further explores the Knowledge Artifact concept from the perspective of time evolving expert systems. In Sect. 4, the Shadow Fact construct is briefly introduced, in order to show how its adoption in KAFKA has allowed to take care computationally of knowledge–bases' variation according to the detection of new information from the domain. Section 5 presents the characteristics of problems and domains addressed by KAFKA: in particular, the role of Knowledge Artifacts and rules is analyzed from both the conceptual and computational point of view, in order to explain how KAFKA allows the user to take care of possible evolutions of the related expert system. In Sect. 6 a case study is presented, illustrating how KAFKA allows the user to interact with the domain expert to complete knowledge bases when new observations are available. Finally, conclusions and directions for further work complete the paper.

2   Related Work

As reported in [4], with the advent of the web and of Linked Data, knowledge sources produced by experts as (*taxonomical*) descriptions of domains' concepts have become strategic assets: SKOS (Simple Knowledge Organizations System) has been recently released as a standard to publish such descriptions in the form of vocabularies.

In this way, knowledge engineering has moved from being a typical centralized activity to being distributed: many experts can share their competencies, contributing to the global growth of knowledge in a given domain. The expert system paradigm has become suitable to solve complex problems exploiting the integration between different knowledge sources in various domains, such as medicine [5] and chemistry [6], and innovative frameworks have been developed to allow non AI experts to implement their own expert systems [7,8].

In this paper, we propose an alternative approach to the development of distributed expert systems, based on the Knowledge Artifact (KA) notion. In Computer Science, artifacts have been widely used in many fields; Distributed Cognition [9] described cognitive artifacts as *"[...] artificial devices that maintain, display, or operate upon information, in order to serve a representational function and that affect human cognitive performance"*. Thus, artifacts are able not only to amplify human cognitive abilities, but also change the nature of the task they are involved into. In CSCW, coordinative artifacts [10] are exploited *"[...] to specify the properties of the results of individual contributions [...], interdependencies of tasks or objects* in a cooperative works setting *[...] a protocol of interaction in view of task interdependencies* in a cooperative work setting *[...]"*, acting as *templates, maps* or *scripts* respectively. In the MAS paradigm [11], artifacts *"[...] represent passive components of the systems [...] that are intentionally constructed, shared, manipulated and used by agents to support their activities [...]"*.

According to the last definition, it is possible to highlight how artifacts are typically considered *passive entities* in literature: they can support or influence human and artificial agents reasoning, but they are not part of it, i.e. they don't specify how a product can be realized or a result can be achieved. In the Knowledge Management research field, Knowledge Artifacts are specializations of artifacts. Cabitza et al. [12] described them as the technological driver that either enables or supports knowledge circulation among people inside organizations. According to Holsapple and Joshi [13], *"A knowledge artifact is an object that conveys or holds usable representations of knowledge"*. Salazar–Torres et al. [14] argued that, according to this definition, KAs are artifacts which represent *"[...] executable-encodings of knowledge, which can be suitably embodied as computer programs, written in programming languages such as C, Java, or declarative modeling languages such as XML, OWL or SQL"*.

Thus, Knowledge Management provides artifacts with the capability to become *active entities*, through the possibility to describe entire decision making processes, or parts of them. In this sense, Knowledge Artifacts can be considered as guides to the development of complete knowledge–based systems. The most relevant case study in addressing this direction is the *pKADS* project [15], that provided a web–based environment to store, share and use knowledge assets within enterprises or public administrations. Each knowledge asset is represented as an XML file and it can be browsed and analyzed by means of an ontological map. Although the reasoning process is not explicitly included in the knowledge asset structure, it can be considered a Knowledge Artifact being machine readable and fully involved in a decision making process development.

3   Motivation: Managing Rapidly Changing Scenarios by Means of Expert Systems

In this paper we are concerned with a rather complex problem, that of modeling time–varying scenarios. In this case, the observed system and its reference environment change in time, passing through a series of macroscopic states, each one characterized by a specific set of relevant rules. Moving from one state to another, the meaning and importance of some events can change drastically, therefore the applicable inferences, as described by the rule set, must change accordingly.

The crucial point from the system point of view is the difficulty for production rules to capture in a precise way the knowledge involved in decision making processes variable in an unpredictable way. The resulting rules' set must be obtained at the end of an intensive knowledge engineering activity, being able to generate new portions of the system effectively and efficiently with respect to the changes in the application domain.

Some examples of these application scenarios can help in clarifying the characteristics of the problems we intend to tackle. A first example is the evolution of the state of an elderly patient affected by a neurologic degenerative disease. Quite often the development of the disease does not proceed in a linear, predictable way; instead long periods of stationary conditions are followed by rapid changes, which lead to another, worse, long lasting state. In this case, the interpretation of some events (such as a fall, or a change in the normal order in which some routine actions are taken) can differ substantially depending on the macro-state of reference. Another case would be an application analyzing urban traffic, with the purpose to help a driver to take the best route to destination. The scenario being analyzed changes significantly with the hour of the day and the day of the week, as well as in response to events modifying the available routes, such as an accident or a street closure due to traffic works.

In these situations, an efficient response of the system is very important, since the computation must be necessarily "real–time", and it is mandatory for the system to check continuously the knowledge-base to understand if it is consistent or not. Here, we present an approach to the development of rule–based systems dynamically changing their behavior according to the evolution, in value and number, of problem variables. The approach is based on the acquisition and representation of Functional Knowledge (FK), Procedural Knowledge (PK) and Experiential Knowledge (EK), with the support of the KA notion.

According to [16], FK is related to the *functional representation* of a product, that *"[...] consists of descriptions of the functionality of components (or (sub-) systems) and the relationship between them."*. To properly capture such relationships, the authors suggest the adoption of ontologies (being able to deal with the semantics of relations); for this reason, ontologies can be defined as Knowledge Artifacts for functional knowledge acquisition and representation.

PK is defined in [17] as the *"[...] understanding of how to apply the concepts learned in any problem solving situations"*. This means that procedural knowledge concerns how to combine concepts to solve a problem. In other words, procedural knowledge is devoted to explain the different steps through which a result is obtained, but it doesn't specify anything on how those steps are implemented.

Finally, some authors defined [18] EK as *"[...] knowledge derived from experience [...]"*. It is important because *"[...] it can provide data, and verify theoretical conjectures or observations [...]"*. Experiential knowledge, that can remain (partly) tacit, allows to describe aspects that procedural knowledge is not able to represent, and opportune tools are needed to capture it; from the Knowledge Artifact definition point of view, this is the reason why the T–Matrix previously cited is provided with a grammar to define the correlations between ingredients and performances: such grammar is the Knowledge Artifact for experiential knowledge representation.

Although functional, procedural and experiential knowledge have been usually treated as separated entities in the past, it is reasonable to assume that they are someway correlated: it should be possible to link the different Knowledge Artifacts involved so as to include in a unique conceptual and computational framework the entire knowledge engineering process, from the requirement analysis (i.e. the identification of all the functional elements to obtain a product or a service) to the implementation of a complete knowledge based system (i.e. the description of the decision making process in a machine–readable form, according to the related experiential knowledge), through the clear and complete specification of all the procedural steps needed to move from inputs to outputs (i.e. which intermediate levels of computation are necessary).

Doing so, the evolution of the expert system from an initial state S_1, characterized by a stable knowledge base with very few details, to a final state S_n, characterized by a fully developed, stable knowledge base, can be continuously checked by the domain expert, with the possibility to add new rules and delete or modify obsolete ones, to include new inputs or outputs and/or to extend the range of values for existing ones according to the domain characteristics. This is the main aim of the KAFKA project, as pointed out in the following sections.

4   Shadow Facts and Their Role in KAFKA

As we know, the typical architecture of a rule–based system is made up of three main components:

- *an inference engine*;
- *a rule base*;
- *a working memory*.

In particular, the working memory is a collection of facts representing all the information the rule based system is working with. In our model, the *input*, *partial output* and *output* node sets are represented as facts in the working memory. Given that Jess has been chosen as the implementation language, three different kinds of facts can be used in KAFKA:

- *ordered facts*, pieces of elementary information stored as (attribute, value) pairs;
- *unordered facts*, record–like structures useful to represent more complex entities, each one composed of one or more simpler elements (i.e. the *slots*);
- *shadow facts*, that are unordered facts whose slots correspond to the properties of a JavaBean.

Being JavaBeans a kind of Java object, shadow facts serve as a connection between the working memory and the Java application running Jess. But, from the KAFKA development point of view, their most interesting feature is the possibility to choose between their *static* or *dynamic* representation in the working memory. A shadow fact is static if its representation changes infrequently or according to an explicit request by the user. On the contrary, a dynamic shadow fact is characterized by a frequent variability over time, with the need for the working memory to keep trace of its changes immediately. Dynamic shadow facts have been used in KAFKA to implement evolution of facts' bases according to the real–time detection of information: for example, let's suppose we have implemented a rule–based system to support users in taking decisions about patients affected by heart diseases.

This kind of application will check heart rate continuously, to be able to recognize possible critical situations. Hopefully, the heart–rate will be normal for the most time, causing the activation and firing of standard rules. But the system should be able to detect immediately possible significant up and down oscillations of the heart rate values, in order to recognize possible critical situations, avoiding them to become irreversible through the execution of proper actions.

The adoption of dynamic shadow facts to represent such kinds of variables enables a KAFKA–based expert system to manage these situations: a state of the system is a collection of shadow facts, whose values can change unpredictably from a time–stamp t_i to the next one t_j. If the current value of the shadow fact is already known, i.e. rules are available in the system to deal with it, proper actions will be taken on time. Otherwise, new actions will be promoted by the expert through the rules' set extension to take care of the new state.

In this way, a state in KAFKA is characterized as a collections of *events* rather than *variables*: the event that causes a shadow fact value change causes a corresponding transition of the system from the state S_i to the state S_j, being sure that it will be properly considered by the expert if unknown.

5   Knowledge Acquisition in Mobile Scenarios: KAFKA Conceptual and Computational Model

5.1   KAFKA Scenario

The typical KAFKA domain is shown in Fig. 1, where two kinds of roles are hypothesized: a KA–User supporting a generic operator solving problems and a KA–Developer, supporting a domain expert in the elaboration of decision making processes. The KA–User is characterized by a state, a collection of quantitative

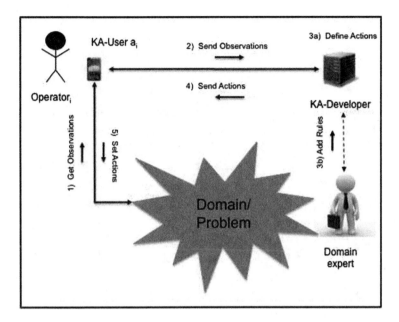

Fig. 1. KAFKA scenario: domain expert and operators virtually communicate to solve a problem in a given domain.

and qualitative parameters (observations on the domain) that can be measured by mobile devices or evaluated by the expert according to a given reasoning. This state can change over time: thus, it is continuously checked by the system in order to discover modifications and take proper actions. The domain expert can interact with the KA–Developer to update the different KA elements.

5.2   KA Definition

As shown in Fig. 2, the Knowledge Artifact model in KAFKA is multi–layered. Given a particular kind of knowledge k_i, $i \in [1, ..., n]$, a Knowledge Artifact for the acquisition of k_i is the pair $KA_{k_i} = \{E, R\}$, where $E = \{e_j\}$, $j \in [1, ..., m]$ and $R = \{r_k\}$, $k \in [1, ..., o]$, are the sets of *elements* and *relationships* among them respectively, with $r : e_i \longrightarrow e_j$, $\forall r \in [r_1, ..., r_o]$.

The elements can be further grouped into subsets according to the nature of knowledge involved: for example, in case of acquiring functional knowledge about a design activity as presented in [16], the elements could be aggregated into *systems* and *sub–systems*. The resulting Knowledge Artifact could be defined as follows: $Ontology = \{\{\{elements\}, \{sub-systems\}, \{systems\}\}, \{is-a, part-of, ...\}\}$.

In order to correlate different KAs, it is necessary to extend the definition above: given a set of kinds of knowledge $K = \{k_1, ..., k_n\}$ and a Knowledge Artifact $KA_{k_i}, \forall i \in [1, ..., n]$ a *Higher–Level Knowledge Artifact* (HLKA) for the acquisition of K is the pair $KA_K^{HL} = \{KA_K, r_K\}$, where $KA_K = \{KA_{k_i}\}$ and

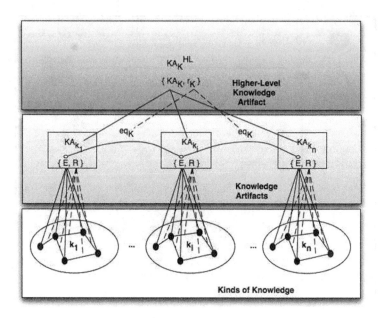

Fig. 2. The multi–layered model of Knowledge Artifact in KAFKA: the first level is the domain knowledge one; the second level concerns the definition of specific Knowledge Artifacts for each of them; the third level unifies Knowledge Artifacts by means of the eq_K relationship.

$r_K : E_{KA_{k_i}} \longrightarrow E_{KA_{k_j}}$, with $i, j \in [1, ..., n]$ is a *correlation* between the element sets of two distinct Knowledge Artifacts. The way this correlation is defined can depend on the knowledge domain: for this reason it could be necessary to specify more than one correlations between two KAs. At the current state of development, there exists only one correlation in KAFKA, namely eq_K (i.e. *equivalence on K*): $eq_K : E_{KA_{k_i}} \longrightarrow E_{KA_{k_j}} | e_1 \equiv e_2$, with $e_1 \in E_{KA_{k_i}}$ and $e_2 \in E_{KA_{k_j}}$. The equivalence notion allows to manage the same entity differently according to the current Knowledge Artifact, but preserving its main features moving from a kind of knowledge to another.

The general model presented here must then be configured on the basis of knowledge involved: the HLKA acts as a library of Knowledge Artifacts, each of them chosen to acquire the related kind of knowledge in the best way. For example, Procedural Knowledge can be captured by many kinds of tools for modeling causal relationships, like e.g. Influence Nets [19], Petri Nets, Causal Nets [20] (C–Nets) or Superposed–Automata Nets [21] (SA–Nets): Influence Nets are particularly suitable for Knowledge Acquisition, since they allow to specify entities and relationships among them in a very intuitive way. Anyway, there are many situation where they could result too few detailed, for example when modeling knowledge domains characterized by temporal dimensions: in those cases, it could be useful to adopt more sophisticated models, like Petri Nets or

Causal Nets to take care more precisely of the event sequentiality. Moreover, Petri Nets and Causal Nets could be not sufficient in case of parallel processes to synchronize occasionally, so that SA–Nets could be more indicated.

The HLKA must then be configured in order to become a complete model for acquiring and relating all the kinds of knowledge involved in the development of knowledge–based systems.

5.3 KA Configuration

As previously stated, KAFKA deals with the acquisition of three kinds of knowledge, i.e. Functional Knowledge (FK), Procedural Knowledge (PK) and Experiential Knowledge (EK). The configuration of KAFKA HLKA begins from the choice of a Knowledge Artifact for each of them:

- FK will be acquired by means of *Taxonomies* (T), characterized by three kinds of elements and two relationships;
- PK will be modeled through *Influence Nets* (IN), characterized by three kinds of elements and two relationship;
- EK will be captured by *Task/Subtask Structures* (TS), characterized by three kinds of elements and one relationship.

Thus, given the kinds of knowledge set $K = \{FK, PK, EK\}$, the KAFKA Higher–Level Knowledge Artifact to deal with K is $KA_K^{HL} = \{\{T, IN, TS\}, eq_K\}$, where $eq_K : T \longrightarrow IN$ and $eq_K : IN \longrightarrow TS$ makes equivalent the element sets of T, IN and TS.

Figure 3 shows the HLKA components: each of them is modeled on a three–tier architecture: *inputs* (I), being the observations necessary to initialize the

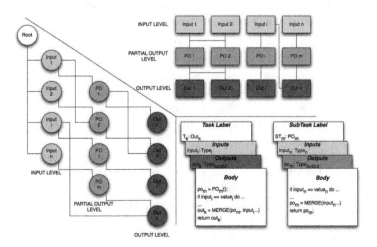

Fig. 3. The relationship existing among Taxonomy (on the left), Influence Net (on the top) and Task/Subtask Structures (on the bottom) in the KAFKA framework.

under–construction system, *partial outputs* (PO), i.e. the results of computations made by the system to reach its goals starting from inputs, and *outputs* (O), the goals of the system.

The first Knowledge Artifact to define is T, whose scope is the clear identification of I, PO and O, and how they are possibly *aggregated to* or *derived from* other elements: $KA_{FK} = T = \{E_T, R_T\}$, with $E_T = \{I \cup PO \cup O\}$ and $R_T = \{is - a, part - of\}$. Elements should be as simple as possible, usually well defined *(attribute, value)* pairs. Anyway, there are many situations in which more complex definitions are needed: the is–a relationship allows to extend the description of an existing input, partial output or output by means of a new value; the part–of relationship allows to aggregate two or more inputs, partial outputs or outputs into a new record–like element.

The Influence Net model is a structured process that allows to analyze complex problems of cause–effect type in order to determine an optimal strategy for the execution of certain actions. The Influence Net is a graphical model that describes the events and their causal relationships. Using information based on facts and experience of the expert, it is possible to analyze the uncertainties created by the environment where actions take place. This analysis helps the developer to identify the events and relationships that can improve or worsen the desired result. The Influence Net can be defined as a 4–tuple (I, PO, O, A), where

- I is the set of *input nodes*, i.e. the information needed by the system to work properly;
- PO is the set of *partial output nodes*, i.e. the collection of new pieces of knowledge and information produced by the system to reach the desired output;
- O is the set of *output nodes*, i.e. the effective answers of the system to the described problem; outputs are values that can be returned to the user;
- A is the set of *arcs* among the nodes: an arc between two nodes specifies that a causal relationship exists between them.

According to the definition above the adopted Knowledge Artifact is $KA_{PK} = IN = \{E_{IN}, R_{IN}\}$, with $E_{IN} = \{I \cup PO \cup O\}$ and $R_{IN} = A = \{affects, affected\text{-}by\}$. The *affects* relationship is defined as follows, *affects*: $I \longrightarrow PO$, *affects*: $I \longrightarrow O$, *affects*: $PO \longrightarrow PO$, *affects*: $PO \longrightarrow O$ and *affects*: $O \longrightarrow O$. The *affected–by* relationship is the vice–versa of *affects*: although it is redundant, it allows an easier description of experiential knowledge by means of Task/Substask Structures.

Task Structures are used to describe how the causal process defined by a given IN is modeled within a rule–based system. Each Task is devoted to define computationally a portion of an Influence Net: in particular, *Subtasks* are procedures to specify how a partial output is obtained, while *Tasks* are used to explain how an output can be derived from one or more influencing partial outputs and inputs. A Task cannot be completed until all the Subtasks influencing it have been finished. In this way, the TS modeling allows to clearly identify all the computational levels of the system. The Task and Subtask bodies are sequences of rules, i.e. $LHS(LeftHandSide) \Rightarrow RHS(RightHandSide)$ constructs.

According to the definition above, the third Knowledge Artifact in KAFKA is $KA_{EK} = TS = \{E_{TS}, R_{TS}\}$, with $E_{TS} = \{I \cup PO \cup O\}$ and $R_{IN} = \{if-do\}$. The *if-do* relationship semantics is the definition of a production rule $if - do : LHS \longrightarrow RHS$, where $LHS \subset I \cup PO \cup O$ and $RHS \subset PO \cup O$.

Each LHS contains the conditions that must be verified so that the rule can be applied: it is a logic clause, which turns out to be a sufficient condition for the execution of the action indicated in the RHS. Each RHS contains the description of the actions to perform as a result of the rule execution.

5.4   KA Implementation

The implementation of the different elements composing the knowledge engineering framework exploits the XML language [22]. A proper schema has been developed for each of them, as well as dedicated parsers to allow the user to interact with them. These XML files contain all the information necessary to compile rule–based systems: as previously stated, the Jess syntax has been chosen to this scope (Fig. 4).

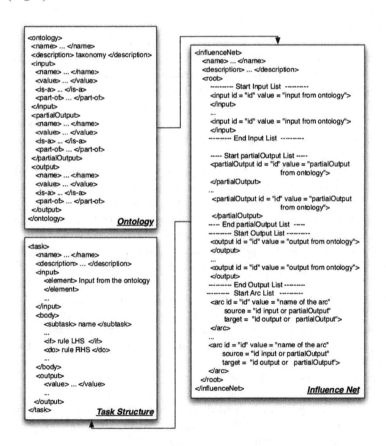

Fig. 4. The XML implementation of the Knowledge Artifact. Relationships among the different elements follow the conceptual model depicted in Fig. 3

Following the conceptual model introduced in the previous section, the first schema is the ontological one: the different components are grouped by the $\langle ontology \rangle$ $\langle /ontology \rangle$ tags, followed by a name and a description to specify which kinds of ontology has been chosen, i.e. *taxonomy* in the current version of KAFKA. Then, the schema presents opportune tags to specify *inputs, partial outputs* and *outputs*. Each element is characterized by a $\langle name \rangle$ and a $\langle value \rangle$ tags. Moreover, it is possible to define $\langle is\text{-}a \rangle$ and $\langle part\text{-}of \rangle$ relationships for each element, as previously stated: the arguments of is–a and part–of tags are the names of elements related to the current one (i.e. the element that generalizes the current one or the element the current one is a component of).

To produce an Influence Net IN, the taxonomy is parsed from inputs to outputs. In this way, different portions of the system under development can be described. Outputs, partial outputs and inputs are tied by arcs which specify the *source* and the *target* nodes.

Each element is associated with a name (inherited by the ontology) and an identifier, useful for building up arcs. Arcs are characterized by a name, i.e. *affects* or *affected–by*. Thus, the IN can be navigated from inputs to outputs, following the chain of *affects* arcs, or backwards, following the sequence of *affected–by* arcs.

Finally, an XML schema for the *Task* (*Subtask* elements of the framework are defined in the same way) can be produced as follows. The parser composes an XML file for each output considered in the Influence Net. The *input* and *subtask* tags allow to define which inputs and partial outputs are needed to the output represented by the Task to be produced. The *body* tag is adopted to model the sequence of rules necessary to process inputs and results returned by influencing Subtasks: a rule is composed of an $\langle if \rangle$ ... $\langle do \rangle$ construct, where the if statement permits to represent the LHS part of the rule, while the do statement concerns the RHS part of the rule.

The XML files introduced above can be incorporated into dedicated decision support systems to guide the user in the design of the underlying Taxonomy, Influence Net and Tasks/Subtasks. Moreover, it is possible to transform the Tasks into a collection of files containing rules written, for instance, in the JESS language. Given the XML code for Task and Subtask Structures, a rule file can be generated by means of opportune parsers. In principle, every language for rule–based system design can be exploited, but the current version of KAFKA adopts Jess. The main reason for this was the possibility to exploit the *shadow fact* construct in the knowledge base to take care of its variability: basically, a shadow fact is an object integrated into the working memory as a fact. For this reason, it is possible to access it for value modifications from every kind of application, and the inference engine will understand the situation, activating a new run of the expert system.

The shadow fact is fundamental to manage the variable scenario in Fig. 1: as shown in Fig. 5 a system transition from state S_i to state S_j can be due to the observation of a not previously considered value for one or more observations. At $State_i$, $Observation_n$ is detected by the KA–User, that was not considered by the current knowledge artifact. A new shadow fact is then generated to take

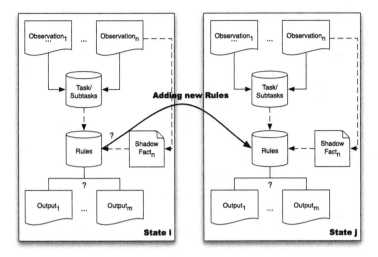

Fig. 5. Transition from a State i to a State j: $ShadowFact_n$ is not recognized in $State_i$, for this reason, new rules are added moving to $State_j$, where $ShadowFact_n$ is known.

care of it (i.e. $ShadowFact_n$), and the KA–Developer is notified about the need for extending the rule set in order to properly manage that value. In this way, the system moves from state S_i, where it is not able to reach a valid solution to the problem, to state S_j, where new rules have been added to fill the gap.

On the other hand, when all the possible values for every observation will be mapped into the set of rules of an expert system, for example in the S_k state, that system will be considered stable, and the related rules' set will be able to generate a solution for every possible configuration of inputs. Thanks to its intrinsically dynamic nature (it is a Java object), the shadow fact is the most suitable technical artifact to take care of such characteristics: by changing its value at run–time, the inference engine will be able to run the current expert system portion whose behavior possibly varies according to that change; in case of no solution, the KA–Developer will be notified about the need for extending both the knowledge artifact and the related set of rules.

5.5   KAFKA Architecture

Every KA–User (i.e. the client in Fig. 6) involved in a problem solving activity is provided with an Android application: this application communicates with the KA–Developer (i.e. the server in Fig. 6) by means of an Internet connection. The KA–User sends data serialized into a JSON[1] object. JSON is an open standard format that uses human–readable text to transmit data objects consisting of attribute–value pairs. For this reason, it is very useful in KAFKA to exchange facts between the client and the server, being sure they are correctly interpreted.

---

[1] JavaScript Object Notation, see http://json.org/.

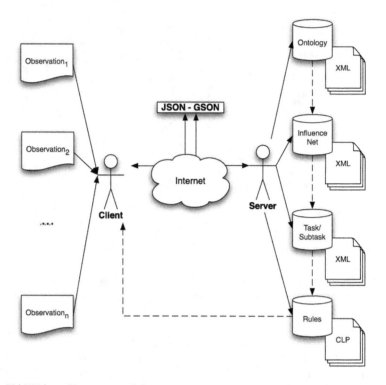

Fig. 6. KAFKA architecture: solid arrows represent concrete flows, whilst dashed ones represent virtual flows.

These data are observations about the conditions of the problem domain. The GSON[2] library has been integrated to convert Java objects (like shadow facts) automatically into their JSON representation.

Then, exploiting the Android primitives, it has been possible to create a stable mechanism for the communication with the server. In particular, the following tools were useful to implement the KA–User in KAFKA:

- *activities*: a class that extends an *Activity* class is responsible for the communication with the user, to support him/her in setting the layout, assigning the listeners to the various widgets (Android's graphical tools) and setting the context menu;
- *listener*: a class that implements the interface *OnClickListener* is a listener. An instance of this object is always associated with a widget;
- *asyncTask*: a class that extends *AsyncTask* is an asynchronous task that performs some operations concurrently with the execution of the user interface (for example the connection to a server must be carried out in an AsyncTask instance, not in an Activity one);

---

[2] See https://sites.google.com/site/gson/gson-user-guide#TOC-Goals-for-Gson.

The typical mechanism to interface the client and the server is the following one: the Activity object prepares the layout and sets the widgets' listeners and a container with the information useful for the server; then, it starts the AsyncTask instance for sending the correct request to the server, passing to it the previously created container. Before starting the asynchronous task, in most cases, the listener activates a dialog window that locks the user interface in waiting for the communication with the server; the AsyncTask predisposes the necessary Sockets for the communication and then performs its request to the server, sending the information about the case study observation enclosed in the container. Before concluding, it closes (dismisses) the waiting dialog window. The KA–Developer creates an instance of the KA model (i.e. the collections of XML files for Ontology, Influence Net and Task/Subtask in Fig. 6) for each active KA–User in communication with it. Then, it executes the related rule–based system (i.e. the collection of .clp files in Fig. 6) and sends answers, serialized into a JSON object, to the KA–User that will be able to take the proper action.

At the current state of development, the rule–based systems generated by a KA–Developer are written in Jess 7.0: this means that they cannot be directly executed by a KA–User, since Jess 7.0 is not fully supported by Android. Thus, the server is responsible for their execution. Anyway, it has been designed to allow the serialization of .clp files too in the future, when Jess will be runnable under Android (i.e. when a stable version of Jess 8.0 will be released). The server, once activated, can accept both requests for the creation of a new system by a domain expert and for the resolution of problems on the basis of existing rule–based systems by a user.

6   Case Study

The case study was inspired by the STOP handbook [23], supplied to the Italian Fire Corps and Civil Protection Department of the Presidency of Council of Ministers for the construction of safety building measures for building struc-tures that have been damaged by an earthquake. In case of a disaster occurring, operators reach the site in order to understand the event consequences and take the proper actions to make safe both human beings and buildings. The case study focused on actions typical of earthquakes, namely *Walls' safety measures*. These actions aims at preventing further rotation or bulging of the wall damaged during an earthquake. It is important to notice that operators are provided with standard equipment to those scopes, i.e. a set of rakers and shores that can be useful in most situations. The main problem is to understand if this equipment can be adopted in case of particularly disrupting events: in fact, the situation found by the operators continuously evolve from a state S_i to a new state S_j according to phenomena like aftershocks. The *STOP App* has been thought for these situations, when operators need more information to e.g. combine rakers in order to sustain walls dramatically damaged by the earthquake or shores to cover very large openings.

The following sections will further explain how the scenario has been effectively translated into a computational system, focusing on how a Knowledge Artifact has been created and instantiated. The main goal of the STOP application has been the possibility to have faster answers by means of a collection of rule–base system that incorporates the STOP handbook knowledge, where the operators can insert the inputs to get outputs in a transparent way. Another important point is the possibility to extend the STOP model when needed, adding rules to the KA–Developer knowledge by means of an opportune interface: as previously stated, the role of shadow facts is crucial to this scope.

6.1   Walls' Safety Measures: Modeling the Knowledge Involved

Rakers are devices adopted to prevent further rotation or bulging of the wall damaged during an earth-quake. There exist two main kinds of rakers: *solid sole* and *flying* [23]. Solid sole rakers can be used when the conditions of the pavement around the damaged walls are good, while flying rakers are useful when rubble is present. Due to their structure, solid sole rakers allow to distribute the wall weight in a uniform manner along the whole pavement, with greater benefits from the wall safety point of view. Anyway, the possibility to concentrate the wall sustain on smaller sections is important too, especially when earthquakes intensity is so strong to break windows or building frontages.

The other two information to fix are *raker class* and *dimensions*. Raker class depends on the distance between the sole and the position of the top horizontal brace on the wall; raker dimensions can be established starting from the raker class, the seismic class related to the earthquake, the wall thickness and the span between the raker shores. The values introduced above constitute the set of system inputs that should be properly used by the KA–Developer to elaborate problem solutions. How these inputs are effectively exploited and which relationships exist among them are also important points to take care of.

This is the goal of the Influence Net depicted in the left part of Fig. 7, which clearly identifies outputs and partial elaborations in order to understand what is the reasoning process that allows to get outputs starting from inputs (in the Figure, light gray rectangles are Inputs, rounded corner rectangles are partial outputs and ovals are outputs). The arcs semantics is *affected–by*. In particular, the type of raker (i.e. Solid Sole or Flying) and its class can be considered as outputs or partial outputs: they have been described as partial outputs, since the final goal of the decision making process is to choose a raker in terms of *name* (that is R1, R2, R3 and so on) and *dimensions*. Raker class and type are characteristics that allow defining the name of the raker, but are not interesting for the user.

The last part of knowledge acquisition and representation is the definition of Tasks and Subtasks, in order to specify how outputs can be obtained from inputs. As previously introduced, an XML file is produced for each output and partial output included into the Influence Net. According to the designed schema, these files contain a description of necessary inputs, expected outputs and the body, i.e. the instructions necessary to transform inputs into outputs.

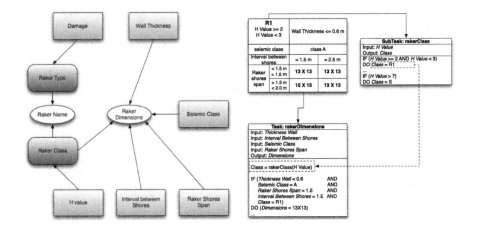

Fig. 7. The Influence Net diagram for Walls Safety case study and a sketch of the decision making process made by the KA–Developer to obtain an output.

These instructions can be $\langle if \rangle \ldots \langle do \rangle$ constructs or invocations of influencing subtasks. The right part of Fig. 7 shows a sketch of the decision making process concerning a portion of the case study Influence Net, starting from the knowledge involved as provided by the STOP handbook: the dashed arrow between *Class* attributes in the Subtask and Task bodies specifies the precedence relationship between them (i.e. the Task must wait for Subtask completion before starting).

The result returned by the subtask is used to value the DIMENSIONS output of the task to be returned to the user as a computational result. The task body is a sequence of IF...DO rules, where different patterns are evaluated in the LHS and the RHS propose an opportune value for the output. The semantics of the rule shown in the Figure is the following: if the *thickness of the wall* to support is *less than 0.6* m and the earthquake *seismic class* is *A* and *raker shores span* is *1.5* m and the *interval between shores* is *1.5* m and the *raker class* is *R1*, the *dimensions* of the raker should be *13 × 13* m^2.

Similar considerations can be made for the other task of the case study, namely *rakerName*, based on the Raker Name node of the Influence Net, and affected by the *rakerClass* and *rakerType* Subtasks.

6.2 The KA–User and the KA–Developer: Two Android Clients

Every operator involved in the emergency procedures to make safe buildings and infrastructures is provided with an Android application on his/her smartphone: this application communicates with the server via the client–server architecture introduced above. Each KA–User sends to the server data about the conditions of the site it is analyzing: according to the STOP handbook, these data allow to make considerations about the real conditions of the building walls and openings after the earthquake, in order to understand which raker or scaffolding to adopt.

Figure 8 presents the GUI for introducing inputs to configure rakers in the first case study: according to the conceptual model described so far, the KA–User guides the user in order to avoid mistakes during parameters' set up.

The KA–User converts the values into GSON instances, which will be sent to the server for elaboration, waiting for answers from it. Values are suggested by the application when available (e.g. in the case of Damaged floor, Seismic class, Sole length and Wall thickness), i.e. when the STOP handbook provides guidelines. Otherwise, the operator measures them and they will be properly interpreted by the server according to the Knowledge Artifact provided by the KA–Developer. This operation mode is sharply different from that of a traditional Expert System: the domain expert associated to the KA-Developer could immediately (i.e. dynamically!) add new rules to the Knowledge Artifact, in order to give suggestions fitting the real conditions observed on-site by the KA–User. This is possible thanks to the adoption of shadow facts for representing observations in KAFKA: when results are provided by the server, the KA–User presents them to the operator. Both outputs present in the case study Influence Net (see Sect. 6.1) are returned, i.e. raker name, that is a combination of partial outputs raker class (value R1) and raker type (value Flying) and dimensions.

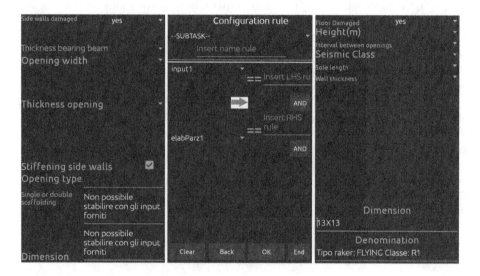

Fig. 8. No output available due to the observations provided by the KA–User: new rules must be added to the knowledge base.

If no output is available, due to the lack of knowledge in the KA, as shown in left part of Fig. 8, the KA–Developer can support the domain expert to complete the knowledge base: the central part of Fig. 8 shows the provided GUI. Then, the rule–based system can be executed again by the KA–User, being sure that valid values will be obtained in output, as shown in right part of Fig. 8.

The system can be run again on different configurations of inputs, producing new outputs according to the KA model: if necessary, new rules can be added moving the system from a stable state S_i to a new stable state S_j. In this way, the rule–based system is iteratively built up according to new discoveries made by the user on the application field.

7   Conclusions and Future Work

This paper addressed the problem of time evolving expert systems design and implementation: the KAFKA approach has been presented from both the theoretical and practical point of view. A unique feature of KAFKA is its development under Android OS, that allows to use it in many contexts characterized by ubiquity of inputs and scalability of problem descriptions.

KAFKA embodies the fundamental idea of a knowledge base which can be extended and modified dynamically following the evolution of the application domain. The framework is intrinsically devoted to develop distributed applications. In this sense, one of the most promising research field is the design and implementation of agent–based systems: according to the literature [24], the KA–Developer could be modeled as a *knowledge-based agent*, whose deliberations depends on the KA model, while KA–Users could be modeled as simple reactive agents, whose scope is observing data on field, sending data to the server and receiving results and/or actions to accomplish on the field.

With respect to existing agent–based platforms developed under Android, like JADE [25] or JaCa [26], the KA–Users and the KA–Developers introduced in our framework are not autonomous at the current state of development: for this reason, we are moving from KAFKA to $KAFKA^2$, by substituting the IN Knowledge Artifact with Bayesian Networks (BN) [27].

More specifically, we know that the Ontology level of KAFKA already represents implicitly all the possible influence relations between the pieces of knowledge that can be collected or derived in a given application scenario. However, the set of rules present in the knowledge base at a specific point in time (the Jess-executable rules) may be insufficient to explain new observations: in this case we need the intervention of the KA-Developer to add new rules (see Sect. 5.4).

In $KAFKA^2$ we are going to add a statistical inference mechanism which allows to generate automatically new rules suitable to the changes occurred in the environment. The process can be informally described as follows.

A Bayesian Network (BN) is built based on the graph structure described at the ontology level; assuming that a probability distribution is assigned to the states and the arcs of the BN, we can derive the rules employed in the expert system as consequences of the most probable combinations of events (or observations). The advantage of the BN is the fact that it extends the functions of the Influence Net in KAFKA by providing a mechanism for generating new rules.

Because we are in a probabilistic setting, a new event/observation that cannot be handled by the expert system is to be considered simply the occurrence of an "not probable LHS". Therefore this new evidence can be fed back to the BN and employed to recompute the state probability distribution. In this way we are able to select new rules to be added to the expert system while the scenario we are observing evolves to a new state.

Two aspects of $KAFKA^2$ require a very careful tuning. The first one is the determination of the initial probability distributions for the BN, which is equivalent to determine which is the initial state S_1 of the observed system (the starting point of the transitions described in Fig. 5). This task can be accomplished by recurring to experimental data analysis or, in some cases, by exploiting the structure of the application domain (in many cases an obvious initial state can be characterized quite easily). A second issue regards the actions to be taken when the not probable events occur. Although we know that, given time and a sufficient number of observations, the expert system will be upgraded to the new state, we are still left with the problem of coping with the "first unexpected events". Depending on the application scenario, they could be safely ignored (no matching LHS means no action) or some sort of ad-hoc action could become necessary. Further research is needed to define the suitable functions capable of modeling correctly the 'cost of no action' in a $KAFKA^2$ system.

References

1. Schreiber, G., Wielinga, B., de Hoog, R., Akkermans, H., Van de Velde, W.: CommonKADS: a comprehensive methodology for KBS development. IEEE Expert **9**, 28–37 (1994)
2. Angele, J., Fensel, D., Landes, D., Studer, R.: Developing knowledge-based systems with MIKE. Autom. Softw. Eng. **5**, 389–418 (1998)
3. Nalepa, G., Ligeza, A.: The HeKatE methodology. Hybrid engineering of intelligent systems. Int. J. Appl. Math. Comput. Sci. **20**, 35–53 (2010)
4. Schreiber, G.: Knowledge acquisition and the web. Int. J. Hum. Comput. Stud. **71**, 206–210 (2013)
5. Yan, H., Jiang, Y., Zheng, J., Fu, B., Xiao, S., Peng, C.: The internet-based knowledge acquisition and management method to construct large-scale distributed medical expert systems. Comput. Meth. Programs Biomed. **74**, 1–10 (2004)
6. Bonastre, A., Ors, R., Peris, M.: Distributed expert systems as a new tool in analytical chemistry. TrAC - Trends Anal. Chem. **20**, 263–271 (2001)
7. Ruiz-Mezcua, B., Garcia-Crespo, A., Lopez-Cuadrado, J.L., Gonzalez-Carrasco, I.: An expert system development tool for non AI experts. Expert Syst. Appl. **38**, 597–609 (2011)
8. Rybina, G.V., Deineko, A.O.: Distributed knowledge acquisition for the automatic construction of integrated expert systems. Sci. Tech. Inf. Process. **38**, 428–434 (2011)
9. Norman, D.A.: Cognitive artifacts. In: Designing Interaction, pp. 17–3 (1991)
10. Schmidt, K., Simone, C.: Mind the gap. Towards a unified view of CSCW. In: COOP, pp. 205–221 (2000)
11. Omicini, A., Ricci, A., Viroli, M.: Artifacts in the A&A meta-model for multi-agent systems. Auton. Agent. Multi-agent Syst. **17**, 432–456 (2008)

12. Cabitza, F., Cerroni, A., Locoro, A., Simone, C.: The knowledge-stream model - a comprehensive model for knowledge circulation in communities of knowledgeable practitioners. In: Proceedings of the International Conference on Knowledge Management and Information Sharing (IC3K 2014), pp. 367–374 (2014)
13. Holsapple, C.W., Joshi, K.D.: Organizational knowledge resources. Decis. Support Syst. **31**, 39–54 (2001)
14. Salazar-Torres, G., Colombo, E., Da Silva, F.C., Noriega, C., Bandini, S.: Design issues for knowledge artifacts. Knowl.-Based Syst. **21**, 856–867 (2008)
15. Butler, T., Feller, J., Pope, A., Emerson, B., Murphy, C.: Designing a core IT artefact for knowledge management systems using participatory action research in a government and a non-government organisation. J. Strateg. Inf. Syst. **17**, 249–267 (2008)
16. Kitamura, Y., Kashiwase, M., Fuse, M., Mizoguchi, R.: Deployment of an ontological framework of functional design knowledge. Adv. Eng. Inform. **18**, 115–127 (2004)
17. Surif, J., Ibrahim, N.H., Mokhtar, M.: Conceptual and procedural knowledge in problem solving. Procedia - Soc. Behav. Sci. **56**, 416–425 (2012)
18. Niedderer, K., Reilly, L.: Research practice in art and design: experiential knowledge and organised inquiry. J. Res. Pract. **6** (2010)
19. Rosen, J.A., Smith, W.L.: Influence net modeling with causal strengths: an evolutionary approach. In: Proceedings of the Command and Control Research and Technology Symposium, pp. 25–28 (1996)
20. Aalst, W., Adriansyah, A., Dongen, B.: Causal nets: a modeling language tailored towards process discovery. In: Katoen, J.-P., König, B. (eds.) CONCUR 2011. LNCS, vol. 6901, pp. 28–42. Springer, Heidelberg (2011). doi:10.1007/978-3-642-23217-6_3
21. de Cindio, F., Michelis, G.D., Pomello, L., Simone, C.: Superposed automata nets. In: Girault, C., Reisig, W. (eds.) Application and Theory of Petri Nets, Selected Papers from the First and the Second European Workshop on Application and Theory of Petri Nets, vol. 52, pp. 269–279. Springer, Heidelberg (1981)
22. Sartori, F., Grazioli, L.: Modeling and understanding time-evolving scenarios. In: Proceedings of the Metadata and Semantics Research - 8th Research Conference, MTSR 2014, Karlsruhe, Germany, 27–29 November 2014, pp. 60–67 (2014)
23. Grimaz, C.S., et al.: Vademecum STOP. Shoring templates and operating procedures for the support of buildings damaged by earthquakes. Ministry of Interior - Italian Fire Service (2010)
24. Russell, S.J., Norvig, P., Canny, J.F., Malik, J.M., Edwards, D.D.: Artificial Intelligence: A Modern Approach, vol. 74. Prentice Hall, Englewood Cliffs (1995)
25. Ughetti, M., Trucco, T., Gotta, D.: Development of agent-based, peer-to-peer mobile applications on android with jade. In: The Second International Conference on Mobile Ubiquitous Computing, Systems, Services and Technologies, UBICOMM 2008. IEEE, pp. 287–294 (2008)
26. Santi, A., Guidi, M., Ricci, A.: JaCa-Android: an agent-based platform for building smart mobile applications. In: Dastani, M., Fallah Seghrouchni, A., Hübner, J., Leite, J. (eds.) LADS 2010. LNCS (LNAI), vol. 6822, pp. 95–114. Springer, Heidelberg (2011). doi:10.1007/978-3-642-22723-3_6
27. Melen, R., Sartori, F., Grazioli, L.: Modeling and understanding time-evolving scenarios. In: Proceedings of the 19th World Multiconference on Systemics, Cybernetics and Informatics (WMSCI 2015), vol. I, pp. 267–271 (2015)

A Property Grammar-Based Method to Enrich
the Arabic Treebank ATB

Raja Bensalem Bahloul[1(✉)], Kais Haddar[1], and Philippe Blache[2]

[1] Multimedia InfoRmation Systems and Advanced Computing Laboratory,
Higher Institute of Computer Science and Multimedia, Sfax, Tunisia
raja_ben_salem@yahoo.com, kais.haddar@yahoo.fr
[2] Laboratoire Parole et Langage, CNRS, Université de Provence,
Marseille, France
pb@lpl.univ-aix.fr

Abstract. We present a method based on the formalism of Property Grammars to enrich the Arabic treebank ATB with syntactic constraints (so-called properties). The Property Grammar formalism is an effectively constraint-based approach that directly specifies the constraints on information categories. This can facilitate the enrichment process. The latter is based on three phases: the problem formalization, the Property Grammar induction from the ATB and the treebank regeneration with a new syntactic property-based representation. The enrichment of the ATB can make it more useful for many NLP applications such as the ambiguity resolution. This allows also the acquisition of new linguistic resources and the ease of the probabilistic parsing process. This enrichment process is purely automatic and independent from any language and source corpus formalism. This motivates its reuse. We obtained good and encouraging experiment results and various properties of different types.

Keywords: Arabic language · Property grammar · Treebank enrichment

1 Introduction

The Property Grammar (GP) formalism [6] is a constraint-based approach that puts the constraint notion at the core of the linguistic analysis. In fact, it does not require the construction of a local structure of the syntactic information before using the constraints it described. Instead, it specifies directly the syntactic information on categories. The other approaches of the same family, however, have other directives. For example, the HPSG (Head-driven phrase structure grammar) needs a local tree [16] and the CDG (Constraint Dependency Grammar) a dependency relation [12]. Moreover, the GP formalism differs from other constraint-based approaches by its simple, direct, local and decentralized representation of linguistic information. Indeed, unlike generative theories, the GP represents, independently, all kinds of information, regardless their position, and even the partial, incomplete or non-canonical information. This promotes its flexibility and robustness. In addition, using the GP formalism can be favorable to parsing process. In effect, many implicit syntactic structures and relations become explicit thanks to this formalism. The specified qualities of the GP formalism encouraged us to

© Springer International Publishing AG 2016
A. Fred et al. (Eds.): IC3K 2015, CCIS 631, pp. 302–323, 2016.
DOI: 10.1007/978-3-319-52758-1_17

use it in the development of a new resource enriched with properties for the Arabic language. Processing this language presents several challenges. These challenges are not only related to certain Arabic specificities to be studied (such as the lack of vowels and diacritic marks and the agglutinative aspect of words), but also to particular linguistic phenomena to be addressed (like the relatives, the anaphora and the coordination). The enrichment task is not easy and direct but requires verification modules of the GP properties in the treebank and matching functions of treated categories. The formalization phase is also challenging. It needs to choose an adequate model and to understand all of the treebank data to succeed the enrichment issue resolution.

The present paper fits in this context. Our goal is to describe the enrichment method of the Arabic treebank ATB with data acquired from a given GP. As a result, we obtain the first Arabic treebank enriched with varied syntactic properties available in variable granularity level according to the user needs. We may also specify the most relevant properties thanks to the frequencies of the treebank categories and properties. This may ease the probabilistic parsing process and evaluate the difficulty of processing cognitive systems. Moreover, new linguistic resources can be obtained from the enriched ATB such as syntactic lexicons and dependency grammars. The proposed enrichment method is based on three phases: the problem formalization, the GP induction from the source treebank ATB and the new treebank generation based on syntactic property. We stared with an empirical phase, a linguistic study of some Arabic sentence structures before testing the method on a large corpus (i.e. the treebank ATB). This task helps us to verify the correctness of the interpretations of the syntactic properties and their validity (satisfaction), and efficiency.

This paper is organized as follows: Sect. 2 is devoted to a brief presentation of related works. Section 3 proposes an Arabic linguistic study within the GP formalism. Section 4 describes the formalization phase. Section 5 explains our enrichment method. Section 6 shows the constraint solver descriptions. Section 7 presents experimental results and discussions. Section 8 gives a conclusion and some perspectives.

2   Related Works

Before quoting some related works, it is necessary to present the main key concepts to use in our contribution: the GP and the ATB. The GP is based on a formalism [6] representing linguistic information through properties (constraints) in a local and decentralized manner. These properties express the relations that may exist between the categories composing the described syntactic structure. Syntactic properties in particular have six types: the linear order ($\prec$), the obligation of co-occurrence ($\Rightarrow$), the interdiction of co-occurrence ($\otimes$), the dependency ($\leadsto$), the interdiction of repetition (Unic) and the head (Oblig).

The ATB [11], is the richest Arabic treebank in reliable annotations (POS tags, syntactic and semantic hashtags), which are also compatible to consensus developed and validated by linguists. Its source documents are relevant, varied and large. They are even converted by several other treebanks into their representations. The ATB grammar is adapted to the Modern Standard Arabic and has a phrase-based representation, which is consistent with the GP hierarchical structure.

Several works are proposed to enrich treebanks in different languages. The contribution of Müller [13] is an instance of converted treebanks. It proposes an annotation of Morphology and NP Structure in the Copenhagen Dependency Treebanks (CDT), which represents different parallel treebanks in many languages such as Danish, English, German, Italian, and Spanish.

The annotations to add in treebanks can be also organized according to well-defined linguistic formalisms. Thus, Oepen et al. [14] followed this directive by developing the Lingo Redwoods, which is a dynamic treebank. This new type of treebank parses analyzed sentences from ERG (English Resource Grammar) according to a precise HPSG formalism. The CCG formalism is another formalism chosen in the treebank enrichment methods. This is particularly the case of the contributions of Çakıcı [7], who created CCGbanks by converting the syntax graphs in the Turkish treebank into CCG derivation trees.

For French, Blache and Rauzy [6] proposed an automatic method, which hybridizes the constituency treebank FTB with constraint-based descriptions using the GP formalism. In addition, this method enriches the FTB with evaluation parameters of the sentence grammaticality.

For Arabic, which is the language that interests us the most, we can find some other works to enrich the ATB. They focus on improving this treebank with new richer annotations or on converting it into new formalisms. The OntoNotes project [10] and the Proposition Bank project (Propbank) for Arabic [15] are some instances of treebank extensions. The latter incorporate semantic level annotations. The contribution of Alkuhlani and Habash [2] provides an enrichment, which adds annotations that models attributes of the functional gender, number and rationality. The work of Abdul-Mageed and Diab [1] has even touched the sentimental level by associating specific annotations to the ATB sentences. There is also the work of Alkuhlani et al. [3], but it enriches the Columbia Arabic Treebank (CATiB) with the most complicated POS tags and lemmas applied in the ATB [11].

As regards the enrichment by employing new formalisms in the treebank source, we can refer to some examples that generates new treebanks: the Habash and Rambow contribution [9] with a TAG grammar, the Tounsi et al. contribution [17] with an LFG grammar and the El-taher et al. contribution [8] with a CCG grammar. Regarding the GP formalism, it was previously hybridized with the French treebank FTB as we have already mentioned.

By inspecting all the works cited above, we may figure out that none of them presents an in-depth formalization phase before proposing the enrichment approach. The absence of this phase can make the establishment of their approaches more difficult due to the lack of pre-specified needed data and the risk of having redundant treatments.

In addition, the enrichment of treebanks can be considered as a Constraint Satisfaction Problem (CSP). In this case, the ATB enrichment processes with new formalisms (TAG, LFG and CCG), which are mentioned above, will be tough. In fact, their representations would require a construction of local structures before referring to the constraints. As already mentioned in Sect. 1, the GP is an approach extremely based on constraint satisfaction. Its application in the ATB enrichment, we can solve these limitations by directly accessing to the variable values of the problem through its categories. An Arabic GP in variable granularity is already available [5]. It is not manually built but automatically generated from an ATB part [4].

3 Arabic Linguistic Study Within the GP Formalism

In order to evaluate the performance of the syntactic properties in the treebank, we followed a linguistic study of some examples of Arabic syntactic structures (phrases). In this study, we present different syntactic relations between these examples to explain the interpretation and the aspects related to each property type in the GP formalism. Therefore, we introduce six examples of Arabic sentences shown in Table 1 below:

Table 1. The examples of the Arabic sentences parsed according to the annotation of the ATB.

S	ظهر ذلك ليس في تونس وحدها بل في خارجها أيضا.	أدى ذلك إلى "تحالف دولي".
T	Zahar+a *'lika layosa fiy tuwnis+a waHod+a-hA, balo fiy xArij+i-hA >ayoD+AF .	->ad~aY *'lika <ilaY " taHAluf+K duwaliy~+K " .
G	This appeared not only in Tunisia but also abroad.	This led to an "international coalition".
P	(S (VP (PV Zahar+a) (NP (PRON *'lika)) (PP (PRT (PART layosa)) (PREP fiy) (NP (NP (NOUN tuwnis+a)) (ADJP (ADJ waHod+a-) (NP (PRON -hA))))) (PUNC ,) (CONJ balo) (PP (PREP fiy) (NP (NOUN xArij+i-) (NP (PRON -hA)))) (ADVP (ADV >ayoD+AF)))) (PUNC .))	(S (VP (PV ->ad~aY) (NP (PRON *'lika)) (PP (PREP <ilaY) (PUNC ") (NP (NOUN taHAluf+K) (ADJ duwaliy~+K)) (PUNC "))) (PUNC .))
S	نجحت تونس سواء عن طريق المساعدات أو من خلال الاستثمارات.	سافر الرجل من تونس إلى مصر.
T	najaHat tuwnis+u sawA'+N Ean Tariyq+i Al+musAEad+At+i CONJ >awo min xilAl+i Al+{sotivomAr+At+i .	sAfar+a Al+rajul+u min tuwnis+a <ilaY miSr+a .
G	Tunisia succeeded, either through aids or through investments.	The man traveled from Tunisia to Egypt.
P	(S (VP (PV najaHat) (NP (NOUN tuwnis+u) (PP (PP (NP (NOUN sawA'+N)) (PREP Ean) (NP (NOUN Tariyq+i) (NP (NOUN Al+musAEad+At+i)))) (CONJ >awo) (PP (PREP min) (NP (NOUN xilAl+i) (NP (NOUN Al+{sotivomAr+At+i))))) (PUNC .))	(S (VP (PV sAfar+a) (NP (NOUN Al+rajul+u)) (PP (PP (PREP min) (NP (NOUN tuwnis+a))) (PP (PREP <ilaY) (NP (NOUN miSr+a))))) (PUNC .))
S	الخطر ليس فقط في أن الاقتصاد ينهار.	تحدث كما كان دائما.
T	Al+xaTar+u layosa faqaT fiy >an~a Al+{iqotiSAd+a ya+nohAr+u .	-taHad~av+a -ka-mA kAn+a dA}im+AF .
G	The danger is not only in the collapsing economy.	He talked as he was always.
P	(S (NP (NOUN Al+xaTar+u)) (VP (PV layosa) (NP (-NONE- *T*)) (PP (ADVP (ADV faqaT)) (PREP fiy) (SBAR (CONJ >an~a) (S (NP (NOUN Al+{iqotiSAd+a)) (VP (IV ya+nohAr+u) (NP (-NONE- *T*)))))))) (PUNC .))	(S (VP (PV -taHad~av+a) (NP (-NONE- *)) (PP (PREP -ka-) (SBAR (WHNP (PRON -mA)) (S (VP (PV kAn+a) (NP (-NONE- *T*)) (NP (ADJ dA}im+AF))))))) (PUNC .))

In Table 1 above, the meanings of the symbols S, T, G and P are respectively as follows: the Arabic Sentences, the Buckwalter Transliterations[1] of the sentences, their Gloss in English language and their Parsing representations according to the ATB annotations [11]. In the following, we present some structure examples extracted from these sentences to explain the application of different syntactic properties of the GP formalism. In particular, we focus, as simple choice, on the PP (Prepositional Phrase) structures to describe each property type.

3.1 The Linearity Property

The relation PREP ≺ NP states that a preposition (PREP) should always precede the Nominal Phrase (NP) in Arabic. Several following examples of the PP structures from Table 1 prove this:

	PREP constituents	NP constituents
1	Fiy	(**NP** (NOUN tuwnis+a)) (**ADJP** (ADJ waHod+a-) (NP (PRON -hA)))
2	Fiy	(**NOUN** xArij+i-) (**NP** (PRON -hA))
3	Ean	(**NOUN** Tariyq+i) (**NP** (NOUN Al+musAEad+At+i))
4	Min	(**NOUN** xilAl+i) (**NP** (NOUN Al+{sotivomAr+At+i))
5	Min	(**NOUN** tuwnis+a)
6	<ilaY	(**NOUN** miSr+a)

For a clearer tree representation, we showed in bold the constituents of each syntactic category, specified in a certain property describing PP. However, we can find from the same table a counter-example, that confutes this relation and shows the NP preceding the PREP, such as the part of PP structure " سواء عن " (sawA'+N Ean / either about):

NP constituents	PREP constituents
(**NOUN** sawA'+N)	Ean

Due to one counter-example, such relation should not be shown in the GP to build.

3.2 The Adjacency Property

Another relation type between the same categories PREP and NP can be noted in the examples of the PP structures presented for the linearity property. This is the adjacency property. Indeed, the PREP is always just after or just before the NP. These two categories can be adjacent if no counter-example is found. However, there is this example of the PP structure below. It shows the Left-Hand Side (LHS) of a rule

---

[1] Arabic Transliteration Table on Tim Buckwalter site: www.qamus.org/transliteration.htm.

example, which is, of course, PP and its Right-Hand Side (RHS), which is the structure
"إلى "تحالف دولي" (<ilaY " taHAluf+K duwaliy~+K " / to an "international coalition"):

LHS	RHS			
PP	**PREP**	**PUNC**	**NP**	**PUNC**
1	<ilaY	"	(**NOUN** taHAluf+K) (**ADJ** duwaliy~+K)	"

This PP structure separates these two categories (PREP and NP) by a punctuation
symbol ("), so the adjacency relation between PREP and NP cannot be valid.

3.3 The Uniqueness Property

This is a unary relation, so it concerns only one category. It induces each unique
constituent in the described syntactic category. In the PP, we have many unique categories such as PREP, SBAR (Subordinate clause), ADVP (Adverbial phrase) and
ADJP (Adjectival Phrase). Some examples of PP structures proving this uniqueness are
shown below for some unique constituents in PP:

Unique constituents	PREP		SBAR	ADVP
LHS	**RHS**			
PP	(**ADVP** (ADV faqaT)) (**PREP** fiy) (**SBAR** (CONJ >an~a) (S (NP (NOUN Al+{iqotiSAd+a)) (VP (IV ya+nohAr+u) (NP (-NONE- *T*)))))			
	(**PREP**-ka-) (**SBAR** (WHNP (PRON -mA)) (S (VP (PV kAn+a) (NP (-NONE- *T*)) (NP (ADJ dA}im+AF)))))			
	(**PREP** fiy) (**NP** (NOUN xArij+i-) (NP (PRON -hA)))			

The first example of the PP structure is common for all specified constituents (PREP,
SBAR, ADVP) because it shows them all unique. The second example is, by contrast,
common for only PREP and SBAR. The NP could have also been unique in the PP
structures if we do not have the PP structure example "سواء عن طريق المساعدات" (sawA'+N
Ean Tariyq+i Al+musAEad+At+i / either through aids) shown in the following:

LHS	RHS
PP	(**NP** (NOUN sawA'+N)) (**PREP** Ean) (**NP** (NOUN Tariyq+i) (NP (NOUN Al +musAEad+At+i)))

3.4 The Obligation Property

This relation is also unary. It concerns the constituents always presented, and which
form a head in the described syntactic category. From the most of the PP structure

examples, we can note that the PREP is a mandotary constituent in the PP. This relation can not be valid for the PP structure example "من تونس إلى مصر" (min tuwnis+a <ilaY miSr+a / from Tunisia to Egypt) shown in the following:

LHS	RHS
PP	**(PP** (PREP min) (NP (NOUN tuwnis+a))) **(PP** (PREP <ilaY) (NP (NOUN miSr+a)))

In this example, the PREP is not directly a mandatory constituent in the derivation subtree of the PP on the top, because this PP structure is composed of two successive PP that each one contains a PREP.

3.5   The Requirement Property

A requirement property indicates that the appearance of the first constituent involves the appearance of the second one. The relation ADVP ⇒ PREP respects this condition in many PP structure examples, such as the structure "فقط في أن الاقتصاد ينهار" (faqaT fiy >an ~ a Al+{iqotiSAd+a ya+nohAr+u / only in the collapsing economy):

LHS	RHS
PP	**(ADVP** (ADV faqaT)) **(PREP** fiy) **(SBAR** (CONJ >an ~ a) (S (NP (NOUN Al +{iqotiSAd+a)) (VP (IV ya+nohAr+u) (NP (-NONE- *T*)))))

However, there are another PP structure example confuting this relation, which is "ليس في تونس وحدها بل في خارجها أيضا" (layosa fiy tuwnis+a waHod+a-hA, balo fiy xArij +i-hA >ayoD+AF / not only in Tunisia but also abroad), shown in the following:

LHS	RHS
PP	**(PP** (PRT (PART layosa)) (PREP fiy) (NP (NP (NOUN tuwnis+a)) (ADJP (ADJ waHod+a-) (NP (PRON -hA))))) **(PUNC** ,) **(CONJ** balo) **(PP** (PREP fiy) (NP (NOUN xArij+i-) (NP (PRON -hA)))) **(ADVP** (ADV >ayoD+AF))

In this example, the ADVP appears without the PREP. Furthermore, if we do not have this counter-example, this relation becomes valid. However, its symmetric relation (PREP ⇒ ADVP) is not necessarily valid. The most famous structure of the PP (PREP NP), in effect, confutes it. This PP structure does not allow NP ⇒ PREP as valid relation. This is because there are PP structures where the PP appears instead of the PREP with the NP.

The relation SBAR ⇒ PREP is, however, always valid, so, in any PP structure example, if we have an SBAR, we find absolutely a PREP.

3.6   The Exclusion Property

The exclusion property is the opposite of the requirement one. It prevents the co-occurrence of two constituents. For the PP, when we observe the categories ADVP and SBAR in many Arabic examples of the PP structures, we can note that they do not appear together in these structures. Here are two examples proving this observation, which are "أيضاً في خارجها بل وحدها في تونس ليس" (layosa fiy tuwnis+waHod+a-hA, balo fiy xArij+i-hA >ayoD+AF / not only in Tunisia but also abroad) and "دائما كان كما" (-ka-mA kAn+a dA}im+AF / as we was always):

	LHS	RHS
Only **ADVP**	**PP**	**(PP** (PRT (PART layosa)) **(PREP** fiy) **(NP** (NP (NOUN tuwnis+a)) (ADJP (ADJ waHod+a-) (NP (PRON -hA))))) **(PUNC** ,) **(CONJ** balo) **(PP** (PREP fiy) (NP (NOUN xArij+i-) (NP (PRON -hA)))) **(ADVP** (ADV >ayoD+AF))
Only **SBAR**		**(PREP** -ka-) **(SBAR** (WHNP (PRON -mA)) (S (VP (PV kAn+a) (NP (-NONE- *T*)) (NP (ADJ dA}im+AF)))))

If we are restricted to these examples, you will see that we have an exclusion relation between the ADVP and the SBAR (ADVP $\otimes$ SBAR). Unlike the requirement properties, this exclusion property is symmetric. Thus, when the SBAR appears, the ADVP do not. However, by expanding our vision on the PP structure example "أن الاقتصاد ينهار فقط في" (faqaT fiy >an $\sim$ a Al+{iqotiSAd+a ya+nohAr+u / only in the collapsing economy) where both SBAR and ADVP appears in the PP structure, this relation becomes invalid.

LHS	RHS
PP	**(ADVP** (ADV faqaT)) **(PREP** fiy) **(SBAR** (CONJ >an $\sim$ a) (S (NP (NOUN Al +{iqotiSAd+a)) (VP (IV ya+nohAr+u) (NP (-NONE- *T*)))))

This linguistic study gives us an idea of the difficulty facing us to determine and enumerate the different valid syntactic relations in the Arabic language. These relations are implicit in the Arabic texts. Making them explicit, thanks to the GP formalism, can be useful for many NLP applications and different domains, such as the ambiguity resolution, the probabilistic parsing and the dependency grammar building.

4   Formalization Phase

As we have already mentioned in Sect. 1, the elaboration of a solid and detailed enrichment method cannot be directly made without modeling the tools to use as input. In our case, this means that we have to generate specific formalizations to the treebank ATB and the GP. This facilitates and clarifies better the enrichment method.

First, we present the description of the CFG (Context-Free Grammar), which is composed of a set of production rules (constructions). The latter are used to produce structures of words. Formally, it is defined by the 4-tuple $G = (N, \Sigma, P, S)$ where: N is a finite set of non-terminal symbols, Σ is a finite set of terminal symbols, P is a finite set of rules formed as $\alpha \to \beta$ with $\alpha \in N$ and $\beta \in (N \cup \Sigma)^*$ and $S \in N$ is the start symbol. The formal language of G is then defined as $L(G) = \{w \in \Sigma^* \mid S \vdash^* w\}$. For each derivation of S, w corresponds to a tree t_w. In natural languages, w corresponds to a sentence Sent, which is associated to a tree t_{Sent} according to the grammar G.

On the one hand, the ATB, as a corpus of manually annotated sentences of a natural language (Arabic) can be seen as a sequence of pairs (Sent, t_{Sent}). So, it is defined by TB = {(Sent, t_{Sent}) | S ⊢* Sent} where Sent is a sequence of Arabic words, giving a complete meaning. So, Sent ∈ M* where M is the set of the treebank words. However, t_{Sent} ∈ $\mathcal{A}$ where $\mathcal{A}$ is the set of trees given by parsing each treebank sentence Sent according to G. As the annotations given by the ATB are extended to several analysis levels (word, phrase and sentence levels), this improves the definition of the ATB to be a 7-tuple TB = (M, Ψ, P, $\mathcal{T}$, Ω, $\mathcal{S}$, $\mathcal{A}$). M is a set of treebank words. Ψ = ψ_1 x ψ_2 x...x ψ_n is an n-tuple of sets ψ_i of information types (morphological, syntactic and semantic). The latter specify the corpus words with the form (c_1, c_2,.., c_n) ∈ Ψ where c_i is an information of a defined type i of n information types (e.g. lexical category, transliteration, gloss). P is the treebank phrase set (a phrase is a sequence of words giving elementary meaning) as p ∈ M*. $\mathcal{T}$ is the elementary tree set t_p given from parsing phrases p ∈ P. Ω = ω_1 x ω_2 x... ω_z is an n-tuple of sets ω_j of information types. The latter specify the corpus phrases with the form (d_1, d_2,.., d_z) ∈ Ω where d_j is an information of a defined type j of z information types (e.g. syntactical category, hashtag). $\mathcal{S}$ is the set of sentences as Sent ∈ M*. $\mathcal{A}$ is the complete tree set obtained from parsing sentences Sent ∈ $\mathcal{S}$.

On the other hand, the GP is a grammar that defines a set of relations between grammatical categories not in terms of production rules (like CFG) but in terms of local constraints (so-called properties). As we specified in the previous section, the syntactic properties describe linguistic phenomena between constituents such as linear precedence (<), mandatory co-occurrence (⇒), restricted co-occurrence (⊗), obligation (oblig), uniqueness (unic) and adjacency (±). Formally, this grammar can be defined by a 3-tuple G' = (N, Σ, R). N is a finite set of syntactic categories. Σ is a finite set of lexical categories. R is a finite set of syntactic properties that links ∀ α ∈ N to ∀ β_1 and β_2 ∈ (N ∪ Σ) in any of the following 6 ways: α: β_1 < β_2, α: β_1 ± β_2, β_1 ∈ unic(α), β_1 ∈ oblig(α), α: β_1 ⇒ β_2, α: β_1 ⊗ β_2. We deduce each of these properties from the set P defined in G.

Now, as the needed tools to use are formally modeled, it is necessary to know how to integrate them to succeed the enrichment method. We may consider this enrichment for the ATB phrases as a satisfaction verification of properties provided from the GP. It can be a Constraint Satisfaction Problem (CSP). Formally, we can model this problem by the 5-uplet TBG = (S(TB), S(G'), Const(TB), Const(G'), Prop(G')) where:

- S(TB) = {p_1, p_2, ...p_n} = P is a finite set of the ATB phrases.
- S(G') = {t_1, t_2, ...t_m} = N is a finite set of the GP syntactic categories.
- Const(TB) = $\bigcup_{i=1}^{n} Const(p_i)$ where Const(p_i) = {c_{i1}, c_{i2}, ...c_{ie}}: set of the words of p_i, label(c_{ix}) is its grammatical category (label(c_{ix}) is equal to $c_1(c_{ix})$ for lexical category or to $d_1(c_{ix})$ for syntactic category).
- Const(G') = $\bigcup_{j=1}^{m} Const(t_j)$ where Const(t_j) = {c_{j1}, c_{j2}, ...c_{jf}}: set of the constituents (grammatical categories) of the syntactic category t_j in the GP.

- $\text{Prop}(G') = \bigcup\limits_{j=1}^{m} Prop(t_j)$ where $\text{Prop}(t_j) = [\text{Prop\_const}(t_j),\ \text{Prop\_unic}(t_j),\ \text{Prop\_}$
  $\text{oblig}(t_j),\ \text{Prop\_lin}(t_j),\ \text{Prop\_adjc}(t_j),\ \text{Prop\_exig}(t_j),\ \text{Prop\_excl}(t_j)]$, for example, $\text{Prop\_lin}(t_j) = \{p_{j1},\ p_{j2},\ \dots p_{jg}\}$ is the linearity property set describing t_j in the GP. Each p_{jk} (with $1 < k < g$) is a relation between two elements c_{jx} and $c_{jy} \in \text{Const}$ (t_j) $(p_{j1} = tj{:}c_{jx} \prec c_{jy})$.

In order to solve this issue, we need, in the first instance, to look in the GP for the syntactic category of each ATB phrase. Formally, for each phrase $p_i \in P$ in the ATB, we search in the GP for its syntactic category $t_j \in N$ where $\text{label}(p_i) = t_j$. The set of properties $\text{Prop}(t_j)$ describing t_j will be used to enrich p_i by verifying the satisfaction of these properties. As a result, this problem would formally be solved.

5   The Enrichment Method of the ATB

Now that the formalization phase is totally accomplished, it became possible to represent, in detail, the other phases of the enrichment method, which would be written in algorithms. Note that these phases are based on the enrichment idea of the French treebank FTB, where the properties were proposed by [6]. For clarity, Fig. 1 shows our ATB enrichment method, which consists of three main phases: formalizing the problem, inducing the GP from the ATB and regenerating the latter with a new syntactic property-based representation.

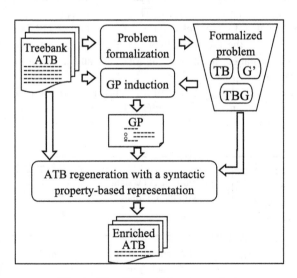

Fig. 1. The ATB enrichment method.

We chose to devote an entire section (the previous one) to explain the first phase, the formalization, as its important role in our enrichment method and particularly in this paper.

The second phase, the GP induction, is already applied by [4], which produced an Arabic GP. In this phase, the GP is constructed automatically from the ATB. This directive is more favorable than building the GP manually. The latter is more challenging and expensive. It needs to use a corpus, which contains all the rules of the Arabic grammar. This increases the GP development time and requires the collaboration of several linguists. Therefore, the obtained GP was automatically induced and independently of the language and source treebank formalism. That is why the GP is not directly generated from the ATB, but rather from a CFG (as shown in Fig. 2).

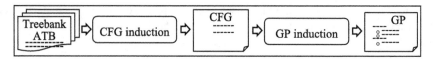

Fig. 2. The induction phase of the GP from the ATB.

For more details, the CFG induction step involves the generation of the set of all the possible assignments for each syntactic category represented in the ATB. This set will be used in the GP induction step to generate the set of properties associated to this syntactic category. All of the GP properties are described except the dependency ones. Adjacency properties are added to this set. They concern the direct order relation between two constituents of the syntactic category. The induction mechanism provides also a control of the granularity level of the categories in order to compromise between quantity and quality of these categories. This control represents each category on feature structures related to hierarchy types. The obtained GP is robust not only because of the power of the GP formalism but also thanks to the qualities inherited from the ATB. For instance, it has rich annotations and a consistent representation structure to the GP one [5].

The obtained GP is used as an input for the ATB regeneration phase with syntactic properties. As mentioned in Fig. 3, this phase is based on many steps: the first is a

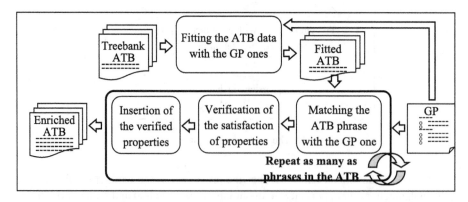

Fig. 3. The ATB regeneration phase with a syntactic property-based representation.

fitting step of the ATB data according to the GP one. The three other steps relate to each phrase in the ATB. Therefore, we need, for all the ATB phrases, to follow a browsing mechanism through the ATB sentences. This is to check for each phrase of each sentence, the satisfaction of the properties describing its syntactic category in the GP. To check this satisfaction, it needs to use previously developed constraint solvers.

In the following sub-sections, we explain the different steps of this phase.

5.1 Fitting the ATB Data with the GP Ones

Since our goal is to enrich the ATB with syntactic properties, it should have a data structure able to host this new information. The parenthesis format "penntree" of the ATB does not provide this structure, and requires preparing its data in a suitable format. The format "xml" can form this structure. To achieve this, we have made a conversion recursive process of encountered open and close parentheses in the format "penntree" to xml tags. In addition, if the ATB category granularity was modified, we would include a verification model of matches in this step to replace the ATB raw categories with categories whose granularity is modified.

The following steps are encapsulated in a browsing mechanism repeated as many as phrases in the ATB. As a result, we will have the fitted ATB as an input, able to host GP syntactic properties. The output is a new version of the ATB, which is enriched with these verified properties as satisfied or not.

5.2 Matching an ATB Phrase with a GP One

The matching between the ATB and the GP consists of browsing the ATB, phrase by phrase, and for each one, searching for the correspondent in the GP of its category. The properties describing the found correspondent will be verified and will enrich the current ATB phrase. Formally, in order to match between an ATB phrase $p_i \in P$ and the correspondent of its category in the GP, we need, for each $\alpha \in N$ in the GP, to look for the case where the p_i category label(p_i) is equal to α (label(p_i) = α).

5.3 Verification of the Satisfaction of the Properties

This step is the heart of the enrichment method. It verifies the satisfaction of the properties, which describes a GP category matched with the ATB phrase. Formally, we just need to verify, for each ATB phrase p_i (with t_j = label(p_i)), the satisfaction of all the properties of Prop(t_j) obtained from the GP. We have used for that a set of methods to check the satisfaction of the properties. Each method, so-called "constraint solver", verifies if a given Arabic phrase tagged with a specific syntactic category respects a given property, which describes this syntactic category in the GP. The solution produced by a solver is the result of this verification (property satisfied or violated).

Then, we associate this solution to the p_i description. As these constraint solvers play an importance role in our method, we have chosen to devote an entire section (the next one) to introduce their descriptions.

5.4 Insertion of the Verified Properties

This task adds to each ATB phrase the result of the verification (either satisfied or not) of the properties that describe its category. The insertion is done by using a new tag that combines the ATB and the GP. However, this enrichment makes too large the new ATB size. It is due to the exponential increase of the number of the new tags with the number of properties in each phrase of the GP.

6 Descriptions of the Constraint Solvers

Let us assume that the matching has been achieved, so the current ATB phrase category is equal to the found GP one. The descriptions of these solvers, introduced in the following, are inspired from the interpretations of [6]. Let us first define some variable definitions to use in these descriptions.

```
p: the given Arabic phrase.
Const(p): constituent set of the ATB phrase p.
Const(t): set of the constituents of the GP category t (where
t=label(p)).
fd: boolean, returns true if c is found.
label(c): grammatical category of the word or the phrase c.
nb_intersect: number of constituents in the intersection
between Const(p) and Const(t)
verif: string ("+" or "-").
nb_occ: number of occurrences of a constituent in Const(p).
type_p: property by type between two constituents cx and cy of
Const(t), type_p contains only cx for unary property type
(uniqueness, obligation).
v_type_p: verified property by type (constituency (const),
linearity (lin), adjacency (adjc), uniqueness (unic), obliga-
tion (oblig), requirement (exig), exclusion(excl)) (has "+" if
satisfied, "-" if not). Firstly, v_type_p is empty (← NIL) and
may remain if the constituents of type_p is not found in p.
verifProp(): method to create a verified property.
```

6.1 The Solver of Constituency Properties

This solver verifies the consistency between the categories of the constituents of the current ATB phrase and the constituents of its GP correspondent. This is to ensure that the intersection of these two sets really includes all the words of the ATB phrase.

```
Input: Const(p), Const(t), nb_intersect ← 0, const_p
Output: v_const_p
for each cₐ in Const(p), do
    for each c_b in Const(t), do
        if label(cₐ)= c_b, then
            nb_intersect ← nb_intersect + 1
if nb_intersect = card(Const(p), then
    v_const_p ← verifProperty(p, Const(p), "+")
else
    v_const_p ← verifProperty(p, Const(p), "+")
return v_const_p
```

This algorithm browses Const(p) and verifies that the category of each element is included in Const(t). The value of nb_intersect is then incremented. If it is equal to the Const(p) cardinal, the property is considered then as satisfied.

6.2   The Solver of the Linearity Properties

This solver checks the current linearity property of the GP syntactic category. This property is satisfied only if the two constituents of this category in that relation are also found in the given phrase and that the first constituent precedes the second one.

```
Input: Const(p), lin_p, v_lin_p ← NIL
Output: v_lin_p
for each cₐ in Const(p), do
    if label(cₐ)= lin_p.cₓ, then
        for each label(c_b) in Const(p), do
            if a≠b and label(c_b)= lin_p.c_y, then
                if a>b, then
                    v_lin_p ← VerifProperty(p, lin_p, "-")
                else
                    v_lin_p ← VerifProperty(p, lin_p, "+")
return v_lin_p
```

This algorithm browses the categories of the Const(p) set elements to search for the two distinct linear constituents c_x and c_y of the GP category t and verifies that the position of the first is not greater than the position of the second one.

6.3   The Solver of the Adjacency Properties

This solver checks the current adjacency property of the syntactic category in the GP. The satisfaction is ensured only if the two adjacent constituents of this category exist in the given phrase and the first is directly before or after the second one.

```
Input: Const(p), adjc_p, v_adjc_p ← NIL
Output: v_adjc_p
for each cₐ in Const(p), do
 if label(cₐ)= adjc_p.cₓ, then
   for each c_b in Const(p), do
    if a≠b and label(c_b)= adjc_p.c_y, then
      if a≠b-1 and a≠b+1, then
        v_adjc_p ← VerifProperty(p, adjc_p, "-")
      else
        v_adjc_p ← VerifProperty(p, adjc_p, "+")
return v_adjc_p
```

This algorithm browses the set Const(p) to look for the two adjacent constituents c_x and c_y of the GP category t and verifies that the second is neither indirectly before nor after the first one. We use the symbol "±" in the adjacency relation.

6.4   The Solver of the Uniqueness Properties

This solver checks the current uniqueness property of the current syntactic category in the GP. The satisfaction is reached if the constituent of this property (in case it has been found) appears only once in the given phrase.

```
Input: Const(p),unic_p,nb_occ←0,v_unic_p ← NIL
Output: v_unic_p
for each cₐ in Const(p), do
 if label(cₐ)= unic_p.cₓ, then
    nb_occ ← nb_occ +1
if nb_occ ≥1, then
 if nb_occ = 1, then
    v_unic_p ← verifProperty(p, unic_p, "+")
 else
    v_unic_p ← verifProperty(p, unic_p, "-")
return v_unic_p
```

This algorithm browses the set Const(p) to search for the constituent unic_p of t and verifies that its cardinality nb_occ is not greater than 1.

6.5   The Solver of the Obligation Properties

This solver checks the current obligation property of the found GP category. This property is satisfied if the constituent "head" of this category is found in the treebank phrase.

```
Input: Const(p),oblig_p,fd ← false,v_oblig_p ← NIL
Output: v_oblig_p
for each cₐ in Const(p), do
  if label(cₐ)= oblig_p.cₓ, then
    fd ← true
    break
if fd = true, then
  v_oblig_p ← verifProperty(p, oblig_p, "+")
else
  v_oblig_p ← verifProperty (p, oblig_p, "-")
return v_oblig_p
```

This algorithm browses Const(p) to search for the obligatory constituent oblig_p of the GP category t. If the algorithm find it, the variable "found" will return true.

6.6   The Solver of the Requirement Properties

This solver checks the current requirement property of the current syntactic category in the GP. The satisfaction is ensured only if, when the constituent involving another in this property, is found in the given phrase, the involved one is also found.

```
Input: Const(s), fd ← false, exig_p, v_exig_p ← NIL
Output: v_exig_p
for each cₐ in Const(s), do
  if label(cₐ)= exig_p.cₓ, then
    fd ← false
    for each c_b in Const(s), do
      if a≠b and label(c_b)= exig_p.c_y, then
        v_exig_p ← verifProperty(s, exig_p, "+")
        fd ← true
      break
    if fd =false then
      v_exig_p ← verifProperty(s, exig_p, "-")
      break
return v_exig_p
```

This algorithm browses the set Const(p) to search for the two constituents c_x and c_y of the GP category t in a requirement relation and verify that, if the first constituent is found in Const(p), then the second one should exist in Const(p).

6.7   The Solver of the Exclusion Properties

This solver checks the current exclusion property of the syntactic category in the GP. This property is satisfied only if its constituents do not appear both in the given phrase.

```
Input: Const(s),excl_p,verif ← "+" ,v_excl_p ← NIL
Output: v_excl_p
a ← search(excl_p.c_x, Const(s))
if a > 0, then
  for each c_b in Const(s), do
    if a≠b and label(c_b)= excl_p.c_y, then
      verif ← "-"
    break
v_excl_p ← verifProperty(s, excl_p, verif)
return v_excl_p
```

This algorithm browses the set Const(p) to look for the constituents c_x and c_y of t in an exclusion relation and mark it as satisfied if they are both not found or only one of them is found in p. So, in all cases there would be a verified property in return. We have used the method search() to search only for the position of c_x in the categories of Const(p).

7   Experimentation and Evaluation

We have tested our method on the ATB corpus (ATB2v1.3 version), which includes 501 stories from the Ummah Arabic News Text. The latter contains 144,199 words before the clitic-separation. As we have already mentioned in the previous section, we need to have the ATB in a "xml" format, as input of the property verification task. Having such format, we had not exempted from preparing a simple version in the fitting step due to the handling difficulty of the available version. We have used more specifically the "penntree" format (the vowelized version) to convert it into "xml". We have induced the GP from only the half of the ATB in order to make the study corpus different to the test one. In what follows, we will present some of the obtained results after citing above the meanings used in the headers of the tables due to lack of space (Table 2):

Table 2. The header meanings.

#	Frequency	#C	Number of possible constituents	XP	Phrase
$\sum$	Total	#R	Number of production rules	#P	Number of properties

First, we have found that the ATB is composed of 841 grammatical categories of which 348 are syntactic (put in 21 phrase groups). Table 3 shows the distribution of the ATB phrase by frequency, the possible constituent number and the production rule number.

From Table 3, we may notice that the most frequent phrase in the ATB is the Nominal Phrase (NP). It even contains large numbers of possible constituents and production rules (equals to 1/3 of all rules). This dominance does not excessively influence the distribution of the properties. According to Table 4 showing information

Table 3. The Distribution of the phrases in the ATB.

XP	#	# C	# R	XP	#	# C	# R	XP	#	# C	# R
NP	110748	299	4824	PRT	2292	13	14	PRN	65	10	20
PP	22100	22	263	ADVP	539	6	68	LST	56	2	2
S	19358	138	1230	NAC	221	18	53	SQ	51	12	26
VP	15947	342	6675	FRAG	178	22	56	CONJP	37	3	2
SBAR	9524	47	380	WHADVP	136	3	34	INTJ	11	1	1
WHNP	4574	3	64	UCP	132	19	88	X	5	4	5
ADJP	3665	88	593	SBARQ	68	19	51	WHPP	3	3	3
								$\sum$		841	14452

Table 4. The distribution of the ATB properties by phrase.

XP	Uniqueness		Linearity		Requirement		Exclusion		$\sum$	
	#P	#	#P	#	#P	#	#P	#	#P	#
NP	22	21686	50	1237	6	125	404	44742192	483	44875988
PP	12	1840	17	1795	15	1947	106	2342600	151	2370282
S	11	539	45	3807	12	89	99	1916442	168	1940235
VP	19	16694	104	26536	16	1493	196	3125612	336	3186282
SBAR	13	5238	30	9053	11	4913	129	1228596	184	1257324
WHNP	5	4574	2	4	2	4	8	36592	18	45751
ADJP	10	88	17	68	12	88	87	318855	127	322764
PRT	12	2290	0	0	0	0	66	151272	79	155854
ADVP	6	559	3	21	4	22	12	6468	26	7609
NAC	9	426	9	189	10	204	33	7293	62	8333
$\sum$	162	54721	346	43096	139	9279	1288	53891401	1958	53998567

about the 10 most frequent phrases (in the lowest granularity level), NP has the greatest numbers of uniqueness (40%) and exclusion (83%) properties. However, the leader in this distribution becomes the VP (Verbal Phrase) for the linearity properties (62%) and the SBAR (subordinate clause) for of the requirement ones (53%). We may also note that dealing with such high frequencies of uniqueness properties for most phrases implies the need to have a unique constituent in each Arabic phrase. The linear order is important to the VPs as to the SBARs. For the constituency properties, we have applied them once for each phrase. Their frequency is then equal to the phrase frequency.

Regardless to the given property distribution, we obtained an important and varied implicit information in such Arabic text. We may give some examples: In the Arabic VP, we have the linearity property IV ≺ PP, which requires that the PP (Propositional Phrase) must never precede the IV (Imperfect Verb). Similarly, we have the requirement property ADJ ⇒ NP, which needs the presence of a NP if an ADJ (adjective) exists.

By focusing on the distribution of the property types, it can be seen that the parts of the obligation and the adjacency properties are virtually zero. The obligation ones have only 3 properties (describing the following 3 phrases: LST, INTJ and WHPP) with 70 occurrences. The adjacency ones do not have any properties. This shows that we do not need to have neither mandatory constituent (head) in the most of the phrases nor any condition about a direct order between constituents of the same phrase. This proves the variety of structuration of the phrase rules in Arabic.

For an overview of all the property types, we can observe their high frequency compared to other enriched treebanks (e.g. the FTB) [6]. Indeed, according to Table 5, only the obligation properties are negligible in the ATB. The others vary from tens of thousands to millions. This large number can go greater if we extend our work to the highest granularity level of grammatical categories in the ATB. This reflects the richness and the variety of the structures in Arabic.

Table 5. The distribution of the properties in treebanks.

Treebanks	Uniqueness	Obligation	Linearity	Requirement	Exclusion	$\sum$
ATB	54721	70	43096	9279	53891401	53998567
FTB	38007	32602	27367	11022	89293	198291

In such phrases, it is also possible to know the most frequent types of properties. The following Fig. 4 represents the distribution of the ATB properties by type (only those with comparable values). Thus, it is important to know that many categories can be absent but not repeated in Arabic phrases. This finding is based on the big difference between uniqueness frequencies of properties and obligation ones. It is also clear that we have a great abundance of the exclusion properties versus a virtual absence of the obligation ones. This wide gap needs to be adjusted by describing new interpretations to these types. The adjacency properties however cannot have another conception because it concerns an order in which information is automatically defined.

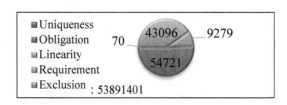

Fig. 4. The distribution of the ATB properties by type.

As already mentioned in the previous section, the verification of the property satisfaction distinguishes those satisfied from those violated. We represent in Table 6 the satisfaction rates of the properties by type of the phrases VP as instance.

Table 6. The satisfaction rates of the VP property types.

Property state	Uniqueness		Linearity		Requirement		Exclusion	
	#	%	#	%	#	%	#	%
Satisfied	15932	99.91	26529	99.99	1491	99.87	3125590	99.99
Violated	15	0.09	7	0.01	2	0.13	22	0.01

According to the obtained results, the number of violated properties is negligible compared to satisfied ones. If the ATB is considered as a large coverage resource, the used GP in the enrichment task inherits also this richness.

The property distribution can be detailed to finer levels by describing individually each property of such phrase and type. This description let us to know which property is more important. We may not have a precise information about the importance of the properties when we are restricted on defining the property distribution by type. Thus, this distribution in a same type can be no homogenous. We present in Fig. 5, for example, the distribution of the VP properties to determine the most frequent ones. The abscise axes of the shown schemes are the property indexes and the ordinate ones are their frequencies. We may consider in that case that the most frequent properties have the highest weight (occurrence number). So, it can be admitted as relevant information. We fix that a strong property have at least 1500 occurrences for the uniqueness and the linearity properties and 500 occurrences for the requirement ones. Table 7 gives us the strong properties of the VP.

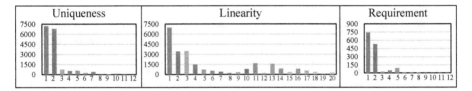

Fig. 5. The distribution of the VP properties.

Table 7. The strong properties of the VP.

	Uniqueness	Linearity	Requirement
Index	1, 2	1, 2; 3; 11, 13	1, 2
Property	PV, IV	PV ≺ {NP, PP}; IV ≺ PP; PRT ≺ {NP, IV}	NOUN ⇒ NP, ADJ ⇒ NP

Using these results, we can automatically measure the property weights in such construction. This information can be included in the GP to ease the parsing process. Indeed, we can check the satisfaction only of the strong properties. The others can be relaxed. This information is also useful to evaluate the difficulty of the processing in cognitive systems since the violation of a strong property implies the most important difficulty.

8   Conclusion and Perspectives

We proposed in the present paper a method of treebank enrichment based on the GP formalism. According to the enrichment process, the efficiency of this formalism is proved both in terms of direct access on constraints as well as in terms of simple and local representation. We started with a linguistic study on Arabic syntactic structures. From these structures, we induced some syntactic properties describing the syntactic categories of these structures. This study verifies also the validity (satisfaction) of the deduced properties. The enrichment method is applied on a large corpus (the ATB) and gave us good results and various properties of different types.

As perspectives, in order to offer a very precise representation of the syntactic information in the ATB, we can enrich or improve the relation set presented in the induced GP. For example, proposing an interpretation of the dependency property or modifying the description of the obligation and exclusion properties. In future works, we can optimize our enrichment method by integrating several control mechanisms into the determination of the syntactic categories and into the verification of their properties. We can go further by applying our enrichment method to other annotated corpora obtained from existing parsers.

References

1. Abdul-Mageed, M., Diab, M.: AWATIF: a multi-genre corpus for modern standard Arabic subjectivity and sentiment analysis. In: Language Resources and Evaluation Conference (LREC 2012), Istanbul, Turkey (2012)
2. Alkuhlani, S., Habash, N.: A corpus for modeling morpho-syntactic agreement in Arabic: gender, number and rationality. In: Association for Computational Linguistics (ACL 2011), Portland, Oregon, USA (2011)
3. Alkuhlani, S., Habash, N., Roth, R.: Automatic morphological enrichment of a morpho-logically underspecified treebank. In: North American Chapter of the Association for Computational Linguistics: Human Language Technologies (HLT-NAACL 2013), pp. 460–470, Atlanta, Georgia, USA (2013)
4. Bensalem, R.B., Elkarwi, M.: Induction d'une grammaire de propriétés à granularité variable à partir du treebank arabe ATB. In: Rencontre des Étudiants Chercheurs en Informatique pour le Traitement Automatique des Langues (RECITAL 2014), pp. 124–135, ATALA, ACL-ontology, Marseille, France (2014)
5. Bahloul, R.B., Elkarwi, M., Haddar, K., Blache, P.: Building an Arabic linguistic resource from a treebank: the case of property grammar. In: Sojka, P., Horák, A., Kopeček, I., Pala, K. (eds.) TSD 2014. LNCS (LNAI), vol. 8655, pp. 240–246. Springer, Heidelberg (2014). doi:10.1007/978-3-319-10816-2_30
6. Blache, P., Rauzy, S.: Hybridization and treebank enrichment with constraint-based representations. In: LREC 2012 - Workshop on Advanced Treebanking, Istanbul, Turkey (2012)
7. Çakıcı, R.: Automatic induction of a CCG grammar for Turkish. In: ACL Student Research Workshop, pp. 73–78, Ann Arbor, Michigan (2005)
8. El-taher, A.I., Abo Bakr, H.M., Zidan, I., Shaalan, K.: An Arabic CCG approach for determining constituent types from Arabic treebank. J. King Saud Univ. Comput. Inf. Sci. 1319–1578 (2014)
9. Habash, N., Rambow O.: Arabic tokenization, part-of-speech tagging and morphological disambiguation in one fell swoop. In: ACL, pp. 573–580, Ann Arbor, Michigan (2005)

10. Hovy, E., Marcus, M., Palmer, M., Ramshaw, L., Weischedel, R.: OntoNotes: the 90% solution. In: North American Chapter of the Association for Computational Linguistics (NAACL 2006), pp. 57–60, USA (2006)
11. Maamouri, M., Bies, A., Buckwalter, T., Mekki, W.: The Penn Arabic treebank: building a large-scale annotated Arabic corpus. In: NEMLAR Conference on Arabic Language Resources and Tools, Cairo, Egypt (2004)
12. Maruyama, H.: Structural disambiguation with constraint propagation. In: ACL 1990 Workshop on Dependency-based Grammars, pp. 31–38. Pittsburgh, Pennsylvania, USA (1990)
13. Müller, H.H.: Annotation of morphology and NP structure in the Copenhagen Dependency Treebanks (CDT). In: International Workshop on Treebanks and Linguistic Theories, pp. 151–162, University of Tartu, Estonia (2010)
14. Oepen, S., Flickinger, D., Toutanova, K., Manning, C.D.: LinGO redwoods - a rich and dynamic treebank for HPSG. In: LREC 2002 - Workshop on Parsing Evaluation, Las Palmas, Spain (2002)
15. Palmer, M., Babko-Malaya, O., Bies, A., Diab, M., Maamouri, M., Mansouri, A., Zaghouani, W.: A pilot Arabic propbank. In: LREC 2008, Marrakech, Morocco (2008)
16. Pollard, C., Sag, I.: Head-driven Phrase Structure Grammars. Chicago University Press, Chicago (1994)
17. Tounsi, L., Attia, M., Van-Genabith, J.: Automatic treebank-based acquisition of Arabic LFG dependency structures. In: The European Chapter of the ACL (EACL) Workshop on Computational Approaches to Semitic Languages, pp. 45–52, Greece (2009)

Driving Innovation in Youth Policies
with Open Data

Domenico Beneventano, Sonia Bergamaschi, Luca Gagliardelli[✉],
and Laura Po

Dipartimento di Ingegneria "Enzo Ferrari",
Università di Modena e Reggio Emilia, Modena, Italy
{domenico.beneventano,sonia.bergamaschi,luca.gagliardelli,
laura.po}@unimore.it

Abstract. In December 2007, thirty activists held a meeting in California to define the concept of open public data. For the first time eight
Open Government Data (OPG) principles were settled; OPG should
be Complete, Primary (reporting data at an high level of granularity),
Timely, Accessible, Machine processable, Non-discriminatory, Non-
proprietary, License-free. Since the inception of the Open Data
philosophy there has been a constant increase in information released
improving the communication channel between public administrations
and their citizens.

Open data offers government, companies and citizens information to
make better decisions. We claim Public Administrations, that are the
main producers and one of the consumers of Open Data, might effectively extract important information by integrating its own data with
open data sources.

This paper reports the activities carried on during a research project
on Open Data for Youth Policies. The project was devoted to explore
the youth situation in the municipalities and provinces of the Emilia
Romagna region (Italy), in particular, to examine data on population,
education and work. We identified interesting data sources both from
the open data community and from the private repositories of local governments related to the Youth Policies. The selected sources have been
integrated and, the result of the integration by means of a useful navigator tool have been shown up. In the end, we published new information
on the web as Linked Open Data. Since the process applied and the tools
used are generic, we trust this paper to be an example and a guide for
new projects that aims to create new knowledge through Open Data.

Keywords: Open government data · Linked Open Data · Youth Policies · Emilia Romagna Region · Municipality of Modena · Data integration · Data visualization

1 Introduction

The Open Data *philosophy* is based on the idea that certain data can be freely
used, modified, and shared by any citizen and shared by anyone for any purpose.

© Springer International Publishing AG 2016
A. Fred et al. (Eds.): IC3K 2015, CCIS 631, pp. 324–344, 2016.
DOI: 10.1007/978-3-319-52758-1_18

This data must be available under an open licence and provided in a convenient and modifiable form that is machine readable. In recent years, the number of international conferences on Open Government and Open Data has been increasing. Showing that the attention of governments to the transparency and the interest of the scientific and economic communities to exploit these data is raising.

Many European countries are developing policies to release their data as Open (Government) Data. OGD has let to several improvements, such as transparency and democratic control, citizen participation, innovation, improved efficiency and effectiveness of government services, impact measurement of policies, the creation of new knowledge from combined data sources and patterns in large data volumes.

The powerful of open data is that many areas can take advantage from their value. And also many different groups of people and organizations can benefit from the availability of open data, including government itself.

This paper describes the activities performed during a one-year research project called "Open Linked Data of the Youth Observatory of the Emilia-Romagna Region", funded by the Municipality of Modena. The main dimensions of analysis of the project concern all the municipalities and provinces of the Emilia Romagna region. The project goals were to identify interesting data sources both from the open data community and from the private repositories of local governments of Emilia Romagna region related to the Youth Policies. In particular, to the topics of youth population, education and employment. The project consists of a integration process aiming at merging together the different information and a visualization and publication process to show up the result of the integration by means of a useful navigator tool and to publish new information as Linked Open Data. The key partners in this project were: the Municipality of Modena, the Department of Culture, Youth and Policies for the Legality of the Emilia Romagna region and the DBGroup[1].

An hight level overview of the project is display below. The remainder of the paper is structured as follows. Section 2 reports the State of the Art of Open Data at national level. Sections 3, 4, 5 and 6 illustrate the set of tools we used to reach the project goals. Finally, Sect. 7 sketches the conclusion and the main difficulties faced during the project.

1.1 Overview of the Project

The project has been developed in four phases (see Fig. 1). The first phase was devoted to a deep and wide analysis of the available data sources (local, regional, national and international) in order to individuate the most relevant ones among the open data a and the proprietary data (such phase will be discussed in Sect. 3). The second phase made use of the open source data integration system, MOMIS[2], to integrate the selected data sources in virtual global views (such phase will be discussed in Sect. 4). The third phase provided an easy-to-use dashboard

---

[1] http://dbgroup.unimo.it/.

[2] http://www.datariver.it/data-integration/momis/.

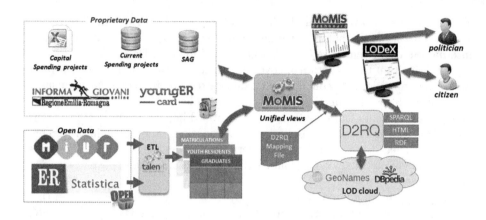

Fig. 1. The high-level workflow of the project.

(i.e. MOMIS dashboard) to visualize the information emerging from aggregated data (such phase will be discussed in Sect. 5). Finally, the fourth phase aimed to make the resulting value-added information, public and searchable on the Web as Linked Open Data (such phase will be discussed in Sect. 6). The third and the fourth phases are not consecutive, they can both be carried out independently on the results of integration process. The results of this projects serve a wide range of users including politicians, who curate government data, and informed citizens who access analytical results from government data.

2   Open Data in Italy

In 2010, the European Commission issued the Digital Agenda for Europe. One of the action points on the Digital Agenda is to "Open up public data resources for re-use". In December 2011, this was made more specific in the form of an "Open Data Strategy". On June 2013, the leaders of the G8 signed an agreement committing to advance open data in their respective countries.

In Italy, in recent years, the open data approach has been adopted by a growing number of Public Administrations and, in some cases, an additional effort has been made to supply Linked Open Data (LOD). At the beginning of June 2015, the open data portal of the italian public administration[3], which since 2011 hosts a catalog of open data published by ministries, regions and local authorities, has been revamped in order to promote transparency, accountability, diffusion and reuse of open data.

In the 2015, the Open Data Barometer project [1] analysed 86 countries across the world that have developed some form of Open Government Data initiative. Italy was ranked in 22nd position, in the same cluster of Spain, Czech Republic, Portugal, Greece, Ireland, and Poland. Common across all these countries is a

---

[3] www.dati.gov.it.

great level of civil society readiness, but a lower level of perceived social impact from open data. In these contexts, OGD initiatives are established, but are progressing relatively slowly when compared to the rest of Europe (UK, Sweden, Austria, France etc.). A picture of the Italian situation showing the impact, the readiness, and the open data is reported in Fig. 2.

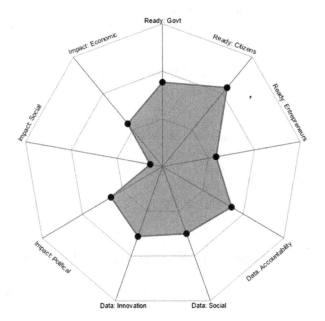

Fig. 2. The Italian situation according to the Open Data Barometer.

Among all the open datasets, the Emilia Romagna Region is present with 642 datasets[4] and 14 open data portals. The open data portals of the Emilia Romagna region include one regional[5], eight municipalities (Piacenza, Bologna, Anzola Emilia, Ferrara, Ravenna, Faenza, Cesena, Rimini), three provinces (Parma, Bologna, Forlì-Cesena) and the public transport company (Passenger Transport Emilia-Romagna)[6].

In Italy, there are very few feedback collected about the consumption of Open Data. The Open Data portals provide indicators such as the number of downloads, but an important indicator should be to figure out how many downloads have generated value. Open Data 200[7] is the first systematic study of Italian companies that use open data in their activities to generate products

---

[4] information available on dati.gov.it at 31st July 2015.

[5] http://dati.emilia-romagna.it/.

[6] The Emilia Romagna Region has nine provinces (Piacenza, Parma, Reggio Nell'Emilia, Modena, Bologna, Ferrara, Forlì-Cesena, Rimini, Ravenna) and 340 municipalities.

[7] http://www.opendata500.com/it/.

and services and create social and economic value. The project is developed by Govlab - New York University in collaboration with Fondazione Bruno Kessler, research institute based in Trento. The project aims at providing a sound basis for evaluating the impact of open data in Italy. The project is still ongoing and the results of the conducted survey are not published yet.

The recent years have seen a great advancement in open data initiatives in Italy at different levels, such as OpenCoesione[8] and Open Bilanci[9] that show the public administration expenses. In these websites the user can view fixed indicators, chosen by the application's and it is not possible for the user to perform customized searches.

Italy is one of the most advanced European countries in terms of smart city initiatives. A report by the European Parliament [2] placed Italy in its top tier of countries ranked by the number of smart city initiatives - over 75% of cities in Italy with a population of over 100,000 have at least one smart initiative; 106 Italian Cities with at least one smart project or initiative [3]. The bricks for building smart cities are the Open Data. Therefore, these public institutions are very active in the distribution of Open Data and boost their use in order to create innovative applications, tools, platforms for different sectors such as e-Government, Health and Wellbeing, Energy Efficiency, Integrated Tourism Services and Mobility, Pollution, Environmental protection.

3   Data Source Selection, Extraction and Cleaning

In the first phase of the project, we focused on the selection of the most relevant data sources w.r.t the main dimensions of analysis of the project, i.e. municipalities and provinces of the Emilia Romagna Region. We have used and analyzed both proprietary and open data sources, the proprietary data sources was provided by the Emilia Romagna region, the sources were:

- **A Database of Current Spending Projects:** it contains information about the fundings provided for projects on youth populations actuated in the different provinces of the region;
- **An Excel File of Capital Spending Projects:** it contains information about funding provided for long period investments (e.g. build structures, equipments);
- **A Database of Social Centers (SAG - "Spazi di Aggregazione Giovanile"):** it lists all the places where young people get together or places where recreational activities for youth are organized;
- **Three Excel Files Related to the Youth Information Project ("Progetto Informagiovani"):** The information centers provide data at local, national and international level on different topics of interest for young people aged 13 to 35 years. The main areas covered are relate to study, work, continuing education, travel and holidays, study and work abroad, leisure,

---

[8] http://www.opencoesione.gov.it/.
[9] http://www.openbilanci.it.

social life and health. These files contain statistics about the number of visitors of the website[10], the number of points on the region and the number of editors;

- **An Excel File About the Young ER Card Project:** YoungERcard[11] is the new card designed by the Emilia Romagna region for young people aged between 14 and 29 residents, students or workers in Emilia Romagna. The card is distributed for free and reserve holders a series of facilities for the enjoyment of cultural and sportive events and discounts at various shops. The file reports information related to the Young Emilia Romagna Card distribution and extensiveness in the region.

Regarding open data sources, several sites and portals that publish information regarding the youth situation have been investigated. Our work of selecting sources and extracting relevant data was driven by few relevant questions arising by the local and regional politicians: "how many funding was provided in any province compared to the number of youth residents (aged 15–34)?", "which is the higher education rate in each province?". These questions have to be answered in an historical perspective, thus monitoring the data of each province over the years. We analysed seven providers of open data (a detailed description of the open data sites is reported in [4]): the *Italian National Institute of Statistics* (ISTAT)[12], the *Alma Graduate* website[13] that yields information on the profile of the graduates and their employment status; the *"Il Mulino" Youth Report* that contains data derived from the survey conducted on a sample of 9000 youth aged from 18 to 29; the *"Oriente" Database of Training Courses*[14] that collects information of all training courses financed or authorized by the Emilia - Romagna Region; the *Emilia Romagna Labor Statistics*[15] that presents some data on the labor market of the Emilia - Romagna region; the *Emilia Romagna Statistical Service*[16] that provides information about several thematics: population, transport, sports, productive sectors, etc. and the *National Student Register*[17], provided by the Italian Ministry of Education, University and Research, that shows the number of matriculates and graduates during the years.

Among the available sources, we selected the ones that supply the number of young residents in each province and their level of education. For our project, we focused on the "population" thematics of the Emilia Romagna Statistical Service and the education data of the MIUR National Student Register.

---

[10] informagiovanionline.it.
[11] https://www.youngercard.it/.
[12] www.istat.it.
[13] www.almalaurea.it/universita/statistiche.
[14] http://orienter.regione.emilia-romagna.it.
[15] http://formazionelavoro.regione.emilia-romagna.it/analisi-sul-mercato-del-lavoro/ approfondimenti/statistiche-sul-lavoro-in-emilia-romagna.
[16] http://statistica.regione.emilia-romagna.it/servizi-online/.
[17] http://anagrafe.miur.it.

The population website of the Emilia Romagna Statistical Service[18] contains data organized on the basis of different measures (age, sex, year of the survey, place ...). The MIUR National Student Register contains information on the number of students and graduates in the various degree courses of Italian universities.

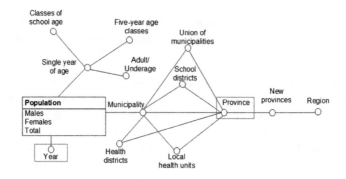

Fig. 3. Conceptual schema of the Regional Statistical Service.

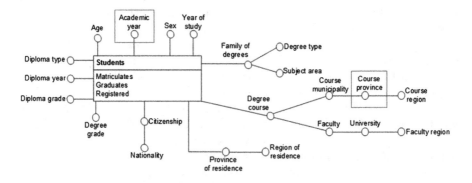

Fig. 4. Conceptual schema of the National Student Register.

The data that was required had to be aggregated by province and year, therefore a preliminary study on which dimensions were provided on these data sources was needed. Figures 3 and 4 show the conceptual schema of the two sources. As you can see, the sources have different dimensions and different level of granularity, thus we needed to select appropriate dimensions to allow a successful integration. From the Emilia Romagna Statistical Service, we extracted the number of male, female and total population with respect to the following dimensions: Year, and Province. From the National Student Register, we

---

[18] http://statistica.regione.emilia-romagna.it/servizi-online/statistica-self-service/pop olazione/popolazione-per-eta-e-sesso.

extracted the number of matriculates, graduates and registered with respect to the following dimensions: Academic Year, and Course Province.

The data available at the web sites was referred to a single year, thus the different files containing annual data were then processed through an ETL tool, the Talend[19] tool, and unified in a single source containing all the years between 2006–2015. During the ETL process, we also applied conversion functions to solve some possible conflicts. Since the academic year is described by a couple of years (e.g. 2012/2013), we applied a conversion function to transform it to a single year. Our decision was to take the first part of the academic year for represent the enrolling year (usually people enroll in the first part of the academic year) and the second part to represent the year of graduation (usually students get their degree in the second part of the academic year). Some naming conflicts occur on the province names. The main problems were found on the provinces of Reggio Nell'Emilia and Forlì-Cesena that were written in different ways, for example "Reggio Emilia", "Reggio-Emilia", "Forlì e Cesena", "Forli-Cesena". For solving this problem, we choose as golden standard the names used on the Regional Statistical Service, namely "Forlì-Cesena" and "Reggio Nell'Emilia", and we converted any other forms to the gold standard.

The data were extracted and saved in a set of MySQL tables. From the Regional Statistical Service, we import data into a table with province, year, number of young residents (aged 15–34 years) attributes; from the National Student Register we import data into two distinct tables with province, year, number of enrolled/graduated students.

4   Data Integration

The data integration process was performed with MOMIS, a data integration system developed by the DBGroup [5,6] of the University of Modena and Reggio Emilia and now distributed by the DATARIVER spin-off[20] as an Open Source tool [7]. In the following, we briefly present the MOMIS's architecture. More information on the MOMIS system and some integration examples can be found on the DATARIVER website[21].

Given a set of heterogeneous and distributed data sources MOMIS generates in a semi-automatic way a unified schema called Global Schema (GS), that allows users to formulate queries on that schema like they are querying a single database. The system performs the integration task, by following an Global-As-View (GAV) approach for creating the mappings between the GS and local schemas of the integrated data sources. MOMIS uses a virtual approach for achieving an integration that preserves the autonomy and security of the local sources.

The integration process is composed of four main phases:

---

[19] https://www.talend.com/.
[20] http://www.datariver.it.
[21] http://www.datariver.it/data-integration/momis/tutorials/.

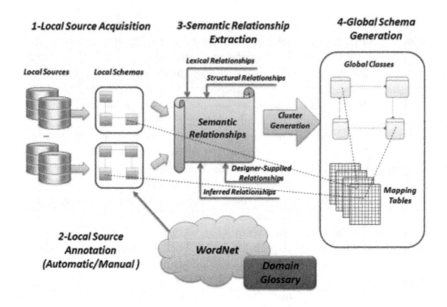

Fig. 5. The MOMIS data integration process.

1. **Local Schema Acquisition:** (Fig. 5-1) the extraction of Local Source Schemas is performed by wrappers that automatically extract the schema of each local source (in this case relational databases or Excel files) and convert it into the common language ODL$_{I3}$.
2. **Local Sources Annotation:** (Fig. 5-2) the designer can perform automatic annotation and/or can manually select a base form and the appropriate Word-Net meaning(s) (i.e. synset(s)) for each term. Moreover, the designer can extend Word-Net with Domain Glossaries. Annotation consists in associating to each class and attribute name and one or more meanings w.r.t. a common lexical reference, i.e. domain glossaries/WordNet[22] [8].
3. **Semantic Relationships Extraction:** (Fig. 5-3) starting from the annotated local schemas, MOMIS derives a set of intra and inter-schema semantic relationships in the form of: synonyms (SYN), broader terms/narrower terms (BT/NT) and related terms (RT) relationships. The set of semantic relationships is incrementally built by adding: structural relationships (deriving from the structure of each schema), lexical relationships (deriving from the element annotations, by exploiting the WordNet semantic network), designer-supplied relationships (representing specific domain knowledge) and inferred relationship (deriving from Description Logics equivalence and subsumption computation).

---

[22] WordNet is a thesaurus for the English language, that groups terms (called lemmas in the WordNet terminology) into sets of synonyms called synsets, provides short definitions (called gloss), and connects the synsets through a wide network of semantic relationships.

4. **GS Generation:** (Fig. 5-4) starting from the discovered semantic relationships and the local sources schemas, MOMIS generates a GS consisting of a set of global classes, plus a corresponding set of Mapping Tables which contain the GAV mappings connecting the global attributes of each global class with the local source attributes. The GS generation is a process where classes describing the same or semantically related concepts in different sources are identified and clustered into the same global class. The designer may interactively refine and complete the proposed integration result through the GUI provided by the Global Schema Designer tool.

At the beginning of the second phase, we had the following data sources:

S1. **Current Spending Projects:** information about funding provided for projects on youth population actuated in the different provinces of the region;

S2. **Capital Spending Projects:** information about funding provided for long period investments (e.g. build structures) organized for each province and year;

S3. **Number of Youth Residents:** number of youth residents in each province of the region Emilia Romagna and for each year;

S4. **Number of Youth Information Centers:** numbers of points of the Youth Information Centers located in each province and year.

S5. **Number of Editors in the Youth Information Centers:** number of member in the editorial staff operating in the Youth Information Centers for each province and year.

S6. **Number of Web Site Visitors Informagiovanionline:** information about the number of the visitors to the website Informagiovanionline for each province and year.

S7. **Young ER Card:** data on the Young ER Card for each province and year.

S8. **Number of Graduates Supply by MIUR:** information about the number of graduated for each province and year.

S9. **Number of Matriculations at the University Supply by MIUR:** information about the number of matriculations for each province and year.

All these sources have two dimensions in common: the year and the province. We conducted two analysis for evaluating the data coverage over the two dimensions. We found no lack of data over the provinces, meaning that the data are spread in all the provinces. Instead, focusing on the entire interval of years (2006–2014), we found a relevant number of missing data, that is reported in Fig. 6.

Since MOMIS allows the creation of more Global Schemas, to avoid a proliferation of null values in the integration result, we created three different Global Schemas:

- **GS-Projects:** this GS considers only S1, S2, S3, S8, S9.
- **GS-Global:** this GS contains the data from all the nine sources;
- **GS-Projects-Informagiovani:** this GS excludes the Young ER Card from the integration;

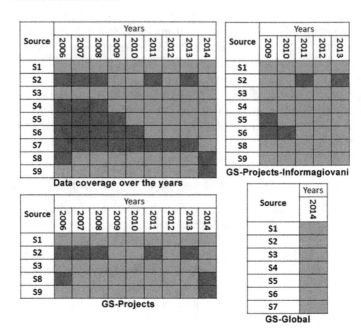

Fig. 6. Data coverage over the years. The red cells represent missing data, the green represent data are present. You can also see the data of each Global Schema. (Color figure online)

As shown in Fig. 6, only few sources includes data over the entire period (2006–2014). In this case, we have taken all data from five of the sources, in order to create charts on the entire period that show the correlation between founded projects, population and education. The GS-Global is the only that includes data on the Young ER Card, it contains all the nine sources, as showed in Fig. 6. Figure 6 highlights that the Young ER Card (S7) have data only for the 2014. Therefore, in the GS-Projects-Informagiovani we included all sources except S7, as showed in Fig. 6. On this integration, we were interested to show the correlations between the data of the Informagiovani project and the data of education and founded projects.

A user can pose queries on the Global Schema by using the Query Manager tool. MOMIS provides another query capability, the Query Manager Web Service, which permits to easily integrate MOMIS with other applications (e.g. Business Intelligence solutions) such as the MOMIS Dashboard to easily visualize synthetic information.

5   Data Visualization

The advantages of using visualization tools to explore the correlation among data are numerous [9]. The analysts do not have to learn any sophisticated method to be able to interpret the resulting graphs. Effective visualizations help users in analyzing and reasoning about data and evidence and make complex data more accessible, understandable and usable. The goal for a good data visualization tool is to unify data mining algorithms and visual user interfaces, to bring the user in direct contact with the data and to make knowledge discovery available to everyone [10].

The integration results of the project have been shown by the MOMIS Dashboard, a web application developed by Datariver20. The MOMIS Dashboard is a interactive visualization tool that offers several views on a set of data. It makes easier to compare data and capture useful information. It allows to filter the data and visualize the results through different charts. In particular, it is possible to display line charts (for showing trends), bar charts, pie charts, bubble charts on a Google Maps, or show the data in a tabular view.

In our project, we designed several charts in order to supply answer to the questions arose from the politicians of the Emilia Romagna region about the Youth Situation. The first questions were focused on the funding compared to the number of youths: "Are the funding provided for each province proportional to the number of youth?", "How do they evolve over time?"

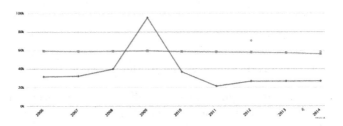

Fig. 7. Fundings and youth population trends in the province of Piacenza from 2006 to 2014. (Color figure online)

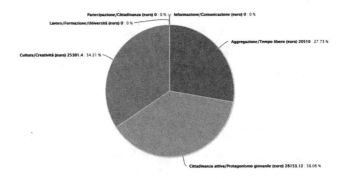

Fig. 8. Funded projects/areas in the province of Modena on 2009.

To visualize the answers of these questions, a bar chart and a bubble chart on Google Maps were created. These charts are activated when the user filters the data by selecting one or more provinces and a single year. Moreover, we created a line chart that is activated in case the user selects one or more provinces and more than one year; these charts show a comparison between the youth population and the investments for each province.

Figure 9 shows the bar chart for the 2014 year, here you see in dark green the fundings for the capital spending projects (in euro), in red the fundings for the current spending projects (in euro) and in light green the number of youth residents.

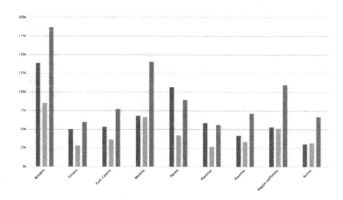

Fig. 9. A bar chart comparing fundings and youth population. (Color figure online)

The same result can be shown with a different visualization in a bubble chart on a map (Fig. 10). In this case, the number of youth is in light green, the fundings for the capital spending projects are displayed in red and the funding for the current spending projects in blue. The map can be zoomed in and out.

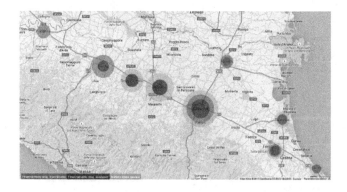

Fig. 10. A map comparing fundings and youth population. (Color figure online)

It is also possible to show over the time how many fundings were provided in one or more provinces in the years, and compare them to the number of youth.

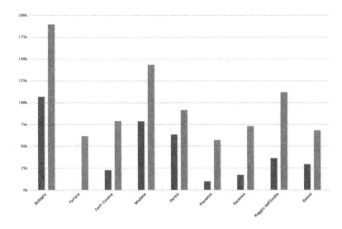

Fig. 11. Number of accesses at the informagiovani.it website (green) compared to the number of youth residents (red) in each province. (Color figure online)

For example, Fig. 7 shows for the province of Piacenza in the period 2006–2014 in light green the number of youth resident, in dark green the founding for capital spending projects (that were provided only in 2012 and 2014), and in red the founding for the current spending projects.

The fundings for projects are assigned on different areas: education, culture, free time, etc. The politicians were interested to see how the fundings were distributed in the different areas. Thus, we devised a pie chart that is activated when the user selects a single province and a single year, that shows the funding provided divided per area. Figure 8 shows the areas that have been funded for the province of Modena on 2009.

Another question that later was arisen is: "How many youth from each province visits the website informagiovani.it compared to the number of residents?". Using statistical data from Google Analytics that was provided by the region, we were able to create the line chart showed in Fig. 11. This chart reports, for the 2013 year, the number of accesses from each province in green, and the number of youth residents in red. As it can be noted, the number of accesses is really low compared to the number of youth, so maybe it is possible to conclude that a more intensive advertisement might increment the visits of the website.

For comparison purposes, we evaluate Tableau[23] and Qlik[24] (cited as a leaders in the Gartner's Magic Quadrant for Business Intelligence and Analytics Platforms [11]). The Tableau Desktop, i.e. the tool to produce visualizations, is very intuitive and the creation of new charts is very quick. The same effectiveness appears in the Tableau reader, i.e. the tool for anyone that consumes the visualizations. Instead, the Qlik platform requires more initial knowledge to design the first chart. Tableau shows to the user different tabs that contain one or more charts. In each tab, some filters that allow the user to change the data

---

[23] http://www.tableau.com.
[24] http://www.qlik.com.

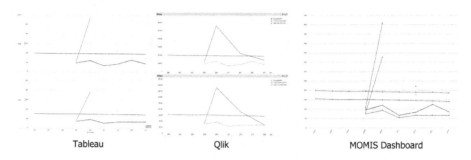

Tableau   Qlik   MOMIS Dashboard

Fig. 12. A comparison of line charts generated with different tools: Tableau, Qlik, MOMIS dashboard. The charts show the current spending projects, capital spending projects and the number of youth residents in each province during the years.

represented in the charts can be defined. The mode of operation of Tableau is opposite to the MOMIS dashboard. In the MOMIS Dashboard, each tab contains only one chart, and the charts are activated after the filters setting (e.g. the selection of more than one year enables the line chart). Both Tableau and Qlik are able to generate the same charts as the one available in the MOMIS dashboard. In the scenario of our application, we observed some limitations of the tools: the maps in Tableau can not show more bubbles representing more measures (differently from the MOMIS Dashboard see Fig. 10); the pie chart can be generated on a single measure, thus the one in Fig. 8 can not be generated with Tableau or Qlik; the line charts can not contain series belonging to different dimensions (in this case, the charts are rendered separately, as shown in Fig. 12).

5.1   MOMIS Dashboard Designer

The MOMIS Dashboard Designer helps the user to create his own charts trough a simple interface, once the charts are created in the Designer they can be showed in the Dashboard. We show an example of use starting from the question "Are the funding provided for each province proportional to the number of youth?". The user that want to answer this question for a specific year need to create a bar chart that shows for each province the number of youth population and the amount of provided fundings. Figure 13 shows the designer interface, on the left side are reported the tables and views of the data source (1), if we select one table/view, the attributes are shown (2). In the right side there are the options that allows the user to create charts and filters (3), and the bottom side shows the created elements (4).

The user start by selecting the appropriate type of charts and the designer shows to the user X/Y fields that have to be filled to create the chart. The user can drag and drop the attributes from the left side, in the example, we selected the number of youth and the fundings. Then, the Designer automatically create the bar chart showed in Fig. 13-5. The Designer also check the number of returned data to see if the generated query is correct for the type of chart. In this particular example, the query is not correct, because, for the bar chart, for each element on the X axis the query have to return zero or one value. This

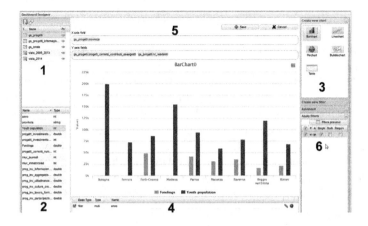

Fig. 13. The MOMIS Dashboard Designer.

probably means that the user have to add a filter to limit the data in the query, so the Designer shows a warning message that suggest to add a filter.

In this example, we wanted to show the data of a single year, so we need to add a filter on the time dimension.

On the right side, using the panel called "Add filters" the user can add a filter to the chart. On this panel there are quick filters (based on the dimensions of the table) that are proposed to the user, so just selecting the attribute "year" the Designer automatically apply a filter on it.

When the filter is applied the warning message disappears and the data are filtered for a single year, the user can also show the value applied on the filter by clicking on the button "Filters preview", as you can see in Fig. 13-6.

6   Linked Open Data Publication and Exploration

In this section we describe the fourth phase of the process, aimed to make the global view containing all the integrated data public and searchable on the Web as Linked Open Data. This phase is composed by two main activities: (1) Linked Open Data Publication: to publish the integrated data as Linked Open Data using D2R Server [12] and map the database schema to RDF using the D2RQ Mapping Language; (2) LOD Exploration and Querying: to explore and query the dataset using the LODeX tool [13,14].

6.1   Linked Open Data Publication

In order to publish in the LOD cloud the project results, we needed a tool for mapping a relational source in RDF. The W3C RDB2RDF Incubator Group [15] had the mission to examine and classify existing approaches to mapping relational data into an RDF source. The tools for automatic mapping generation define a set of mappings between RDB and RDF namely: an RDB record is a RDF node, the column name of an RDB table is a RDF predicate and an RDB table cell is a value. Among these tools, we selected D2RQ [12] as it allows users to define customized mappings.

The tool allows to access relational databases as virtual, read-only RDF graphs, therefore it avoids the replication of information into an RDF store. We used the D2RQ Platform[25] for publishing the dataset in RDF. D2RQ provides a SPARQL access to the content of the database as Linked Data over the Web, and several other opportunities like RDF dumps, API calls, HTML views. The declarative D2RQ mapping language is used to define a set of mappings between the database schema and the RDFS vocabulary or OWL ontology.

Another key aspect in publication of LOD is the identification of vocabularies that can be used to describe the dataset. In our case, we recognized two potential vocabularies: GeoNames and DBpedia. An example of D2RQ mapping to an external vocabulary is reported below:

province	dbpedia	geonames
Bologna	http://dbpedia.org/resource/Province_of_Bologna	http://sws.geonames.org/3181927/
Ferrara	http://dbpedia.org/resource/Province_of_Ferrara	http://sws.geonames.org/3177088/
Forlì-Cesena	http://dbpedia.org/resource/Province_of_Forlì-Cesena	http://sws.geonames.org/3176745/
Modena	http://dbpedia.org/resource/Province_of_Modena	http://sws.geonames.org/3173330/
Parma	http://dbpedia.org/resource/Province_of_Parma	http://sws.geonames.org/3171456/
Piacenza	http://dbpedia.org/resource/Province_of_Piacenza	http://sws.geonames.org/3171057/
Ravenna	http://dbpedia.org/resource/Province_of_Ravenna	http://sws.geonames.org/3169560/
Reggio nell'Emilia	http://dbpedia.org/resource/Province_of_Reggio_Emilia	http://sws.geonames.org/3169524/
Rimini	http://dbpedia.org/resource/Province_of_Rimini	http://sws.geonames.org/6457404/

Fig. 14. References to instances of external sources.

```
map:Statistics_anno a d2rq:PropertyBridge;
    d2rq:belongsToClassMap map:Statistics;
    d2rq:property dbpedia-owl:Year;
    d2rq:propertyDefinitionLabel"anno";
    d2rq:column"gs_totale.anno";
    d2rq:datatype xsd:integer; .
```

Here the column *gs_totale.anno* is mapped to the property *dbpedia-owl:Year;*.

We also use of the D2RQ mappings to link instances of the global view (GS-Global) with instances of the local sources, as in the following example.

```
map:Statistics_informagiovani
    a d2rq:PropertyBridge;
    d2rq:belongsToClassMap map:Statistics;
    d2rq:property vocab:Informagiovani;
    d2rq:refersToClassMap map:Informagiovani;
    d2rq:join"gs_totale.provincia =
                informagiovani.provincia";
    d2rq:join"gs_totale.anno =
        informagiovani.anno"; .
```

---

[25] http://d2rq.org.

A specific procedure is needed to connect the instances of our source with instances of other sources in the LOD cloud. In this case, we wanted to add geographical information to our dataset, thus we linked to the Emilia Romagnas provinces defined in DBpedia and in GeoNames. We create a table (see Fig. 14) containing references to instances of external sources and we have included this table in our dataset. By using these data, each time we refer to a province in this table, we link the instance to the specific province in DBpedia and in GeoNames. Once the dataset is selected and the mappings defined, D2RQ automatically creates a SPARQL access to the LOD source.

6.2 LOD Exploration and Querying

Once a LOD dataset is available, a tool to navigate, explore and query it is necessary. We exploited LODeX [13], a tool able to provide a summary of a LOD source starting from scratch, thus supporting users in exploring and understanding the contents of a dataset. Moreover, LODEX provides a visual query interface [14] to easily compose queries, that are automatically translated in Sparql and executed on a LOD source.

By using tools for browsing and querying a LOD source, we can explore a graphical representation of the source and exploit a convenient visual query panel to extract information from the dataset. With LODeX, the user can take advantage of the Schema Summary that represents the selected source (classes, properties and other statistical information) and by picking graphical elements out of the Schema Summary, he/she can create a visual query. The tool also supports the user in browsing the results and, eventually, refining the query.

The prototype has been evaluated on the SPARQL endpoint of the GS-Global and the visualization is shown in Fig. 15 and is available online at http://dbgroup.unimo.it/lodex2/ok#!/schemaSummary/999.

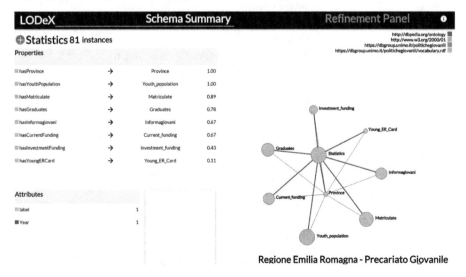

Fig. 15. The visualization of the LOD youth policies dataset.

7   Conclusion

This paper has exemplified how a Public Administrations can benefit from the use of Open Data and can effectively extract new and important information by integrating its own datasets with open data sources.

We consider our work can be helpful for future open government projects aiming to exploit and publish open data. To outline some guidelines, we identified some issues as important factors that may seriously affect the entire process

Table 1. Main issues handled during the project and their criticality (expressed through a three star rating).

Phase	Operation	Criticality
Data selection, extraction and cleaning	Requirements definition of the datasets to be searched for	***
	Discovery of relevant open data sources	**
	Investigation on how open data sources are structured and what information they contain (most of the time the open data sources do not provide a schema nor an high level view of the data)	***
	Discovery of the overlaps among different sources (heterogeneity in the format and in the granularity of the available data might affect this task)	**
	Data cleaning	**
Data integration	Selection of the dimensions for the integration	*
	Availability of data in relation to the chosen dimensions	**
Data visualization	Deciding of what data to show and to compare	**
	Selection of the more suitable chart formats to display the data	**
	Selection of filters to be applied on the data	**
Linked Open Data publication and exploration	Build an ontology for describing the data	*
	Selection of tools to convert data into LOD format	**
	Choose whether it is convenient to publish data in a static rdf file or to convert on the fly data located in a relational database	*
	Looking for similar instances in external datasets (e. Geonames or DBpedia) and building links	**

of exploration, selection, consumption, integration till the publication of open data. Table 1 summarizes the main problems handled during the project and their criticality.

Open data from the Italian Government is an important national resource, serving as fuel for innovation and scientific discovery. It is central to a more efficient, transparent, and collaborative democracy and it might be of benefit to different groups of people, industries, and communities. Citizens could map demographic, income, delinquency rate, and school data to determine which area is the best place to live. Patients could find information on the effectiveness and quality of treatments and waiting time in the hospitals, nursing homes, and physicians, empowering them to make smarter health care choices.

Moreover, publishing Open Data can create a return. As reported in the study of impact of re-use of Public Data Resources [16], over a certain period of time, increased re-use of Open Data will generate new products and services and better return via taxes to government budget.

An important future direction of this project is the possibility to treat data at different level of granularity. Our intention is to add the possibility for the users to link and access data at different levels of the structural granularity (e.g. city, province, region for the geographical dimension, or month, year, decande for the time dimension).

Acknowledgements. This project has been realized thanks to the collaboration of the Department of Culture, Youth and Policies for the Legality of the Emilia Romagna region, the Municipality of Modena (Italy), and the DBGROUP of the University of Modena and Reggio Emilia.

We are thankful to Sergio Ansaloni, Paola Francia, Giulio Guerzoni, Marina Mingozzi, Fabio Poggi, Antonio Volpone for many fruitful discussions on the various aspects of integrating and releasing open government data covered in this paper.

References

1. Davies, T., Sharif, R.M., Alonso, J.M.: Open data barometer global report, (second edition). Technical report, World Wide Web Foundation (2015)
2. Manville, C., Cochrane, G., Cave, J., Millard, J., Pederson, J.K., Thaarup, R.K., Liebe, A., Wissner, M., Massink, R., Kotterink, B.: Mapping smart cities in the eu. Technical report, European Union (2014)
3. Dameri, R.P., Rossignoli, C., Bonomi, S.: How to govern smart cities? Empirical evidences from Italy. In: Proceedings of the 15th European Conference on eGovernment 2015 (ECEG 2015), p. 61. Academic Conferences Limited (2015)
4. Beneventano, D., Bergamaschi, S., Gagliardelli, L., Po, L.: Open data for improving youth policies. In: Proceedings of the International Conference on Knowledge Engineering and Ontology Development (KEOD 2015), Part of IC3K 2015, Lisbon, Portugal, 12–14 November 2015, pp. 118–129 (2015)
5. Beneventano, D., Bergamaschi, S., Guerra, F., Vincini, M.: Synthesizing an integrated ontology. IEEE Internet Comput. **7**, 42–51 (2003)
6. Bergamaschi, S., Castano, S., Vincini, M., Beneventano, D.: Semantic integration of heterogeneous information sources. Data Knowl. Eng. **36**, 215–249 (2001)

7. Bergamaschi, S., Beneventano, D., Corni, A., Kazazi, E., Orsini, M., Po, L., Sorrentino, S.: The open source release of the MOMIS data integration system. In: Proceedings of the Nineteenth Italian Symposium on Advanced Database Systems (SEBD 2011), Maratea, Italy, 26–29 June 2011, pp. 175–186 (2011)

8. Miller, G.A., Beckwith, R., Fellbaum, C., Gross, D., Miller, K.: Wordnet: an on-line lexical database. Int. J. Lexicogr. **3**, 235–244 (1990)

9. Tufte, E.R.: The Visual Display of Quantitative Information. Graphics Press, Cheshire (1986)

10. Fayyad, U.M., Wierse, A., Grinstein, G.G.: Information Visualization in Data Mining and Knowledge Discovery. Morgan Kaufmann, San Francisco (2002)

11. Sallam, R.L., Hostmann, B., Schlegel, K., Tapadinhas, J., Parenteau, J., Oestreich, T.W.: Magic quadrant for business intelligence and analytics platforms. Technical report, Gartner. Gartner Research Note G00270380 (2015)

12. Bizer, C., Cyganiak, R.: D2rq-lessons learned. In: W3C Workshop on RDF Access to Relational Databases, p. 35 (2007)

13. Benedetti, F., Bergamaschi, S., Po, L.: Lodex: a tool for visual querying linked open data. In: Proceedings of the International Semantic Web Conference ISWC 2015 Posters and Demo Track, Bethlehem, PA, USA, 11 October 2015 (2015)

14. Benedetti, F., Bergamaschi, S., Po, L.: Visual querying LOD sources with lodex. In: Proceedings of the 8th International Conference on Knowledge Capture, K-CAP 2015, Palisades, NY, USA, 7–10 October 2015, pp. 12:1–12:8 (2015)

15. Sahoo, S.S., Halb, W., Hellmann, S., Idehen, K., Thibodeau Jr., T., Auer, S., Sequeda, J., Ezzat, A.: A survey of current approaches for mapping of relational databases to RDF. Technical report, W3C RDB2RDF Incubator Group Report (2009)

16. Carrara, W., Chan, W.S., Fischer, S., van Steenbergen, E.: Creating value through open data. Technical report, European Commission (2015)

OntologyLine: A New Framework for Learning Non-taxonomic Relations of Domain Ontology

Omar El idrissi esserhrouchni[1(✉)], Bouchra Frikh[2],
and Brahim Ouhbi[1]

[1] LM2I Lab ENSAM, Moulay Ismaïl University,
Marjane II, B.P. 4024, Meknès, Morocco
omar.elid@gmail.com, ouhbib@yahoo.co.uk
[2] LTTI Lab ESTF, Sidi Mohamed Ben Abdellah University,
B.P. 2427, Route d'imouzer, Fès, Morocco
bfrikh@yahoo.com

Abstract. Domain Ontology learning has been introduced as a technology that aims at reducing the bottleneck of knowledge acquisition in the construction of domain ontologies. However, the discovery and the labelling of non-taxonomic relations have been identified as one of the most difficult problems in this learning process. In this paper, we propose OntologyLine, a new system for discovering non-taxonomic relations and building domain ontology from scratch. The proposed system is based on adapting Open Information Extraction algorithms to extract and label relations between domain concepts. OntologyLine was tested in two different domains: the financial and cancer domains. It was evaluated against gold standard ontology and was compared to state-of-the-art ontology learning algorithm. The experimental results show that OntologyLine is more effective for acquiring non-taxonomic relations and gives better results in terms of precision, recall and F-measure.

Keywords: Ontology learning · Cancer ontology · Financial ontology · Non-taxonomic relations extraction · Knowledge acquisition · Open information extraction

1 Introduction

In computer science, ontologies are defined as a formal, explicit specification of a shared conceptualization [1]. They provide several potential benefits in representing and processing knowledge, including the sharing of common knowledge of information among human and software agents [1], and the reuse of domain knowledge for a variety of applications [2]. Nowadays, ontologies are widely used in many areas such as finance [3], e-business [4], medicine [5] and biomedical informatics [6].

However, manual construction of such domain ontologies is time consuming and a costly task that requires an extended knowledge of the domain [7]. Due to those limitations, over the last decade, automatic and semi-automatic ontology learning methods have been developed to support the automatic engineering of domain

© Springer International Publishing AG 2016
A. Fred et al. (Eds.): IC3K 2015, CCIS 631, pp. 345–364, 2016.
DOI: 10.1007/978-3-319-52758-1_19

ontologies. Ontology learning consists of three main phases, the lexical term identification, the taxonomic relations extraction and the non-taxonomic relations learning [8]. This paper is focused on learning non-taxonomic relations for domain ontology, which is a hot and challenging research topic at present.

In the literature, several approaches have been proposed for learning domain ontologies. Nevertheless, most of them address only the way to learn the lexical and the taxonomical part of domain ontologies. However, the domain specific semantics relations are ignored. The phase of non-taxonomic relation extraction is recognized as one of the most difficult and least tackled problems [9]. This phase includes two different problems:

(a) Discovering the existence of relevant arbitrary relations between concepts.
(b) Labeling these relations according to their semantic meaning.

In general, a non-taxonomic relation models the relation between concepts, expressed by a verb, or a verbal phrase, that doesn't constitute a hierarchy among concepts. Compared with taxonomic relations, non-taxonomic relations have less explicit manifestation as well as more diverse expression. For instance, considering the financial domain, the relations between two concepts "bank" and "credit" can be presented by the following expressions: "banks offer credit" or "credit can be got from banks". Thus, non-taxonomic relations learning needs to resolve the issue caused by implicity and diversity. Moreover, in the ontology learning process, non-taxonomic relations identification layer uses background knowledge from domain taxonomy in order to extract non-taxonomic relations of the domain [10]. However, the quality of the extracted non-taxonomic relations will highly depend on the accuracy of its source domain taxonomy.

In previous works [11, 12], we covered the learning of taxonomic relations for domain ontology and introduced an efficient method that builds domain taxonomy more accurately than other benchmark algorithms. In this paper, we present our new framework, named OntologyLine[1], for acquiring non-taxonomic relations. It is an extension of our previous work [13]. The final goal of our research is the automatic construction of domain ontologies from scratch.

The OntologyLine system is able to discover and label automatically non-taxonomic relationships of domain ontology. The proposed method is based on integrating and adapting Open Information Extraction (Open IE) algorithms to extract and label domain relations between concepts. Open IE approaches were introduced recently by Banko et al. in 2007 [14]. So there are only a small number of projects using it. To the best of our knowledge, they were never used as a part of the process of learning domain ontology. This work adapts these types of algorithms to learn domain ontology. We integrate our system with three well-known Open IE tools: Reverb [15], Ollie [16] and ClausIE [17].

To evaluate our ontology learning framework, the experiment was performed in two different domains, the financial and the cancer domains. The extracted non-taxonomic relations by OntologyLine for each domain was compared to Sanchez

---

[1] An online version is available at www.ontologyline.com.

and Moreno (S-M) algorithm [18] using gold standard ontology. The precision, recall and F-measure of our method were observed to be considerably higher, in both domains, than those of S-M methods for non-taxonomic relations learning.

The rest of the paper is organized as follows. Section 2 discusses an overview of previous research performed in the non-taxonomic learning area. Section 3 presents the general steps for learning domain ontology. Then in Sect. 4, the OntologyLine algorithm is detailed. In Sect. 5, the architecture of the proposed framework is presented. Section 6, describes the experiments and the evaluation results. Finally, conclusions and future work are defined in Sect. 7.

2 Related Works

Some techniques have already been proposed for learning non-taxonomic relationships of domain ontology. Most of them combine different levels of machine learning and linguistic analysis.

Maedche and Staab [19] propose a semi-automatic method for learning domain ontology. It uses generalized association rules to identify relationships between pairs of words and proposes the best possible level in the hierarchy where to add the relationships. Nevertheless, this technique does not address the problem of labeling relationships. As consequence, users have to complete this task manually and without any assistance. In the same context, Villaverde et al. [9] use the strength of the association between concepts and verb given by POS-tagging rules to suggest multiple labels relationship between concepts. They identify non-taxonomic relationships by the presence of two concepts of the taxonomy in the same sentence with a verb between them within a window of N terms. One limitation of this method is that the authors refer to the verbs and the concepts as single words when in fact, in most cases; they appear in the form of verb phrases. However, the accuracy will be reduced and the recall will be low.

Alternatively, Jiang and Tan [21] attempt to acquire non-taxonomic relationships between concepts using regular expressions and natural language processing. Their algorithm performs a full parsing of the entire corpus using the Berkeley Parser [22] and inspects the entire corpus to identify instances of patterns that indicate non-taxonomic relationships. However, the use of regular expressions limits the identification of the relations in the corpus. It may lack relevant relationships that are not recognized by the used patterns. Another important element to note is that their method involves a full parsing from text, which allows better extraction of concepts, but it is very expensive in time.

Other works use the web as a source for learning domain ontologies and discovering non-taxonomic relations. Sanchez and Moreno [18] proposed a method for discovering non-taxonomic relations based on using search engines. They developed a technique for learning domain patterns using domain-relevant verb phrases extracted from web pages provided by search engines. These domain patterns are then used to extract and label the non-taxonomic relations using linguistic and statistical analysis. However, to avoid the natural language complexity, they apply some restriction, for example: only sentences with present tenses are used, verb phrases containing modifiers in the form of adverbs are rejected. Such restrictions limit the discovery of domain relations; which negatively impact on the recall of the extracted relations. Wong et al. [25] proposed a hybrid approach based on techniques of lexical simplification, word disambiguation and association inference for acquiring non-taxonomic relations by using search engines' page count. However, such kind of methods largely depends on the quality of the search engine results and their accessibility.

In more recent works, Serra et al. [24] use on their work Natural Language Processing (NLP) and statistics to discover associated concept pairs and verbs, and then employed the verbs to label non-taxonomic relations. The system provides three types of extraction rules: The Sentence Rule (SR), the Sentence Rule with Verb Phrase (SRVP) and the Apostrophe Rule (AR). The importance of the extracted relations is measured using co-occurrence frequency. Punuru and Chen [26] utilized a statistical method with log likelihood ratio to discover non-taxonomic relations. The relation label is suggested by measuring the verb frequency inverse concept frequency (VF-ICF) value for the relations. However, the recall value obtained by that methodology was poor.

In the sequel of this paper, we propose a new method to overcome all the limitations listed below and to learn domain ontology from scratch. We incorporate, for the first time in the literature, Open IE algorithms in the domain ontology extraction process. Thus, we benefit from the performance and the experience of these algorithms in the area of non-taxonomic relations extraction.

3 Non-taxonomic Relations Learning Steps

This section introduces the non-taxonomic relations learning process, describing the main steps for discovering automatically and from scratch non-taxonomic relations of domain ontology. Buitelaar et al. [10] define the main steps of learning non-taxonomic relations for domain ontology as follows (see Fig. 1):

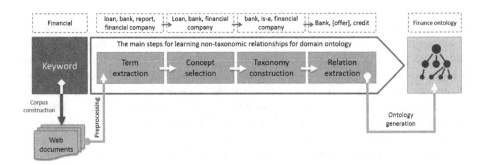

Fig. 1. The main steps for learning non-taxonomic relations for domain ontology.

Step 1 - term extraction: It's needed to identify terms that are likely related to the studied domain. This step constitutes a principal prerequisite for unsupervised concept acquisition. In our work, the term extraction is based on the neighborhood of an initial keyword that characterizes the input corpus. In instance, for the financial domain

(Fig. 1), the initial keyword given as input for building the ontology is *"Financial"* and the terms *"bank, loan, report, financial company"* are examples of the extracted terms in this phase of the ontology learning process.

Step 2 - Concept selection: This step aims to select the most relevant terms from those previously extracted. The filtered terms constitute the concepts of the studied domain. Concepts selection can be performed using statistic-based techniques [28, 29], linguistic-based techniques or a hybrid one [30].

In our work, we use a statistical measure that combines between Chir statistic and Conditional Mutual Information measure to select the most relevant domain concepts (see Sect. 4). In our example of the financial ontology, the word *"report"*, extracted in the first step, does not belong to the financial domain. Using our combined method, previously cited, that noisy word will be excluded in this second stage of the learning process of ontology. Therefore, only the terms *"bank, loan, financial company"* will be retained as relevant concepts of the financial domain.

Step 3 - Taxonomy construction: Learning or extracting taxonomic relations means finding hyponyms between the selected concepts with the goal of constructing a concept hierarchy. It can be performed in various ways such as using predefined relations from existing background knowledge [31], using hierarchical clustering [28], relying on semantic relatedness between concepts [32], or using linguistic and logical rules or patterns [21].

Our proposed algorithm uses the term structure through string matching of the immediate posterior and anterior words of a keyword concept to extract their hierarchies. In our example, we extract the hyponym relation between the two concepts *"bank"* and *"financial company"*.

Step 4 - Relation extraction: Discovering and labeling non-taxonomic relations are mainly reliant on the analysis of the structure and dependencies of candidate sentences. In this phase, verbs are good indicators for non-taxonomic relations and are used to label such relations.

Our main contribution in this work is related to this step of the ontology construction process by introducing a novel technique based on Open IE algorithms to extract and label non-taxonomic relations of domain ontology. In our financial domain example, the relation between *"bank"* and *"credit"* was identified and labeled with the verb *"offer"*.

4   OntologyLine Framework

In this section we present our framework for learning non-taxonomic relations of domain ontology. An overview of the OntologyLine process is shown in Fig. 1, which illustrates the sequence of the different data processing steps, as well as the input and output of this process.

Based on the review presented in the previous Sect. 3, the OntologyLine system consists of four main steps. At first, term extraction takes place to identify terms that are likely related to the studied domain from a corpus of web documents. Then, these terms are filtered based on several criteria (as it will be seen in Sect. 4.2) and selected in the concept selection step. The selected terms constitute the concepts of the studied domain. After concept selection, the taxonomic relations between the extracted concepts are determined to create the concept hierarchy. Next, the non-taxonomic relations of each concept of the constructed taxonomy are discovered. Finally, the obtained result is generated in an OWL ontology file to form the constructed domain ontology.

This section continues by describing each of the previously discussed OntologyLine steps. First, the term extraction is explained. Then, the approach used for domain concept selecting will be discussed. Next, the method for building the concept hierarchy is presented. Finally, the process for learning non-taxonomic relations is detailed.

4.1   Term Extraction

As shown in Sect. 3, the first step of the process of learning non-taxonomic relations of domain ontology is to extract from the input corpus the terms that represent the domain. OntologyLine is based on the neighborhood of the introduced keyword that is fairly representative of the studied domain to extract relevant terms (candidate concepts). The algorithm selects the anterior and the posterior nouns of the introduced keyword as relevant terms for the next step of concept selection. To get nouns from the corpus, we use a part-of-speech tagger that tags words that appear in the corpus.

4.2   Concept Selection

The most relevant terms for a specific domain need to be selected from the previously extracted terms to identify domain concepts. In order to achieve this goal, we used the Conditional Score-Measure already defined in our previous work [11]. As we have shown in that work, Conditional Score-Measure statistics is more efficient than other benchmark algorithms for selecting domain concepts. It combines between the Chir statistic and a semantic similarity based on Conditional Mutual Information. For each candidate concept, a score is calculated on the basis of this measure. This score is used to determine the relevance of the extracted terms.

In the next three subsections, first, we present the used Chir statistic. Then, we describe the semantic similarity that is based on Conditional Mutual Information. Finally, we present the formula of the introduced conditional score-measure to select relevant domain concepts.

4.2.1   Chir Statistic

The Chir statistic proposed by Li et al. [34] is an extended variant of the χ^2 statistic to measure the degree of dependency between a term w and a category c of documents. The χ^2 statistic is defined as:

$$\chi^2_{w,c} = \sum_i \sum_j \frac{(O(i,j) - E(i,j))^2}{E(i,j)} \tag{1}$$

Where $O(i, j)$ is the observed frequency of documents that belongs to category j and contains w. $E(i, j)$ is the expected frequency of category j and term i.

Li et al. showed that their method could improve the performance of text clustering by selecting the words that help to distinguish documents in different groups. Indeed, when the χ^2 statistic measure the lack of independence between the terms in the category [35], the Chir statistic selects only relevant terms that have strong positive dependency on certain categories in the corpus and remove the irrelevant and redundant terms. To define the term goodness of a term w in a corpus with m classes, Li et al. use a combining formula of χ^2 and a category dependency measure $R(w,\ c)$ defined by:

$$R_{w,c} = \frac{O(w,c)}{E(w,c)} \tag{2}$$

Where $O(w,\ c)$ is the number of documents that are in the category c and contain the term w, and $E(w,\ c)$ is the expected frequency of the category c to contain the term w. If there is a positive dependency, then $R(w,\ c)$ should be larger than 1. If there is negative dependency, $R(w,\ c)$ should be smaller than 1. In the case of the no-dependency between the term w and the category c, the term category dependency measure $R(w,\ c)$ should be close to 1. In summary, when $R(w,\ c)$ is larger than 1, the dependency between w and c is positive, otherwise, the dependency is negative.

The final formula of the Chir statistic was defined by:

$$r_{\chi^2}(w) = \sum_{j=1}^{m} p\left(R_{w,c_j}\right) \chi^2_{w,c_j} \ With \ R_{w,c_j} > 1 \tag{3}$$

Where

$$p\left(R_{w,c_j}\right) = \frac{R_{w,c_j}}{\sum_{i=1}^{m} R_{w,c_i}} \ With \ R_{w,c_i} > 1 \tag{4}$$

is the weight of $\chi^2_{w,\,c}$ in the corpus in terms of $R_{w,\,c_j}$.

A bigger $r_{\chi^2}(w)$ value indicates that the term is more relevant to the domain. When there is a positive dependency between the term w and the category c_j.

4.2.2 Conditional Information Measure

In our ontology learning process, conditional mutual information is used to measure dependency between two terms w and w' conditioned by the occurrence of a parent term w_p. On the basis of Brun et al. [36] work, two terms are considered similar if their mutual information with all terms in the vocabulary is nearly the same. Thus, by introducing conditional information in the similarity measure defined in their work, the new measure is described by:

$$Sim(w, w'/w_P) = \frac{1}{2|V|} \sum_{i=1}^{|v|} \left(\frac{\min\left(I(zi, w/w_P), I(zi, w'/w_P)\right)}{\max\left(I(zi, w/w_P), I(zi, w'/w_P)\right)} + \frac{\min\left(I(w, zi/w_P), I(w', zi/w_P)\right)}{\max\left(I(w, zi/w_P), I(w', zi/w_P)\right)}\right) \tag{5}$$

where V is the vocabulary, and $I(z_i, w/w_p)$ is the Conditional Mutual Information of terms z_i and w conditioned by the presence of the term w_p. The Conditional Mutual Information formula is given by Zhang et al. [37] as following:

$$I(z_i, w/w_P) = P_d(z_i, w, w_P) log \frac{P(w_P)P_d(z_i, w, w_P)}{P_d(z_i, w_P)P_d(w, w_P)} \quad (6)$$

Where d is the withdrawal, $P(w_p)$ is the probability of the term w_p. $P_d(z_i, w_p)$ and $P_d(w, w_p)$ are the probability of succession of the terms z_i and w_p, w and w_p respectively in the window observation. $P_d(z_i, w, w_p)$ is the probability of succession of the terms z_i, w and w_p in the window observation. This probability can be estimated by the ratio of the number of times that the term z_i is followed by the terms w and w_p within the window, with the cardinal of the vocabulary:

$$P(z_i, w, w_P) = \frac{f_d(z_i, w, w_P)}{|V|} \quad (7)$$

Where $f_d(z_i, w, w_p)$ is the number of times that the term z_i is followed by the terms w and w_p.

4.2.3 Conditional Score-Measure

To identify the relevant concepts from those extracted, we have defined a hybrid measure based on the weighting model combining the Chir-statistic and the similarity measure using conditional information. This new scoring measure was defined as:

$$S(w_c/w_p) = \lambda * r_{\chi^2}(w_c) + (1 - \lambda) * sim(w_c, w_k/w_p) \quad (8)$$

Where λ is a weighting parameter between 0 and 1.

Since relevant concepts have strong dependency with the studied domain and convey semantically similar information with respect to their parent concept and to the initial keyword, the candidate concepts having strong score are likely to be relevant and should be integrated into the extracted taxonomy.

4.3 Taxonomy Construction

In this section, the algorithm used to discover and select representative concepts and construct the domain taxonomy is described.

The OntologyLine algorithm is based on the analysis of a large number of web corpus files in order to find relevant concepts and to identify their taxonomic relationships. The concept hierarchy extraction is based on analyzing the neighborhood of an initial keyword that characterizes the domain. It is a particularity of the English Language where the immediate posterior and anterior words of a keyword express a semantic specialization between them [38]. Theses candidate concepts are processed in order to select the most relevant ones by performing our statistical algorithm based on the Conditional Mutual Information. The selected classes are finally incorporated into the ontology. The process is repeated recursively in order to find new terms and build the hierarchy of concepts.

As shown in Fig. 2, this algorithm is described by the following steps:

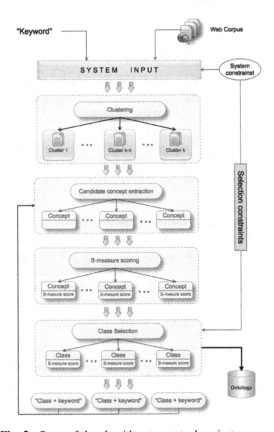

Fig. 2. Steps of the algorithm to create domain taxonomy.

1. Perform a k-means clustering algorithm on the set of all documents and get initial clusters.
2. Start with a keyword that has to be representative enough for the domain and a set of parameters that constrain the search and the concept selection.
3. Extract all the candidate concepts by analyzing the neighborhood of the initial keyword; select the anterior words and posterior words as candidate concepts.
4. For each candidate concept, calculate its score $S(w_c/w_p)$ measure by using (8).
5. Sort the terms in descending order of their $S(w_c/w_p)$ measure.
6. Select the top l terms from the list.
7. The l concepts extracted are incorporated as classes or instances in the taxonomy.

8. For each concept incorporated in the taxonomy, a new keyword is constructed joining the new concept and the initial one. This process is repeated recursively until no more results are found.
9. Finally, a refinement process is performed in order to obtain a more compact taxonomy and avoid redundancy.

4.4   Non-taxonomic Relations Extraction

In what follows, our methodology for non-taxonomic relation extraction is based on the integration of Open IE in the process of learning non-taxonomic relationships for domain ontology. Thus, we benefit from the performance and the experience of these algorithms in the area of relations extraction.

Open IE approaches are relatively a recent paradigm for extracting relations from unstructured documents. They were introduced by Banko et al. [14]. Open IE systems facilitate domain independent discovery of relations. They extract all possible relations without any prerequisite or restriction. In recent years, many systems for Open IE have been developed. For our non-taxonomic learning process, we integrate the most recent ones, mainly: Reverb [15], Ollie [16] and ClausIE [17].

Reverb uses shallow syntactic parsing to identify relations expressed by verbs. The system takes a sentence as input, identifies a candidate pair of noun phrase arguments *(arg1, arg2)* from the sentence, and then uses the learned extractor to label each word between the two arguments as part of the relation phrase or not.

In Ollie system, the authors use context analysis to extract relations in a given sentence. It extracts not only relations expressed via verb phrases, but also relations mediated by adjectives and nouns.

ClausIE is the most recent Open IE system. It differs from Reverb and Ollie approaches in that it separates the detection of useful information expressed in a sentence from their representation in terms of extractions. ClausIE exploits linguistic knowledge to first detect clauses in an input sentence and to subsequently identify the type of each clause according to the grammatical function of its constituents.

However, in our domain extraction context, due to its open-domain and open-relation, gross use of Open IE algorithms is unable to relate the extracted relations to domain ontology. Subsequently, an adaptation of the used Open IE algorithms is necessary to overcome this limitation.

To address this limitation, we have implemented a solution based on the concepts of the taxonomy already extracted in the previous stage. The proposed process for learning non-taxonomic domain relationships with Open IE tools is performed in three steps:

Step1: For each concept of the taxonomy, we extract from the corpus all the sentences where the concept c is found.

Step2: For each extracted sentence, we discover all possible relations using one of the proposed Open IE algorithms. As an output, we obtain a set of relational tuples *<Arg1, Rel, Arg2>* that describes the sentence verb relation *(Rel)* and its arguments *(Arg1 and Arg2)*.

Step3: Finally, we judged each tuple as related to the studied domain or not, based on whether it contains the concept *c* in one of the extracted arguments. The selected relations are incorporated into the result ontology.

This process is repeated until all concepts of the taxonomy are processed.

5   OntologyLine Architecture

In this section, we present the architecture of our proposed framework for discovering non-taxonomic relations and learning domain ontology. The OntologyLine system consists of five basic components, namely: Corpus Pre-processing Component (CPC), Data Pre-processing Component (DPC), Algorithm Library Component (ALC), Open IE Component (OIEC) and Ontology Generation Component (OGC) (Fig. 3).

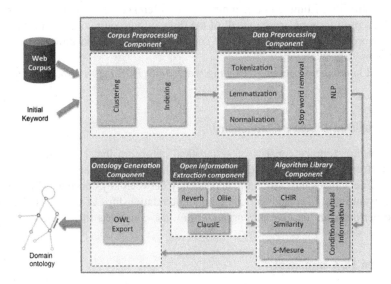

Fig. 3.   Ontologyline architecture.

This section goes on to provide a detailed description of each component of the OntologyLine system architecture. First, the Corpus Preprocessing component is discussed. Then, the Data Preprocessing Component is presented. After that, the Algorithm Library Component is introduced. Next the Open IE Component is described. Finally, the Ontology Generation Component is exposed.

5.1   Corpus Pre-processing Component

This component aims to import the corpus into the system and prepare it for processing. In a first stage, the k-means clustering algorithm [39] is applied. It partitions the corpus into *k* clusters so that two documents within the same cluster are more closely related than two documents from two different clusters. In a second step, an indexing process is executed to index the full contents of the clustered documents. Based on the neighborhood of an initial keyword to extract domain concepts and the relations that exist between them, the indexation allows an efficient and fast retrieval of this information.

5.2   Corpus Pre-processing Component

In the Data Preprocessing Component, text information is filtered and cleaned. The preprocessing includes the following elements:

- Tokenization: Splitting strings into their component words based on delimiters.
- Normalization: Elimination of stylistic differences due to capitalization, punctuation, word order, and characters not in the Latin alphabet.
- Lemmatization: Elimination of grammatical differences due to verb tense, plurals, etc. It's used to improve the recall of the domain ontology concepts in the corpus.
- Stop word removal: Remove of very common words. The Glasgow stop word [40] list is used in this work.
- Natural language processing: Tagging words with POS tags and parsing sentences.

5.3   Algorithm Library Component

The Algorithm Library consists of a statistical algorithm that (1) extracts candidate concepts from a collection of documents and stores the sentences where they are located; (2) evaluates the taxonomic dependency between key concepts; (3) selects the most relevant ones; (4) discovers non-taxonomic relationships of the selected concepts (5) and performs an iterative mining algorithm that constructs the ontology.

5.4   Open IE Component

The Open IE component proposes three algorithms to extract non-taxonomic relations: Reverb, Ollie and ClausIE. It communicates with the previous component "Algorithm Library" in input/output mode. According to the Open IE algorithm chosen at the beginning of the process, and from the sentences provided by the "Algorithm Library", the Open IE component identifies non-taxonomic relationships of each concept of the taxonomy and return verbs and concepts that describe the identified relationships.

5.5   Ontology Generation Component

In this final phase, the resulted domain ontology is generated in an OWL file format by using the Jena toolkit. The resulting file can be visualized using an ontology editor such as Protegé or OntoEdit.

6 Evaluation

The evaluation process is an essential step that should be performed in any ontology learning approach. It's particularly important in automatic approaches as the present work. In this section, we evaluate the performance of our methodology for learning domain ontology against the S-M algorithm [18]. The evaluation is performed in two different domains: financial and cancer. The assessment includes a comparison of the extracted non-taxonomic relations by each algorithm using gold standard ontology. However, the evaluation of the lexical and the taxonomical level of the obtained ontology, in both domains, were already presented in previous works [11, 12, 33]. The comparison was made against S-M algorithm and others related works, and showed that our method was more efficient in extracting relevant domain concepts and learning concept hierarchies.

Furthermore, due to the lack of an evaluation corpus and ontology officially released that covers non-taxonomic relations of the studied domains, we build our own corpuses and gold standard ontologies. Indeed, most existing ontologies are not linked to a precise document collection and they result from a concerted effort of domain experts based on their background knowledge of the domain and their experience. It is therefore unfair to compare an automatically built ontology from a specific corpus with such expert ontology as the learned ontology will depend heavily on the corpus from which it is constructed.

In the next subsections, the implementation of the algorithms is briefly described. Then, the used corpuses and the gold standard otology are given. Next, the measures used for the evaluation are presented. Finally, the experimental results are discussed.

6.1 Implementation of the Algorithms

We developed our OntologyLine system as an online web application using JEE 7. For the integration of the three Open IE algorithms (Reverb, Ollie and ClausIE) in the domain ontology learning process, we used the original Java source code published by their owners and we adapted it to extract domain relations.

Concerning Sanchez and Moreno algorithm, in the absence of its source code, and to perform the comparison with our algorithm, we developed their algorithm from scratch by referring to its description in their paper.

6.2 Domain Corpuses Application

In order to prepare our test corpuses automatically, we developed a Java program to retrieve text documents from the web. The financial corpus was constructed from the financial news Website "Yahoo! Finance". We were based on a period of six months (January through Jun 2015), to retrieve randomly new articles. 10760 news documents[2]

---

[2] Available at http://www.ontologyline.com/OntoLineProjects/financial10760-OntologyLine/corpus/corpus.zip.

were retrieved for that period. They represent our financial corpus. Regarding the construction of the cancer corpus, we used the "PubMed" library of Medicine's search to extract all Medline abstracts articles that contain the term "cancer" in the title. The search query was applied for the same period used to build the financial corpus. The obtained cancer corpus contains 3056 abstracts[3].

6.3   Gold Standard Ontology

The gold standard ontology was developed for each corpus with a collaboration of domain specialists. It was developed in two steps in order to give equal chances to all the tested algorithms. In the first step, all the candidate algorithms were launched on each corpus. Then, the resulting ontologies were automatically merged distinctly into a single one using the Protégé ontology merging functionality [20]. Redundant concepts and relations were exported only once. Our aim is to build for each corpus, a reference ontology based on the previous resulting ontology of each algorithm. In the second step, the constructed ontologies were presented to the domain specialist for validation. He performs a cleanup of the constructed ontologies and removes erroneous concepts and invalid relations. The final ontologies were used as reference ontologies for our evaluation process. The number of the extracted non-taxonomic relationships in the financial and the cancer corpuses is shown respectively in Tables 1 and 2. Tables 3 and 4 summarize the number of non-taxonomic relationships in the Gold standard ontology of each dataset after the specialist validation and the refinement of the merged ontology.

Table 1. Number of generated non-taxonomic relations from the financial corpus.

Source	OntologyLine using Reverb	OntologyLine using Ollie	OntologyLine using ClausIE	Sanchez & Moreno	Merged ontology
Number of extracted non-taxonomic relations	113	102	97	85	145

Table 2. Number of generated non-taxonomic relations from the cancer corpus.

Source	Our method using Reverb	Our method using Ollie	Our method using ClausIE	Sanchez & Moreno	Merged ontology
Number of extracted non-taxonomic relations	167	132	107	109	239

---

[3] Available   at   http://www.ontologyline.com/OntoLineProjects/cancer3056-OntologyLine/corpus/corpus.zip.

Table 3. Number of non-taxonomic relations in the financial gold standard ontology.

Source	Gold standard ontology
Number of non-taxonomic relations	114

Table 4. Number of non-taxonomic relations in the cancer gold standard ontology.

Source	Gold standard ontology
Number of non-taxonomic relations	157

6.4   Evaluation Metrics

Previous studies have used several evaluation methods [18, 23, 27] to evaluate the non-taxonomic relations layer of a constructed ontology. The aim of this evaluation is to determine whether the learned relations are correct and belong to the target domain.

To determine the performance of each algorithm at this layer we used the precision, recall and F-measure metrics. The *Precision* (9) measures to which extend incorrect relations can be rejected. It is computed as the ratio between the number of correct selected relations ($r_{relevant}$) and the number of all extracted relations (r_{all}).

$$Precision = \frac{r_{relevant}}{r_{all}} \tag{9}$$

The Recall (10) measures how much of the existing relations in the gold standard ontology are extracted in the learned ontology. It is computed as the ration of the number of relevant extracted relations ($r_{relevant}$) divided by the total number of relevant relations in the gold standard ontology ($g_{relevant}$).

$$Recall = \frac{r_{relevant}}{g_{relevent}} \tag{10}$$

Finally, the F-Measure (11) provides the weighted harmonic mean of precision and recall, summarizing the global performance of the algorithm in learning non-taxonomic relations.

$$F - Measure = \frac{2 * Precision * Recall}{Precision + Recall} \tag{11}$$

6.5   Experimental Results

In the interest of a fair comparison, all the algorithms were executed under the same conditions. The test was performed on a Mac OS X machine, with 8 GB of RAM.

Table 5. Metric results of non-taxonomic relations obtained on the financial corpus.

Algorithm	Precision	Recall	F-measure
OntologyLine with Reverb Open IE[a]	72,57%	71,93%	72,25%
OntologyLine with Ollie Open IE[b]	72,55%	64,91%	68,52%
OntologyLine with ClausIE Open IE[c]	70,10%	59,65%	64,45%
Sanchez & Moreno (S-M)	62,35%	46,49%	53,27%

[a]The result ontology is available at http://www.ontologyline.com/
buildOntology/financial10760-OntologyLineReverb/financial
[b]The result ontology is available at http://www.ontologyline.com/
buildOntology/financial10760-OntologyLineOllie/financial
[c]The result ontology is available at http://www.ontologyline.com/
buildOntology/financial10760-OntologyLineClausIE/financial

Table 6. Metric results of non-taxonomic relations obtained on the cancer corpus.

Algorithm	Precision	Recall	F-measure
OntologyLine with Reverb Open IE[a]	76,65%	81,53%	79,01%
OntologyLine with Ollie Open IE[b]	72,73%	61,15%	66,44%
OntologyLine with ClausIE Open IE[c]	74,77%	50,96%	60,61%
Sanchez & Moreno (S-M)	66,97%	46,50%	54,89%

[a]The result ontology is available at http://www.ontologyline.com/
buildOntology/cancer3056-OntologyLineReverb/cancer
[b]The result ontology is available at http://www.ontologyline.com/
buildOntology/cancer3056-OntologyLineOllie/cancer
[c]The result ontology is available at http://www.ontologyline.com/
buildOntology/cancer3056-OntologyLineClausIE/cancer

The evaluation results illustrated in Tables 5 and 6 show that our method, used with the three Open IE algorithms, outperforms the S-M method in both corpora. Indeed, OntologyLine reaches its best performance when using Reverb as an external open IE tool. In both financial and cancer domains, it has much higher precision and recall than the other algorithms. It achieves a precision that is 16% higher than S-M algorithm and is nearly identical to that when using OntologyLine with Ollie or ClausIE. In term of recall, OntologyLine with Reverb is 10% higher than when our method is used with Ollie, 20% when it's used with ClausIE and is more than double the recall of S-M. The performance of OntologyLine with Reverb is even greater in the cancer domain, achieving a recall 75% higher than S-M algorithm.

Therefore, OntologyLine proves to be quite effective in discovering non-taxonomic relations of domain ontology, especially when it's combined with Reverb.

Tables 7 and 8 show some examples of the learned non-taxonomic relations using OntologyLine with Reverb respectively in the financial and the cancer domains.

Table 7. Examples of the learned non-taxonomic relations from the financial corpus using OntologyLine with Reverb.

The extracted relations		
Subject (NP)	Verb (VP)	Object (NP)
Financial company	Provide	Loan
Financial market	Include	Oil market
Financial investor	Offer	Aircraft financing loan
Company	Do business with	Consumer financial provider
Company	Claim	Financial service
Dealer	Be affiliate with	Financial company
Investor	Be interested in	Financial stock
Bank	Be a provider of	Financial service
Device	Can improve access to	Financial service

Table 8. Examples of the learned non-taxonomic relations from the cancer corpus using OntologyLine with Reverb.

The extracted relations		
Subject (NP)	Verb (VP)	Object (NP)
Breast cancer	Be observe for	Woman
Breast cancer	Treated with	Trastuzumab
Cancer patient	Receive	Chemotherapy
Breast cancer patient	Receive	Radiation therapy
Physical exercise	Be beneficial to	Breast cancer patient
Drug	Be associate with	Cancer risk
Prostate cancer	Accelerate	Disease
Prostate cancer	Treat with	Radiotherapy
Hypercholesterolemia	Promote	Prostate cancer
Lung cancer	Cause by	Smoke

7 Conclusion

In this article, we have presented OntologyLine, our new system for discovering non-taxonomic relations and building domain ontology from scratch. One of the main difficulties of domain ontology learning process. The novelty of our approach in this area is that it is based on adjusting Open IE algorithms to extract domain relations between concepts. The other contribution is that it addresses the problem of labeling non-taxonomic relations with high precision, recall and F-measure. OntologyLine integrates three well-known Open IE algorithms, namely: Reverb, Ollie and ClausIE. The system was tested in two different domains: the financial and cancer domains. It was evaluated against gold standard ontology and was compared to state-of-the-art ontology learning algorithm Sanchez and Moreno. The obtained results prove that using Open IE tools in the process of learning domain ontology is an interesting way to extract ontological relations with higher degree of precision and recall.

As future work we would like to enhance the proposed system as follows. We will improve the nature of the extracted relations by integrating the automatic extraction of axioms in the OntologyLine process. So the system can build a more complete and richer ontology. Also, we think validate the result ontology by integrating it in a decision-support system, such as for financial decision making or investment recommendations. This would allow us to evaluate our system in being used in such area.

References

1. Gruber, T.R.: A translation approach to portable ontology specifications. Knowl. Acquis. **5** (2), 199–220 (1993)
2. Noy, N.F., McGuinness, D.L.: Ontology development 101: a guide to creating your first ontology. Stanford Knowledge Systems Laboratory. Technical report KSL-01-05 and Stanford Medical Informatics Technical report SMI-2001-0880 (2001)
3. Du, J., Zhou, L.: Improving financial data quality using ontologies. Decis. Support Syst. **54** (1), 76–86 (2012)
4. Zhao, Y., Li, Z., Wang, X., Halang, W.A.: Decision support in e-business based on assessing similarities between ontologies. Knowl.-Based Syst. **32**, 47–55 (2012)
5. Wen, Y.-X., Wang, H.-Q., Zhang, Y.-F., Li, J.-S.: Ontology-based medical data integration for regional healthcare application. In: Li, S., Jin, Q., Jiang, X., Park, J.J. (eds.) Frontier and Future Development of Information Technology in Medicine and Education. LNEE, vol. 269, pp. 1667–1672. Springer, Heidelberg (2014). doi:10.1007/978-94-007-7618-0_191
6. Paul, S., Maji, P.: Gene ontology based quantitative index to select functionally diverse genes. Int. J. Mach. Learn. Cybern. **5**(2), 245–262 (2014)
7. Wong, M.K., Abidi, S.S.R., Jonsen, I.D.: A multi-phase correlation search framework for mining non-taxonomic relations from unstructured text. Knowl. Inf. Syst. **38**(3), 641–667 (2014)
8. Maedche, A., Staab, S.: Ontology learning for the semantic web. IEEE Intell. Syst. **16**(2), 72–79 (2001)
9. Villaverde, J., Persson, A., Godoy, D., Amandi, A.: Supporting the discovery and labeling of non-taxonomic relationships in ontology learning. Expert Syst. Appl. **36**(7), 10288–10294 (2009)
10. Buitelaar, P., Cimiano, P., Magnini, B.: Ontology learning from text: an overview. In: Ontology Learning from Text: Methods, Evaluation and Applications, pp. 3–12. IOS Press, Amsterdam (2005)
11. El idrissi esserhrouchni, O., Frikh, B., Ouhbi, B.: HCHIRSIMEX: an extended method for domain ontology learning based on conditional mutual information. In: Third IEEE International Information Science and Technology (CIST), pp. 99–95 (2014)
12. Frikh, B., Djaanfar, A.S., Ouhbi, B.: A new methodology for domain ontology construction from the web. Int. J. Artif. Intell. Tools **20**(06), 1157–1170 (2011)
13. El idrissi esserhrouchni, O., Frikh, B., Ouhbi, B.: Learning non-taxonomic relationships of financial ontology. In: KEOD, pp. 479–489 (2015)
14. Banko, M., Cafarella, M.J., Soderland, S., Broadhead M., Etzioni O.: Open information extraction from the Web. In: Proceedings of the 20th International Joint Conference on Artifical Intelligence (IJCAI), pp. 2670–2676 (2007)

15. Fader, A., Soderland, S., Etzioni, O.: Identifying relations for open information extraction. In: Proceedings of the Conference on Empirical Methods in Natural Language Processing, pp. 1535–1545. Association for Computational Linguistics (2011)
16. Schmitz, M., Bart, R., Soderland, S., Etzioni, O.: Open language learning for information extraction. In: Proceedings of the 2012 Joint Conference on Empirical Methods in Natural Language Processing and Computational Natural Language Learning, pp. 523–534. Association for Computational Linguistics (2012)
17. Del Corro, L., Gemulla, R.: ClausIE: clause-based open information extraction. In: Proceedings of the 22nd International Conference on World Wide Web, pp. 355–366. International World Wide Web Conferences Steering Committee (2013)
18. Sánchez, D., Moreno, A.: Learning non-taxonomic relationships from web documents for domain ontology construction. Data Knowl. Eng. **64**(3), 600–623 (2008)
19. Maedche, A., Staab, S.: Semi-automatic engineering of ontologies from text. In: Proceedings of the 12th International Conference on Software Engineering and Knowledge Engineering, pp. 231–239 (2000)
20. Protégé: Protégé v5.0 (2015). http://protege.stanford.edu. Accessed 11 Feb 2016
21. Jiang, X., Tan, A.H.: Mining ontological knowledge from domain-specific text documents. In: Proceedings of the Fifth IEEE International Conference on Data Mining. IEEE (2005)
22. Petrov, S., Barrett, L., Thibaux, R., Klein, D.: Learning accurate, compact, and interpretable tree annotation. In: Proceedings of the 21st International Conference on Computational Linguistics and the 44th Annual Meeting of the Association for Computational Linguistics, pp. 433–440. Association for Computational Linguistics (2006)
23. Xu, Y., Li, G., Mou, L., Lu, Y.: Learning non-taxonomic relations on demand for ontology extension. Int. J. Softw. Eng. Knowl. Eng. **24**(08), 1159–1175 (2014)
24. Serra, I., Girardi, R., Novais, P.: PARNT: a statistic based approach to extract non-taxonomic relationships of ontologies from text. In: Proceedings of the 10th International Conference on Information Technology. IEEE (2013)
25. Wong, W., Liu, W., Bennamoun, M.: Acquiring semantic relations using the web for constructing lightweight ontologies. In: Theeramunkong, T., Kijsirikul, B., Cercone, N., Ho, T.-B. (eds.) PAKDD 2009. LNCS (LNAI), vol. 5476, pp. 266–277. Springer, Heidelberg (2009). doi:10.1007/978-3-642-01307-2_26
26. Punuru, J., Chen, J.: Learning non-taxonomical semantic relations from domain texts. J. Intell. Inf. Syst. **38**(1), 191–207 (2012)
27. Sánchez, D., Batet, M., Martínez, S., Domingo-Ferrer, J.: Semantic variance: an intuitive measure for ontology accuracy evaluation. Eng. Appl. Artif. Intell. **39**, 89–99 (2015)
28. Maedche, A., Volz, R.: The ontology extraction and maintenance framework text-to-onto. In: Proceedings of the ICDM 2001 Workshop on Integrating Data Mining and Knowledge Management (2001)
29. Makrehchi, M., Kamel, M.S.: Automatic taxonomy extraction using google and term dependency. In: Proceedings of the IEEE/WIC/ACM International Conference on Web Intelligence, pp. 321–325. IEEE Computer Society (2007)
30. Meijer, K., Frasincar, F., Hogenboom, F.: A semantic approach for extracting domain taxonomies from text. Decis. Support Syst. **62**, 78–93 (2014)
31. Lee, S., Huh, S.Y., McNiel, R.D.: Automatic generation of concept hierarchies using wordnet. Expert Syst. Appl. **35**(3), 1132–1144 (2008)
32. Pekar, V., Staab, S.: Taxonomy learning: factoring the structure of a taxonomy in to a semantic classification decision. In: 19th International Conference on Computational Linguistics, vol. 1, pp. 1–7. Association for Computational Linguistics (2002)

33. El Idrissi Esserhrouchni, O., Frikh, B., Ouhbi, B.: Building ontologies: a state of the art, and an application to finance domain. In: Fifth International Conference on Next Generation Networks and Services (NGNS), pp. 223–230. IEEE (2014)
34. Li, Y., Luo, C., Chung, S.M.: Text clustering with feature selection by using statistical data. IEEE Trans. Knowl. Data Eng. **20**(5), 641–652 (2008)
35. Saengsiri, P., Meesad, P., Wichian, S.N., Herwig, U.: Comparison of hybrid feature selection models on gene expression data. In: IEEE International Conference on ICT and Knowledge Engineering, pp. 13–18 (2010)
36. Brun, A., Smaïli, K., Haton, J.P.: WSIM: une méthode de détection de thème fondée sur la similarité entre mots. Actes de TALN, pp. 145–154 (2008)
37. Zhang, Y., Zhang, Z.: Feature subset selection with cumulate conditional mutual information minimization. Expert Syst. Appl. **39**(5), 6078–6088 (2012)
38. Grefenstette, G.: Short query linguistic expansion techniques: palliating one-word queries by providing intermediate structure to text. In: Pazienza, M.T. (ed.) SCIE 1997. LNCS, vol. 1299, pp. 97–114. Springer, Heidelberg (1997). doi:10.1007/3-540-63438-X_6
39. Jain, A.K., Murty, M.N., Flynn, P.J.: Data clustering: a review. ACM Comput. Surv. (CSUR) **31**(3), 264–323 (1999)
40. The Glasgow stop word. http://ir.dcs.gla.ac.uk/resources/linguistic_utils/stop_words. Accessed 11 Feb 2016

Software Is Part Poetry, Part Prose

Iaakov Exman[1]([⊠]) and Alessio Plebe[2]

[1] Software Engineering Department,
The Jerusalem College of Engineering – JCE - Azrieli, Jerusalem, Israel
`iaakov@jce.ac.il`
[2] Department of Cognitive Science, University of Messina, Messina, Italy
`alessio.plebe@unime.it`

Abstract. Software is part Poetry, part Prose. But it has much more in common with both forms of natural language, than usually admitted: software concepts, rather than defined by syntactic oriented computer programming languages, are characterized by the semantics of natural language. This paper exploits these similarities in a two-way sense. In one way the software perspective is relevant to the analysis of natural language forms, such as poems. In the other way round, this paper uses properties of both Poetry and Prose to facilitate a deeper understanding of highest-level software abstractions. Running software or poetry leads to understanding of the meaning conveyed by the conceptual structure. Refactoring embeds the understanding obtained by running software or poetry, into their modified conceptual structure.

Keywords: Software · Conceptual · Runnable · Understanding · Poetry · Prose

1 Introduction

We claim that before proposing a theory of software engineering one must understand the nature of software itself. This work focuses on theoretical implications of natural language aspects of highest-level software abstractions. It was triggered by a dialogue of the authors during a festive dinner at a conference in Rome last year and continued by electronic means. The dialogue below was partially transcribed and slightly edited in the Galileo style [12], serving as an introduction to the issues in this work.

1.1 Galilean Dialogue on Two Software Views

Alessio – I agree that in some sense software is deeper than Chomskian theories. Chomsky sharply divided between syntax and semantics, moving almost all the burden of language on the syntax side. This is not the way natural language works. Indeed most exponents of the cognitive linguistics enterprise challenged the syntactocentricsm of generative grammar. One of the first was George Lakoff, a former student of Chomsky, who, looking for linguistic expressions supporting the alleged syntax autonomy, found so many counterexamples, to become convinced of the contrary [17]. He turned into one of the leading exponents of cognitive linguistics, together with

© Springer International Publishing AG 2016
A. Fred et al. (Eds.): IC3K 2015, CCIS 631, pp. 365–380, 2016.
DOI: 10.1007/978-3-319-52758-1_20

Langacker [18], Fauconnier [9] and others. But maybe you have in mind other reasons why the Chomskian account is limited regarding software.

Iaakov – Since I like *gedanken* experiments [5], I ask you to imagine the following experiment. Assume that from a person's birth until the age of fifteen one is supposed to learn the mother tongue and use it strictly according to grammatical, syntactic rules. From the age of fifteen until the age of twenty one gradually uses words with the same meaning as before, but more and more liberated from grammatical rules. From the age of twenty onwards one is totally free to speak poetry instead of prose. In the world of such experiment, as the meaning of a word does not follow from grammar, but is dependent on a context, say an ontology, Chomskian theories would be unimportant. So is software, less grammar, more conceptual.

Alessio – Your second issue is the analogy with poetry, I must say I didn't caught it in Rome, maybe now I can understand a bit more. Repeating a poem to myself (mentally or aloud) corresponds somehow to *executing* it. That's interesting. As the execution of software affects hardware components, registers, memory and so on, the *execution* of a poem will elicit responses, in emotional brain centers, recall long-term memories, activate semantic networks. In both cases the meaning of the code (poetic or software) is in the activation resulting from its execution. It sounds fine. Of course, one may raise several possible objections. For example: what is special for poetry in this analogy? Wouldn't be similar when reading a novel, or a newspaper article? Maybe I'm still far from catching completely the intents of your analogy.

Iaakov – Poetry is paradoxical. On the one hand, it is more constrained by structures. On the other hand, it is less grammatical, more audacious. This is like object oriented software, more structured, and freer in conceptual terms. But we are also aware of other software assets which are prose-like.

Alessio – Now, I come to my point. What I shared with you was a thought I had since long, but never articulated in detail: the possibility that the road taken by computation toward software in the 60's has been the result of the influence of Chomsky. It had the consequence of a paradigm shift from mathematics to linguistics. In the early years of computation, it was entirely mathematical, with central concepts like Dedekind's recursive functions. Turing devised its foundational machine in 1937 as a contribution to Hilbert's Entscheidungsproblem [31]. Even the introduction of *compilation* by Hopper in 1953 [16] was totally unrelated with linguistics. It was only after the publication of Chomsky's "Syntactic Structures" in 1957 [6], and his huge success, that Backus [1] and McCarthy [20] launched the concept of *programming languages*, hinging at large inside the Chomskian tools: generative grammar, tokenization, parsing, translation, and so on.

Iaakov – I get your point.

Alessio – I remember you objecting that history does not go with alternatives, it has no sense in imagining a different destiny in different contingencies. But my point is not historical, it is ontological. There is a widespread assumption, that software is actually a sort of language, one that follows the same rules of natural languages, because it *is* a language in its essence. I'm not saying that this is a wrong belief. I'm just saying that it is rooted in historical contingencies – see also Nofre [25], for a sociological account of these contingencies which could be possibly right or wrong.

Iaakov – It is easier for me to agree with an ontological point of view than with a historical one. I prefer to state that software proper – the runnable part, not the requirements and other assets – is in essence a complex semantic structure, rather than a language. It is runnable meaning.

Alessio – The two alternatives do not affect anything with regard to how efficient is to treat computation using linguistic tools. But it is clearly important when dealing with the ontological status of software.

Iaakov – I would add a cautious caveat. Efficiency has more to do with the underlying machine, than with the runnable meaning itself.

Alessio – Good, I will stop for now… Let me just add that this sort of conversation is quite new for me. But I'm interested in continuing, and it touches two sides of my interests: from one side, the philosophy of computing and on the other side linguistic meaning. I'll do it with pleasure.

1.2 Paper Organization

The remaining of the paper is organized as follows. Section 2 deals with poetry as software, Sect. 3 with refactoring Poetry, Sect. 4 with software as Prose, Sect. 5 with Software as conceptual constructs. The paper ends with a discussion.

2 Poetry as Software

Here we point out features that poetry has in common with software. We display poems in diagrams as if we were describing a kind of software. See [11] for another paper analyzing poetry, having in mind software.

2.1 Poetry Has Structure

From the earliest to most modern samples, poems have structure. Figure 1 displays a modern sonnet by St. Vincent Millay [22, 23]. It has four stanzas, with respectively 4, 4, 3 and 3 verses, and classical rhymes, e.g. in the 1st stanza, *ended* rhymes with *extended*, and *all* rhymes with *fall*.

To make the comparison of poetry with software more concrete, we treat this poem as a piece of software, providing its UML "class diagram". Each stanza is assumed to be a different class. This is seen in Fig. 2. If the reader is not familiar with UML [2, 26], one can think it as an ontology graph containing concepts (classes).

Structure reflects meaning, thus class names were chosen as the most meaningful word in each stanza. Class attributes are significant nouns, and the class functions are the significant verbs. Inheritance links classes with related themes. Association links a class with the previous one, of which it is aware. The overall sonnet class diagram resembles a typical software design pattern – like Observer or Mediator [13].

> **Sonnet -** *E. St. Vincent Millay*
>
> Only until this cigarette is ended,
> A little moment at the end of all,
> While on the floor the quiet ashes fall,
> And in the firelight to a lance extended,
>
> Bizarrely with the jazzing music blended,
> The broken shadow dances on the wall,
> I will permit my memory to recall
> The vision of you, by all my dreams attended.
>
> And then, adieu,—farewell!—the dream is done.
> Yours is a face of which I can forget
> The colour and the features, every one,
>
> The words not ever, and the smiles not yet;
> But in your day this moment is the sun
> Upon a hill, after the sun has set.

Fig. 1. SONNET by E. St. Vincent Millay – classical structure of a modern poem with four stanzas, with 4, 4, 3 and 3 verses, with classical rhymes at the end of verses.

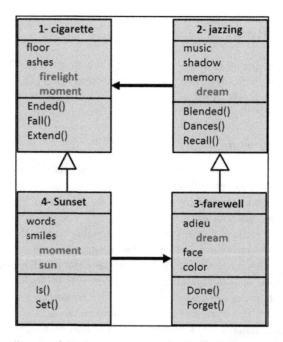

Fig. 2. UML class diagram of the above sonnet – each class (yellow rectangle) fits to one stanza and is numbered by stanzas order. The 1st one is "cigarette". The class middle part contains "nouns" (attributes). The lowest part contains "verbs" (functions). Vertical white triangle arrowheads denote inheritance, i.e. similar themes. Bold red words are common to pairs of classes. Black horizontal arrows denote associations, i.e. a class aware of the previous one. (Color figure online)

2.2   Poetry Has Metaphors

A metaphor is a figure of speech in which a term refers to an object or action that it does not literally denote, implying a resemblance. Metaphors are a common way to generate new meanings of words, indeed new words and idioms.

A neutral dictionary definition of cigarette is just a smoking device: a small roll of finely cut tobacco for smoking, enclosed in a wrapper of thin paper. But if one reads the 1st stanza in St. Vincent Millay's sonnet, a cigarette is a peculiar device to measure time, by its gradual shortening due to the falling ashes. It also implies momentary memories of an ending affectional relation, in the 3rd and 4th stanzas.

Similarly with the so to speak sunset, of the faraway sun, the closest star to planet earth which is actually rotating around the sun. Here the sun represents a less peculiar device to measure a short time and perhaps human vanity.

2.3   Poetry Is Runnable

Running a poem is to read it in one's head or aloud, once and again, to understand its contents. Running means understanding (Fig. 3).

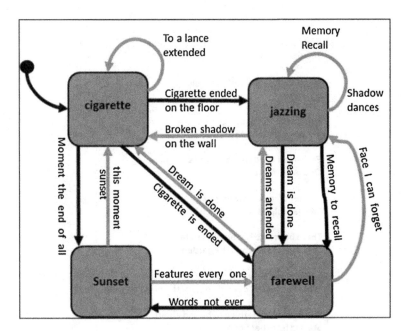

Fig. 3. Running (reading in one's head) the Sonnet by E. St. Vincent Millay – in each of the four states one reads the respective stanza (class). One may proceed to the next state, to a related state or return to a previous state. The reason for a specific transition is written close to the transition arrow: this reason may be either a difficulty to be solved or just an associative link. For instance, in the upper transition from *cigarette* to *jazzing* one has "cigarette ended on the floor". Here "on the floor" is associatively related to "on the wall" in the opposite direction transition.

3 Refactoring Poetry as if It Were Software

In this section we analyze poems which illustrate cases of analogy to software in which one needs to refactor [10, 21], i.e. change the structure of classes.

3.1 Interrupted Stanzas

Poems may have shorter than expected stanzas, which convey meaning by their very interruptions. Figure 4 displays a poem, named Edge, written by Plath [29]. This poem has modern characteristics, as quite free structure and the absence of rhyme.

Plath's poem structurally has 10 very short stanzas of 2 verses each. But after some "poem running" in one's head, one perceives that many of the sentences of the poem are broken into consecutive verses.

For instance let us look at the sentence:

Her bare
Feet seem to be saying:

Edge – *Sylvia Plath*

The woman is perfected.
Her dead

Body wears the smile of accomplishment,
The illusion of a Greek necessity

Flows in the scrolls of her toga,
Her bare

Feet seem to be saying:
We have come so far, it is over.

Each dead child coiled, a white serpent,
One at each little

Pitcher of milk, now empty.
She has folded

Them back into her body as petals
Of a rose close when the garden

Stiffens and odors bleed
From the sweet, deep throats of the night flower.

The moon has nothing to be sad about,
Staring from her hood of bone.

She is used to this sort of thing.
Her blacks crackle and drag.

Fig. 4. "Edge" poem by Sylvia Plath – it has ten very short stanzas of only two verses. One perceives broken sentences, which rather link stanzas into three groups.

It starts in the 2nd verse of the 3rd stanza (*Her bare*) and continues in the 4th stanza (*Feet seem to be saying:*). There is a whole blank line between the stanzas strongly suggesting a deliberate interruption, as part of the poem significance, despite the fact that the word *Feet* begins with a capital letter insinuating that it starts a new sentence.

The structure of the poem, in particular the many interruptions, conveys semantics. Further examination of the poem links these interruptions to its main meaning.

Searching the Web [33] one finds that *Edge* was written a short time before Plath's relatively young age suicide, a drastic interruption of her life. Given this information, the title *Edge*, its broken sentences, and the overall poem turn meaningful.

3.2   Refactoring Poetry into Longer Classes

If one persists in the one-to-one correspondence between stanzas and UML classes, one would obtain ten classes for Plath's *Edge* poem class diagram.

In Fig. 5 one can see the corresponding class diagram of Plath's *Edge* poem. It contains just three classes, named by their most significant words, with attributes and functions given by considerations similar to those that lead to the diagram in Fig. 2.

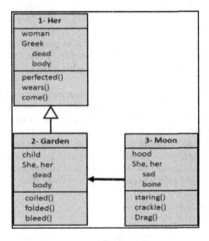

Fig. 5. UML class diagram of the Edge poem – each class (same conventions as in Fig. 2) fits to a group of stanzas (see text) and is numbered by the groups order. The 1st one is "Her".

Why three instead of ten classes? Resuming the "poem running" in one's head, one finds, in spite of broken sentences, a continuity among stanzas. Interruptions rather link between consecutive stanzas. One then divides the latter into three groups.

The 1st group has stanzas 1 to 4. Its subject is the woman and her body. It is perfected, an accomplishment, and a Greek tragedy brings it beyond the *Edge*, it is over. The 2nd group has stanzas 5 to 8. The garden may be associated with a kindergarten, with the Garden of Eden (the serpent, the woman), the flowers and odors. The 3rd group has two stanzas 9 and 10, with a distant detached moon is staring at the tragedy, but "nothing is to be sad about", she (the moon? the woman?) is used to this sort of thing.

So, instead of many too short stanzas, we refactor the poem classes into just three consistent ones. Refactoring classes, see e.g. [10, 21] is a software technique based upon semantics and efficiency considerations. Its aim is to facilitate comprehension of the software system, for purposes, such as maintenance, reuse, and so on. More details about this technique are given in Subsect. 5.3.

3.3 Refactoring Poetry into Shorter Classes

Sometimes poems may have longer than expected stanzas, conveying meaning by their extended continuity. They may be refactored by dividing them into smaller structures. An example is the poem "Borges and I" by Jorge Luis Borges, whose English translation [3] is seen in Fig. 6. In order to understand it, one may "run it" in one's head or hear the poet Borges himself, reading the original in Spanish ref. [4].

Borges and I – *Jorge Luis Borges*

The other one, Borges, is the one to whom things happen. I wander through Buenos Aires, and pause, perhaps mechanically nowadays, to gaze at an entrance archway and its metal gate; I hear about Borges via the mail, and read his name on a list of professors or in some biographical dictionary. I enjoy hourglasses, maps, eighteenth century typography, etymology, the savour of coffee and Stevenson's prose: the other shares my preferences but in a vain way that transforms them to an actor's props. It would be an exaggeration to say that our relationship is hostile; I live, I keep on living, so that Borges can weave his literature, and that literature justifies me. It's no pain to confess that certain of his pages are valid, but those pages can't save me, perhaps because good writing belongs to no one, not even the other, but only to language and tradition. For the rest, I am destined to vanish, definitively, and only some aspect of me can survive in the other.

Little by little, I will yield all to him, even though his perverse habit of falsifying and exaggerating is clear to me. Spinoza understood that all things want to go on being themselves; the stone eternally wishes to be stone, and the tiger a tiger. I am forced to survive as Borges, not myself (if I am a self), yet I recognise myself less in his books than in many others, less too than in the studious strumming of a guitar. Years ago I tried to free myself from him, and passed from suburban mythologies to games of time and infinity, but now those are Borges' games and I will have to think of something new. Thus my life is a flight and I will lose all and all will belong to oblivion, or to that other.

I do not know which of us is writing this page.

Fig. 6. The "Borges and I" poem by Jorge Luis Borges – it has two long stanzas, followed by a third one containing just a single sentence.

The overall meaning of the poem is the complex interaction between Borges the person and the Borges the poet. Borges the person wishes to live eternally, but the only possibility to somehow accomplish this goal is to be submissive to Borges the poet. On the other hand, Borges the poet has no life independent of Borges the person. This is the conflict between two entangled personae in the same body.

A tentatively refactored "Borges and I" poem can be seen in Fig. 7. The main idea is to disentangle the stanzas referring exclusively to the "other" Borges, from those of the person writing the poem. But, by the last verse in the poem, disentanglement fails.

The class diagram corresponding to Fig. 7 is seen in Fig. 8. It shows:

- two classes (1 and 3) explicitly referring to "the other" Borges;
- one class (2) referring just to the poem writer persona;

- one class (4) dealing with the interactions between the two personae;
- two classes (5 and 6) whose subject is survival and eternity;
- one short class (7) admitting failure in the separation of the two personae.

Borges and I – *Jorge Luis Borges*

The other one, Borges, is the one to whom things happen.

I wander through Buenos Aires, and pause, perhaps mechanically nowadays, to gaze at an entrance archway and its metal gate; I hear about Borges via the mail, and read his name on a list of professors or in some biographical dictionary. I enjoy hourglasses, maps, eighteenth century typography, etymology, the savour of coffee and Stevenson's prose:

the other shares my preferences but in a vain way that transforms them to an actor's props. It would be an exaggeration to say that our relationship is hostile;

I live, I keep on living, so that Borges can weave his literature, and that literature justifies me. It's no pain to confess that certain of his pages are valid, but those pages can't save me, perhaps because good writing belongs to no one, not even the other, but only to language and tradition.

For the rest, I am destined to vanish, definitively, and only some aspect of me can survive in the other. Little by little, I will yield all to him, even though his perverse habit of falsifying and exaggerating is clear to me. Spinoza understood that all things want to go on being themselves; the stone eternally wishes to be stone, and the tiger a tiger.

I am forced to survive as Borges, not myself (if I am a self), yet I recognise myself less in his books than in many others, less too than in the studious strumming of a guitar. Years ago I tried to free myself from him, and passed from suburban mythologies to games of time and infinity, but now those are Borges' games and I will have to think of something new. Thus my life is a flight and I will lose all and all will belong to oblivion, or to that other.

I do not know which of us is writing this page.

Fig. 7. The "Borges and I" poem by Jorge Luis Borges refactored after running it in our heads – it has five longer stanzas, enclosed by two stanzas with just one sentence.

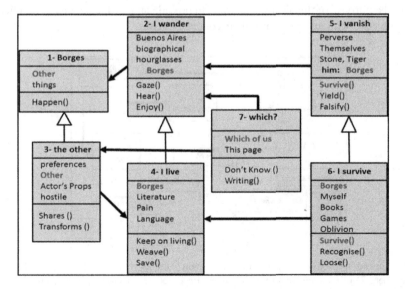

Fig. 8. Class diagram of the "Borges and I" refactored poem by Jorge Luis Borges – compare with poem in Fig. 7 and the paper text for an interpretation of these classes.

4 Prose Is Easily Readable

Software assets other than runnable programs, e.g. requirements, specification documents or user's guides, are essentially prose. By prose we mean all texts, literary, professional, excluding a few types: poetry, text containing formulas (mathematical or chemical), very specialized texts like philosophical ones. The immediate quality of prose is to be easily readable, as people are used to speak and write in prose.

4.1 But Prose also Has Metaphors

On the other hand, prose texts have varying degrees of sophistication. Thus all the complex characteristics of natural language, say ambiguity, synonymy, use of metaphors etc., may appear in ordinary prose, demanding deeper comprehension.

5 Software Systems Are Conceptual Constructs

Software systems are hierarchical. Moving from bottom to top, one meets assembly language near to a machine, next a high-level language as Java, then UML and on top a set of system domain ontologies. From the ontology concepts one can derive UML classes. The highest abstraction level is conceptual content, closest to human beings.

5.1 Conceptual Content of Abstract Factory

In order to illustrate the idea of conceptual content of software, we use a software design pattern class diagram as a case study.

The class diagram of the *Abstract Factory* software design pattern [13] is seen in Fig. 9. The purpose of this design pattern is to provide an interface to create families of related objects, without specifying their concrete classes.

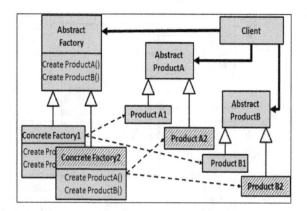

Fig. 9. Abstract factory design pattern class diagram – the abstract factory class may have any number (here 2) of concrete factory sub-classes. Each concrete factory has the same family of products (here Product A and Product B). Concrete Factory 2 and its products have class names with hatched background.

A general observation, when looking at this class diagram, is that none of its concepts is part of the vocabulary (the reserved words) of a programming language. So they are not intrinsic software artefacts. Let us now specifically examine some of these concepts. We start with "*Abstract Factory*". Both words "abstract" and "factory" appear in the English dictionary as independent words with specific meanings, having no necessary relation to software. The word "abstract" has been incorporated with a special meaning into software. It denotes a certain kind of class.

5.2   Software Has Metaphors

The concept "*Abstract Factory*" is a metaphorical usage of natural language terms absorbed into software. Looking carefully at the Abstract Factory class diagram, one sees that *all* its concepts result from metaphorical usage. "Create" is another such example, meaning construction of an object in an object oriented programming language. A "factory" is thus a class whose main purpose is to construct "objects".

Moreover, "*concrete factory*" is a typical example of ambiguity. It is not a factory that fabricates concrete or itself made of concrete (in the dictionary: *concrete* is a strong construction material made of sand, conglomerate gravel, in a cement matrix).

Even the term "family" of related objects explaining the purpose of the abstract factory design pattern metaphorically extends the meaning of a non-software word.

Summarizing our claim, the highest-level abstraction of software is a set of concepts that may be extracted from domain ontologies and included in UML class diagrams. Since these concepts are ordinary natural language words, all the complexities of natural language, such as ambiguity, synonymy, figures of speech as metaphors, are fundamental for the understanding, development and maintenance of software, and therefore should be part of a basic theory of software.

Are these complexities so fundamental for software? The immediate answer is that if software itself should automatically manipulate software, these complexities should be taken into account.

5.3   Refactoring Software

The purpose of refactoring software is to rearrange the software system structure, at the UML level and above. The software engineering literature has several refactoring kinds (e.g. [10, 21]). It is done by the following actions:

(1) *move methods* – from a class to another, without changing class numbers;
(2) *break down* – classes into smaller ones;
(3) *coalesce* – a few classes (at least two of them) into a bigger one;
(4) *add new* – not previously existing classes;
(5) *delete* – classes considered superfluous.

Here we are interested in the conceptual aspects of refactoring. It can be succinctly summarized as:

(a) run the software system in an IDE (Integrated Development Environment) or if the system is quite simple, run it in your head;

(b) identify the class and function names, which most commonly are natural language concepts, or technical terms found in suitable ontologies;

(c) choose the appropriate refactoring from the five above mentioned actions;

(d) perform the chosen refactoring.

5.4   A Conceptual Refactoring Example

To illustrate the idea, we give a conceptual refactoring example – in this case an addition of a new class, with its necessary functions. Assume one has diverse types of documents – pure text, HTML text, Microsoft Word and Pdf – each type being represented by a class. In addition one has some operations that can be executed on any type of document – e.g. to print, to display on a screen, or to compress the document before transmitting it by some communication message.

The previously existing classes are the documents and operations. All of these were linked to the document parent class: the document types by inheritance and the operations by an association. The software engineer is supposed to get the insight that document types are conceptually different from operation devices. Thus, in order to facilitate understanding of the software system, the engineer is expected to add a new class, say a *visitor*, to differentiate the operation devices from the document types.

A *visitor* class – itself a metaphor – is a well-known software component. It signals to any posterior reader of the software that the *visitor* software design pattern – one of the patterns in the "Gang of Four" (GoF) book [13] – has been used, pointing out that its operation device sub-classes can be applied to any document type in the relevant system. The corresponding system class diagram after refactoring is shown in Fig. 10. All the class names are natural language words or technical terms, and none is a reserved word in a programming language.

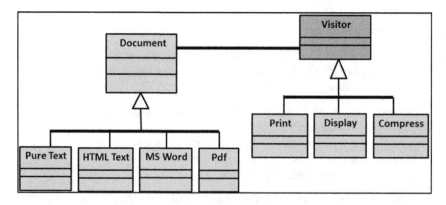

Fig. 10. Class diagram of a *Visitor* operating on documents – this is the refactored system after the addition of the visitor class (green background), to the previous existing classes (yellow background). (Color figure online)

5.5 A Metaphor Design Pattern

Given that metaphors are so common in higher abstraction levels of software, we propose a generic *Metaphor* design pattern. The purpose of this design pattern is to easily change or add a new context for a given term with multiple meanings.

This proposed design pattern is modelled after the *Strategy* pattern [13], with a role inversion. In the Strategy pattern a context is fixed and strategies are variable. In the Metaphor pattern a term is fixed and its meaning providing contexts are variable.

The class diagram of this proposed Metaphor pattern is shown in Fig. 11. Its generic classes are:

- *Metaphor* – it contains the several meanings of a given term; it receives a term meaning as input and sets its specific context;
- *Term* – it is an (abstract) interface declaring a SetContext() function and corresponding Actions();
- *Contexts* – these are concrete classes with different domains, each say given by an ontology and its specific actions.

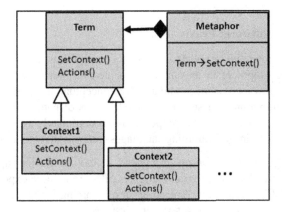

Fig. 11. Metaphor design pattern class diagram – in a metaphor a single term is fixed and its various meanings are set by variable contexts. In this diagram only two contexts are shown, but they imply that any number of contexts can be added.

An example of a fixed term is "*bridge*". This term has numerous different but metaphorically related meanings given by the respective contexts: e.g. civil engineering, odontology, card games, and even design patterns.

6 Discussion

For the purposes of this discussion, instead of referring in separate specifically to the similarity of software either to poetry or to prose, we jointly refer to both under the rubric of software *natural language* conceptual issues.

6.1   Foundational Issues

Significant issues have been opened in the literature concerning software conceptual contents. Is software semantics intrinsic (inner) or extrinsic (outer)? In other words, is software semantics just given by the inner workings of the computing machine or is it deeply related to the human conceptual (outer) world?

White [32] mentions obstacles to solve this symbol grounding problem, e.g. the difficulty to assign a clear boundary between inside and outside of the computer system. Piccinini [28] argues for "computation without representation", i.e. meaning of symbols and states is given by functional properties of computational systems. According to Smith's [30] "participatory computation", any physical computing system is inherently situated in its environment such that its processes extend beyond the system boundaries, which stand in semantic relations to distal states of affairs.

Another issue is *universality*. Do conceptual contents affect only restricted kinds of software systems? Absolutely not: characteristics of natural languages – ambiguity, synonymy, metaphors for new word invention, or metonymy in the context of software design pattern functionality [24] – are widespread at a given time, and also expected to persist along time. They are inherent to the vitality of natural languages.

6.2   Conceptual Software: Praxis

From the praxis viewpoint, explicit consideration of software conceptual contents enables system development in an ontology-oriented fashion [7, 27]. There are also tools to improve software system modularity by conceptual analysis [8, 15, 19].

The practical importance of semantic considerations of natural language found in higher software abstraction levels, as opposed to the dominantly syntactic concerns in lower levels, viz. code in programming languages, refer to diverse aspects:

(a) *Natural language for non-programmers* – mobile device applications are increasingly used by non-programmers, i.e. software is more exposed and should be understood by people not "speaking" programming languages.

(b) *Software systems complexity* – software systems are growing in complexity and criticality, with potentially life-threatening situations, e.g. autonomous vehicles, remote surgery, and automatic power stations. Complex systems design is done in higher abstraction levels to enable design comprehension.

(c) *Relation between structure and behavior* – structure conveys part of the software meaning; exercising software behavior (running it) leads to understanding; thus refactoring – changing structure – effectively embeds the obtained understanding in the modified structure.

6.3   Conceptual Software: Theory

FCA (Formal Concept Analysis) [14, 15] is a well-developed formalism dealing with concepts. It involves lattice theory and related algebraic domains of mathematics. Besides its theoretical importance, it has a variety of applications.

One raises the issue of boundaries of the formalism applicability: are there software systems for which this formalism is insufficient? We encourage exploration beyond these boundaries, eventually leading to new discoveries.

7 Conclusion

The main contribution of this work is raising issues concerning the importance of conceptual analysis for software theory – which follows from inherent characteristics of natural languages, rather than from programming languages.

Acknowledgements. The authors wish to acknowledge significant suggestions by two anonymous referees. I.E. is grateful to Yedidia Argaman for actual experiments with "running" poetry.

References

1. Backus, J.W.: The syntax and semantics of the proposed international algebraic language of Zurich ACM-GAMM conference. In: Proceedings of the International Conference on Information Processing, Paris (1959)
2. Booch, G., Rumbaugh, J., Jacobson, I.: The Unified Modeling Language User Guide, 2nd edn. Addison-Wesley, Boston (2005)
3. Borges, J.L.: Selected Poems. English translation by A.S. Kline (2013). http://www.poetry intranslation.com/PITBR/Spanish/Borges.htm#_Toc192667921
4. Borges, J.L.: Borges y yo. Original text and audio in Spanish by the poet. http://palabravirtual.com/index.php?ir=ver_voz1.php&wid=725&p=Jorge%20Luis%20Borges&t=Borges%20y%20yo
5. Brown, J.R., Fehige, Y.: Thought experiments. In: Zalta, E.N. (ed.) Stanford Encyclopedia of Philosophy (2011). http://plato.stanford.edu/archives/fall2011/entries/thought-experiment/
6. Chomsky, N.: Syntactic Structures. Mouton, Berlin (1957)
7. Exman, I., Yagel, R.: ROM: an approach to self-consistency verification of a runnable ontology model. In: Fred, A., Dietz, Jan, L.,G., Liu, K., Filipe, J. (eds.) IC3K 2012. CCIS, vol. 415, pp. 271–283. Springer, Heidelberg (2013). doi:10.1007/978-3-642-54105-6_18
8. Exman, I., Speicher, D.: Linear software models: equivalence of modularity matrix to its modularity lattice. In: Proceedings of the 10th ICSOFT International Joint Conference on Software Technologies, Colmar, France, pp. 109–116 (2015). doi:10.5220/0005557701090116
9. Fauconnier, G.: Mappings in Thought and Language. Cambridge University Press, Cambridge (1997)
10. Fowler, M., Beck, K., Brant, J., Opdyke, W., Roberts, D.: Refactoring: Improving the Design of Existing Code. Addison-Wesley, Boston (1999)
11. Gabriel, R.P.: Designed as designer. In: Proceedings of the 23rd ACM SIGPLAN Conference on Object Oriented Programming Systems, Languages and Applications, OOPSLA 2008 (2008). ACM SIGPLAN Not. **43**(10), 617–632. doi:10.1145/1449955.1449813
12. Galilei, G.: Dialogue Concerning the Two Chief World Systems (1632). Italian, Translated by Stillman Drake. University of California Press, Berkeley, CA (1953)
13. Gamma, E., Helm, R., Johnson, R., Vlissides, J.: Design Patterns. Addison-Wesley, Boston (1995)

14. Ganter, B., Wille, R.: Formal Concept Analysis: Mathematical Foundations. Springer, New York (1999)
15. Ganter, B., Stumme, G., Wille, R.: Formal Concept Analysis - Foundations and Applications. Springer, Berlin (2005)
16. Hopper, G.M.: Compiling routines. Comput. Autom. **2**(4), 1–5 (1953)
17. Lakoff, G.: A principled exception to the coordinate structure constraint. In: Proceedings of the 21st Regional Meeting Chicago Linguistic Society. Chicago Linguistic Society (1986)
18. Langacker, R.W.: Foundations of Cognitive Grammar. Stanford University Press, Stanford (1987)
19. Lindig, C., Snelting, G.: Assessing modular structure of legacy code based on mathematical concept analysis. In: ICSE 1997 Proceedings of the 19th International Conference on Software Engineering, pp. 349–359. ACM (1997). doi:10.1145/253228.253354
20. McCarthy, J.: Recursive functions of symbolic expressions and their computation by machine. Commun. ACM **3**(4), 184–195 (1960). doi:10.1145/367177.367199
21. Mens, T., Tourwe, T.: A survey of software refactoring. IEEE Trans. Softw. Eng. **32**, 126–139 (2004). doi:10.1109/TSE.2004.1265817
22. St. Vincent Millay, E.: Unnamed Sonnets I-XII, pp. 97–110. Mitchell Kennerley, New York (1921)
23. Modern American Poetry, Online Poems. www.english.illinois.edu/maps/poets/m_r/millay/online_poems.htm
24. Noble, J., Biddle, R., Tempero, E.: Metaphor and metonymy in object-oriented design patterns. In: ACSC 2002 Proceedings of the 25th Australasian Conference on Computer Science, vol. 4, pp. 187–195 (2002). doi:10.1145/563857.563823
25. Nofre, D., et al.: When technology became language, the origins of the linguistic conception of computer programming, 1950–1960. Technol. Cult. **55**(1), 40–75 (2014). doi:10.1353/tech.2014.0031
26. OMG (Object Management Group): Unified Modeling Language (UML) Specification version 2.5 (2015). http://www.omg.org/spec/UML/2.5
27. Pan, J.Z., et al. (eds.): Ontology-Driven Software Development. Springer, Heidelberg (2013)
28. Piccinini, G.: Computation without representation. Philos. Stud. **137**(2), 205–241 (2008). doi:10.1007/s11098-005-5385-4
29. Plath, S.: Edge Poem (1963). http://www.poetryfoundation.org/poem/178970
30. Smith, B.C.: The foundations of computing. In: Scheutz, M. (ed.) Computationalism: New Directions, pp. 23–58. MIT Press, Cambridge (2002)
31. Turing, A.: On computable numbers with an application to the entscheidungsproblem. Proc. London Math. Soc. **s2-42**, 230–265 (1937). doi:10.1112/plms/s2-42.1.230
32. White, G.: Descartes among the robots – computer science and inner/outer distinction. Mind. Mach. **21**(2), 179–202 (2011). doi:10.1007/s11023-011-9232-4
33. Plath, S.: Wikipedia (2015). https://en.wikipedia.org/wiki/Sylvia_Plath

Automatic Pattern Generator of Natural Language Text Applied in Public Health

Anabel Fraga(✉), Juan Llorens, Eugenio Parra, and Valentín Moreno

Computer Science Department, Carlos III of Madrid University,
Av. Universidad 30, Leganés, Madrid, Spain
{afraga,llorens,vmoreno}@inf.uc3m.es

Abstract. At the moment, a huge amount of scientific articles is available, referring to a wide variety of topics like medicine, technology, economics, finance, and so on. Scientific papers show results of scientific interest and also present the evaluation and interpretation of relevant arguments. Due to the fact that these papers are created with a high frequency it is feasible to analyze how people write in a given domain. Within the discipline of natural language processing there are different approaches to analyze large amounts of text corpus. Identification patterns with semantic elements in a text, let us classify and examine the corpus to facilitate interpretation and management of information through computers. At the moment, a semiautomatic or automatic way to generate natural language patterns is not available or quite complicated. In the paper, it is shown how a tool developed for this research is tested in a domain of public health. The results obtained – by means of a tool and aided by graphs – provide groups of words that are used (to determine if they come from a specific vocabulary), most common grammatical categories, most repeated words in a domain, patterns found, and frequency of patterns found. A domain of public health has been selected containing 800 papers concerning different topics referring to genetics. The topics include mutations, genetic deafness, DNA, trinucleotide, suppressor genes, among others. An ontology of public health has been used to provide the basis of the study.

Keywords: Indexing · Ontologies · Knowledge · Patterns · Reuse · Retrieval · Public health

1 Introduction

There are huge quantities of scientific articles referring to any topic like medicine, technology, economics, and finance, among others. Christopher Manning states in his book that: "People write and say lots of different things, but the way people say things - even in drunken casual conversation - has some structure and regularity." [11].

The main issue in this paper is: how do people write? Nowadays, researchers conduct investigations using natural language processing tools, generating indexing and semantic patterns that help to understand the structure and relation of how writers communicate through their papers.

This project will use a natural language processing system which will analyze a corpus of papers acquired by Ramon and Cajal Hospital from Madrid. The documents

© Springer International Publishing AG 2016
A. Fred et al. (Eds.): IC3K 2015, CCIS 631, pp. 381–395, 2016.
DOI: 10.1007/978-3-319-52758-1_21

will be processed by the system and will generate simple and composed patterns. These patterns will give us different results which we can analyze and conclude the common aspects the documents have even though they are created by different authors but are related to the same topic [1–3]. The study uses as center of the study. an ontology created in a national founded project for Oncology and it has been extended with general terms of public health.

The reminder of this paper is organized as follows: Sect. 2 presents a state of the art of the main topics of the paper; Sect. 3 includes the summary of our approach to information processing; Sect. 4 discusses the results, and finalizes with conclusions.

2   State of the Art

2.1   Information Reuse

Reuse in software engineering is present throughout the project life cycle, from the conceptual level to the definition and coding requirements. This concept improves the quality and optimization of the project development, but it has difficulties in standardization of components and combination of features. Also, the software engineering discipline is constantly changing and updating, which quickly turns obsolete the reusable components [10].

At the stage of system requirements, reuse is implemented in templates to manage knowledge in a higher level of abstraction, providing advantages over lower levels and improving the quality of the project development. The patterns are fundamental reuse components that identify common characteristics between elements of a domain and can be incorporated into models or defined structures that can represent the knowledge in a better way.

2.2   Natural Language Processing

The need for implementing Natural Language Processing techniques arises in the field of the human-machine interaction for many tasks such as text mining, information extraction, language recognition, language translation, and text generation, particularly for fields that require lexical, syntactic and semantic analysis to be recognized by a computer [5]. The natural language processing consists of several stages which take into account the different techniques of analysis and classification supported by the current computer systems [6]:

(1) Tokenization: The tokenization corresponds to a previous step on the analysis of the natural language processing, and its objective is to demarcate words by their sequences of characters grouped by their dependencies, using separators such as spaces and punctuation [13]. Tokens are items that are standardized to improve their analysis and to simplify ambiguities in vocabulary and verbal tenses.

(2) Lexical Analysis: Lexical analysis aims to obtain standard tags for each word or token through a study that identifies the turning of vocabulary, such as gender, number and verbal irregularities of the candidate words. An efficient way to

perform this analysis is by using a finite automaton that takes a repository of terms, relationships and equivalences between terms to make a conversion of a token to a standard format [8]. There are several additional approaches that use decision trees and unification of the databases for the lexical analysis but this not covered for this project implementation [4, 19].

(3) <u>Syntactic Analysis</u>: The goal of syntactic analysis is to explain the syntactic relations of texts to help a subsequent semantic interpretation [12], and thus using the relationships between terms in a proper context for an adequate normalization and standardization of terms. To incorporate lexical and syntactic analysis, in this project were used deductive techniques of standardization of terms that convert texts from a context defined by sentences through a special function or finite automata.

(4) <u>Grammatical Tagging</u>: Tagging is the process of assigning grammatical categories to terms of a text or corpus. Tags are defined into a dictionary of standard terms linked to grammatical categories (nouns, verbs, adverb, etc.), so it is important to normalize the terms before the tagging to avoid the use of non-standard terms. The most common issues of this process are about systems' poor performance (based on large corpus size), the identification of unknown terms for the dictionary, and ambiguities of words (same syntax but different meaning) [14, 20]. Grammatical tagging is a key factor in the identification and generation of semantic index patterns, in where the patterns consist of categories not the terms themselves. The accuracy of this technique through the texts depends on the completeness and richness of the dictionary of grammatical tags.

(5) <u>Semantic and Pragmatic Analysis</u>: Semantic analysis aims to interpret the meaning of expressions, after on the results of the lexical and syntactic analysis. This analysis not only considers the semantics of the analyzed term, but also considers the semantics of the contiguous terms within the same context. Automatic generation of index patterns at this stage and for this project does not consider the pragmatic analysis.

2.3   Information System: RSHP Model as Universal Schema

RSHP is a model of information representation based on relationships that handles all types of artifacts (models, texts, codes, databases, etc.) using a unified scheme. This model is used to store and link generated pattern lists to subsequently analyze them using specialized tools for knowledge representation [9, 10]. Within the Knowledge Reuse Group at the University Carlos III of Madrid the RSHP model is used for projects relevant to natural language processing. [1, 7, 17, 18] The information model is presented in Fig. 1. An analysis of sentences and basic patterns are shown in Fig. 2.

3   Information Process

Several steps have been done to start analyzing the set of documents. The processing steps are mentioned below:

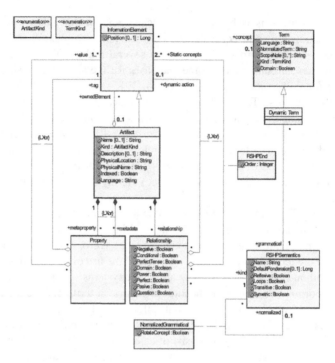

Fig. 1. RSHP information representation model. (Llorens et al. 2005).

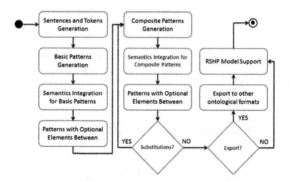

Fig. 2. Analysis of sentences and basic patterns.

1. *Document Choice* – Choose the documents to analyze out of the eight hundred. These have been picked randomly.
2. *Document Formatting* – Due to the fact that the system for analyzing text and discovery patterns [15–17] only analyzes text documents, these papers had to be converted since they were given in a PDF format. The conversion of these documents has been manual. Headers, footers, page numbers, references have been removed because the tool analyzes sentences of each document and the ones that

have been avoided are not relevant at the time of the analysis. The system extended the Information schema described in Fig. 1, it is shown in Fig. 3. The extension consists of a corpus of documents indexed, the basic patterns generated, the additional patterns containing sub-patterns, and the new semantics generated for the domain; the core is always the general schema RSHP.

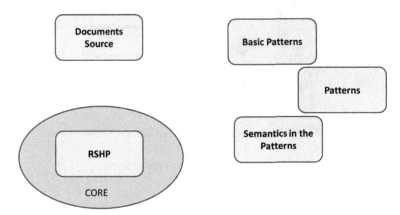

Fig. 3. Extended version (in a graphical mode) of the information schema for the system to be used in the study.

3. *Pattern Creation* – After the documents have been inserted in the tool (also called system, it is a software that allows the creation of the patterns in a semiautomatic manner [17]), you proceed to create basic patterns and patterns with all the text documents. The minimum of frequency can be changed. A pattern is a set of slots that allows the Natural Language Processing of text by pattern matching techniques. Basic patterns are those with two sides of simple kind, it means without using sub-patterns or another patterns as left or right side in the binary structure.

4. *Analysis* – Different scenarios will be used in order to analyze the different results and have a final conclusion. The domain of Public health is based on a previously existing oncology ontology, the ontology at the end of the process increased in 60570 terms, the ontology structure is shown in Fig. 4. These scenarios will be described below.

The scenarios selected for the study are as follows (a summary is shown in Fig. 5):
Scenario I: Use all grammatical categories with a minimum frequency of 1 and without a difference in semantics.
This scenario has the following characteristics:

1. The basic patterns for the text documents in the pattern analysis system (Fig. 5) were generated in this scenario.
2. Generate all patterns for these documents using all grammatical categories located in Create patterns tab in pattern analysis system.
3. Minimum of frequency to create a pattern is 1.

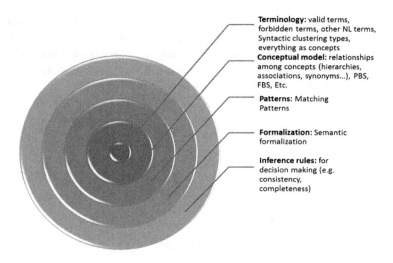

Fig. 4. Ontology structure created in layers.

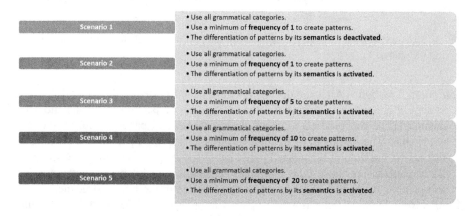

Fig. 5. Summary of the steps followed.

Scenario II: Use all grammatical categories with a minimum frequency of 1 and differ patterns by its semantics.

This scenario has the following characteristics:

1. Since basic patterns were already created in scenario one, it is not necessary to create them again because basic patterns analyze each of the documents and with the result of these is how patterns can be created. In this case the steps shown in Scenario I (step 2 and 3) are the same.

Scenario III: Use all grammatical categories with a minimum of frequency of 5 and differ patterns by its semantics.

This scenario has the following characteristics:

1. In this case the steps shown in Scenario I (step 2 and 3) are the same, but applied to a frequency of 5.

Scenario IV: Use all grammatical categories with a minimum frequency of 10 and differ patterns by its semantics.

This scenario has the following characteristics:

1. In this case the steps shown in Scenario I (step 2 and 3) are the same, but applied to a frequency of 10.

Scenario V: Use all grammatical categories with a minimum frequency of 20 and differ patterns by its semantics.

This scenario has the following characteristics:

1. In this case the steps shown in Scenario I (step 2 and 3) are the same, but applied to a frequency of 20.

The patterns are frequency pattern creation is showed in Figs. 6 and 7. The objective of this paper is to discuss the results comparing the scenarios, but in an extended version a detailed presentation of each scenario might be of interest to the audience (Table 1).

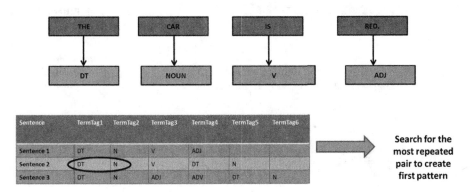

Fig. 6. Example of frequency of patterns at first level.

The frequency pattern creation is showed in Figs. 6 and 7. The objective of this paper is to discuss the results comparing the scenarios, but in an extended version a detailed presentation of each scenario might be of interest to the audience.

Figures 6 and 7, show the creation of patterns and sub-patterns by frequency of appearance. It means for the example shown, that P1 is the most repeated basic pattern, and (P2, P1) is the most repeated pattern with sub-patterns. Nouns are the center in all the cases, a structured manner of writing was found in most of the papers.

Each pattern might be differentiated by semantics as shown in Fig. 8. The following table shows the patterns that have semantics on both sides. Since the difference of pattern by its semantics has been activated you can see in the table that there are not multiple lines of one pattern with the same name. In this case, the patterns are differed by the semantics and they have a distinct identifier Pattern (Table 2).

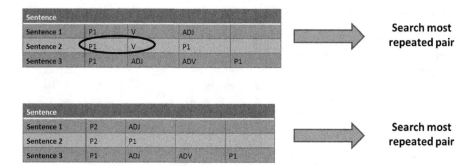

Fig. 7. Example of frequency of patterns at the following level.

Table 1. Patterns created in all scenarios.

Number of patterns created				
S 1	S 2	S 3	S 4	S 5
9188	9818	2145	1171	650

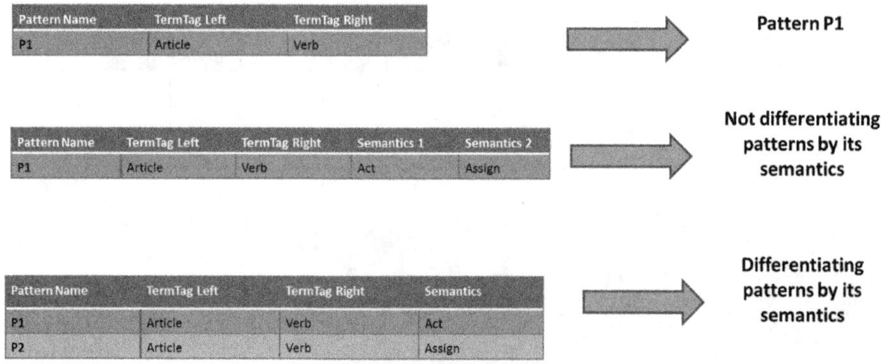

Fig. 8. Example of patterns by its semantics.

Table 2. Patterns with semantics in both sides.

Pattern	Semantic left	Semantic right
P666	- RANGE <= (MAXIMUM) (OK)	- RANGE ALL (OK)
P1402	- RANGE <= (MAXIMUM) (OK)	- RANGE LITTLE-FEW-SOME (OK)
P2105	- RANGE <= (MAXIMUM) (OK)	- RANGE MUCH-MANY (OK)
P702	- Deny (OK)	- RANGE <= (MAXIMUM) (OK)
P465	- MODAL OPTIONAL (OK)	- Deny (OK)
P1587	- MODAL OPTIONAL (OK)	- Provide (OK)
P2116	- RANGE ALL (OK)	- RANGE <= (MAXIMUM) (OK)
P1273	- RANGE MUCH-MANY (OK)	- RANGE > MINIMUM (OK)
P2114	- RANGE MUCH-MANY (OK)	- RANGE < MAXIMUM (OK)

It is interesting to notice that most recognized semantic is the one assigned to ranges, modal verbs and negative sentences.

4   Results

Basic Patterns

After the basic patterns were created, all the sentences from the text documents were analyzed and to each of the words (known in the database as token text) a termtag or syntactic tag was assigned with the help of the tables Rules Families and Vocabulary in the Requirements Classification database.

You may find the most repeated words in the domain of documents in the Basic patterns table. The most repeated words in grammatical categories such as nouns, verbs and nouns coming from the ontology we used.

Patterns

The name of patterns is different depending on the requirements you use when you generate them. In this case there are 5 scenarios which have different results because each of them had different characteristics. When the minimum frequency to create patterns is higher the patterns will be less.

Scenario 1 and 2 have the same minimum of frequency (1) but the difference is that scenario 2 has the differentiation of patterns by its semantics activated; the result of differentiating is that more patterns are created due to the fact that there are patterns with distinct identifiers. There has been the decision to use this option for the rest of scenarios (scenarios 3, 4, 5), this way we can analyze the maximum number of patterns created and also the semantics of them can be easier to understand (Fig. 9).

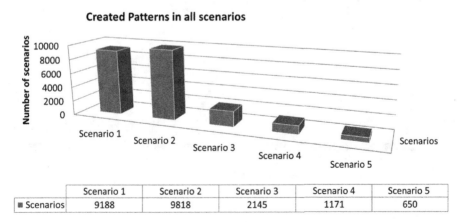

	Scenario 1	Scenario 2	Scenario 3	Scenario 4	Scenario 5
■ Scenarios	9188	9818	2145	1171	650

Fig. 9. Created patterns in all scenarios.

Patterns Created with Same Termtags

Among all the scenarios there has been one pattern in common. This is composed of two same termtags on the left and right side. This pattern has the same identifier and name for all scenarios it is pattern P1 (Table 3).

Table 3. Patterns with same termtag.

Pattern name	Term tag left	Term tag right
P1	Unclassified noun	Unclassified noun

Unclassified nouns are the most common termtags in all the text documents used. Some words that are unclassified nouns are abbreviations, some words in another language, slang language, uncommon symbols, and scientific terms.

Patterns Created with Two Different Termtags

There are patterns which have a different termtags on the left and right side of a pattern. After comparing the results of the five scenarios it has been observed that the termtags with higher frequency are common between all scenarios for each side.

An example of a pattern generated presented as a binary tree is shown in Fig. 10 (Table 4).

For all scenarios the 20 patterns with higher frequency as termtags on each side have been shown in graphs. Also, it has been shown how these patterns are composed.

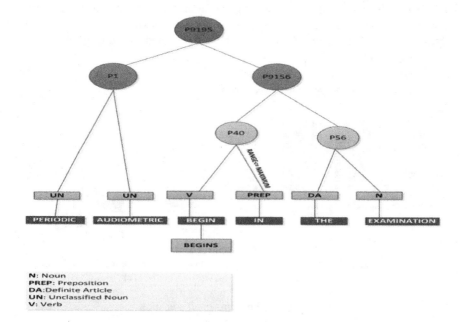

N: Noun
PREP: Preposition
DA: Definite Article
UN: Unclassified Noun
V: Verb

Fig. 10. Binary tree representation of a sentence.

Table 4. Most common termtags.

Most common termtags in all scenarios	
Left side	Right side
Unclassified noun	Unclassified noun
Adverb	Adverb
Noun(ontology oncology)	Noun (ontology oncology)
Noun	Noun
Verb	Verb
Adverb	Adverb
Preposition	Preposition

The composition of these patterns is common between all scenarios. The termtags for these patterns are unclassified nouns, noun, prepositions, verbs, nouns (oncology ontology), adverbs, adjectives. These termtags are the ones that also are the most common in all text documents.

Semantic Patterns

Some of the created patterns in all scenarios have semantics assigned on the left or right side, one side, or some patterns do not have.

In scenario one, there are patterns with different semantics but they have the same identifier and name, this is because the option to differentiate them by semantics was inactive. The rest of patterns for the other scenarios are not repeated and they are unique. Observing the results, the semantics for patterns in different scenarios is similar.

The change is noticeable in the pattern name and ids. In this case, they are different because in every scenario the number of patterns was different due to the minimum of frequency used. While the minimum of frequency is higher, the number of created patterns is less.

- The higher minimum of frequency used for this project has been 20.
- With the help of the system used for the creation of patterns has been successful. Processing documents at a time for basic patterns was not issue for the tool.
- 11% of the term names found in the documents were from the ontology added.
- Writers about genetic deafness use a similar vocabulary and appropriate terms.
- Studying patterns will facilitate the search of documents in search engines or databases.
- It can also assist the writing for any user that is not a researcher or scientist. With the help of patterns documents can be written (see Fig. 11).

As expected, the frequency of term tags is higher in the terms already existing or new terms coming from the new domain to be included, these are called unclassified terms nouns as shown in Figs. 12 and 13.

Looking at the graphs, it is evident that unclassified nouns is the number one term tag most used in all text documents with a frequency of 32.2% (64745 terms). The system defines these as nouns that can't be classified due to the fact that they can be a word from a complex vocabulary, abbreviation, and slang language, among others.

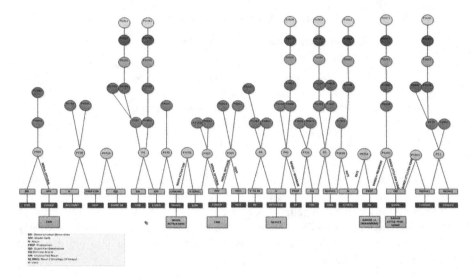

Fig. 11. Segment of the graph with patterns and semantics assigned.

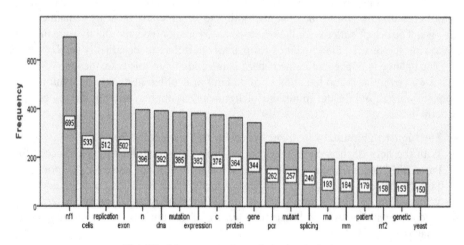

Fig. 12. Most repeated words in the analysis.

Since we added an ontology with term names referring to oncology, 11% of the words come from this vocabulary. This means, that if we haven't added the ontology there will be around 44% of unclassified nouns. This could have decreased our results in the patterns creation because most of the patterns would have been created with only unclassified nouns. However, adding this ontology improves it. Symbols, prepositions and definite articles are the termtags categories that follow; this is reasonable since these are termtags that we use in our everyday speaking or writing.

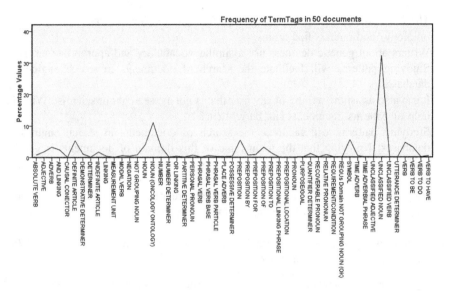

Fig. 13. Frequency of Tern Tags in the analysis.

5 Discussion and Conclusion

Using a domain of documents related to genetic engineering and making different scenarios to create patterns it has been possible to conclude the following:

1. First of all, a semiautomatic or automatic way to generate natural language patterns was not available or quite complicated. It is shown how a tool developed is tested in a domain of public health.
2. A minimum of frequency of 1, 5, 10, 20 has been used to create patterns. In the results, we can confirm that while the minimum of frequency is higher the number of patterns created is less.
3. The higher minimum of frequency used for this project has been 20. This means that every 20 pairs of basic patterns a pattern has been created. After analyzing results, the domain of documents has more than 20 pairs repeated, meaning that more scenarios can be made in a future.
4. Even though the documents are created by different authors, the patterns created are common among all scenarios. The patterns can have a different name but the composition of them is consistent in the different tests made in this project.
5. With the help of a system for generating patterns in an automatic way the creation of patterns has been successful. Processing documents at a time for basic patterns was not a problem for the tool.
6. After finishing all scenarios and analyzing results, we can conclude that authors writing papers about a same topic (in this case, genetic deafness) have similarity in how they write. 11% of the term names found in the documents were from the ontology added. This improved the creation of frequency patterns to observe the compositions of the nouns from the oncology and any other grammatical category.

The most common combination for a pattern in all scenarios was noun (oncology ontology) and unclassified nouns.

7. Writers about genetic deafness use a similar vocabulary and appropriate terms.
8. Studying patterns will facilitate the search of documents in search engines or databases.
9. It can also assist the writing of any user that is not a researcher or scientist. With the help of patterns documents can be written.
10. Studying patterns will facilitate the search of documents in search engines or databases. Knowing that the main topic of this domain of documents is genetic deafness, we can select the most used words to create a sentence and look for it in any search engine. One example can be searching "Mutation AND Deafness", since we now know that these articles are related because they use similar vocabulary, our results will be papers, magazines, books, webpages regarding this same topic.
11. It can also assist the writing for any user that is not a researcher or scientist. With the help of patterns documents can be written.

After ending all scenarios and analyzing results, we suggest some recommendations for this study:

- Expand the existing vocabulary table in the database. More ontologies can be included, symbols, slang languages, words in different language, among others.
- Differentiate patterns by its semantics to maximize the creation and different compositions of them.
- Using different minimum of frequencies at the moment of creating patterns will help to compare and analyze results.
- Expand the existing vocabulary table in the database. More ontologies can be included, symbols, slang languages, words in different language, among others.
- Differentiate patterns by its semantics to maximize the creation and different compositions of them.
- Using different minimum of frequencies at the moment of creating patterns will help to compare and analyze results.
- Expand the existing vocabulary table in the database. More ontologies can be included, symbols, slang languages, words in different language, among others. Differentiate patterns by its semantics to maximize the creation and different compositions of them.

Acknowledgements. The Authors Thank the AGO2 Project, Founded by the Ministry of Education of Spain for Aiding the Author in the Research and Production of This Paper.

The Research Leading to These Results Has Received Funding from the European Union's Seventh Framework Program (FP7/2007-2013) for Crystal – Critical System Engineering Acceleration Joint Undertaking under Grant Agreement No 332830 and from Specific National Programs and/or Funding Authorities.

References

1. Abney, S.: Part-of-speech tagging and partial parsing. In: Young, S., Bloothooft, G. (eds.) Corpus-Based Methods in Language and Speech Processing. An ELSNET book. Bluwey Academic Publishers, Dordrecht (1997)
2. Alonso, L.: Herramientas Libres para Procesamiento del Lenguaje Natural. Facultad de Matemática, Astronomía y Física. UNC, Córdoba, Argentina. 5tas Jornadas Regionales de Software Libre. 20 de noviembre de 2005. http://www.cs.famaf.unc.edu.ar/ ~ laura/freeNLP
3. Amsler, R.A.: A taxonomy for English nouns and verbs. In: Proceedings of the 19th Annual Meeting of the Association for Computational Linguistic, Stanford, California, pp. 133–138 (1981)
4. Carreras, X., Márquez, L.: Phrase recognition by filtering and ranking with perceptrons. In: Proceedings of the 4th RANLP Conference, Borovets, Bulgaria, September 2003
5. Cowie, J., Wilks, Y.: Information Extraction. In: Dale, R. (ed.) Handbook of Natural Language Processing, pp. 241–260. Marcel Dekker, New York (2000)
6. Dale, R.: Symbolic approaches to natural language processing. In: Dale, R. (ed.) Handbook of Natural Language Processing. Marcel Dekker, New York (2000)
7. Gómez-Pérez, A., Fernando-López, M., Corcho, O.: Ontological Engineering. Springer, London (2004)
8. Hopcroft, J.E., Ullman, J.D.: Introduction to Automata Theory, Languages and Computations. Addison-Wesley, Reading (1979)
9. Llorens, J., Morato, J., Genova, G.: RSHP: an information representation model based on relationships. In: Damiani, E., Jain, L.C., Madravio, M. (eds.) Soft Computing in Software Engineering. Studies in Fuzziness and Soft Computing Series, vol. 159, pp. 221–253. Springer, Heidelberg (2004)
10. Llorens, J.: Definición de una Metodología y una Estructura de Repositorio orientadas a la Reutilización: el Tesauro de Software. Universidad Carlos III (1996)
11. Christopher, Manning: Foundations of Statistic Natural Language Processing, p. 81. Cambridge University, Cambridge (1999)
12. Martí, M.A., Llisterri, J.: Tratamiento del lenguaje natural, p. 207. Universitat de Barcelona, Barcelona (2002)
13. Moreno, V.: Representación del conocimiento de proyectos de software mediante técnicas automatizadas. Anteproyecto de Tesis Doctoral. Universidad Carlos III de Madrid, Marzo (2009)
14. Poesio, M.: Semantic Analysis. In: Dale, R. (ed.) Handbook of Natural Language Processing. Marcel Dekker, New York (2000)
15. Rehberg, C.P.: Automatic pattern generation in natural language processing. United States Patent. US 8,180,629 B2, 15 May 2012, January 2010
16. Riley, M.D.: Some applications of tree-based modeling to speech and language indexing. In: Proceedings of the DARPA Speech and Natural Language Workshop. Morgan Kaufmann, California, pp. 339–352 (1989)
17. Suarez, P., Moreno, V., Fraga, A., Llorens, J.: Automatic generation of semantic patterns using techniques of natural language processing. In: SKY, pp. 34–44 (2013)
18. Thomason, R.H.: What is Semantics? Version 2. 27 March 2012. http://web.eecs.umich.edu/ ~ rthomaso/documents/general/what-is-semantics.html
19. Triviño, J.L., Morales Bueno, R.: A Spanish POS tagger with variable memory. In: Proceedings of the Sixth International Workshop on Parsing Technologies (IWPT-2000). ACL/SIGPARSE, Trento, Italia, pp. 254–265 (2000)
20. Weischedel, R., Metter, M., Schwartz, R., Ramshaw, L., Palmucci, J.: Coping with ambiguity and unknown through probabilistic models. Comput. Linguist. 19 369–382

Knowledge Management and Information Sharing

Active Integrity Constraints: From Theory to Implementation

Luís Cruz-Filipe[1]([⊠]), Michael Franz[1], Artavazd Hakhverdyan[1],
Marta Ludovico[2], Isabel Nunes[2], and Peter Schneider-Kamp[1]

[1] Department of Mathematics and Computer Science, University of Southern
Denmark, Campusvej 55, 5230 Odense M, Denmark
{lcf,petersk}@imada.sdu.dk, mf@bfdata.dk, artavazd19@gmail.com
[2] Faculdade de Ciências da Universidade de Lisboa, Campo Grande,
1749-016 Lisboa, Portugal
marta.al.ludovico@gmail.com, in@fc.ul.pt

Abstract. The problem of database consistency relative to a set of integrity constraints has been extensively studied since the 1980s, and is still recognized as one of the most important and complex in the field. In recent years, with the proliferation of knowledge repositories (not only databases) in practical applications, there has also been an effort to develop implementations of consistency maintenance algorithms that have a solid theoretical basis.

The framework of active integrity constraints (AICs) is one example of such an effort, providing theoretical grounds for rule-based algorithms for ensuring database consistency. An AIC consists of an integrity constraint together with a specification of actions that may be taken to repair a database that does not satisfy it. Both denotational and operational semantics have been proposed for AICs. In this paper, we describe **repAIrC**, a prototype implementation of the algorithms previously proposed targeting SQL databases, i.e., the most prolific type of databases. Using **repAIrC**, we can both validate an SQL database with respect to a given set of AICs and compute possible repairs in case the database is inconsistent; the tool is able to work with the different kinds of repairs that have been considered, and achieves optimal asymptotic complexity in their computation. It also implements strategies for parallelizing the search for repairs, which in many cases can make untractable problems become easily solvable.

1 Introduction

Databases are among the most prolific software items in today's world, being components of virtually every non-trivial software system used in practice – from mobile apps to enterprise management systems. Besides facilitating efficient storage and retrieval of data, one of the main tasks of database management systems is ensuring data integrity, i.e., guaranteeing semantic relationships between data not captured by the syntactic structure of the database hold at all times.

© Springer International Publishing AG 2016
A. Fred et al. (Eds.): IC3K 2015, CCIS 631, pp. 399–420, 2016.
DOI: 10.1007/978-3-319-52758-1_22

Typical database management systems allow the user to specify integrity constraints on the data as logical statements that are required to be satisfied at any given point in time. The typical database consistency tasks include guaranteeing that such constraints still hold after updating databases [1], and determining what repairs have to be made when the constraints are violated [13], without making any assumptions about how the inconsistencies came about. Repairing an inconsistent database is a highly complex process; also, it is widely accepted that human intervention is often necessary to choose an adequate repair [10]. That said, every progress towards automation in this field is nevertheless an important contribution, and criteria to choose among different possible repairs allow for a reasonable level of semi-automation.

In particular, the framework of active integrity constraints [5,11] was introduced more recently with the goal of giving operational mechanisms to compute repairs of inconsistent databases. This framework has subsequently been extended to consider preferences [3] and to find "best" repairs automatically [7] and efficiently [6]. Active integrity constraints (AICs) are expressive enough to encompass the majority of integrity constraints that are typically found in practice, and they allow the definition of preferred ways to calculate repairs, through specific actions to be taken in specific inconsistent situations.

To the best of our knowledge, no real-world implementation of an AIC–enhanced database system exists today. This paper presents a prototype tool that implements the tree–based algorithms for computing repairs presented in [5,7]. While not yet ready for productive deployment, this implementation can work successfully with virtually any database management system supporting access through SQL, and is readily extendible to other (nearly arbitrary) database management systems thanks to its modular design.

This paper is structured as follows. We summarize related work in Sect. 1.1. Section 2 recapitulates previous work on active integrity constraints and repair trees. Section 3 introduces our tool, repAIrC, and describes its implementation, focusing on the new theoretical results that were necessary to bridge the gap between theory and practice. Section 4 then discusses how parallel computation capabilities are incorporated in repAIrC to make the search for repairs more efficient. The discussion in these two sections is illustrated by a non-trivial running example. Section 5 summarizes our achievements and gives a brief outlook into future developments. This paper extends the work previously described in [8].

1.1   Related Work

The problem of maintaining data consistency when changing a database has been the focus of intensive research for over three decades. The survey paper [1], which extensively describes the state of the art in 1988, remains actual in its characterization of the concept of "good" update, and identification of three main change operations: insertion of new facts, deletion of existing facts, and modification of information.

Database changes can be caused by two distinct scenarios, which lead to the distinct notions of database update and database revision [10,13]. A *data-*

base update occurs whenever the world changes and the database needs to be updated to reflect this fact; a *database revision* happens when new knowledge is obtained about a world that did not change. This distinction is especially relevant in deductive databases and open-world knowledge bases, where the known information is not assumed to be complete. In spite of their conceptual difference, updates and revisions can in practice be addressed by similar techniques. In particular, they both often demand changes that conflict with the integrity constraints, and the database must be repaired in order to regain consistency.

Over the years, several authors have proposed alternative approaches to the problem of how to repair an inconsistent database. One possibility is to read integrity constraints as rules that suggest possible actions to repair inconsistencies [1]; another is to express database dependencies through logic programming, namely in deductive databases [14,16,17]. A more algorithmic approach uses event-condition-action rules [19,20], where actions are triggered by specific events, for which a procedural semantics has been defined. This paper focuses on the formalism of active integrity constraints [11], which will be described in more detail in the next section.

Several algorithms for computing repairs of inconsistent databases have been proposed and studied throughout the years, focusing on the different ways integrity constraints are specified and on the different types of databases under consideration [12,14,15,17]. This multitude of approaches is not an accident: deciding whether an inconsistent database can be repaired is typically a Π_2^P- or co-Σ_2^P- complete problem, and there is no reason to believe in the existence of general-purpose algorithms for this problem, but one should rather focus on developing specific algorithms for particular interesting cases [10]. The formalism of active integrity constraints also establishes a hierarchy of database (weak) repairs, which can be used to define preferences among these and obtain more automation in the process [5].

2   Active Integrity Constraints

Active integrity constraints (AICs) were introduced in [11] and further explored in [4,5], which define the basic concepts and prove complexity bounds for the problem of repairing inconsistent databases. These authors introduce declarative semantics for different types of repairs, obtaining their complexity results by means of a translation into revision programming. In practice, however, this does not yield algorithms that are applicable to real-life databases; for this reason, a direct operational semantics for AICs was proposed in [7], presenting database-oriented algorithms for finding repairs. The present paper describes a tool that can actually execute these algorithms in collaboration with an SQL database management system.

2.1   Syntax and Declarative Semantics

For the purpose of this work, we can view a database simply as a set of atomic formulas over a typed function-free first-order signature Σ, which we will assume throughout to be fixed. Let $\mathcal{A}t$ be the set of closed atomic formulas over Σ. A database $\mathcal{I}$ *entails* literal L, $\mathcal{I} \models L$, if $L \in \mathcal{A}t$ and $L \in \mathcal{I}$, or if L is not a with $a \in \mathcal{A}t$ and $a \notin \mathcal{I}$.

An integrity constraint is a clause

$$L_1, \ldots, L_m \supset \bot$$

where each L_i is a literal over Σ, with intended semantics that $\forall (L_1 \wedge \ldots \wedge L_m)$ should not hold. As is usual in logic programming, we require that if L_i contains a negated variable x, then x already occurs in $L_1, \ldots, L_{i-1}$. We say that $\mathcal{I}$ *satisfies* integrity constraint r, $\mathcal{I} \models r$, if, for every instantiation θ of the variables in r, it is the case that $\mathcal{I} \not\models L\theta$ for some L in r; and $\mathcal{I}$ satisfies a set η of integrity constraints, $\mathcal{I} \models \eta$, if it satisfies each integrity constraint in η.

If $\mathcal{I} \not\models \eta$, then $\mathcal{I}$ may be updated through *update actions* of the form $+a$ and $-a$, where $a \in \mathcal{A}t$, stating that a is to be inserted in or deleted from $\mathcal{I}$, respectively. A set of update actions $\mathcal{U}$ is *consistent* if it does not contain both $+a$ and $-a$, for any $a \in \mathcal{A}t$; in this case, $\mathcal{I}$ can be updated by $\mathcal{U}$, yielding the database

$$\mathcal{U}(\mathcal{I}) = (\mathcal{I} \cup \{a \mid +a \in \mathcal{U}\}) \setminus \{a \mid -a \in \mathcal{U}\}.$$

The problem of database repair is to find $\mathcal{U}$ such that $\mathcal{U}(\mathcal{I}) \models \eta$.

Definition 1. *Let $\mathcal{I}$ be a database and η a set of integrity constraints. A* weak repair *for $\langle \mathcal{I}, \eta \rangle$ is a consistent set $\mathcal{U}$ of update actions such that: (i) every action in $\mathcal{U}$ changes $\mathcal{I}$; and (ii) $\mathcal{U}(\mathcal{I}) \models \eta$. A* repair *for $\langle \mathcal{I}, \eta \rangle$ is a weak repair $\mathcal{U}$ for $\langle \mathcal{I}, \eta \rangle$ that is minimal w.r.t. set inclusion.*

The distinction between weak repairs and repairs embodies the standard principle of *minimality of change* [21].

The problem of deciding whether there exists a (weak) repair for an inconsistent database is NP-complete [5]. Furthermore, simply detecting that a database is inconsistent does not give any information on how it can be repaired. In order to address this issue, those authors proposed active integrity constraints (AICs), which guide the process of selection of a repair by pairing literals with the corresponding update actions.

In the syntax of AICs, we extend the notion of update action by allowing variables. Given an action α, the literal corresponding to it is $\mathsf{lit}(\alpha)$, defined as a if $\alpha = +a$ and not a if $\alpha = -a$; conversely, the update action corresponding to a literal L, $\mathsf{ua}(L)$, is $+a$ if $L = a$ and $-a$ if $L = $ not a. The *dual* of a is not a, and conversely; the dual of L is denoted L^D. An *active integrity constraint* is thus an expression r of the form

$$L_1, \ldots, L_m \supset \alpha_1 \mid \ldots \mid \alpha_k$$

where the L_i (in the *body* of r, $\mathsf{body}(r)$) are literals and the α_j (in the *head* of r, $\mathsf{head}(r)$) are update actions, such that

$$\{\mathsf{lit}(\alpha_1)^D, \ldots, \mathsf{lit}(\alpha_k)^D\} \subseteq \{L_1, \ldots, L_m\}.$$

The set $\mathsf{lit}(\mathsf{head}\,(r))^D$ contains the *updatable* literals of r. The *non-updatable* literals of r form the set $\mathsf{nup}(r) = \mathsf{body}\,(r) \setminus \mathsf{lit}\,(\mathsf{head}\,(r))^D$.

The natural semantics for AICs restricts the notion of weak repair.

Definition 2. *Let $\mathcal{I}$ be a database, η a set of AICs and $\mathcal{U}$ be a (weak) repair for $\langle\mathcal{I},\eta\rangle$. Then $\mathcal{U}$ is a* founded *(weak) repair for $\langle\mathcal{I},\eta\rangle$ if, for every action $\alpha \in \mathcal{U}$, there is a closed instance r' of $r \in \eta$ such that $\alpha \in \mathsf{head}\,(r')$ and $\mathcal{U}(\mathcal{I}) \models L$ for every $L \in \mathsf{body}\,(r') \setminus \{\mathsf{lit}(\alpha)^D\}$.*

The problem of deciding whether there exists a weak founded repair for an inconsistent database is again NP-complete, while the similar problem for founded repairs is Σ_2^P-complete. Despite their natural definition, founded repairs can include circular support for actions, which can be undesirable; this led to the introduction of justified repairs [5].

We say that a set $\mathcal{U}$ of update actions is *closed* under r if $\mathsf{nup}(r) \subseteq \mathsf{lit}(\mathcal{U})$ implies $\mathsf{head}\,(r) \cap \mathcal{U} \neq \emptyset$, and it is closed under a set η of AICs if it is closed under every closed instance of every rule in η. In particular, every founded weak repair for $\langle\mathcal{I},\eta\rangle$ is by definition closed under η.

A closed update action $+a$ (resp. $-a$) is a *no-effect* action w.r.t. $(\mathcal{I},\mathcal{U}(\mathcal{I}))$ if $a \in \mathcal{I} \cap (\mathcal{U}(\mathcal{I}))$ (resp. $a \notin \mathcal{I} \cup (\mathcal{U}(\mathcal{I}))$). The set of all no-effect actions w.r.t. $(\mathcal{I},\mathcal{U}(\mathcal{I}))$ is denoted by $\mathsf{ne}_{\mathcal{I}}(\mathcal{U}(\mathcal{I}))$. A set of update actions $\mathcal{U}$ is a justified action set if it coincides with the set of update actions forced by the set of AICs and the database before and after applying $\mathcal{U}$ [5].

Definition 3. *Let $\mathcal{I}$ be a database and η a set of AICs. A consistent set $\mathcal{U}$ of update actions is a* justified action set *for $\langle\mathcal{I},\eta\rangle$ if it is a minimal set of update actions containing $\mathsf{ne}_{\mathcal{I}}(\mathcal{U}(\mathcal{I}))$ and closed under η. If $\mathcal{U}$ is a justified action set for $\langle\mathcal{I},\eta\rangle$, then $\mathcal{U} \setminus \mathsf{ne}_{\mathcal{I}}(\mathcal{U}(\mathcal{I}))$ is a* justified weak repair *for $\langle\mathcal{I},\eta\rangle$.*

In particular, it has been shown that justified repairs are always founded [5]. The problem of deciding whether there exist justified weak repairs or justified repairs for $\langle\mathcal{I},\eta\rangle$ is again a Σ_2^P-complete problem, becoming NP-complete if one restricts the AICs to contain only one action in their head (*normal* AICs).

2.2 Operational Semantics

The declarative semantics of AICs is not very satisfactory, as it does not capture the operational nature of rules. In particular, the quantification over all no-effect actions in the definition of justified action set poses a practical problem. Therefore, an operational semantics for AICs was proposed in [7], which we now summarize.

Definition 4. *Let $\mathcal{I}$ be a database and η be a set of AICs.*

- *The* repair tree *for $\langle\mathcal{I},\eta\rangle$, $T_{\langle\mathcal{I},\eta\rangle}$, is a labeled tree where: nodes are sets of update actions; each edge is labeled with a closed instance of a rule in η; the root is $\emptyset$; and for each consistent node n and closed instance r of a rule in η, if $n(\mathcal{I}) \not\models r$ then for each $L \in \mathsf{body}\,(r)$ the set $n' = n \cup \{\mathsf{ua}(L)^D\}$ is a child of n, with the edge from n to n' labeled by r.*

- *The* founded repair tree *for* $\langle \mathcal{I}, \eta \rangle$, $T^f_{\langle \mathcal{I}, \eta \rangle}$, *is constructed as* $T_{\langle \mathcal{I}, \eta \rangle}$ *but requiring that* $\mathsf{ua}(L)$ *occur in the head of some closed instance of a rule in* η.
- *The* well-founded repair tree *for* $\langle \mathcal{I}, \eta \rangle$, $T^{wf}_{\langle \mathcal{I}, \eta \rangle}$, *is also constructed as* $T_{\langle \mathcal{I}, \eta \rangle}$ *but requiring that* $\mathsf{ua}(L)$ *occur in the head of the rule being applied.*
- *The* justified repair tree *for* $\langle \mathcal{I}, \eta \rangle$, $T^j_{\langle \mathcal{I}, \eta \rangle}$, *has nodes that are pairs of sets of update actions* $\langle \mathcal{U}, \mathcal{J} \rangle$, *with root* $\langle \emptyset, \emptyset \rangle$. *For each node n and closed instance r of a rule in* η, *if* $\mathcal{U}_n(\mathcal{I}) \not\models r$, *then for each* $\alpha \in \mathsf{head}\,(r)$ *there is a descendant n′ of n, with the edge from n to n′ labeled by r, where:* $\mathcal{U}_{n'} = \mathcal{U}_n \cup \{\alpha\}$; *and* $\mathcal{J}_{n'} = (\mathcal{J}_n \cup \{\mathsf{ua}(\mathsf{nup}(r))\}) \setminus \mathcal{U}_n$.

The properties of repair trees are summarized in the following results, whose detailed proofs can be found in [7].

Theorem 1. *Let* $\mathcal{I}$ *be a database and* η *be a set of AICs. Then:*

1. $T_{\langle \mathcal{I}, \eta \rangle}$ *is finite.*
2. *Every consistent leaf of* $T_{\langle \mathcal{I}, \eta \rangle}$ *is labeled by a weak repair for* $\langle \mathcal{I}, \eta \rangle$.
3. *If* $\mathcal{U}$ *is a repair for* $\langle \mathcal{I}, \eta \rangle$, *then there is a branch of* $T_{\langle \mathcal{I}, \eta \rangle}$ *ending with a leaf labeled by* $\mathcal{U}$.
4. *If* $\mathcal{U}$ *is a founded repair for* $\langle \mathcal{I}, \eta \rangle$, *then there is a branch of* $T^f_{\langle \mathcal{I}, \eta \rangle}$ *ending with a leaf labeled by* $\mathcal{U}$.
5. *If* $\mathcal{U}$ *is a justified repair for* $\langle \mathcal{I}, \eta \rangle$, *then there is a branch of* $T^j_{\langle \mathcal{I}, \eta \rangle}$ *ending with a leaf labeled by* $\mathcal{U}$.
6. *If* η *is a set of normal AICs and* $\langle \mathcal{U}, \mathcal{J} \rangle$ *is a leaf of* $T^j_{\langle \mathcal{I}, \eta \rangle}$ *with* $\mathcal{U}$ *consistent and* $\mathcal{U} \cap \mathcal{J} = \emptyset$, *then* $\mathcal{U}$ *is a justified repair for* $\langle \mathcal{I}, \eta \rangle$.

Not all leaves will correspond to repairs of the desired kind; in particular, there may be weak repairs in repair trees. Also, both $T^f_{\langle \mathcal{I}, \eta \rangle}$ and $T^j_{\langle \mathcal{I}, \eta \rangle}$ typically contain leaves that do not correspond to founded or justified (weak) repairs – otherwise the problem of deciding whether there exists a founded or justified weak repair for $\langle \mathcal{I}, \eta \rangle$ would be solvable in non-deterministic polynomial time. The leaves of the well-founded repair tree for $\langle \mathcal{I}, \eta \rangle$ correspond to a new type of weak repairs, called *well-founded weak repairs*, not considered in the original works on AICs.

2.3 Parallel Computation of Repairs

The computation of founded or justified repairs can be improved by dividing the set of AICs into independent sets that can be processed independently, simply merging the computed repairs at the end [6]. Here, we adapt the definitions given therein to the first-order scenario. Two sets of AICs η_1 and η_2 are independent if the same atom does not occur in a literal in the body of a closed instance of two distinct rules $r_1 \in \eta_1$ and $r_2 \in \eta_2$. If η_1 and η_2 are independent, then repairs for $\langle I, \eta_1 \cup \eta_2 \rangle$ are exactly the unions of a repair for $\langle \mathcal{I}, \eta_1 \rangle$ and $\langle \mathcal{I}, \eta_2 \rangle$.

When one considers founded, well-founded or justified repairs, this notion can be made stronger. Since those semantics use the information in the heads of the rules, rather than in the bodies, we can obtain similar results by considering *strong independence*: η_1 and η_2 are strongly independent if, for each $r_1 \in \eta_1$ and $r_2 \in \eta_2$, there is no atom occurring both in the head of a closed instance of r_1 and in the body of a closed instance of r_2, or conversely.

If an atom occurs in a literal in the body of a closed instance of a rule in η_2 and in an action in the head of a closed instance of a rule in η_1, but not conversely, then we say that η_1 *precedes* η_2. Founded/justified (but not well-founded) repairs for $\eta_1 \cup \eta_2$ can be computed in a stratified way, by first repairing $\mathcal{I}$ w.r.t. η_1, and then repairing the result w.r.t. η_2.

Splitting a set of AICs into independent sets or stratifying it can be solved using standard algorithms on graphs, as we describe in Sect. 4.

3   The Tool

The tool `repAIrC` is implemented in Java, and its simplified UML class diagram can be seen in Fig. 1. Structurally, this tool can be split into four main separate components, centered on the four classes marked in bold in that figure.

- Objects of type `AIC` implement active integrity constraints.
- Implementations of interface `DB` provide the necessary tools to interact with a particular database management system; currently, we provide functionality for SQL databases supported by JDBC.
- Objects of type `RepairTree` correspond to concrete repair trees; their exact type will be the subclass corresponding to a particular kind of repairs.
- Class `RunRepairGUI` provides the graphical interface to interact with the user.

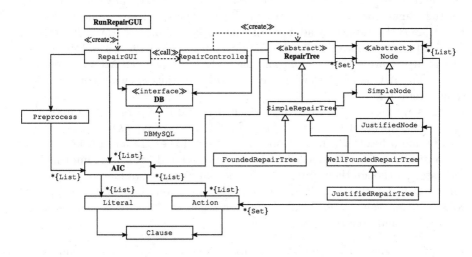

Fig. 1. Class diagram for `repAIrC`.

An important design aspect has to do with extensibility and modularity. A first prototype focused on the construction of repair trees, and used simple text files to mimick databases as lists of propositional atoms, in the style of [5, 7]. Later, parallelization capabilities were added (as explained in Sect. 4), requiring changes only to `RepairController` – the class that controls the execution of the whole process. Likewise, the extension of `repAIrC` to SQL databases and the addition of the stratification mechanism only required localized changes in the classes directly concerned with those processes.

The next subsections detail the implementation of the four classes `AIC`, `DB`, `RepairTree` and `RunRepairTreeGUI`.

3.1 Representing Active Integrity Constraints

In the practical setting, it makes sense to diverge a little from the theoretical definition of AICs.

- Real-world tables found in DBs contain many columns, most of which are typically irrelevant for a given integrity constraint.
- The columns of a table are not static, i.e., columns are usually added or removed during a database's lifecycle.
- The order of columns in a table should not matter, as they are identified by a unique column name.

To deal pragmatically with these three aspects, we will use a more database-oriented notation to write down atoms, namely allowing the arguments to be provided in any order, but requiring that the column names be provided. The special token $ is used as first character of a variable. So, for example, the literal `hasInsurance(firstName=$X, type='basic')` will match any entry in table `hasInsurance` having value `basic` in column `type` and any value in column `firstName`; this table may additionally have other columns. Negative literals are preceded by the keyword `NOT`, while actions must begin with + or -. Literals and actions are separated by commas, and the body and head of an AIC are separated by ->. The AIC is finished when; is encountered, thus allowing constraints to span several lines.

AICs are provided in a text file, which is parsed by a parser generated automatically using JavaCC and transformed into objects of type `AIC`. These contain a body and a head, which are respectively `List<Literal>` and `List<Action>`; for consistency with the underlying theory, `Literal` and `Action` are implemented separately, although their objects are isomorphic: they contain an object of type `Clause` (which consists of the name of a table in the database and a list of pairs column name/value) and a flag indicating whether they are positive/negated (literals) or additions/removals (actions).

Example 1. Consider an employee database containing tables empl (employee), categ (category), supvsr (supervisor), unsup (unsupervised) and insured, among others. This database includes the following AICs.

$$\text{empl}(X), \text{empl}(Y), \text{supvsr}(X, Y), \text{supvsr}(Y, X) \supset -\text{supvsr}(X, Y) \qquad (r_1)$$
$$\text{empl}(X), \text{empl}(Y), \text{supvsr}(X, Y), \text{unsup}(Y) \supset -\text{unsup}(Y) \qquad (r_2)$$
$$\text{empl}(X), \text{cat}(X, \text{junior}), \text{unsup}(X) \supset -\text{cat}(X, \text{junior}) \qquad (r_3)$$
$$\text{empl}(X), \text{not insured}(X, \text{basic}) \supset +\text{insured}(X, \text{basic}) \qquad (r_4)$$

Intuitively, (r_1) states that two employees cannot supervise each other, and the preferred way to correct this error is by changing the supvsr table. Rule (r_2) states that employees who have a supervisor are not unsupervised, and its head assumes that the information in the supvsr table is more correct. Rule (r_3) states that junior employees cannot be unsupervised, and the last rule states that all employees must have a basic insurance.

These AICs are written in the concrete text-based syntax of the `repAIrC` tool as

```
empl(id=$X), empl(id=$Y),
  supvsr(master=$X,slave=$Y), supvsr(master=$Y,slave=$X)
  -> - supvsr(master=$X,slave=$Y);

empl(id=$X), empl(id=$Y),
  supvsr(master=$X,slave=$Y), unsup(empId=$Y)
  -> - unsup(empId=$Y);

empl(id=$X), cat(type=junior, empId=$X), unsup(empId=$X)
  -> - cat(type='junior',empId=$X);

empl(id=$X), NOT insured(empId=$X, type='basic')
  -> + insured(empId=$X, type='basic');
```

respectively, assuming the corresponding column names for the attributes. Note that, thanks to our usage of explicit column naming, the order of the columns in each table is not important.

3.2   Interfacing with the Database

Database operations (queries and updates) are defined in the DB interface, which contains the following methods.

- `getUpdateActions(AIC aic)`: queries the database for the instances of aic not satisfied in its current state, and returns a `Collection <Collection<Action>>` that contains the corresponding instantiations of the head of aic.
- `update(Collection<Action> actions)`: applies all update actions in the collection `actions` to the database (void).
- `undo(Collection<Action> actions)`: undoes the effect of all update actions in `actions` (void).
- `aicsCompatible(Collection<AIC> aics)`: checks that all the elements of aics are compatible with the structure of the database.

– `disconnect()`: disconnects from the database (void). The connection is established when the object is originally constructed.

Some of these methods require more detailed comments. The construction of the repair tree also requires that the database be changed interactively, but upon conclusion the database should be returned to its original state. In theory, this would be achievable by applying the `update` method with the duals of the actions that were used to change the database; but this turns out not to be the case for deletion actions. Since the AICs may underspecify the entries in the database (because some fields are left implicit), the implementation of `update` must take care to store the values of all rows that are deleted from the database. In turn, the `undo` method will read this information every time it has to undo a deletion action, in order to find out exactly what entries to re-add.

The method `aicsCompatible` is necessary because the AICs are given independently of the database, but they must be compatible with its structure – otherwise, all queries will return errors. Including this method in the interface allows the AICs to be tested before any queries are made, thus significantly reducing the number of exceptions that can occur during program execution.

Currently, `repAIrC` includes an implementation `DBMySQL` of DB, which works with SQL databases. The interaction between `repAIrC` and the database is achieved by means of JDBC, a Java database connectivity technology able to interface with nearly all existing SQL databases. In order to determine whether an AIC is satisfied by a database, method `getUpdateActions` first builds a single SQL query corresponding to the body of the AIC. This method builds one `SELECT` statement for the positive literals in the body of the AIC, and another for the negative literals, if they occur. Each time a new variable is found, the table and column where it occurs are stored, so that future references to the same variable in a positive literal can be unified by using inner joins. If there is a `SELECT` statement for the negative literals, it is then connected to the other one using a `WHERE NOT EXISTS` condition. Variables in the negative literals must necessarily appear first in a positive literal in the same AIC; therefore, they can then be connected by a `WHERE` clause instead of an inner join.

Example 2. The bodies of the two last integrity constraints in Example 1 generate the following SQL queries.

```
SELECT * FROM empl t0          SELECT * FROM empl t0
  INNER JOIN cat t1            WHERE NOT EXISTS
  ON t0.id=t1.empId           (SELECT * FROM insured t1
  INNER JOIN unsup t2            WHERE t1.empId=t0.id
  ON t0.id=t2.empId             AND t1.type='basic')
  WHERE t1.type = 'junior'
```

3.3 Implementing Repair Trees

The implementation of the repair trees directly follows the algorithms described in Sect. 2. Different types of repair trees are implemented using inheritance, so that most of the code can be reused in the more complex trees. The trees are constructed in a breadth-first manner, and all non-contradictory leaves that are found are stored in a list. At the end, this list is pruned so that only the minimal elements (w.r.t. set inclusion) remain – as these are the ones that correspond to repairs.

While constructing the tree, the database has to be temporarily updated and restored. Indeed, to calculate the descendants of a node, we first need to evaluate all AICs at that node in order to determine which ones are violated; this requires querying a modified version of the database that takes into account the update actions in the current node.

In order to avoid concurrency issues, we use a transaction-style methodology to perform these updates, where we first change the database, then perform the necessary SQL queries, and finally rollback to the original state, guaranteeing that other threads interacting with the database during this process neither see the modifications nor lead to inconsistent repair trees. This becomes of particular interest when the parallel processing tools described in Sect. 4 are put into place. Although this adds some overhead to the execution time, at the end of that section we discuss why scalability is not a practically relevant concern.

After finding all the leaves of the repair tree, a further step is needed in the case one is looking for founded or justified repairs, as the corresponding trees may contain leaves that do not correspond to repairs with the desired property. This step is skipped if all AICs are normal, in view of the results from [7]. For founded repairs, we directly apply the definition: for each action α, check that there is an AIC with α in its head and such that all other literals in its body are satisfied by the database.

For justified repairs, the validation step is less obvious. Directly following the definition requires constructing the set of no-effect actions, which is essentially as large as the database, and iterating over subsets of this set. This is obviously not possible to do in practical settings. Therefore, we use some criteria to simplify this step.

Lemma 1. *If a rule r was not applied in the branch leading to $\mathcal{U}$, then $\mathcal{U}$ is closed under r.*

Proof. Suppose that r was never applied and assume $\mathsf{nup}(r) \subseteq \mathsf{ne}_\mathcal{I}(\mathcal{U})$. Then necessarily $\mathsf{head}\,(r) \cap \mathsf{ne}_\mathcal{I}(\mathcal{U}) \neq \emptyset$, otherwise r would be applicable and $\mathcal{U}$ would not be a repair.

By construction, $\mathcal{U}$ is clearly also closed for all rules applied in the branch leading to it.

Let $\mathcal{U}$ be a candidate justified weak repair. In order to test it, we need to show that $\mathcal{U} \cup \mathsf{ne}_\mathcal{I}(\mathcal{U})$ is a justified action set (see [7]), which requires iterating over all subsets of $\mathcal{U} \cup \mathsf{ne}_\mathcal{I}(\mathcal{U})$ that contain $\mathsf{ne}_\mathcal{I}(\mathcal{U})$. Clearly this can be achieved by iterating over subsets of $\mathcal{U}$.

But if $\mathcal{U}^* \subseteq \mathcal{U}$, then $\mathsf{nup}(r) \cap \mathcal{U}^* = \emptyset$; this allows us to simplify the closedness condition to: if $\mathsf{nup}(r) \subseteq \mathsf{ne}_\mathcal{I}(\mathcal{U})$, then $\mathcal{U}^* \cap \mathsf{head}\,(r) = \emptyset$. The antecedent needs then only be done once (since it only depends on $\mathcal{U}$), whereas the consequent does not require consulting the database.

The following result summarizes these properties.

Lemma 2. *A weak repair $\mathcal{U}$ in a leaf of the justified repair tree for $\langle \mathcal{I}, \eta \rangle$ is a justified weak repair for $\langle \mathcal{I}, \eta \rangle$ iff, for every set $\mathcal{U}^* \subseteq \mathcal{U}$, if $\mathsf{nup}(r) \subseteq \mathsf{ne}_{\mathcal{I}}(\mathcal{U})$, then $\mathcal{U}^* \cap \mathsf{head}\,(r) = \emptyset$.*

The different implementations of repair trees use different subclasses of the abstract class `Node`; in particular, nodes of `JustifiedRepairTrees` must keep track not only of the sets of update actions being constructed, but also of the sets of non-updatable actions that were assumed. These labels are stored as `Set<Action>` using `HashSet` from the Java library as implementation, as they are repeatedly tested for membership everytime a new node is generated.

For efficiency, repair trees maintain internally a set of the sets of update actions that label nodes constructed so far as a `Set<Node>`. This is used to avoid generating duplicate nodes with the same label. Since this set is used mainly for querying, it is again implemented as a `HashSet`. Nodes with inconsistent labels are also immediately eliminated, since they can only produce inconsistent leaves.

3.4   Interfacing with the User

The user interface for `repAIrC` is implemented using the standard Java GUI widget toolkit `Swing`, and is rather straightforward. On startup, the user is presented with the dialog box depicted in Fig. 2.

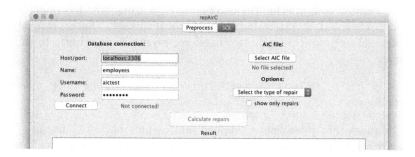

Fig. 2. The initial interface screen for `repAIrC`, providing the user with options to connect to a database, load a file with AICs, choose the desired type of repairs, and compute them.

The user can then provide credentials to connect to a database, as well as enter a file containing a set of AICs. If the connection to the database is successful and the file is successfully parsed, `repAIrC` invokes the `aicsCompatible` method required by the implementation of the DB interface (see Sect. 3.2) and verifies that all tables and columns mentioned in the set of AICs are valid tables and columns in the database. If this is not the case, then an error message is generated and the user is required to select new files; otherwise, the buttons for configuration and computation of repairs become active.

Once the initialization has succeeded, one can check the database for consistency and obtain different types of repairs, computed using the repair tree described above. As it may be of interest to obtain also weak repairs, the user is given the possibility of selecting whether to see only the repairs computed, or all valid leaves of the repair tree – which typically include some weak repairs. In both cases the necessary validations are performed, so that leaves that do not correspond to repairs (in the case of founded or justified repairs) are never presented. A typical step in the interaction is shown in Fig. 3.

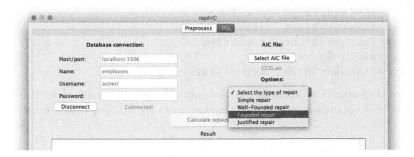

Fig. 3. The interface screen for `repAIrC` after the connection to the database has been successfully established. The drop-down menu is expanded, illustrating the different possibilities.

Example 3. We illustrate the usage of `repAIrC` with the set of AICs from Example 1, over a database in the following state.

EMPL
jane
john
mark

CAT	
jane	boss
john	clerk
mark	junior

SUPVSR	
jane	john
john	jane
john	mark

UNSUP
jane
mark

INSURED	
jane	gold
john	basic

Observe the inconsistencies in this database: john and jane mutually supervise each other; mark is both supervised by john and marked as unsupervised; he is also a junior employee marked as unsupervised. Finally, neither jane nor mark have a basic insurance.

An example output screen after successful computation of the founded repairs for this database can be seen in Fig. 4. The four repairs are:

```
{+insured(empId=$jane, type='basic'),
 -supvsr(master=$john, slave=$jane),
 -unsup(empId=$mark),
```

```
 +insured(empId=$mark, type='basic')}
{+insured(empId=$jane, type='basic'),
 -cat(type='junior', empId=$mark),
 -supvsr(master=$john, slave=$jane),
 -supvsr(master=$john, slave=$mark),
 +insured(empId=$mark, type='basic')}

{+insured(empId=$jane, type='basic'),
 -supvsr(master=$jane, slave=$john),
 -unsup(empId=$jane),
 -unsup(empId=$mark),
 +insured(empId=$mark, type='basic')}

{+insured(empId=$jane, type='basic'),
 -cat(type='junior', empId=$mark),
 -supvsr(master=$jane, slave=$john),
 -supvsr(master=$john, slave=$mark),
 -unsup(empId=$jane),
 +insured(empId=$mark, type='basic')}
```

Observe that constants included in the original AICs are marked with quotes, whereas those obtained by instantiating variables are marked with a dollar sign.

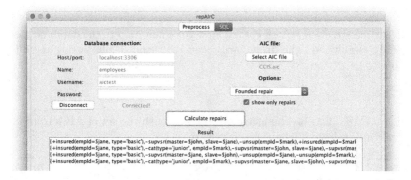

Fig. 4. Possible founded weak repairs of the inconsistent database w.r.t. the AICs from Example 1.

4 Parallelization and Stratification

As described in Sect. 2.3, it is possible to parallelize the search for repairs of different kinds by splitting the set of AICs into independent sets; in the case of founded or justified repairs, this parallelization can be taken one step further by also stratifying the set of AICs. Even though finding partitions and/or stratifications is asymptotically not very expensive (it can be solved in linear time by the well-known graph algorithms described below), it may still take noticeable time if the set of AICs grows very large.

Since, by definition, partitions and stratifications are independent of the actual database, it makes sense to avoid repeating their computation unless the set of AICs changes. For this reason, parallelization capabilities are implemented in `repAIrC` in a two-stage process. Inside `repAIrC`, the user can switch to the `Preprocess` tab, which provides options for computing partitions and stratifications of a set of AICs. This results in an annotated file which still can be read by the parser; in the main tab, parallel computation is automatically enabled whenever the input file is annotated in a proper manner. Figure 5 shows the base view of the preprocessing tab.

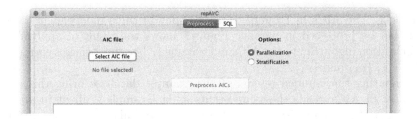

Fig. 5. Interface for computing parallelization and stratification of AICs.

4.1   Parallelization

Computing optimal partitions in the spirit of [6] is not feasible in a setting where variables are present, as this would require considering all closed instances of all AICs – but it is also not desirable, as it would also result in a significant increase of the number of queries to the database. Instead, we work with the definition of strong independency given in Sect. 2. Given a set of AICs, `repAIrC` constructs the adjacency matrix for the undirected graph whose nodes are AICs and such that there is an edge between r_1 to r_2 iff r_1 and r_2 are not independent. A partition is then computed simply by finding the connected components in this graph by a standard graph algorithm. The pseudo-code for the corresponding method `partition`, which takes a set of AICs and returns the set of its partition into independent subsets, is given in Fig. 6.

The partitions computed can then be written to a file. Each partition begins with the line

`#PARTITION_BEGIN_[NO]#`

where `[NO]` is the number of the current partition, and ends with

`#PARTITION_END#`

and the AICs in each partition are inserted in between, in the standard format.

```
procedure partition(aics)
   visited[v] ← ⊥ (∀ v∈aics)
   partitions ← []
   foreach v ∈ aics do
     if ¬visited[v] then
       part ← []
       findCC(v,part,aics,visited)
       partitions += [part]
```

```
procedure findCC(u,part,
                      aics,visited)
   part += [u]
   visited[u] ← ⊤
   foreach v ∈ aics do
     if ¬visited[v] and u≾v then
       findCC(v,part,aics,visited)
```

Fig. 6. Algorithm for partitioning a set of AICs into independent subsets. The test u ≾ v is done by reading the previously computed adjacency matrix.

Example 4. The AICs in Example 1 can be split in three independent classes, which can be processed in parallel. Figure 7 shows the result of applying the parallelization algorithm to the AICs in Example 1. The three classes correspond to $\{r_1, r_2\}$, $\{r_3\}$ and $\{r_4\}$.

Computing the founded repairs of our example database using this parallelized set of AICs yields the same results as in Example 3, albeit in a different order, but takes approx. 0.43 s, whereas the unparallelized version required approx. 6 s.

Fig. 7. Parallelization of the set of AICs from Example 1.

As pointed out in Sect. 2, we can only split a set of AICs into strongly independent sets if we are computing founded, well-founded or justified repairs. Therefore, if repAIrC is given a set of partitioned AICs and asked to compute all repairs, it will produce a warning message and ignore the parallelization.

4.2 Stratification

To compute the partitions for stratification, we need to find the strongly connected components of a similar graph. This is now a directed graph where there is an edge from r_1 to r_2 if r_1 precedes r_2. The implementation is a variant of Tarjan's algorithm [18], adapted to give also the dependencies between the connected components; the pseudo-code is given in Fig. 8.

```
procedure stratify(aics)
    visited[v] ← ⊥ (∀ v∈aics)
    lowlink ← empty map
    dependencies ← ∅
    partitions ← []
    i ← 0
    foreach v ∈ aics do
        if ¬visited[v] then
            findSCC(v,aics,i,lowlink,visited,partitions,[])
    foreach u,v ∈ aics do
        if u ⪯ v and p[u] ≠ p[v] then
            dependencies ← dependencies ∪ (p[u],p[v])

procedure findSCC(v,aics,i,lowlink,visited,partitions,stack)
    visited[v] ← ⊤
    lowlink[v] ← i
    i++
    push v into stack
    isRoot ← ⊤
    foreach w ∈ aics do
        if v ⪯ w and ¬visited[w] then
            findSCC(w,aics,i,lowlink,visited,partitions,stack)
        if lowlink[v] > lowlink[w] then
            lowlink[v] ← lowlink[w]
            isRoot ← ⊥
    if isRoot then
        part ← []
        do
            pop x from stack
            part += [x]
            lowlink[x] ← ∞
        while x ≠ v
        partitions += [part]
```

Fig. 8. Algorithm for stratifying a set of AICs. In procedure stratify, the notation p[u] denotes the (only) element of partitions containing u.

The computed stratification is then presented in a similar syntax to the previous one, to which a dependency section is added, between the special delimiters #DEPENDENCIES_BEGIN# and #DEPENDENCIES_END #, and it can again be written to a file. The dependencies are included in this section as a sequence of strings X -> Y, one per line, where X and Y are the numbers of two partitions and Y precedes X.

Example 5. The two AICs r_1 and r_2 in Example 1 cannot be parallelized, as was seen in Example 4, since they both use the supvsr table, which can be changed by r_1. They can however be stratified, as r_2 only changes unsup, which is not used by r_1. Preprocessing this example by repAIrC returns the output in Fig. 9. Now each AIC is indeed in a separate set, and there is a dependency $r_1 \prec r_2$ – meaning that we can repair the database w.r.t. r_1 before considering r_2.

Computing the founded repairs of our example database using this stratified set of AICs now takes only approx. 0.07 s.

Fig. 9. Partitions computed by repAIrC for the AICs in Example 1.

These examples illustrate the practical speedup obtained by splitting the set of AICs. Indeed, the independence of the several AICs in Example 1 is very clear in the repairs computed in Example 3, as they all share common parts corresponding to repairing r_3 and r_4. By processing all AICs separately we drastically reduce the size of the trees we need to build: parallelization allows us to build three trees instead of one single tree whose branches are all possible interleavings of the branches in the three (necessarily smaller) trees. Likewise, stratification again replaces one tree by two smaller ones, eliminating some interleavings of branches. In general, by stratifying AICs, we get an exponential decrease on the size of the repair trees being built – and therefore also on the total runtime.

However, as mentioned in Sect. 2, stratification only works when computing founded or justified repairs. If repAIrC is fed a stratified set of AICs and asked to compute e.g. well-founded repairs, it will warn the user that these options are incompatible and ignore the stratification.

In addition to alleviating the exponential blowup of the repair trees, parallelization and stratification also allow for a multi-threaded implementation, where repair trees are built in parallel in multiple concurrent threads. To ensure that the dependencies between the partitions are respected, the threads are instructed to wait for other threads that compute preceding partitions. In Example 5, the thread processing partition 2 would be instructed to first wait for the thread processing partition 1 to finish.

Our example showed that significant speedups were observable even when processing small parallelizable sets of AICs. For larger sets of AICs, parallelization and stratification are necessary to obtain feasible runtimes. In one test case, which allowed for 15 partitions to be processed independently, the stratified version computed the founded repairs in approximately 1 s, whereas the sequential version did not terminate within a time limit of 15000 s. This corresponds to a speedup of at least four orders of magnitude, demonstrating the practical impact of the contributions of this section.

4.3   Practical Assessment

In the theoretical worst case, parallelization and stratification will have no impact on the construction of the repair tree, as it is possible to construct a set of AICs with no independent subsets. However, the worst case is not the general case, and it is reasonable to expect that real-life sets of AICs will actually have a high parallelization potential.

Indeed, integrity constraints typically reflect high-level consistency requirements of the database, which in turn capture the hierarchical nature of relational databases, where more complex relations are built from simpler ones. Thus, when specifying *active* integrity constraints there will naturally be a preference to correct inconsistencies by updating the more complex tables rather than the most primitive ones.

Furthermore, in a real setting we are not so much interested in repairing a database once, but rather in ensuring that it remains consistent as its information changes. Therefore, it is likely that inconsistencies that arise will be localized to a particular table. The ability to process independent sets of AICs separately guarantees that we will not be repeatedly evaluating those constraints that were not broken by recent changes, focusing only on the constraints that can actually become unsatisfied as we attempt to fix the inconsistency.

For the same reason, scalability of the techniques we implemented is not a relevant issue: there is no practical need to develop a tool that is able to fix hundreds of inconsistencies efficiently simultaneously, since each change to the database will likely only impact at most a few AICs.

5   Conclusions and Future Work

We described a prototype implementation of repAIrC, a tool to check the integrity of real-world SQL databases with respect to a given set of active integrity constraints. Furthermore, repAIrC implements a set of previously published algorithms to compute repairs of inconsistent databases, being able to deal with the different semantics for active integrity constraints that have been proposed so far. It can also split a set of AICs using known results on parallelization and stratification and perform parallel computations of independent repairs, thereby achieving a practical improvement that can reach several orders of magnitude. The theoretical soundness of repAIrC follows from results previously published in [3,5–7,11]. We believe that it is the first step towards an implementation of consistency maintenance features in database management systems based on a strong theoretical background.

We could go one step further in the automation process and instruct repAIrC to apply repairs to the database automatically. However, this does not seem a good strategy: in general, there are several possible repairs, and it has long been pointed out [10] that there will always be some instances where human intervention is necessary to sort out among the different possibilities. On the contrary, the design of repAIrC is such that the computation of repairs is isolated from and transparent to other concurrent uses of the database. This is accomplished by using standard SQL transaction and rollback mechanisms.

For real-world applications, the next logical step is to move beyond databases into more generic reasoning systems. There are currently several models for heterogeneous knowledge management systems, of which the framework of heterogeneous nonmonotonic multi-context systems [2] is one of the most general that have been positively received by the community. Multi-context systems are nowadays used in practice, as implementations already exist, and they are flexible enough to adapt to several different usage scenarios. A subset of the authors of this paper is currently working on defining integrity constraints for multi-context systems [9], which we plan to extend to active integrity constraints in the near future. Extending repAIrC to this more encompassing framework would then be the natural next step. We believe the modularity embedded in its design will be a key ingredient towards this task.

Finally, on the more technical side, we also intend to increase repAIrC's performance by means of the integration of a local database cache. In this way, repAIrC will be able to execute the repeated update/undo actions required during the construction of the different repair trees without interacting with the external database, thereby reducing the significant overhead introduced by that connection.

Acknowledgements. This work was supported by the Danish Council for Independent Research, Natural Sciences, and by FCT/MCTES/PIDDAC under centre grant to BioISI (Centre Reference: UID/MULTI/04046/2013). Marta Ludovico was sponsored by a grant "Bolsa Universidade de Lisboa / Fundação Amadeu Dias". The authors would also like to thank Graça Gaspar and Patrícia Engrácia for many interesting discussions on the topic of active integrity constraints.

References

1. Abiteboul, S.: Updates, a new frontier. In: Gyssens, M., Paredaens, J., Gucht, D. (eds.) ICDT 1988. LNCS, vol. 326, pp. 1–18. Springer, Heidelberg (1983). doi:10. 1007/3-540-50171-1_1
2. Brewka, G., Eiter, T.: Equilibria in heterogeneous nonmonotonic multi-context systems. In: AAAI, pp. 385–390. AAAI Press (2007)
3. Caroprese, L., Greco, S., Molinaro, C.: Prioritized active integrity constraints for database maintenance. In: Kotagiri, R., Krishna, P.R., Mohania, M., Nantajeewarawat, E. (eds.) DASFAA 2007. LNCS, vol. 4443, pp. 459–471. Springer, Heidelberg (2007)
4. Caroprese, L., Greco, S., Zumpano, E.: Active integrity constraints for database consistency maintenance. IEEE Trans. Knowl. Data Eng. 21(7), 1042–1058 (2009)
5. Caroprese, L., Truszczyński, M.: Active integrity constraints and revision programming. Theory Pract. Logic Prog. 11(6), 905–952 (2011)
6. Cruz-Filipe, L.: Optimizing computation of repairs from active integrity constraints. In: Beierle, C., Meghini, C. (eds.) FoIKS 2014. LNCS, vol. 8367, pp. 361–380. Springer, Heidelberg (2014)
7. Cruz-Filipe, L., Engrácia, P., Gaspar, G., Nunes, I.: Computing repairs from active integrity constraints. In: Wang, H., Banach, R. (eds.) TASE, pp. 183–190. Piscataway, IEEE (2013)
8. Cruz-Filipe, L., Franz, M., Hakhverdyan, A., Ludovico, M., Nunes, I., Schneider-Kamp, P.: repAIrC: a tool for ensuring data consistency by means of active integrity constraints. In: Fred, A., Dietz, J., Aveiro, D., Liu, K., Filipe, J. (eds.) IC3K, vol. 3, pp. 17–26. SCITEPRESS, Setúbal (2015)
9. Cruz-Filipe, L., Nunes, I., Schneider-Kamp, P.: Integrity constraints for general-purpose knowledge bases. In: Gyssens, M., Simari, G. (eds.) FoIKS 2016. LNCS, vol. 9616, pp. 235–254. Springer, Heidelberg (2016)
10. Eiter, T., Gottlob, G.: On the complexity of propositional knowledge base revision, updates, counterfactuals. Artif. Intell. 57(2–3), 227–270 (1992)
11. Flesca, S., Greco, S., Zumpano, E.: Active integrity constraints. In: Moggi, E., Warren, D.S. (eds.) PPDP, pp. 98–107. ACM, New York (2004)
12. Kakas, A.C., Mancarella, P.: Database updates through abduction. In: McLeod, D., Sacks-Davis, R., Schek, H.-J. (eds.) VLDB, pp. 650–661. Morgan Kaufmann, San Francisco (1990)
13. Katsuno, H., Mendelzon, A.O.: On the difference between updating a knowledge base and revising it. In: Allen, J.F., Fikes, R., Sandewall, E. (eds.) KR, pp. 387–394. Morgan Kaufmann, San Francisco (1991)
14. Marek, V.W., Truszczyński, M.: Revision programming, database updates and integrity constraints. In: Gottlob, G., Vardi, M.Y. (eds.) ICDT 1995. LNCS, vol. 893, pp. 368–382. Springer, Heidelberg (1995)
15. Mayol, E., Teniente, E.: A survey of current methods for integrity constraint maintenance and view updating. In: Chen, P.P., Embley, D.W., Kouloumdjian, J., Liddle, S.W., Roddick, J.F. (eds.) ER 1999. LNCS, vol. 1727, pp. 62–73. Springer, Heidelberg (1999). doi:10.1007/3-540-48054-4_6
16. Naqvi, S.A., Krishnamurthy, R.: Database updates in logic programming. In: Edmondson-Yurkanan, C., Yannakakis, M. (eds.) PODS 1988, pp. 251–262. ACM, New York (1988)
17. Przymusinski, T.C., Turner, H.: Update by means of inference rules. J. Log. Program. 30(2), 125–143 (1997)

18. Tarjan, R.E.: Depth-first search and linear graph algorithms. SIAM J. Comput. **1**(2), 146–160 (1972)
19. Teniente, E., Olivé, A.: Updating knowledge bases while maintaining their consistency. VLDB J. **4**(2), 193–241 (1995)
20. Widom, J., Ceri, S. (eds.): Active Database Systems: Triggers and Rules for Advanced Database Processing. Morgan Kaufmann, San Francisco (1996)
21. Winslett, M.: Updating Logical Databases. Cambridge Tracts in Theoretical Computer Science. Cambridge University Press, Cambridge (1990)

Predictive Analytics in Business Intelligence Systems via Gaussian Processes for Regression

Bruno H.A. Pilon[1]([✉]), Juan J. Murillo-Fuentes[2], João Paulo C.L. da Costa[1],
Rafael T. de Sousa Júnior[1], and Antonio M.R. Serrano[1]

[1] Department of Electrical Engineering,
University of Brasilia (UnB), Brasilia, DF, Brazil
bhernandes@gmail.com
[2] Department of Signal Theory and Communications,
University of Sevilla, Seville, Spain

Abstract. A Business Intelligence (BI) system employs tools from several areas of knowledge to deliver information that supports the decision making process. Throughout the present work, we aim to enhance the predictive stage of the BI system maintained by the Brazilian Federal Patrimony Department. The proposal is to use Gaussian Process for Regression (GPR) to model the intrinsic characteristics of the tax collection financial time series that is kept by this BI system, improving its error metrics. GPR natively returns a full statistical description of the estimated variable, which can be treated as a measure of confidence and also be used as a trigger to classify trusted and untrusted data. In our approach, a bidimensional dataset reshape model is used in order to take into account the multidimensional structure of the input data. The resulting algorithm, with GPR at its core, outperforms classical predictive schemes in this scenario such as financial indicators and artificial neural networks.

Keywords: Business Intelligence · Predictive analytics · Gaussian processes · Kernel methods

1 Introduction

In corporations and governmental institutions, the ability to obtain fast, reliable and comprehensive business information is strategic to efficiently manage business and institutional operations and to support the decision making process [4]. In this domain of expertise, Business Intelligence (BI) has evolved as an important field of research. Furthermore, outside of the academic world, BI has been recognized as a strategic initiative and a key enabler for effectiveness and innovations in several practical applications in the business universe.

In this context, advances in technology have massively increased the volume of electronic data available, with about 2.5 EB of digital data being created each day in the world, a number which is doubling every 40 months approximately [12]. On the other hand, a great part of this new data lacks structure.

© Springer International Publishing AG 2016
A. Fred et al. (Eds.): IC3K 2015, CCIS 631, pp. 421–442, 2016.
DOI: 10.1007/978-3-319-52758-1_23

Organizing and analyzing this rising volume of raw digital data and finding meaningful and useful information in its content are key points in BI systems.

In this work[1], we address an existing BI system of the Brazilian federal government with the objective of improving the performance of its predictive analytics stage. The BI system used in this work, maintained by the Brazilian Federal Patrimony Department (SPU)[2], contains data regarding the monthly tax collection of that federal department. This BI system was designed also to perform fraud and irregularities detection such as tax evasion.

Fraud techniques are a perpetually evolving enterprise. When a new fraud detection scheme becomes public domain, criminals are likely to use this information to evade themselves from the new detection method, a fact that contributes to limiting the public exchange of ideas regarding this topic [3]. Just in the year of 2012, global credit, debit and prepaid card fraud losses reached $11.27 billion [21]. Of that, card issuers lost 63% and merchants lost the other 37% [21].

Predictive fraud detection approaches have been used in [9], where an Artificial Neural Network (ANN) is used for fraud detection in credit card operations. In [24], an ANN based predictor is used in real world BI data for forecasting and a set of heuristics based on error metrics decides if the predicted data is possibly fraudulent or regular. In [6], the decision making process of identifying frauds in bank transactions is performed using decision trees to classify information gathered using the CRISP-DM management model of data mining in large operational databases logged from internet bank transactions. In [14], supported vector machines and genetic algorithms are used to identify electricity theft.

Our proposal is to enhance the predictive module of the BI system that forecasts the amount of tax to be collected by the SPU and improve its error metrics. The model chosen as the core predictor is based on Gaussian process for regression (GPR), a widely used family of stochastic process schemes for modeling dependent data. GPR exhibit two essential properties that dictate the behavior of the predicted variable [20]. First, a Gaussian process is completely determined by its mean and covariance functions, which reduces the amount of parameters to be specified since only the first and second order moments of the process are needed. Second, they belong to the family of nonparametric methods, i.e., the predicted values are a function of the observed values, where all finite-dimensional distributions sets have a multivariate Gaussian distribution [18].

GPR has several major advantages. GPR returns a complete statistical description of the predicted variable. In a BI environment this can add confidence to the final results and help the evaluation of the performance. The statistical description can be also be obtained in a full classification framework [27], or used as a trigger to easily label a pair of classes. Also, GPR can be learned in a principled manner, avoiding cross-validation approaches, even with

---

[1] A preliminary version of this paper was presented in the 7th International Joint Conference on Knowledge Discovery, Knowledge Engineering and Knowledge Management (IC3K), Lisbon, Portugal, November 12–14, 2015 [19].

[2] In Portuguese, *Secretaria do Patrimônio da União*.

multidimensional data. GPR can be independently modeled in each dimension, which adds flexibility for data sets with different degrees of correlation among its dimensions.

This paper is organized as follows. In Sect. 2, a theoretical foundation on the relevant topics of GPR is presented. In Sect. 3, the considered BI system and dataset model is introduced. In Sect. 4, a predictive algorithm based on a unidimensional GPR is developed. In Sect. 5, a method for reshaping the original data set is proposed, allowing the application of GPR in a bidimensional data set. In Sect. 6, a technique for optimizing the hyperparameters of the GPR's covariance function is presented and the resulting experimental prediction is included. Finally, in Sect. 7, conclusions and considerations are drawn.

2   Gaussian Process for Regression

Gaussian processes belong to the family of stochastic processes that can be used for modeling dependent data observed over time and/or space [20]. In this paper, the main interest is on supervised learning, which can be characterized by a function that maps the input-output relationship learned from empirical data, *i.e.* a training data set. In this study, the output function is the amount of tax to be collected at any given month by SPU, and hence a continuous random variable.

In order to make predictions based on a finite data set, a function h needs to link the known sets of the training data with all the other possible sets of input-output values. The characteristics of this underlying function h can be defined in a wide variety of ways [1], and that is where Gaussian processes are applied. Stochastic processes, as the Gaussian process, dictate the properties of the underlying function as well as probability distributions govern the properties of a random variable [20].

Two properties make Gaussian processes an interesting tool for inference. First, a Gaussian process is completely determined by its mean and covariance functions, requiring only the first and second order moments to be specified, which makes it a non parametric model whose structure is fixed and completely known. Second, the predictor of a Gaussian process is based on a conditional probability and can be analytically solved with simple linear algebra, as shown in [8].

2.1   Gaussian Process

Multivariate Gaussian distributions are useful for modeling finite collections of real-valued random variables due to their analytical properties. *Gaussian processes* extend this scenario, evolving from distributions over random vectors to distributions over random functions.

A stochastic process is a collection of random variables, *e.g.* $\{h(\mathbf{x}) : \mathbf{x} \in \mathcal{X}\}$, defined on a certain probability space and indexed by elements from some set [5]. Just as a random variable assigns a real number to every outcome of a random experiment, a stochastic process assigns a sample function to every outcome of a random experiment [5].

A Gaussian process is a stochastic process where any finite subcollection of random variables has a multivariate Gaussian distribution. In other words, a collection of random variables $\{h(\mathbf{x}) : \mathbf{x} \in \mathcal{X}\}$ is a Gaussian process with mean function $m(\cdot)$ and covariance function $k(\cdot, \cdot)$ if, for any finite set of elements $\{x_1, x_2, \ldots, x_n \in \mathcal{X}\}$, the associated finite set of random variables $h(\mathbf{x})$ have a distribution of the form

$$
\mathcal{N}\left(\begin{bmatrix} m(x_1) \\ \vdots \\ m(x_n) \end{bmatrix}, \begin{bmatrix} k(x_1, x_1) & \cdots & k(x_1, x_n) \\ \vdots & \ddots & \vdots \\ k(x_n, x_1) & \cdots & k(x_n, x_n) \end{bmatrix}\right). \tag{1}
$$

The notation for defining $h(\mathbf{x})$ as a Gaussian process is

$$
h(\mathbf{x}) \sim \mathcal{GP}(m(\mathbf{x}), k(\mathbf{x}, \mathbf{x}')), \tag{2}
$$

for any $\mathbf{x}$ and $\mathbf{x}' \in \mathcal{X}$. The mean and covariance functions are given, respectively, by:

$$
m(\mathbf{x}) = \mathbb{E}[\mathbf{x}],
$$
$$
k(\mathbf{x}, \mathbf{x}') = \mathbb{E}[(\mathbf{x} - m(\mathbf{x}))(\mathbf{x}' - m(\mathbf{x}'))]; \tag{3}
$$

also for any $\mathbf{x}$ and $\mathbf{x}' \in \mathcal{X}$.

Intuitively, a sample function $h(\mathbf{x})$ drawn from a Gaussian process can be seen as an extremely high dimensional vector obtained from an extremely high dimensional multivariate Gaussian, where each dimension of the multivariate Gaussian corresponds to an element x_k from the index $\mathcal{X}$, and the corresponding component of the random vector represents the value of $h(x_i)$ [20].

2.2 Regression Model and Inference

Let $S = \{(\mathbf{x}_i, y_i)\}_{i=1}^{m}, \mathbf{x} \in \mathbb{R}^n$ and $y \in \mathbb{R}$, be a training set of independent identically distributed (iid) samples from some unknown distribution. In its simplest form, GPR models the output nonlinearly by [18]:

$$
y_i = h(\mathbf{x}_i) + \nu_i; \quad i = 1, \ldots, m \tag{4}
$$

where $h(\mathbf{x}) \in \mathbb{R}^m$. An additive iid noise variable $\nu \in \mathbb{R}^m$, with $\mathcal{N}(0, \sigma^2)$, is used for noise modeling. Other noise models can be seen in [13]. Assume a prior distribution over function $h(\cdot)$ being a Gaussian process with zero mean:

$$
h(\cdot) \sim \mathcal{GP}(0, k(\cdot, \cdot)), \tag{5}
$$

for some valid covariance function $k(\cdot, \cdot)$ and, in addition, let $T = \{(\widehat{\mathbf{x}}_i, \widehat{y}_i)\}_{i=1}^{\widehat{m}}$, $\widehat{\mathbf{x}} \in \mathbb{R}^n$ and $\widehat{y} \in \mathbb{R}$, be a set of iid testing points drawn from the same unknown distribution S.

Given the training data distribution, S; the prior distribution, $h(\cdot)$; and the test inputs, the columns of the matrix $\widehat{\mathbf{X}} \in \mathbb{R}^{n \times m}$; the use of standard tools of Bayesian statistics such as the Bayes' rule, marginalization and conditioning allows the computation of the posterior predictive distribution over the testing outputs $\widehat{\mathbf{y}} \in \mathbb{R}^m$ [20].

Deriving the conditional distribution of $\widehat{\mathbf{y}}$ results in the predictive equations of GPR. We refer to [20] for further details:

$$\widehat{\mathbf{y}}|\mathbf{y}, \mathbf{X}, \widehat{\mathbf{X}} \sim \mathcal{N}(\boldsymbol{\mu}, \boldsymbol{\Sigma}), \tag{6}$$

where

$$\boldsymbol{\mu} = \mathbf{K}(\widehat{\mathbf{X}}, \mathbf{X})[\mathbf{K}(\mathbf{X}, \mathbf{X}) + \sigma^2 \mathbf{I}]^{-1}\mathbf{y},$$
$$\boldsymbol{\Sigma} = \mathbf{K}(\widehat{\mathbf{X}}, \widehat{\mathbf{X}}) + \sigma^2 \mathbf{I} - \mathbf{K}(\widehat{\mathbf{X}}, \mathbf{X})[\mathbf{K}(\mathbf{X}, \mathbf{X}) + \sigma^2 \mathbf{I}]^{-1}\mathbf{K}(\mathbf{X}, \widehat{\mathbf{X}}).$$

Since a Gaussian process returns a distribution over functions, each of the infinite points of the function $\widehat{\mathbf{y}}$ have a mean and a variance associated with it. The expected or most probable value of $\widehat{\mathbf{y}}$ is its mean, whereas the confidence about that value can be derived from its variance.

2.3   Covariance Functions

In the previous section, it was assumed that the covariance function $k(\cdot, \cdot)$ is known, which is not usually the case. In fact, the power of the Gaussian process to express a rich distribution on functions rests solely on the shoulders of the covariance function [25], if the mean function can be set or assumed to be zero. The covariance function defines similarity between data points and its form determines the possible solutions of GPR [18].

A wide variety of families of covariance functions exists, including squared exponential, polynomial, etc. See [20] for further details. Each family usually contains a number of free parameters, the so-called hyperparameters, whose value also need to be determined. Therefore, choosing a covariance function for a particular application involves the tuning of its hyperparameters [20].

The covariance function must be positive semi-definite, given that it represents the covariance matrix of a multivariate Gaussian distribution [18]. It is possible to build composite covariance functions by adding simpler covariance functions, weighted by a positive hyperparameter, or by multiplying them, as adding and multiplying positive definite matrices results in a positive definite matrix [18].

One of the most commonly used covariance function in GPR is the squared exponential kernel given by (7), which reflects the prior assumption that the latent function to be learned is smooth [2].

$$k(\mathbf{x}, \mathbf{x}') = \sigma^2 \cdot \exp\left(-\frac{(\mathbf{x} - \mathbf{x}')}{2\theta^2}\right). \tag{7}$$

In a nutshell, the hyperparameter σ controls the overall variance of the kernel function and the hyperparameter θ controls the distance from which two points will be uncorrelated, both of them presented in (7). These free parameters allow a flexible customization of the problem at hand [2], and maybe selected by inspection or automatically tuned by maximum likelihood (ML) using the training data set.

The covariance function in GPR plays the same role as the kernel function in other approaches such as support vector machines (SVM) and kernel ridge regression (KRR) [17]. Typically, these kernel methods use cross-validation techniques to adjust its hyperparameters [18], which are highly computational demanding and essentially consists of splitting the training set into k disjoint sets and evaluate the performance of the hyperparameters [20].

GPR can infer the hyperparameters from samples of the training set using the Bayesian framework [18]. The marginal likelihood of the hyperparameters of the kernel given the training data set can be defined as:

$$p(\mathbf{y}|\mathbf{X}) = \int p(\mathbf{y}|\mathbf{h}, \mathbf{X})p(\mathbf{h}|\mathbf{X})d\mathbf{h}. \tag{8}$$

Recalling that $\mathbf{X}$ is dependent of the hyperparameter's set, [28] proposes to maximize the marginal likelihood in (8) in order to obtain the optimal setting of the hyperparameters. Although setting the hyperparameters by ML is not a purely Bayesian solution, it is fairly standard in the community and it allows using Bayesian solutions in time sensitive applications [18]. More detailed information regarding practical considerations about this topic will be presented in Subsect. 6.1 and can be found in [11].

3   BI System and Dataset

From a process point of view, BI systems can be divided into two primary activities: insert data into the system and extract information and knowledge out of the system [26]. The traditional architecture of the key components in a generic BI systems is shown in Fig. 1. We refer to [7,15] for a background regarding the components and the architecture of BI systems.

The BI system addressed in this works belongs to SPU, which is a Brazilian federal agency legally responsible for managing, supervising and granting permission to use federal real estate properties in Brazil. The monthly revenue managed by this agency comes mainly from taxes and other associated fees collected by its Department of Patrimony Revenue Management[3] [22].

The tax collected by SPU is based on a massive amount of federal legislation spread out among the Constitution of Brazil, laws, decrees and executive orders. A BI system designed for SPU was first implemented by [23], where a predictive analytics module based on artificial neural network forecasted the monthly amount of tax to be collected.

Figure 2 shows the architecture of the current BI system of SPU. We refer to [23] for further details.

---

[3] In portuguese, *Departamento de Gesto de Receitas Patrimoniais.*

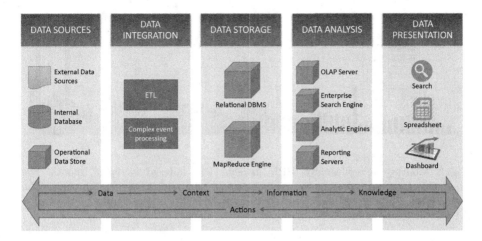

Fig. 1. A traditional architecture and components of a generic BI system.

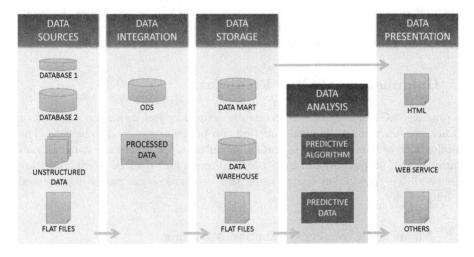

Fig. 2. Architecture of the current state-of-the-art BI system of SPU.

The data set used in this entire work is a financial time series: the monthly tax collection of SPU. The series ranges from years 2005 to 2010. The amount collected, expressed in *reais*[4] (R$), is treated as a random variable indexed by the xth month, where x ranges from 1 to 72. Thus, $x = 1, \ldots, 12$ is related to the first year's collection (2005); $x = 13, \ldots, 24$ is related to the second year's collection (2006), and so forth.

For comparison purposes, only the first 60 months of the data (ranging from 2005 to 2009) were used to build the covariance matrix and estimate the hyperparameters of the Gaussian process. The data regarding the year 2010 was

---

[4] *Reais* is the Brazilian currency.

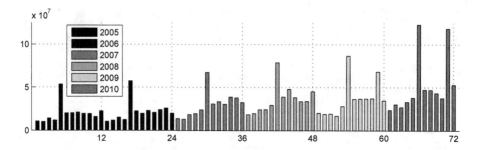

Fig. 3. Monthly tax collected by SPU, in *reais* (R\$), indexed by the xth month. The gray scale bars, representing the years between 2005 and 2009, were chosen as the training set, and the red bars, representing the year 2010, were chosen as the target set. (Color figure online)

exclusively used to evaluate the performance of the proposed predictor by error measurement. Therefore, the first five years of data will be referred as the training data set, and the sixth year of data will be referred as the target data set. Figure 3 shows a bar plot of the data model used in this work.

4   Unidimensional GPR Predictor

In practice, a Gaussian process can be fully defined by just its second moment, or covariance function, if the mean function can be set or assumed to be zero. The implications of this approach is studied in Subsect. 4.1, where the data normalization and a unidimensional model for the mean and covariance functions are discussed. The prediction results using this unidimensional model is presented in Subsect. 4.2.

4.1   Mean and Covariance Function Modeling

Considering the training SPU data set in Fig. 3, a preprocessing stage normalized that data set by a mean subtraction - transforming it into a zero mean data set - and an amplitude reduction by a factor of one standard deviation. Thus, the mean function in (3) can be set to zero and the focus of the GPR modeling can be fully relied on the covariance function.

Some features of the training data are noticeable by visual inspection, such as the long term rising trend and the periodic component regarding seasonal variations between consecutive years. Taking those characteristics into account, a combination of some well known covariance functions is proposed in order to achieve a more complex one, which is able to handle those specific data set characteristics.

The uptrend component of the data set was modeled by the following linear covariance function:

$$k_1(\mathbf{x}, \mathbf{x}') = \mathbf{x}^T \mathbf{x}'. \tag{9}$$

A closer examination of the data set reveals that, yearly, there is a peak in the tax collection. Additionally, for the years of 2005 and 2006, the peak occurred in the fifth month (May), whereas from 2007 to 2010 the peak occurred in the sixth month (June). The shift of this important data signature makes the seasonal variations not to be exactly periodic. Therefore, the periodic covariance function

$$k_{2,1}(\mathbf{x}, \mathbf{x}') = \sigma_1^2 \, \exp\left(-\frac{2\sin^2[\frac{\pi}{\theta_2}(\mathbf{x} - \mathbf{x}')]}{\theta_1^2}\right)$$

is modified by the squared exponential covariance function

$$k_{2,2}(\mathbf{x}, \mathbf{x}') = \exp\left(-\frac{(\mathbf{x} - \mathbf{x}')}{2\theta_3^2}\right),$$

resulting in the following covariance function to model the seasonal variations:

$$k_2(\mathbf{x}, \mathbf{x}') = k_{2,1} \cdot k_{2,2}. \tag{10}$$

Finally, the sum of the characteristic components in (9) and (10), also with a measured noise assumed to be additive white Gaussian with variance σ_n^2 leads to the proposed noisy covariance function:

$$k(\mathbf{x}, \mathbf{x}') = k_1(\mathbf{x}, \mathbf{x}') + k_2(\mathbf{x}, \mathbf{x}') + \sigma_n^2 \mathbf{I}. \tag{11}$$

In (11), the hyperparameter σ_1 gives the magnitude, or scaling factor, of the covariance function k_2. The θ_1 and θ_3 give the relative length scale of periodic and squared exponential functions, respectively, and can be interpreted as a "forgetting factor". The smaller these values, the more uncorrelated two given observations x and x' are. On the other hand, θ_2 controls the cycle of the periodic component of the covariance function, forcing that underlying function component to repeat itself with a period given by θ_2 time indexes.

As an example of the individual contributions of each component of the covariance function to the final prediction, Fig. 4 shows the decomposed product function $k_2(\mathbf{x}, \mathbf{x}')$ of (10) in terms of the periodic and the squared exponential components. The input observed data is the normalized SPU data set in Fig. 3.

The plots of Fig. 4 were obtained with the hyperparameters $\sigma_1^2 = 1$; $\theta_1 = 0.3$; $\theta_2 = 12$; $\theta_3 = 60$ and $\sigma_n^2 = 0.1$.

The magnitude σ_1^2 was set to 1 not to distort the resulting function regarding the training set; the θ_1 was set to 0.3 month due to the poor month-to-month correlation that the data presents; the θ_2 was set to 12 months due the periodicity of the data; the θ_3 was set to 60 months to ensure all data points are taken into account in the final prediction results and, at least, the σ_n^2 was set to 0.1 to add some white Gaussian noise on the observation set. At this point, it is important to remember that the initial choice of hyperparameters have only taken into consideration the characteristics of the original data set. Later, on Subsect. 6.1, we present a optimization method for tuning them.

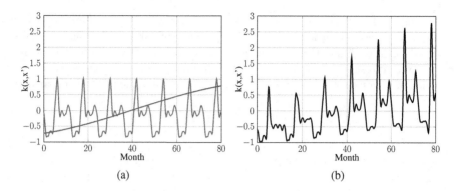

Fig. 4. Normalized plot of the posterior inference of the Gaussian process, indexed by a continuous time interval $\mathcal{X} = [0, 80]$, obtained using the covariance function (a) $k_{2,1}(\mathbf{x}, \mathbf{x}')$ in red (the periodic component) and $k_{2,2}(\mathbf{x}, \mathbf{x}')$ in blue (the squared exponential component); (b) $k_2(\mathbf{x}, \mathbf{x}')$ in black (the product of both components). (Color figure online)

4.2   Unidimensional Prediction Results

With the covariance function defined in (11) and a set of training points given by the first 60 months of the normalized SPU data of Fig. 3, it is possible to formulate a GPR with time as input.

The GPR's characteristic of returning a probability distribution over a function enables the evaluation of the uncertainty level of a given result. For each point of interest, the Gaussian process can provide the expected value and the variance of the random variable, as shown in Fig. 5.

It is noticeable that, for the twelve month prediction using the proposed model, two predicted months fell off the confidence band that delimitates the 95% certainty interval - June and November. These two months have a high contribution on the overall prediction error on this initial approach.

5   Bidimensional Data Reshape

In this section, we propose a pre-processing stage based on the cross-correlation profile of the original data set. This profile is used to separate highly correlated months into one dimension and poor correlated months into a different dimension, leading to a two dimensional structure. Subsect. 5.1 shows an analysis of the time cross-correlation results and implications on the proposed model, and Subsect. 5.2 shows the proposed reshaped data set.

5.1   Time Cross-Correlation

Although the uptrend and the periodic seasonal characteristics are prominent in our data set, some important features of the data are not visible at first sight.

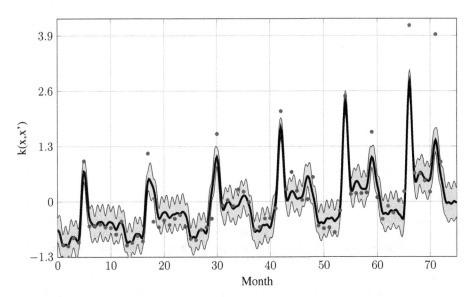

Fig. 5. Prediction results from conditioning the posterior Gaussian jointly distribution at a continuous time interval $\mathcal{X} = [0, 75]$. The blue dots are the training data, the red dots are the target data, the black tick line is the expected value at a time index and the gray band represents the 95% confidence interval (two standard deviations above and below the expected value). (Color figure online)

Considering that the covariance function used to define the GPR is based on a measure of distance, where closer pairs of observation points tend to have a strong correlation and distant pairs of points tend to have a weak correlation, a measure of month-to-month correlation in SPU data can reveal the accuracy of that approach.

The cross-correlation between two any infinite length, real-valued and discrete sequences $\mathbf{a}$ and $\mathbf{b}$ is given by $\mathbf{R_{ab}}(i) = \mathbb{E}[\mathbf{a}_j \mathbf{b}_{j-i}]$ [16]. In practice, sequences $\mathbf{a}$ and $\mathbf{b}$ are likely to have a finite length, therefore the true cross-correlation needs to be estimated since only partial information about the random process is available. Thus, the estimated cross-correlation, with no normalization, can be calculated by [16]:

$$\hat{\mathbf{R}}_{\mathbf{ab}}(i) = \sum_{j=0}^{J-i-1} \mathbf{a_{j+i}} \, \mathbf{b_j}, \quad \text{for } i \geq 0. \tag{12}$$

Figure 6 shows a plot of the absolute cross-correlation of the entire SPU data as sequence $\mathbf{a}_j$, and the last year's target data as sequence $\mathbf{b}_j$. The smaller sequence, $\mathbf{b}_j$, was zero-padded to give both sequences the same length. The resulting cross-correlation was also normalized to return 1.0 exactly where the lag i matches the last year's target data month-by-month.

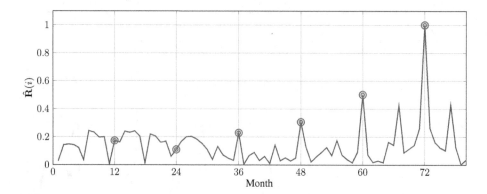

Fig. 6. Estimated absolute normalized cross-correlation between the target data and the whole SPU data set. The sequence was trimmed due to the zero-padding, and the red circles highlight where the lag i is a multiple of 12 months. (Color figure online)

The cross-correlation between the target data and the rest of the sequence exhibited a couple of interesting features about the data. First, it can be noted that the first two years are poorly correlated with the last year. Second, there are some clear peaks on the cross-correlation function where the lag i is a multiple of 12.

Some important conclusions arise from those features. First one is that there is not much information about the last year on the first two years of data, and the amount of information rises as it gets closer to the target. This complies with the distance based correlation function previously proposed.

Also, the peaks pattern shows that the month-to-month correlation is poor, since we only get high correlation values when comparing January of 2010 with January of 2009, 2008, 2007; February of 2010 with February of 2009, 2008, 2007 and so forth. Although some secondary order correlation peaks can be noted, their correlation are smaller than the noisy first two years, leading to the assumption that they do not provide much information.

5.2   Dataset Reshape

With the objective of incorporating the knowledge obtained from the time cross-correlation showed in the previous subsection, some changes were made in the overall modeling proposed. An exponential profile shows a good approximation for modeling the cross-correlation peaks, although the vicinity of the peaks demonstrates a very low correlation with the target data.

In spite the fact that an exponential profile is the main characteristic of the squared exponential covariance function, for it to be a good approximation the exponential profile is required to be present at all times. In this case, the cross-correlation profile shows that the tax collected 12 months before the prediction is more correlated than the tax collected on the previous month of the prediction.

In order to take advantage of the squared exponential covariance function in translating the peaks correlation profile and, at the same time, to carry the characteristics of the original data, this section proposes to convert the original one dimensional SPU data into a two dimensional array, with the first dimension indexed by month $M = 1, 2, \ldots, 12$ and the second dimension indexed by year $Y = 1, 2, \ldots, 6$. This leads to a reshape of the 1D data of Fig. 3 into the 2D data array presented at Fig. 7.

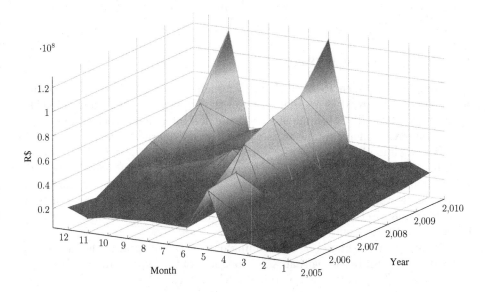

Fig. 7. Plot of the SPU data set converted in a 2D array.

With this new array as the input of our Gaussian process, we can now separate the mean and the covariance function in a two dimensional structure, with different hyperparameters for it in each dimension. Considering the cross-correlation profile of our data shown in Subsect. 5.1, we will assume that only the amount of tax collected on January of 2005, 2006, 2007, 2008 and 2009 will influence the predictive quantity of tax collected in January of 2010, and analogously to the other months. In other words, the information used by the predictor will be obtained mainly from the highlights of Fig. 6. Therefore, from this point forward, the selected approach is to apply the final covariance function showed in (11) exclusively in the monthly dimension.

6   Optimization and Results

This section describes the technique used to optimize the hyperparameters of the proposed covariance function and the resulting prediction using the optimum settings. In addition, we describe preliminary proposals for a classification stage aimed at future studies. In Subsect. 6.1, the knowledge of the cross-correlation profile is applied into the covariance function model and the hyperparameters evaluation. In Subsect. 6.2, the bidimensional resulting prediction is shown and in Subsect. 6.3 a series of performance measurements and error comparisons are made with the previously obtained results, including comparisons with a similar approach using Neural Networks proposed in the literature and a usual financial estimating technique. In Subsect. 6.4, a classification stage based on the statistical description of GPR is discussed, labeling the data into regular or possibly fraudulent.

6.1 Hyperparameters Tuning

Regarding the initial choice of the hyperparameters and its tuning, that learning problem can be viewed as an adaptation of the hyperparameters to a collection of observed data. Two techniques are usual for inferencing their values in a regression environment: (i) the cross-validation and (ii) the maximization of the marginal likelihood.

Since our observed data possess a trend, splitting it would require some detrending approach in the pre-processing stage. Also, the number of training data points in this work is small, and the use of cross-validation would lead to an even smaller training set [20]. Therefore, the marginal likelihood maximization was chosen to optimize the hyperparameter's set.

The marginal likelihood of the training data is the integral of the likelihood times the prior (8). Recalling that $\mathbf{X}$ is dependent of the hyperparameter's set $\boldsymbol{\Theta}$, it is shown in [20] that the log marginal likelihood can be stated as:

$$\log p(\mathbf{y}|\mathbf{X},\boldsymbol{\Theta}) = -\frac{1}{2}\mathbf{y}^{\mathbf{T}}\mathbf{K}_y^{-1}\mathbf{y} - \frac{1}{2}\log|\mathbf{K}_y| - \frac{n}{2}\log 2\pi. \tag{13}$$

In (13), $\mathbf{K}_y = \mathbf{K}_h + \sigma_n^2\mathbf{I}$ is the covariance matrix of the noisy targets $\mathbf{y}$ and $\mathbf{K}_h$ is the covariance matrix of the noise-free latent function $\mathbf{h}$. To infer the hyperparameters by maximizing the marginal likelihood in (8), [20] shows a numerically stable algorithm that seeks the partial derivatives of the logarithmic marginal likelihood in (13) with respect to the hyperparameters.

In order to apply the technique presented in [20], we must address an important restriction in our case. Our final covariance function in (11) possess an hyperparameter θ_2, one of the periodic covariance function's hyperparameters, that dictates the overall period of that function. As seen in Subsect. 5.1, the optimum periodicity of the covariance function should be within a finite set of multiples of 12, leading to $\hat{\theta}_2 = \{12, 24, 36, 48, 60\}$. Taking that restriction into consideration, the optimization results achieved with this approach are shown in Table 1.

The results above described were obtained with algorithm proposed by [20] and the following computational workflow:

Table 1. Optimized set of hyperparameters Θ, σ_1^2 and σ_n^2 after 100 iterations, using the marginal likelihood with the kernel in (11).

Predicted month	θ_1	θ_2	θ_3	σ_n^2	σ_1^2
01. January	0.9907	12	1019	0.2653	0.6935
02. February	0.9360	12	1361	0.2670	0.6552
03. March	0.9151	12	71.12	0.3952	0.6406
04. April	0.8792	12	46.87	0.3662	0.6154
05. May	1.0012	12	23.02	1.5523	0.7008
06. June	1.0000	24	6465	0.5056	0.7000
07. July	0.8919	12	90.50	0.4273	0.4273
08. August	0.7594	12	48.60	0.5075	0.5315
09. September	0.8613	12	88.59	0.3749	0.6029
10. October	0.8994	12	39.55	0.4587	0.6296
11. November	1.0000	24	1252	0.3934	0.7000
12. December	0.8705	12	77.79	0.4636	0.6094

- Define the initial values of the hyperparameter's set Θ. In our case, $\Theta = \{1; 12; 60\}$. The initial magnitude $\sigma_1^2 = 0.7$ and initial noise variance $\sigma_n^2 = 0.1$ were also treated as hyperparameters and, therefore, optimized together with the set Θ;
- Evaluate the marginal likelihood of the periodic component among the finite set of θ_2, keeping the other hyperparameters fixed at their initial values;
- Choose the periodic hyerparameter with the maximum marginal likelihood;
- Evaluate the marginal likelihood of the resting hyperparameters, keeping the periodic hyperparameter fixed;
- Choose the final set of hyperparameters with the maximum marginal likelihood.

Two main considerations regarding the above approach are stated in [20]. The first one is that the likelihood distribution is multimodal, *i.e.* is dependent of the initial conditions of Θ. Also, the inversion of the matrix $\mathbf{K}_y$ is computationally complex.

6.2 Bidimensional Prediction Results

Figure 8 shows a plot of the predicted values using the optimized hyperparameters, where it can be seen that the uncertainty of May's prediction is quite high, mainly because the tax collection profile changed drastically in the training data. This behavior contradicts the linear increasing trend that were used to model the covariance function, since the linear regression of this specific month shows a clear downtrend. However, in spite of the uncertainty level, the prediction of this month turned out to be precise.

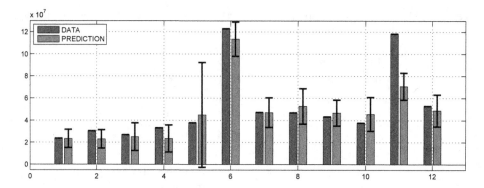

Fig. 8. Plot of the Gaussian process prediction in blue, target SPU data in red. The error bars corresponds to a confidence interval of two standard deviations with respect to the predictive mean (around 95% of confidence). (Color figure online)

Also, it can be noted that November was the only month whose target value fell off the uncertainty predictive interval delimited in this section. In spite the fact that the predicted value is larger than the last year's value for this month, the rate of growth from 2009 to 2010 could not be estimated by this model based only on the information of the training data.

6.3 Prediction Comparison and Error Metrics

The resulting prediction obtained in Subsect. 6.2 will be evaluated by comparison with other predictive techniques and analyzed by different error metrics between the target data and the predictive data. The comparative evaluation will be made month-by-month with two other predictive approaches, one using an artificial neural network and another using an economical indicator. Also, an yearly comparison will be made with the projected tax collection, a revenue estimation made by the Brazilian federal government and published by SPU.

The approach proposed by [24] addressed the same problem, where an artificial neural network is used to predict the SPU tax collection for the year of 2010. On the other hand, a pure financial approach consists of projecting the annual tax collection of SPU by readjusting the previous year's collection by an economic indicator. In this case, the chosen indicator to measure the inflation of the period is the National Index of Consumer's Prices (IPCA), consolidated by the Brazilian Institute of Geography and Statistics (IBGE). In 2009, the twelve month accumulated index was 4.31% [10].

The error metrics used in this subsection aim to evaluate the goodness of fit between the predicted and the testing data set for all the predictive approaches, using the normalized root mean squarred error (NRMSE), the mean absolute relative error (MARE), the coefficient of determination (d) and the coefficient of efficiency (e). The descriptive formulas of each metric is described in Appendix A.

All the predictive approaches, including the one proposed in this work, have their prediction error calculated with respect to the target data and the results are resumed in Table 2.

Table 2. Performance comparison by several error metrics.

Error metric	Optimum value	Gaussian process	Art. neural network	Inflation
NRMSE	0	0.44833	0.46320	0.56246
MARE	0	0.14830	0.31021	0.23222
d	1	0.82107	0.89463	0.92603
e	1	0.78072	0.7659	0.67730

It is important to notice that the overall error in the Gaussian process prediction showed in Table 2 is mainly concentrated in November. Removing this month from the error measurements would lead to NRMSE $= 0,22644$, MARE $= 0,12524$, d $= 0,94$ and e $= 0,94359$.

Figure 9 shows a comparative plot among the target data and all the predictive approaches side by side.

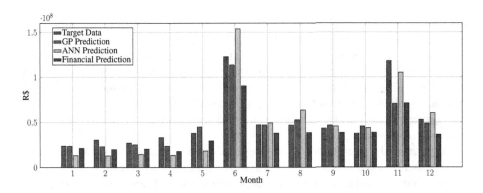

Fig. 9. Monthly plot of target data and predictive results, in *Reais* (R$).

Finally, it is interesting to note that the Brazilian government revenue estimation, published by SPU on its annual report [22], projects an amount of tax collection by SPU in 2010 of R$ 444, 085, 000.00, whereas the total amount collected that year was R$ 635, 094, 000.00 - a gross difference of 38.48% between the estimated and the executed amount of tax collection.

The GPR approach presented in this work, in a yearly basis, projected a total tax collection amount of R$ 620, 703, 197.42, resulting in a gross difference of 2.27% between the projected and executed amounts.

6.4   Classification Stage Proposals

The statistical description of the estimated variable, natively given by Gaussian processes in the regression stage, can be used to build heuristics to classify a predicted dataset into regular or possibly fraudulent. Here, we propose two different heuristics that are suitable to fraud detection scenarios. However, given the limited information publicly available from SPU regarding the dataset used in this work, the evaluation of the proposed schemes is incomplete and deserve to be better investigated in future studies.

The resulting regression obtained through GPR, presented in Fig. 8, shows the variance of the estimated variable as a measure of confidence by translating it into error bars. Since the chosen confidence range can be as large or as small as we desire it to be, it is possible to optimize a classification stage based on this information and, hence, build a trigger where high error bars means high probability of fraud and vice versa. In our case, without any doubt this system would classify May (month number 5) as a possibly fraudulent one. Despite the high uncertainty level of the prediction of this month, the prediction showed to be accurate when compared to the target data.

Another classification approach using the variance information can be build simply by confronting the predicted confidence interval with the real data, when it becomes available. In our case, this system would classify November (month number 11) as a possibly fraudulent one. SPU's annual report [22] states that an extraordinary revenue of R\$ 73, 759, 533.99 happened in 2010, but it is not possible to precise in which month it happened. In november, the difference between the predicted value and the actual revenue was R\$ 55, 015, 235.13.

Whereas the first proposed system returns the classified data in advance, together with the predicted values in the regression stage, the second system needs the real revenue data in order to classify it. On the other hand, the second approach seeks for samples that are most dissimilar from the norm, whereas the first approach needs to be optimized in order to learn the norm and distinguish anomalous behaviors.

As previously mentioned, it is not possible to evaluate the performance of these classification stage proposals due to the limited information regarding our dataset, but the preliminary results using the statistical description of the estimated variable showed in this section encourages further studies on this topic.

7   Conclusions

BI is one of the most challenging and active fields of research nowadays. The multidisciplinary aspect of BI, not rarely embracing knowledge from exact and social sciences, makes its overall development not trivial. Business executives, end users, customers, CIO's and engineers are a few examples of people that must be addressed by a BI solution in an organization. Often, a BI system must be broken into smaller portions to allow an expert suitable approach for each one of its parts.

Governmental BI systems aimed at fraud and irregularities detection are a continuously evolving topic due to the changing nature of fraudulent behavior. In addition, the exchange of ideas regarding fraud detection is limited in the public domain, as publishing information about fraud detection schemes ends up helping the circumvention of these schemes. It is not by chance that the technology and development process of major BI systems focused on fraud detection in financial institutions, banks, credit card issuers, governmental organizations, etc., are rarely made public.

In this work, we aimed to enhance the predictive stage performance of a real world governmental BI system maintained by SPU, an agency of the Brazilian federal government. In order to achieve that, a predictive algorithm based on GPR was developed, aimed to model the intrinsic characteristics of a specific financial series. A unidimensional model for the covariance function of the Gaussian process was proposed, and a pre-processing stage reshaped the original data set based on its cross-correlation profile. That approach empowered the use of a unidimensional GPR model in a bidimensional environment by isolating high correlated months in one dimension and poor correlated months in another dimension.

Although Neural Networks are known for their flexibilities and reliable results when used for regression of time series, GPR are a transparent environment, with a parametric covariance function and no hidden layers, which can be an advantage when evaluating different components of a time series. The hyperparameters of the covariance function of the Gaussian process were optimized with a ML approach, *i.e.* the proposed model let the data speak for itself by learning the hyperparameters only with information obtained from the data. It is relevant to notice that the optimization algorithm can converge to a local minimum, making the initial choice of hyperparameters a critical part of the optimization task [19].

Another positive point of GPR is related to the complete statistical description of the predicted data, which resulted in a powerful tool of confidence. Using this feature, a classification stage can be built to trigger trusted and possibly fraudulent tax collection data based on the confidence interval of the prediction [19].

The regression results outperformed some classical predictive approaches such as ANN and financial indicator by several error metrics. In a yearly basis, the difference between the estimated and the real tax collection for 2010 using the approach proposed in this work was of 2.27%, whereas that difference reached 38.48% with the Brazilian government own estimation method [19].

The approach explored in this work showed to be particularly useful for a small number of training samples, since the covariance function chosen to model the series results in a strong relationship for closer training points and a weak relationship for distant points. On the other hand, adding more training years before 2005 should not make a substantial difference in the prediction result using this method.

Acknowledgements. The authors wish to thank the Spanish Government (TEC 2012-38800-C03-02), the European Union (FEDER), the Brazilian research, development and innovation agencies FAPDF (TOA 193.000.555/2015), CAPES (FORTE 23038.007604/2014-69) and FINEP (RENASIC/PROTO 01.12.0555.00), as well as the Brazilian Ministry of Planning, Budget and Management, for their support to this work.

A   Error Metric Formulas

For notation purposes, consider $\mathbf{t} \in \mathbb{R}^n$ as a target vector with desired values and $\mathbf{y} \in \mathbb{R}^n$ as an output vector of a regression model; $\mu_\mathbf{t}$ and $\mu_\mathbf{y}$ as the mean of $\mathbf{t}$ and $\mathbf{y}$, respectively; and $\sigma_\mathbf{t}^2$ as the variance of $\mathbf{t}$. The goodness of fit between $\mathbf{t}$ and $\mathbf{y}$ will be given in terms of:

1. Normalized Root Mean Squared Error (NRMSE):

$$\sqrt{\frac{1}{n} \frac{\sum_{i=1}^{n}(t_i - y_i)^2}{\sigma_\mathbf{t}^2}}$$

2. Mean Absolute Relative Error (MARE):

$$\frac{1}{n} \sum_{i=1}^{n} \left| \frac{t_i - y_i}{t_i} \right|$$

3. Coefficient of Determination (d):

$$\left(\frac{\sum_{i=1}^{n}(t_i - \mu_\mathbf{t})(y_i - \mu_\mathbf{y})}{\sqrt{\sum_{i=1}^{n}(t_i - \mu_\mathbf{t})^2}\sqrt{\sum_{i=1}^{n}(y_i - \mu_\mathbf{y})^2}} \right)^2$$

4. Coefficient of Efficiency (e):

$$1 - \frac{\sum_{i=1}^{n}(t_i - y_i)^2}{\sum_{i=1}^{n}(t_i - \mu_\mathbf{t})^2}$$

References

1. Bernardo, J., Berger, J., Dawid, A., Smith, A., et al.: Regression and classification using Gaussian process priors. Bayesian Stat. **6**, 475 (1998)
2. Blum, M., Riedmiller, M.: Optimization of Gaussian process hyperparameters using Rprop. In: European Symposium on Artificial Neural Networks, Computational Intelligence and Machine Learning (2013)
3. Bolton, R.J., Hand, D.J.: Statistical fraud detection: A rev. Stat. Sci. pp. 235–249 (2002)
4. Cheng, H., Lu, Y.C., Sheu, C.: An ontology-based business intelligence application in a financial knowledge management system. Expert Syst. Appl. **36**(2), 3614–3622 (2009)

5. Cinlar, E.: Introduction to Stochastic Processes. Courier Dover Publications, New York (2013)
6. Da Rocha, B.C., de Sousa Júnior, R.T.: Identifying bank frauds using CRISP-DM and decision trees. Int. J. Comput. Sci. Inf. Technol. (IJCSIT) $2(5)$, 162–169 (2010)
7. Davenport, T.H., Harris, J.G.: Competing on Analytics: The New Science of Winning. Harvard Business Press, Brighton (2007)
8. Davis, R.A.: Gaussian process. In: Brillinger, D. (ed.) Encyclopedia of Environmetrics, Section on Stochastic Modeling and Environmental Change. Willey, New York (2001)
9. Dorronsoro, J.R., Ginel, F., Sánchez, C., Cruz, C.: Neural fraud detection in credit card operations. IEEE Trans. Neural Netw. $8(4)$, 827–834 (1997)
10. IBGE: Historical Series of IPCA (2013). www.ibge.gov.br/home/estatistica/indicadores/precos/inpc_ipca
11. MacKay, D.J.: Information Theory, Inference and Learning Algorithms. Cambridge University Press, Cambridge (2003)
12. McAfee, A., Brynjolfsson, E.: Big data: the management revolution. Harv. Bus. Rev. 90, 60–66 (2012)
13. Murray-Smith, R., Girard, A.: Gaussian process priors with ARMA noise models. In: Irish Signals and Systems Conference, pp. 147–152, Maynooth (2001)
14. Nagi, J., Yap, K., Tiong, S., Ahmed, S., Mohammad, A.: Detection of abnormalities and electricity theft using genetic support vector machines. In: TENCON 2008-2008 IEEE Region 10 Conference, pp. 1–6. IEEE (2008)
15. Negash, S.: Business intelligence. Commun. Assoc. Inf. Syst. $13(1)$, 54 (2004)
16. Orfanidis, S.J.: Optimum Signal Processing: An Introduction. McGraw-Hill, New York (2007). ISBN 0-979-37131-7
17. Pérez-Cruz, F., Bousquet, O.: Kernel methods and their potential use in signal processing. Signal Proc. Mag. $21(3)$, 57–65 (2004)
18. Pérez-Cruz, F., Vaerenbergh, S., Murillo-Fuentes, J.J., Lázaro-Gredilla, M., Santamaria, I.: Gaussian processes for nonlinear signal processing. IEEE Signal Proc. Mag. $30(4)$, 40–50 (2013)
19. Pilon, B.H.A., Murillo-Fuentes, J.J., da Costa, J., de Sousa Jr., R.T., Serrano, A.M.R.: Gaussian process for regression in business intelligence: a fraud detection application. In: Proceedings of the 7th International Joint Conference on Knowledge Discovery, Knowledge Engineering and Knowledge Management (IC3K), pp. 39–49 (2015)
20. Rasmussen, C.E., Williams, C.K.I.: Gaussian Processes for Machine Learning. The MIT Press, Cambridge (2006). ISBN 0-262-18253-X
21. Robertson, D.: Global card fraud losses reach \$11.27 billion in 2012. Nilson Rep. (1023) 6 (2013)
22. Secretaria de Patrimonio da União (SPU): Relatório de gestão 2010 (2011). http://www.planejamento.gov.br/secretarias/upload/Arquivos/processo_contas/2010/SPU/1_SPU2010_Relatorio_de_Gestao.pdf
23. Serrano, A.M.R.: Study and implementation of a predictive analytics module for the business intelligence system of the Brazilian ministry of planning, budget and management. Universitat Politcnica de Catalunya (2012)
24. Serrano, A.M.R., da Costa, J., Cardonha, C.H., Fernandes, A.A., de Sousa Jr., R.T.: Neural network predictor for fraud detection: a study case for the federal patrimony department. In: Proceeding of the Seventh International Conference on Forensic Computer Science (ICoFCS) 2012, pp. 61–66. ABEAT, Brasília, Brazil (2012). ISBN 978-85-65069-08-3

25. Snoek, J., Larochelle, H., Adams, R.P.: Practical Bayesian optimization of machine learning algorithms. In: NIPS, pp. 2960–2968 (2012)
26. Watson, H.J., Wixom, B.H.: The current state of business intelligence. Computer **40**(9), 96–99 (2007)
27. Williams, C.K., Barber, D.: Bayesian classification with Gaussian processes. IEEE Trans. Pattern Anal. Mach. Intell. **20**(12), 1342–1351 (1998)
28. Williams, C.K., Rasmussen, C.E.: Gaussian Processes For Regression (1996)

Semantically Enhancing Recommender Systems

Nuno Bettencourt[1]([✉]), Nuno Silva[1], and João Barroso[2]

[1] GECAD and Instituto Superior de Engenharia,
Instituto Politécnico do Porto, Porto, Portugal
{nmb,nps}@isep.ipp.pt
https://www.isep.ipp.pt/
http://www.gecad.isep.ipp.pt

[2] INESC TEC and Universidade de Trás-os-Montes e Alto Douro,
Vila Real, Portugal
jbarroso@utad.pt
http://www.utad.pt
https://www.inesctec.pt

Abstract. As the amount of content and the number of users in social relationships is continually growing in the Internet, resource sharing and access policy management is difficult, time-consuming and error-prone. Cross-domain recommendation of private or protected resources managed and secured by each domain's specific access rules is impracticable due to private security policies and poor sharing mechanisms. This work focus on exploiting resource's content, user's preferences, users' social networks and semantic information to cross-relate different resources through their meta information using recommendation techniques that combine collaborative-filtering techniques with semantics annotations, by generating associations between resources. The semantic similarities established between resources are used on a hybrid recommendation engine that interprets user and resources' semantic information. The recommendation engine allows the promotion and discovery of unknown-unknown resources to users that could not even know about the existence of those resources thus providing means to solve the cross-domain recommendation of private or protected resources.

Keywords: Recommendation · Access policy · Unknown-Unknown

1 Introduction

The Internet has recently grown to over three billion users. On certain social networks, more than two hundred thousand photographs are uploaded every minute. Such rate of content generation and social network building make the task of sharing resources more difficult for users.

Standard resource sharing in the Internet is achieved by granting users with access to resources, but they are commonly restricted to resources hosted on a single domain. Access policies are consequently issued to users registered on the same domain. Sharing resources with users that are not registered on the same

© Springer International Publishing AG 2016
A. Fred et al. (Eds.): IC3K 2015, CCIS 631, pp. 443–469, 2016.
DOI: 10.1007/978-3-319-52758-1_24

domain has proven insecure or difficult to achieve. Referencing and accessing resources protected by access policies in other web domains (apart from where they are hosted) is practically unsupported by existing web applications.

In cross-domain sharing, such difficulties encourage: (i) the cloning of the resource to different domains; (ii) the multiplication of users' internal and social identity.

The goal of this work is to provide a seamless cross-web-domain infrastructure that provides secure, rich and supportive resource managing and sharing processes. It proposes a distributed and decentralised architectural model by fostering cross-web-domain resource sharing, resource dereferencing and access policy management. It adopts the principles of the Web and of World Wide Web Consortium (W3C) standards or recommendations.

In order to support user management of access policies, a recommendation provider capable of recommending access policies to users is included in the architecture (Sect. 2). The proposed recommender system features a hybrid engine consisting on the combination of different filtering techniques that exploit user profiles, their social networks, resources content, (distributed) provenance and traceability information (Sect. 3).

A prototype to demonstrate the infrastructure's feasibility was designed and implemented to prove that the architecture model can be deployed in a real world scenario. The hybrid recommendation process was tested using an available data set where information was interpreted to simulate human behaviour in the system (Sects. 4 and 5).

Finally, the last section gives an overview of the proposed solution.

2   Background Knowledge

This work allows the recommendation of access policies to resource authors. This section provides an insight about recommendation processes, resources that are currently not easily shared because of access policy restrictions and Authentication, Authorisation and Accountability (AAA) architectures.

2.1   Recommendation

Recommendation is something that has become part of everyone's daily lives. To reduce uncertainty and help coping with information overload when trying to choose among various alternatives, people usually rely on suggestions given by others, which can be given directly by recommendation texts, opinions of reviewers, books, newspapers, etc. [17]. Users are willing to follow others' recommendations and to give back to the community.

When deciding between which product to buy, users want to be able to read opinions from other buyers [8] and tend to follow them as they are considered experienced users [22].

Currently, recommendation is widely used in electronic commerce [1,7,16]. In e-commerce web applications, trust is based on the feedback of previous online interactions between members as shown by the authors in [13,14].

In the Internet perspective, there are other areas in which recommendation is also relevant, such as resource recommendations on websites (*e.g.* Pinterest), documents (*e.g.* Slideshare, Pocket) and users (*e.g.* LinkedIn, Facebook, Google+). With the Internet's continual evolution, recommender systems have also evolved. While initially recommendation was only used in e-commerce websites for recommending similar or most bought items to users, nowadays the process of recommendation has improved such that the recommendation of friendship and/or relationship between users of a social network has become a quite common task on typical social web applications.

Recommender systems rely on two basic elements:

User/Item Actions. Represents user actions upon items and may include a possible rating.

Item Similarities. Represents the associations between users or between items. Most recommender systems provide off the shelve algorithms to calculate item similarity during the recommendation process.

The output of a recommender system is a scored list of recommended items to users. The maximum number of retrieved recommendations is specified by the value of AT.

A systematisation of the user's consciousness about resources is presented next, which will be helpful to perceive the importance of the recommendation process in the scope of this work.

2.2   Known-Known, Unknown-Unknown

A user's consciousness about something can be characterised according to two dimensions: perception of reality and reality of perception.

Applying such rationale to resources' location and users' knowledge awareness of those, a particular resource can be classified as (*cf.* Fig. 1):

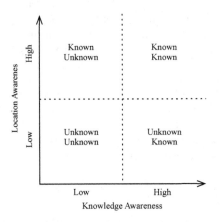

Fig. 1. Location awareness *vs.* Knowledge awareness.

Known-Knowns. These are resources whose existence and location are known by the user *e.g.* a photograph is taken of a person, and the person knows about its existence and its location.

Known-Unknowns. These are resources a user recognises he/she knows nothing about until he/she finds them *e.g.* a person finds a photo by chance on which he/she appears, knows its location but was not aware of its existence.

Unknown-Knowns. These are resources that the user does not know how to find, but knows about their existence, *e.g.* a photo is taken of a person, the person knows about its existence but does not know about its location. With time and searching investment the person might get to its location.

Unknown-Unknowns. These are resources whose existence the user is not even aware of *e.g.* a photo is taken of a person but the person does not know its existence or where to locate it. These types of resources would only come up on searches related to the user if contextual information is used.

This classification emphasises the fact that the same existing information is perceived differently by users. There are different reasons for these different perceptions, including (i) access policy restrictions and (ii) information overload. Recommender systems are conceptually fit to help users perceive resources as (useful) known-knowns.

Access policy restrictions prevent users to access resources that would be of their interest. The recommender system mediates between the owner (that has the resource and can grant access to it) and the beneficiary (that is interested in the resource). Recommender systems will:

- recommend the owner with access policies to grant access permissions to another user upon the resource;
- recommend the beneficiary to request access permissions for a certain resource that is not accessible and that is a known-unknown, unknown-known or unknown-unknown to the reader.

Information overload "occurs when the amount of input to a system exceeds its processing capacity" [18]. In this context, information overload occurs because the owner is not able to match the large number of his/her protected resources with the potentially large number of interested readers. In that sense, recommender systems will:

- recommends the owner with suggestions of potentially interested users that are not able to access the resources;
- recommends the beneficiary, which is overloaded by the quantity of users that he/she would have to contact to request access to known-unknown, unknown-known or unknown-unknown resources.

2.3   Conceptual Architecture

Based on the nomenclature and responsibilities proposed by the Internet Engineering Task Force's reference architecture for AAA in the Internet [20], this section describes the architecture for a system capable of accomplishing the envisaged goal.

When included as part of a multi-domain decentralised AAA system, the conceptual architecture sets the stage for defining protocol requirements between engaged systems.

Commonly accepted names for the various entities involved in the architecture are:

- Policy Enforcement Point (PEP) [12,20,23,24];
- Policy Decision Point (PDP) [12,20,23,24];
- Policy Information Point (PIP) [12,20];
- Policy Retrieval Point (PRP) [9,20];
- Policy Administration Point (PAP) [3,12,19].

Other existing architectures use the concept of an IDentity Provider (IdP) that provide features for creating and maintaining users' identity.

The decentralised structure is capable of providing authentication, authorisation, access control management and recommendation based on resources, users, provenance and traceability information in a distributed and decentralised system, by promoting the usage of action sensors, metadata generators and semantic rules (*cf.* Fig. 2). This architecture is novel in respect to the following aspects:

- despite most of the components maintaining the same names as in typical architectures, their responsibilities and features are enhanced to address the defined requirements;
- adds a recommendation component that is responsible for the recommendation of access policies;
- boosts these components by replacing legacy and traditional non-standard formats and procedures with new data representation by using semantic web standards, capable of a better and explicit knowledge and information description.

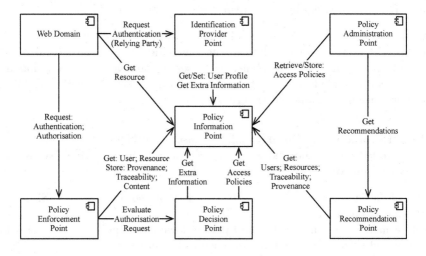

Fig. 2. Conceptual architecture.

In particular, managed and exploited information is a cornerstone of this work:

- resource, *i.e.* anything in the world that can be referred, either physical or virtual, that is identified;
- user, a special kind of resource representing a human or artificial agent in the system.

The conceptual architecture uses components that have been previously addressed by the authors in [2] and proposes the adoption of a new component which is the focus of the work being presented. Each component has specific responsibilities and features (*cf.* Fig. 2).

In [10] the author suggests a typical operation pattern for providing resource authorisation. In this operation pattern, the PEP is responsible for intercepting access requests sent from the user to perform some type of action upon a resource. The PEP, on behalf of the user, requests authorisation for accessing the resource. This request is forward to the PDP, which is the entity that has the engine for evaluating access policies. It uses the information provided by the PEP and the specified access policies to determine if the user should be allowed or denied access to the resource.

The PDP uses the PRP and PIP to retrieve policies and attributes referenced in the policies. When the PDP finishes the evaluation of access policies, it returns an answer to the PEP stating whether access has been granted or denied to the user. If access is granted, the resource is retrieved from the hosting server. The PAP is the system entity used for managing the access policies. For that it uses the features of PRP to retrieve existing policies and store changes to those. Some of the component features are described next.

Identification Provider Point. Provides users with a new identity and appropriate credentials. The following features are enhanced or added:

- allows identity generation and credentials creation to new users;
- allows managing each user's internal and social identity in the virtual world;
- provides an authentication relying party service that allows legacy domains that do not provide Friend-Of-A-Friend + Secure Sockets Layer (FOAF+SSL) authentication to validate users credentials.

Policy Enforcement Point. Enforces user's authentication and guarantees controlled and authorised access to resources. The following features are enhanced or added:

- typical basic authentication methods are replaced by FOAF+SSL cross-domain authentication;
- enforcement is no longer achieved by using local access policies, but instead it is replaced by a distributed and decentralised method;
- action sensors capture User-Generated Content (UGC) and actions.

Policy Decision Point. Evaluates access policies in order to decide if a user should or not be granted access to a resource. The following features are added or changed:

- replaces traditional role or attribute based authorisation mechanisms by an authorisation mechanisms capable of handling semantic, declarative and expressive access policy languages;
- provides decentralised access policy evaluation that is used in a cross-domain perspective;
- obtains, if necessary, semantic information from the PIP for evaluating a particular policy;
- offers reasoning capabilities over more expressive access policy rules that exploit the system's semantics.

Policy Information Point. Manages the information needed for the authentication, authorisation and recommendation processes. The following features are added or enhanced:

- information management of:
  - resources' content, including their type, attributes/properties and preferred hosting domain;
  - provenance and traceability information over UGC and content;
- generating and publishing information according to an explicit and public semantic specification (*i.e.* ontology).

Policy Administration Point. Enables users to manage access policies over existing resources. The following features are added or enhanced:

- access policies are specified by rules instead of directly assigning users to resources or placing users in particular roles;
- proprietary access policies over resources imposed by closed domains are replaced by far more flexible and expressive rules that capture the rationale behind a particular access policy beyond current approaches;
- provides and promotes the means to create access policies based not only on user attributes and relationships, but also on resource attributes;
- provides and promotes the means to define more complex access policies through semantic reasoning over contextual information and meta-information.

Policy Recommendation Point. This component is a novelty in AAA systems. It recommends access policies that are applied to users and resources. These are some of the envisaged responsibilities and features:

- recommend access policies by combining collaborative, social content and semantic filtering methods, allowing the recommendation of known-unknown and unknown-unknown resources to users;
- allow customising the recommendation process, namely the weights for each filtering method.

This component proposal is addressed in the next section.

3   Proposal

The Policy Recommendation Point recommends known-unknown and unknown-unknown resources to users in a cross-domain perspective, by exploring the information gathered by the system, namely user profiles, social network relationships, provenance and traceability information.

Having an access policy definition based on similarities between resources, users or domain knowledge, is being half-way to enabling an automatic recommendation system based on information such as FOAF profiles, interest topics and contexts to provide the sharing of resources.

Traditionally, the responsibility of sharing resources always comes down to the resource's author, based on his/hers restricted perception/knowledge of the whole network of users and resources. Resource access policy recommendation is a process that is introduced to widen that vision by which a system notifies the resource author when other users would probably benefit or rejoice from having access to a particular resource.

The access policy recommendation process aids resource authors in granting or denying access to existent resources by making use of similarity factors between resources and social relationships, suggesting which users should be given access to each resource.

It also eases resource authors' task of sharing resources by finding similar access policies that could be reapplied to similar resources. It is envisaged that recommendation can aid users in the access policy management process regarding their resources, and give other users access to resources that would not have previously been accessible to them.

This is achieved by enriching and enhancing the access policy recommendation process with existing users' and resources' meta-information, and creating a hybrid recommendation method capable of understanding not only the concepts of users and resources but also provenance and traceability annotations gathered from user actions.

A resource context is produced by the analysis of each resource's content and meta-information, while a relationship context is created based on the existing relationship depth between users [21], each user's profile, linked resources and consequent relationships.

One of the outcomes of this proposal is the creation of semantic rules that match similarities between contexts [5]. Therefore, for every resource or relationship, a context is generated and multiple contexts may exist for the same resource.

This PRP is responsible for:

- the implementation of a hybrid recommendation engine;
- guiding users through the resource-sharing process by suggesting access policies for their resources:
  - by evaluating feedback actions regarding the acceptance or rejection of recommended resource sharing;
  - avoiding rejected recommendations from being recommended again;
- recommending known-unknown and unknown-unknown resources.

3.1   Hybrid Recommendation Engine

When an application responsible for ensuring access control is aware of all users' resources and social relationships, such application is capable of recommending resources to new users that have recently became part of the resource author's social network. This already happens on typically closed applications (*e.g.* Slideshare, ResearchGate, etc.) but is still not being used in a cross-domain perspective for all user resources. Contrary to such closed environments, this proposal consists on performing such task in a cross-domain perspective.

The recommendation is enhanced with semantic information for cross-domain web applications relying on an open and distributed social network based on FOAF profiles, provenance and traceability information.

Users' public resources are used in the recommendation process to enable associations between users, between resources or between users and resources. Despite already being publicly accessible, recommendation of publicly accessible resources is performed because other users that do not know of their existence can eventually have interest in them.

The proposed recommendation process consists of a hybrid approach accomplished by the combination of users' profiles, resources' meta-information, traceability and provenance annotations, social network analysis and domain knowledge as depicted in Fig. 3.

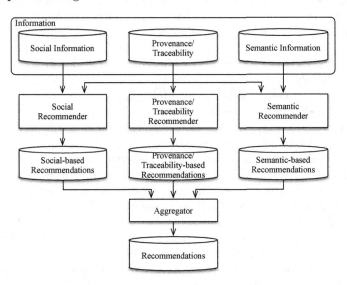

Fig. 3. Hybrid recommendation information.

The semantic filtering relates to problems as recommending known-unknown and unknown-unknown resources that users had little or even no knowledge about. The recommendation service is built on top of three filtering methods that are capable of dealing with different sets of information as displayed in Table 1.

Table 1. Filtering methods input information.

Input information		Filtering method		
		Content	Collaborative	Context
Resources	URI	–	✓	–
	Attributes	✓	–	✓
	Content	✓	–	✓
	Meta-information	✓	–	✓
Users	URI	–	✓	–
	Topic preferences	✓	–	✓
	Social network	–	–	✓
	Actions	–	✓	✓
Semantic	Provenance	–	–	✓
	Traceability	–	–	✓
	Other ontologies	–	–	✓

The following methods are therefore suggested for the PRP:

Content-based Filtering Method. Recommends existing resources by comparing resource attributes, content and meta-information to the user's profile attributes and topic preferences in order to verify the resource's relevancy to the user. This relevancy is given by the similarity between resource attributes and the user's topic preferences. The content-based filtering method is enriched mainly by exploiting resources' content, resources' generated meta-information and users' interest topic preferences.

Collaborative Filtering Method. It recommends resources based on the following pairs of connections: (users, users), (resources, resources) and (users, resources). This process is content-agnostic, meaning that it only recommends resources based such these collaboration patterns, where similarities between users linked to resources are used to infer other new possible connections between users and resources. The collaborative filtering method uses information that associates users' actions to resources.

Context Filtering Method. Recommends resources that match the proposed user's topic preferences or semantically related topics. This filtering method expands the capabilities of the content-based filtering method by introducing reasoning over knowledge concepts. When the user context and resource context match, the recommender system recommends that resource to the user. Interest topics are semantically described, providing not only hierarchical relations between topics but also a graph of other connections between semantic information. Contexts are obtained through the usage of ontologies and semantic rules that provide grounding to this filtering method. The filtering method is enriched by semantic information derived from multiple domains, that include users' FOAF profiles, topic interests, social network graphs, resources' meta-information, provenance and traceability annotations.

While the recommendation process runs continually, it is triggered by several changes in the system, namely:

User-Generated Content. When users create, edit or change an existing resource, resource content is analysed by specific meta-information generators that generate semantic information.

User-Generated Actions. When users perform actions over resources, they are implicitly building their profile. When their profile changes, it is necessary to trigger a recommendation process because a change in a user profile might suggest access to other resources as it influences the collaborative and context-filtering method.

Access Policy Modification. When users create, change or remove access policies, the recommendation process is triggered because other users may now have access to resources that they did not have before, which also influences the collaborative-filtering method.

Social Network Changes. Whenever a user becomes part of or is removed from another user's social network. In fact, this process is quite similar to the addition of new resources because a new user is actually a special case of a new resource that is identified by a corresponding Uniform Resource Identifier (URI).

The proposed hybrid recommender system is depicted in Fig. 4. The recommendation process differs from typical recommendation processes because it

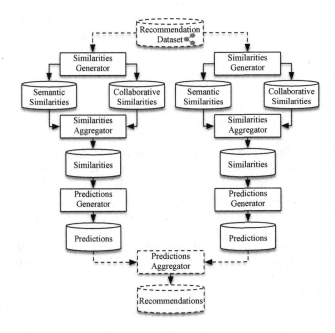

Fig. 4. Recommender system overall process.

supports similarities aggregation (for each individual recommendation process) and predictions aggregation from different individual recommendation processes. The dashed lines represent components that are run only once, independently of how many recommendation processes configurations are used in the system, while the solid lines represent processes that are run singularly for each one of the individual configurations.

The following processes take place during the recommendation process:

- recommendation dataset, maps the system ontology to the recommendation dataset;
- similarities generator, generates similarities between users or between items from the recommendation dataset;
- similarities aggregator, aggregates similarities obtained from different similarity generators;
- predictions generator, predicts the user's interests in available items;
- predictions aggregator, aggregates prediction results;
- recommendations, provides resource author's recommendations.

Similarities Generator. When using collaborative filtering techniques, most recommender engines only evaluate the similarities between users and between items. This phase occurs before applying the prediction algorithms and influences predictions' outcomes. The similarity generator process allows the creation of similarity sets by either using existing similarity algorithms or obtaining them from a semantic model, *e.g.* by using a property that relates concepts. Semantic information is used in the recommender system through the usage of semantic item similarities, obtained through reasoning from the system's ontology.

The similarity generation process is divided in two processes:

Collaborative Similarities Generator. This process generates similarities between users and between items by using existing off-the-shelf user and item similarity algorithms (*e.g.* Tanimoto Coefficient Similarity and Log-Likelihood Similarity). An example is shown in Table 2.

Semantic Similarities Generator. This process is responsible for mapping domain knowledge from the ontology (*e.g.* user preferences) for the recommendation process. In order to semantically infer similarities between concepts of the ontology, properties or rules are used to derive the semantic associations between different concepts. An example is shown in Table 3.

Similarities Aggregator. A similarity aggregator is proposed to merge the similarities provided by the recommenders' similarity generator and the ones provided from an external source (*i.e.* semantic). The similarity aggregator process is proposed for recommender systems to aggregate different similarity sets originated from different similarity generating processes, as depicted in Fig. 5.

Table 2. Generic similarities [Weightless].

Item	Item
1	1
1	2
1	3
2	1
1	3
...	...

Table 3. Semantic similarities [Weightless].

Item	Item
1	2
1	2
1	3
1	5
2	1
3	1
...	...

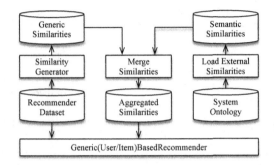

Fig. 5. Similarity aggregation process example.

The similarities aggregator process aggregates the similarities resulting from the above distinct processes. If required, the aggregation process allows assigning different weights for each of the different generated similarities. Semantic associations *per-se* do not have a specific scoring being classified as boolean. These either exist or not, *i.e.* have a value of one or are not present on the ontology. Yet, multiple records for the same association can appear (*cf.* Table 3) and in different orders The similarity aggregator is composed by two inner processes:

- value normalisation for different similarity sets. In this case, different similarity sets (*cf.* Tables 2, 3) can be aggregated using two different approaches:
  - using a boolean similarity approach by deleting duplicate similarities. The resulting sets are shown in Tables 4 and 5;
  - using a weighted similarity approach by counting the number of equally repeating similarity pairs. The resulting sets are shown in Tables 6 and 7.
- average calculation for the different values of the same similarity pair is performed. This happens after values have been normalised *e.g.* using simple linear conversion (*cf.* Tables 8, 9). The similarity aggregator calculates the average between the recommender-based (*cf.* Table 6) and the semantic-based (*cf.* Table 7) similarities by using:
  - a union average approach, by calculating the average of common similarities *i.e.* even if a similarity only appears on one of the generators. The resulting set is shown in Table 10;

Table 4. Generic similarities [Boolean].

Item	Item	Weight
1	1	1
1	2	1
1	3	1
...	...	...

Table 5. Semantic similarities [Boolean].

Item	Item	Weight
1	2	1
1	3	1
1	5	1
...	...	...

Table 6. Generic similarities [Weighted].

Item	Item	Weight
1	1	12.00
1	2	10.00
1	3	18.00
...	...	...

Table 7. Semantic similarities [Weighted].

Item	Item	Weight
1	2	3
1	3	2
1	5	1
...	...	...

Table 8. Generic similarities [Normalised weight].

Item	Item	Weight	Weight [0–1]
1	1	12.00	0.25
1	2	10.00	0.00
1	3	18.00	1.00
...	...	...	...

Table 9. Semantic similarities [Normalised weight].

Item	Item	Weight	Weight [0–1]
1	2	3.00	1.00
1	3	2.00	0.50
1	5	1.00	0.00
...	...	...	...

Table 10. Similarities aggregation [Union average].

Item	Item	Weight
1	1	0.13
1	2	0.50
1	3	0.75
1	5	0.00
...	...	...

Table 11. Similarities aggregation [Intersection average].

Item	Item	Weight
1	2	0.50
1	3	0.75
...	...	...

- using an intersection average approach, by calculating the average of all common similarities, discarding similarities that are only present on one of the similarity generators. The resulting set is shown in Table 11.

Predictions Generator. Most recommender systems present two categories of prediction algorithms: user-based or item-based, not allowing both to be used in the same execution. The result of this process is a list of tuples combining user, item and a score. The higher the score, the more relevant it is to the user.

Predictions Aggregator. To overcome the problem of only being able to use one type of predictions for each recommender execution, a process for aggregating predictions from different executions of different recommenders is proposed. The predictions aggregator process is responsible for producing a recommendation list consisting in tuples (user, item, score), based on the previously calculated predictions.

Prediction aggregators allow the combination of multiple configurations, executed in parallel, into one set of recommendations. As there are different predictions provided from different configurations (*i.e.* originated from item or user-based recommender) with different input similarities (*i.e.* recommendation-based similarities, collaborative similarities and semantic similarities), the recommender engine must aggregate all those predictions.

It aggregates multiple prediction results from different configurations into one new prediction result, where prediction scores are normalised to a given range. By default, the proposed predictions aggregator, normalises the aggregation to values on a scale from zero to one by using a simple linear conversion expressed in Eq. 1.

$$NewValue = \left(\frac{(OldValue - OldMin) * NewRange}{OldRange} \right) + NewMin \quad (1)$$

The aggregation function can be configured and may adopt several functions. An example of an average score between different prediction generators is shown in Eq. 2, where each prediction can be favoured over another by giving a different weight to each prediction set.

$$score(item) = \frac{score(p_1) * w_1 + score(p_2) * w_2}{w_1 + w_2} \quad (2)$$

Tables 12 and 13 demonstrate how the normalisation of values might affect recommendation results. In the first table, when prediction values are not normalised, item number one is the least recommended to user number one, while it becomes the second most recommended item when using normalised values.

Table 12. Predictions without normalised scoring.

User	Item	P1 score	P2 score	Average score	Weighted score (P1=0.8, P2=0.2)
1	1	2.00	25.00	13.50	6.60
1	2	1.00	35.00	18.00	7.80
1	3	2.00	32.00	17.00	8.00

Table 13. Predictions with normalised scoring.

User	Item	P score	P2 score	Average score	Weighted score (P1=0.8, P2=0.2)
1	1	1.00	0.00	0.50	0.80
1	2	0.00	1.00	0.50	0.20
1	3	1.00	0.70	0.85	0.94

Results are scored and the final list is filtered by the top AT value *i.e.* the number of maximum recommendations per user, defined in the system configuration. This is necessary because the user considers only a few of the first recommendations [4, 6].

3.2   Notifications and Feedback

When the recommendation process succeeds in recommending access to resources, the resource author is notified with a message containing:

- the resource to be shared;
- the user to whom the resource is being shared;
- an explanation of why the resource is being recommended.

When a resource sharing is recommended, the system checks with the resource's author if he wishes to assign the access privilege to the proposed user. If the author wants to assign the privilege to the proposed user, the PRP takes the necessary actions to notify the proposed user. When authors accept resource sharing recommendations, these are translated into access policies over resources.

When authors accept resource sharing recommendations, these are translated into access policies over resources.

The author may receive recommendation notifications of access policies granting access to users that may not be part of his/hers social network. When sharing is recommended to users outside the author's social network, the inclusion of that user in the author's social network must be achieved prior to the sharing act, otherwise sharing is not permitted. To this end, the inclusion of a new relationship is proposed. If accepted, the author's FOAF profile is changed accordingly.

The proposed user who should be given access to the resource also receives a notification message stating:

- that a resource exists that might be suitable for the user;
- an explanation of why the resource is being recommended.

Each user receives a list of resources that were shared with him/her, and a request to express whether or not that resource is relevant to him/her, thus providing feedback to the recommender system. This feedback is captured in the form of traceability information and will be used as supporting information.

3.3   Known-Unknown and Unknown-Unknown Resources

In order for a system to be able to recommend the sharing of known-unknown and unknown-unknown resources, it must be possible to establish associations between resources, between users and between users and resources that are not possible to establish by means of content or collaborative analysis. The semantic-filtering method uses ontologies to map existing information and allow the inference of new knowledge by providing associations between resources that would not have been associated before.

In order to better understand how one could benefit from obtaining access to an unknown-unknown resource, let's consider the use case of resources (*e.g.* photographs) that are generated at an event that takes place on a public space. Let's also consider that those resources have been annotated for having recognised but not identified people that may or not be part of the resource author's social network.

On the same place where the event takes place, other passers-by could appear on some of those resources but would never have access to, because no connection exists between those passers-by and the social event except that both co-existed in the same place for a certain period of time.

This unidentified person on each of the resources represents any possible user that may be interested in that resource, not because of any relationships with the user but because that person was at the same time and place where some photographs were taken and could eventually appear in one or more.

By enriching the system with ontologies capable of performing deductive reasoning about events, space and time (from multiple and different sources of information), the architecture infers that the passerby was located in the same place and time the social event took place and therefore presumes a possible association between passers-by and the social event resources. For this, it is possible to narrow down the possibilities of people that could be passers-by at that location and time if the recognised but unidentified person is in the same context on which the photos were taken, and as a result recommend the resource sharing to that unidentified user, by using the following information:

- user profile;
- user contextual information:
  - users' geo-referenced position;
  - users' geo-referenced position's time;

- resource creation time and location;
- provenance and traceability information from user actions:
  - event records of their physical performance while practicing sports.

When the system discovers which unidentified users were at the same time and place, by comparing their location at a given time with the resource time and location, the resource's author is notified in order to share those resources with those particular users.

This type of recommendation can only be derived if different resources' contexts are matched. In this situation, time and location create the context for the presented resources. Nevertheless, this is just an example of a possible context. The conditions for specifying contexts can be fully captured by ontologies and semantic rules, thus being easily extended and reused by multiple recommendation system.

4   Experiments

The aim of the experiments was to prove that even with a large dataset of information, semantic information would improve existing algorithms. For that, a larger set of information and a recommender system are required.

The recommender engine should feature a hybrid mechanism that makes use of collaborative, content and semantic filtering techniques. Yet, these features are not natively supported by mainstream recommender systems.

Mahout recommender engine is a framework that provides advanced expansion features and makes use of collaborative filtering but it does not provide content or semantic filtering techniques, as these must use domain-specific approaches [11].

In order to provide this support with content and semantic filtering techniques, Mahout's recommendation process was modified to enable the aggregation of similarities between items and between users, together with Mahout's similarities generation.

Conducting the evaluation in a real world would be time-consuming and would hence face cold-start problems typically associated with collaborative filtering techniques. For these reasons, it was decided that the system should be evaluated according to an existing dataset.

Several datasets used on the Second International Workshop on Information Heterogeneity and Fusion in Recommender Systems (Hetrec'2011) were analysed in order to prove their appropriateness to the desired evaluation.

After a careful inspection of the content of the LastFM dataset it was clear that it would provide more useful information than the one in the Delicious Dataset or MovieLens, thus promoting the content and semantic filtering. For this reason, LastFM was the chosen dataset for the experiments as it suits the evaluation needs, considering a carefully planned interpretation and mapping to the ontology used in the system. The LastFM dataset is further enhanced with data from the Freebase and Music Brainz datasets.

Figure 6 depicts the entities and associations from LastFM, Freebase and Music Brainz dataset. It is worth noticing that Music Brainz's Musical Artist is used for the single purpose of data integration between LastFM and Freebase datasets.

Due to the lack of integration and explicit semantics of the source datasets, it is necessary to derive and integrate the implicit semantics from the existing datasets into a domain ontology. The mapping stage is responsible for converting the source datasets into a domain ontology. This mapping process is depicted in Fig. 3. The dotted lines represent mappings from the source datasets to the domain ontology.

Each lastfm:User individual/instance gives origin to a domain:User individual. Listen and Tag actions are combined into the general domain's Action because Mahout recommender system does not distinguish between different types of user actions. Each LastFM Musical Artist individually originates a domain's Musical Artist.

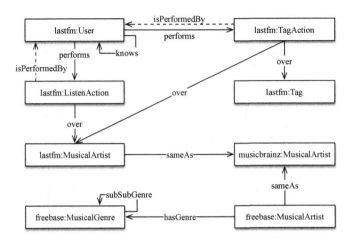

Fig. 6. Source datasets.

The original lastfm:Tag individuals are interpreted as domain Musical Genres' individuals. This is the result of users' manual tagging of each Musical Artist. Yet, while these users' actions complement the Musical Artists with associations to Musical Genres, the original LastFM dataset does not provide information about each Musical Artist and their related Musical Genres. In order to simulate the generation of semantic information, when UGC is captured, an enrichment process is performed for providing an association between domain:MusicalArtist and domain:MusicalGenre.

In order to enrich the domain dataset with users' preferences for musical genres it is necessary to transform the information in the source dataset (i.e. Tags) into semantic content by performing the mapping represented as mapping a) in Fig. 7.

Domain's Musical Genre individuals are obtained by the union of any Freebase Musical Genre:

- whose description matches LastFM's Tag's value by using a reconcile process. In the end of this process, 4698 of the initial 11946 tags were correctly reconciled to their semantic equivalent domain Musical Genre;
- that are tagged against the Musical Artist. Freebase's and LastFM's Musical Artist are not directly associated. Nevertheless, when a Music Brainz Musical Artist is the same for both Freebase and LastFM, one may conclude they are the same.

A transitive property "hasSubGenre" is added to the domain ontology to relate sub-genres. This "hasSubGenre" relation provides the necessary information for semantic filtering recommendation.

The process of generating the recommendation's dataset consists in obtaining the following sets of information from the system's ontology, to comply with the recommendation model presented in Fig. 7.

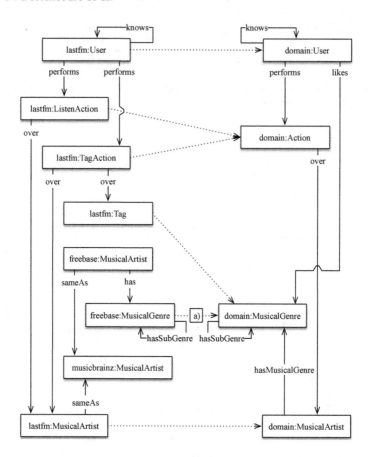

Fig. 7. Source datasets to domain ontology mapping.

The specific domain ontology is translated to a generic ontology that is used by the system. The mapping between both ontologies is depicted in Fig. 8.

Mahout's recommendation process recognises users, items, and similarities between users or between items, user actions and their weights.

Because Mahout's recommender system does not recognise or handle ontologies, a mapping between the system's ontology and Mahout's recommender model is necessary. It converts the system's ontology data into a format that the recommendation engine can use (*cf.* Fig. 9).

According to Fig. 9, it is possible to derive the following concepts:

User. Derived from the foaf:Person concept. Each "foaf:Person" from the system's ontology is mapped to "rec:User" in the recommendation dataset.

Item. Derived from the "prv:DataItem concept". Each "prv:DataItem" is mapped from the system's ontology to the "rec:Item" concept in the recommendation dataset.

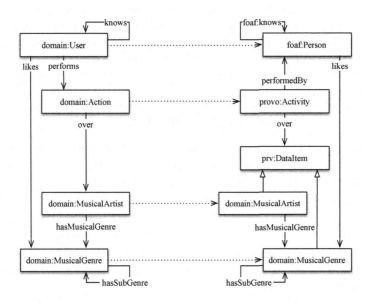

Fig. 8. Domain ontology to system ontology mapping.

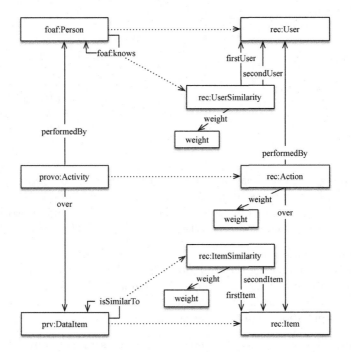

Fig. 9. System ontology to recommendation dataset mapping.

Actions. Derived from the "provo:Activity" concept. Each "provo:Activity" from the system's ontology is mapped into a rec:Action in Mahout's. For each mapped activity, respective relationships with the user ("performedBy" property) and items ("over" property) are created.

User/User Similarities. Derived from the "foaf:knows" property. Each "foaf:knows" property originates a "rec:UserSimilarity" individual.

Item/Item Similarities. Derived from the "isSimilarTo" property. This similarity set is the outcome of the semantic filtering approach.

The weight of each user action, item similarity and user similarity is obtained by the number of repetitions that occur during the mapping process. The resulting dataset represents the input data for the recommender system.

5   Evaluation

This evaluation suite gathers measurements of the recommendation evaluation execution under different runtime configurations. Some of the most relevant baseline configurations are shown in Table 14.

Table 14. Baseline configurations results.

Configuration ID	Similarity		Prediction		Measures			
	Log-Likelihood	Tanimoto	User/Item-based	Boolean/Weighted	AT	Precision	Recall	F1
C1	L	-	I	B	25	0,0814	0,4969	0,1399
C2	-	T	I	B	25	0,0774	0,4726	0,1331
C4	-	T	U	B	25	0,0754	0,4652	0,1297
C7	L	-	U	W	25	0,0471	0,3173	0,0820
C8	-	T	U	W	25	0,0473	0,3180	0,0824
C10	L	-	U	B	25	0,0730	0,4541	0,1258

Each configuration evaluation consists in the calculation of average precision, recall and f1. Experiment's results are compared to those of an initial baseline experiment that is obtained by using the dataset with the simplest possible configuration. Experiments are characterised according to the following dimensions:

– the recommendation dataset;
– the process of generating the training model and relevant items;
– the process of generating and aggregating similarities;
– the process of generating and aggregating predictions;
– the recommendation engine configurations (*e.g.* AT).

Each experiment has its own configuration of these dimensions. The experiments were conducted for a top AT of 25, 50 and 150. Each configuration evaluation consists in the calculation of average precision, recall and f1.

Table 15. Similarities aggregation configurations and results.

Configuration ID	Similarity		Prediction		Aggregation				Measures			
	Log-Likelihood	Semantic	User/Item-based	Boolean/Weighted	Normalised/Std	Union	Union average	Intersection average	AT	Precision	Recall	F1
C1	L	-	I	B	-	-	-	-	25	**0,0814**	**0,4969**	**0,1399**
C104	L	S	I	B	S	UN	-	-	25	+0,0605	+0,2874	+0,1005
C105	-	S	I	B	-	-	-	-	25	−0,0018	−0,0427	−0,0045
C109	L	S	I	B	N	-	UA	-	25	+0,0456	+0,2011	+0,0750
C110	L	S	I	B	N	-	-	IA	25	+0,0455	+0,2000	+0,0748

Table 16. Baseline configurations aggregated predictions results.

	Configurations		Measures				F1 Comparison	
Config-uration ID	Config-uration 1	Config-uration 2	AT	Precision	Recall	F1	Config-uration 1	Config-uration 2
C11	C1	C4	25	0,0821	0,5021	0,1412	+0,0013	+0,1115
C12	C1	C7	25	0,0475	0,3194	0,0826	−0,0573	+0,0006
C13	C1	C8	25	0,0479	0,3218	0,0834	−0,0565	+0,0010
C14	C1	C10	25	0,0790	0,4851	0,1359	−0,0040	+0,0101
C15	C2	C4	25	0,0787	0,4808	0,1353	+0,0022	+0,0056
C16	C2	C7	25	0,0475	0,3188	0,0826	−0,0505	+0,0006
C17	C2	C8	25	0,0475	0,3188	0,0826	−0,0505	+0,0002
C18	C2	C10	25	0,0737	0,4566	0,1269	−0,0062	+0,0011

Experiment's results are compared to those of an initial baseline experiment that is obtained by using the dataset with the simplest possible configuration.

Baseline configurations were created using Mahout's algorithms without injecting any extra similarities in the process, as depicted in Table 15 as configuration C1.

The configurations derived from the C1 baseline configuration are configured with an item-based boolean recommender that uses the Log-Likelihood Similarities algorithm as shown in Table 15.

By using the baseline recommender configuration solely with semantic similarities (*i.e.* C105), precision and recall values drop when compared to the baseline (*i.e.* C1).

Yet, when aggregating both the recommender system similarities and the semantic similarities, using an approach without averaging both similarity sets weights (*i.e.* C104), it produces much better results: precision is about six per cent higher, recall around twenty-nine per cent and f1 about ten per cent higher than the baseline and the normalised averaged approaches (*i.e.* C109 and C110) with union or average intersection.

Using a normalised approach with intersection provides worse results than a non-normalised union of all results.

Table 16 shows the normalised results of aggregation predictions from different configurations. As observed, the results are generally worse than running each configuration independently.

6   Conclusions

The access policy recommendation process aids resource authors in granting or denying access to existing resources by making use of similarity factors between resources and social relationships and suggesting which users should be given access to each resource.

As any other recommendation process, the one proposed is based upon three main parts: users, resources and associations between users and resources. Yet, provenance and traceability annotations, users' social awareness, list of interest topics, resources' and users' context are used in the recommendation process to infer users' interest in resources.

Captured provenance and traceability information are used together with the user's social networks and resources' contents as to automatically propose which access policies should be added to a certain resource.

For providing recommendations, the proposal for generating and aggregating semantic similarities and predictions are described and evaluated.

The results demonstrate that by introducing similarities calculated from content and semantic information into a collaborative filtering technique – either focusing on social networking, user profiles or resource content – it is possible to improve recommendation results.

In [15] the authors used the MovieLens dataset for measuring the system's recommendation performance, using a mean average precision measure. Their precision values for an AT of 50 vary from 0.0272 to 0.0687, which are on par with the values obtained by the experiments conducted with the baseline configuration for the same AT (0.0255 to 0.0345). In the conducted experiments precision barely drops below 0.0500 hitting a maximum of around 0.0800 for a top AT value of 25, which is better than the best values (0.0699, AT = 5) observed by the authors in [15]. This proves that precision measures produce quite small results yet good enough for providing comparison between different systems in an evaluation phase.

The usage of similarities produced from semantic content injected in collaborative-filtering techniques proved to enhance the results in all the configuration scenarios, showing that precision values higher than ten per cent are easily achievable. This was achieved by using only a minimal subset of information that a semantic system can have. Provenance and traceability information, together with enriched semantic information, can indeed make the resource recommendation better.

From the evaluation results, it is possible to conclude that for the used dataset, item-based recommender systems provides better results than user-based recommenders. Furthermore, weighted analysis of actions provided worse results than using a boolean approach (cf. Table 14). This may be due to existing disparities in the values of each action in the original dataset. Aggregating similarities from different sources in the same recommendation process produces better results than aggregating predictions for the same configurations when run in parallel, considering only one of the similarity sets (cf. Table 16).

In summary, the proposed system architecture provides the following features and functionalities:

- the resource author is recommended with new access policies that would facilitate sharing resources with other users;
- it allows discovering resources that users did not even known existed;

- users are given a list of resources which match their interests or contexts, even though specific names and content are not shown unless the author gives them permission to access it, *i.e.* the resource author will know which user is requesting access but the requesting user does not know who is the author;
- semantically enhanced recommendations, allowing the creation of contexts (*e.g.* time and space) for resources and users;
- is possible to combine several similarity algorithms in each individual recommender system run;
- it is possible to aggregate the predictions of different recommender systems running in parallel by coalescing all the results into a single final recommendation list.

The adoption of a hybrid access-policy-recommendation engine enables the enrichment of access policy recommendations by using additional information provided by the system.

Acknowledgements. This work is supported through FEDER Funds, by "Programa Operacional Factores de Competitividade - COMPETE" program and by National Funds through "Fundaçãopara a Ciência e a Tecnologia (FCT)" under the project Ambient Assisted Living for All (AAL4ALL – QREN 13852).

References

1. Adomavicius, G., Alexander, T.: Context-Aware Recommender Systems. In: Ricci, F., Rokach, L., Shapira, B., Kantor, P.B. (eds.) Media, chapp. 7. Springer, US (2011)
2. Bettencourt, N., Silva, N.: Recommending access to web resources based on user's profile and traceability. In: the Tenth IEEE International Conference on Computer and Information Technology, CIT 2010, IEEE, Bradford, UK, June 2010
3. Convery, S.: Network authentication, authorization, and accounting: part one: concepts, elements and approaches. Internet Protoc. J. **10**(1), 2–11 (2007)
4. Duhamel, T., Cooreman, G., De Vuyst, P.: MC DC 2009 - UNITE Report. Technical report, IAB Europe (2009). http://www.iabeurope.eu/files/7513/6852/2734/mc-dc-2009-iab-unite-report.pdf. Accessed 3 May 2015
5. Ghita, S., Nejdl, W., Paiu, R.: Semantically rich recommendations in social networks for sharing, exchanging and ranking semantic context. Soc. Netw. **3729**, 293–307 (2005)
6. Kimberley, S.: European Web Users Stop Searching After First 10 Results (2009). http://www.mediaweek.co.uk/article/974179/european-web-users-stop-searching-first-10-results-report-reveals. Accessed 3 May 2015
7. Linden, G., Smith, B., York, J.: Amazon.com recommendations: item-to-item collaborative filtering. IEEE Internet Comput. **7**(1), 76–80 (2003)
8. MacKinnon, K.A.: User generated content vs. advertising: do consumers trust the word of others over advertisers? Elon J. Undergrad. Res. Commun. **3**(1), 14–22 (2012)
9. Nair, S.: XACML reference architecture (2013). https://www.axiomatics.com/blog/entry/xacml-reference-architecture.html. Accessed 3 May 2015

10. Nimmons, S.: Policy enforcement point pattern (2012). http://www.stevenimmons. org/2012/02/policy-enforcement-point-pattern/. Accessed 4 May 2015
11. Owen, S., Anil, R., Dunning, T., Friedman, E.: Mahout in Action. Manning, Manning (2011)
12. Parducci, B., Lockhart, H.: eXtensible Access Control Markup Language (XACML) version 3.0. Technical report, OASIS, January 2013
13. Resnick, P., Kuwabara, K., Zeckhauser, R., Friedman, E.: Reputation systems. Commun. ACM **43**(12), 45–48 (2000)
14. Ruohomaa, S., Kutvonen, L., Koutrouli, E.: Reputation management survey. In: Second International Conference on Availability, Reliability and Security, ARES 2007, Vienna, Austria, April 2007
15. Said, A., Kille, B., De Luca, E.W., Albayrak, S.: Personalizing tags: a folksonomy-like approach for recommending movies. In: Proceedings of the Second International Workshop on Information Heterogeneity and Fusion in Recommender Systems, HetRec 2011, pp. 53–56. ACM, Chicago, October 2011
16. Schafer, J.B., Konstan, J.A., Riedl, J.: E-commerce recommendation applications. Data Min. Knowl. Discov. **5**(1), 115–153 (2001)
17. Shardanand, U., Maes, P.: Social information filtering: algorithms for automating "Word of Mouth". In: the Proceedings of the ACM Conference on Human Factors in Computing Systems, CHI 1995, vol. 1. ACM Press/Addison-Wesley Publishing Co. (1995)
18. Speier, C., Valacich, J.S., Vessey, I.: The influence of task interruption on individual decision making: an information overload perspective. Decis. Sci. **30**(2), 337–360 (1999)
19. Stephen, L., Dettelback, W., Kaushik, N.: Modernizing Access Control with Authorization Service. Oracle - Developers and Identity Services, November 2008
20. Vollbrecht, J.R., Calhoun, P.R., Farrell, S., Gommans, L., Gross, G.M., de Bruijn, B., de Laat, C.T., Holdrege, M., Spence, D.W.: AAA Authorization Framework [RFC 2904], the Internet Society (2000)
21. Wasserman, S., Faust, K.: Social Network Analysis: Methods and Applications, Structural Analysis in the Social Sciences, 1st edn. Cambridge University Press, Cambridge (1994)
22. Wasserman, S.: The amazon effect, May 2012. http://www.thenation.com/print/article/168125/amazon-effect. Accessed 27 Apr 2015
23. Westerinen, A., Schnizlein, J.: Terminology for policy-based management [RFC 3198], the Internet Society (2001)
24. Yavatkar, R., Pendarakis, D., Guerin, R.: A Framework for Policy-based Admission Control [RFC2753], the Internet Society (2000)

Analyzing Social Learning Management Systems for Educational Environments

Paolo Avogadro$^{(\boxtimes)}$, Silvia Calegari, and Matteo Dominoni

University of Milano-Bicocca, v.le Sarca 336/14, 20126 Milano, Italy
{paolo.avogadro,calegari,dominoni}@disco.unimib.it

Abstract. A Social Learning Management System (Social LMS) is an instantiation of an LMS where there is an inclusion and strong emphasis of the social aspects mediated via the ICT. The amount of data produced within the social educational network (since all the students are potential creators of material) outnumbers the information of a normal LMS and calls for novel analysis methods. At the beginning, we introduce the architecture of the social learning analytics required to manage the knowledge of a Social LMS. At this point, we adapt the Kirkpatrcik-Phillips model for scholastic environments in order to provide assessment and control tools for a Social LMS. This requires the definition of new metrics which clarify aspects related to the single student but also provide global views of the network as a whole. In order to manage and visualize these metrics we suggest to use modular dashboards which accommodate for the different roles present in a learning institution.

Keywords: Social Learning Management System · Social learning analytics · KirckPatrick-Phillips model · Key performance indicators · Dashboard · e-Learning platform

1 Introduction

The integration of social aspects in a learning environment can be defined as an exchange of information and ideas, supplemented by interactions with a personal or professional network of users. A social learning program must provide users with immediate access to relevant content and to seasoned experts who can impart their wisdom [7,13,14]. Social ICT, for example, renders collaboration very natural. The core components of this technology are: social networks, wikis, chat rooms, forums, blogs, expert directories and expertise location, content libraries with content ranked for relevance, shared communities of interest, online coaching and mentoring, etc. By the addition of the gamification element, players can also develop and test their skills and learn complex subjects involving multiple roles and relationships. In the last decade, the goal of many researchers has been to define a Social LMS in order to built a complete "learning environment" that provides a support network, as well as the ability to collaborate, and share information to solve problems. In the modern world, learning organizations (e.g., schools and academies) are expected to go beyond the disciplines of

© Springer International Publishing AG 2016
A. Fred et al. (Eds.): IC3K 2015, CCIS 631, pp. 470–491, 2016.
DOI: 10.1007/978-3-319-52758-1_25

building content for use in the classroom or online. They should provide context and pathways through which people can learn but also improve as learners and become better creators of content. For this reason, social learning environments have to comprise both formal and informal learning elements in an augmented vision of the blended learning paradigm [11]. While most formal training can be done in classes or in isolation, social learning systems may offer the ability to develop learning communities in which groups of learners and trainers share information and collaborate on their learning experience. Thus, a Social LMS integrates social networking, collaboration and knowledge sharing capabilities, as well as interactive elements that enable users to rate contents. In this paper, we propose a model for social network learning which is based on the idea that the act of learning should be an organic process comprising many subjects (students, teachers, specialists, families, etc.). The students remain the most important subjects of the e-learning environment since their development is the final goal of the school, however the interplay with the other subjects cannot be underestimated since a static world where only the students evolve while the school or the families do not change is clearly unrealistic. For a better insight of the process, all the subjects need analytic tools returning important information and automatic analysis [1,3,16]. The information provided is not to be intended as a replacement of the specific capabilities of the teachers, specialists etc. At the contrary, it is thought as an additional tool for these subjects which allows them to have a good understanding of the success or failure of their initiatives. With this information they are able to adopt measures which can improve the quality of the learning process. The platform must also include metrics which demonstrate real value for the learning organization. Learning management systems are designed to measure the performance of a learner on assessments, but they are less efficient at measuring the effectiveness of content [3,9]. Since social learning is dependent on content generated from a variety of sources, a good social learning solution should measure who are the most reliable content providers and which content is not being accessed at all. As a result one can prioritize what is most effective. The goal is to connect users using social tools to accelerate learning.

The standard Kirkpatrick-Phillips model, which is used as a benchmark for learning assessment in organizations, can be also adopted for a Social LMS with a different look. The Kirkpatrick-Phillips model is defined by 5 levels that are: Reaction, Learning, Behavior, Results, and ROI. In this paper, novel key performance indicators (KPI) for each Kirkpatrick-Phillips's level are defined in order to evaluate the impact of the Social LMS on the student's improvements. It is important to take into account that each role is in need of different information in order to improve. A general model of interactions among these subjects is beyond the scope of this paper and pertains more to the social studies, however we want to provide tools which help to give quantitative assessments. At the moment, this paper focuses the work on students, but we are planning to extend our model on the other subjects.

The paper is organized as follows. The architecture for the social learning analytics is presented in Sect. 2. Section 3 presents the revised Kirckpatrick-Phillips models for Social LMS, Sect. 4 defines the metrics for each *new* Kirckpatrick-Phillips's level, Sect. 5 introduces how the metrics can be applied to the forum and chat room social components, Sect. 6 gives an insight on how a dashboard for students can be defined that considers the *new* metrics. Finally, in Sect. 7 some conclusions and future works are stated.

2   An Innovative Architecture for Social Learning Analytics

This section presents an architecture dedicated to the analysis of data produced in an e-learning platform. The data are provided by heterogeneous sources of information by making more complicated the process of learning analytics within an educational context. The description of the architecture is an extension of the one defined in [20]. The innovative aspects are related to the addition of the social elements that, in our opinion, are assuming an increasing importance in a Social LMS. The architecture presented in [20] is the outcome of a workshop held in Paris in February 2015 where the team of experts have followed the guidelines provided by the Apereo Learning Analytics Initiative. Figure 1 shows the core elements of the new architecture where the added components are: (1) Social LMS, (2) Educational Social Network - EduSN, and (3) Social Alert and Intervention System. The focus of this work is to provide an evidence of the importance of the data from a Social LMS that are complementary to the student's formal grades and that allow to have a complete vision of the student's academic path. The difficulty lies in obtaining suited indicators to quantify the informal activities of a student and aggregating the formal academic information with the informal one (see Sects. 3 and 4). In detail:

- Library Systems: it takes information about the permanence of a student in library and the academic material borrowed.
- Student Information System: it considers information related to a student such as ID number, formal grades, absences, group to which they belong (see Sect. 4), etc.
- VLE - Virtual Learning Environment: it monitors the student's actions when he/she interacts with the learning platform in order to capture his/her interests by analyzing, for example, the academic material printed, saved, downloaded, viewed (video, images, chapters of book, ...), etc.
- Social LMS: it considers all the students' activities related to the ability to collaborate and share information to solve a problem such as the information generates within a chat or a forum, a thread discussion within a social wall, etc. In addition, a key role is assumed by the possibility to track the students' preferences on the information created by the others with the support of social judgments such as like/not-like, hearts, smiles, stars, etc.
- Educational Social Network: the interactions arising from the Social LMS component allow to implicitly create virtual linkages among the students according

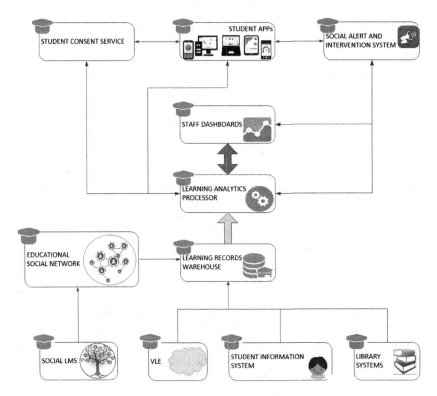

Fig. 1. An architecture for social learning analytics.

to the knowledge shared (see Sect. 4). The analysis of these links creates groups of students based on common interests. In addition, groups can also be defined by considering the physical spaces where students learn such as classes, laboratories, gyms, etc. All these groups define the so called Educational Social Network, a network of learners who act together for solving a problem.

- Learning Records Warehouse: it stores the information acquired by the previous components within ad-hoc data structure/format.
- Learning Analytics Processor: it is the core element of the architecture and its aim is to provide evidence of the effectiveness of the improvement of a training path when a student uses the learning components previously explained. The task is to analyze the data acquired and with the adoption of suited key performance indicators to analyses the learning trend. In this work, we make a detailed analysis on the definition of new metrics devoted to the integration of knowledge from a Social LMS with standard academic knowledge such as formal grades, absences, etc. The social learning analytics arising from these new metrics allow to analyses the formal and informal activities of the students in order to have a complete perspective of what the learner's interests are. In case of poor academic performances, if the teachers can access many indicators, this can help them to undertake proper actions to correct the problem.

- Staff Dashboard: it displays to the users (e.g., teachers, students) the results of the Learning Analytics Processor component with the consequence of having a graphic perception of what the training of a student is. At this step, the correct graphic representation is selected (see Sect. 6) to provide optimal insights of each key performance indicator.
- Social Alert and Intervention System: it comprises the workflow processes to report to the users a note about events or problems. Within a social context, it can be useful to be informed about a social event that comprises students belonging to a same group (e.g., team sport, meeting, classwork, etc.). It can also report a bad academic performance such as low formal grades, a great number of absences, a poor participation to the class activities (e.g., chat, forum), etc.
- Student Consent Service/Student Apps: if a student gives his/her approval then, the services developed by the Staff Dashboard to show the metrics and the Social Alert and Intervention System are installed on the student's devices.

3   The Kirckpatrick-Phillips Model for a Social LMS

The classic Kirkpatrick-Phillips model (KP) [8,10,12] is widely used to evaluate the quality of a training for companies [18]. According to the KP there are 5 levels which are used to evaluate the whole learning process: Reaction, Learning, Behavior, Results and Return on Investment (ROI), respectively. Summarizing, *Reaction* is an indicator which allows to understand how the training was received by the participants. It deals with impressions and tastes regarding the participants - Did they like it? How was the environment? Was the content relevant? Although a positive reaction does not guarantee learning, a negative reaction almost certainly reduces its effectiveness. *Learning* refers to the idea of assessing how much of the information has been understood and retained by the participants of the training. If possible, participants take tests before the training (pretest) and after training (post test) to determine the amount of learning that has occurred. In a company, however, the quality of the work after a training is expected to be better than before, this quantity is measured in the *Behavior* level. The evaluation issue, at this level, attempts to answer questions such as - are the newly acquired skills, knowledge, or attitude being used in the everyday environment of the learner? How did the performance change due to the training provided? For many trainers this level represents the best assessment of a program's effectiveness. A final assessment on the training comprising all the parts just mentioned is necessary, in the KP model this is usually referred to as *Results*. Good results are achieved if some indicators are improved in the organization such as the increased efficiency, decreased costs, increase revenue, improved quality, etc. The *ROI* indicator clarifies if the training was beneficial to the company once its cost is taken into account, with the objective that the effect should be worth the cost. Are we achieving a reasonable return on investment? The ROI formula [12] is calculated as $ROI = (Benefit - Cost)/Cost \times 100$.

Companies use the KP model to assess the investment in organizational learning and development although from the literature it emerges a clear difficulty in measuring with suited metrics the 5 levels just described. It is rarely possible to have data that allow to measure the effectiveness of each level by considering the evaluations of the tangible and intangible benefits in relation with the results of the investment as described in Sect. 1. Due to these problems, in [18] a complementary approach for enterprise training program management is proposed with the intent to overcome the barriers that companies can have when adopting the KP model, that are: (1) isolation of the participant as a factor that has impact on corporate results, (2) lack of standard metrics within the adopted LMS, and (3) lack of standardized data to be used as a benchmark for comparing the defined training functions.

A difficulty in adopting the KP model is that the information from each prior level paves the way for the evaluation of the next level. Thus, each successive level represents a more precise measure of the effectiveness of the training program, but at the same time requires a more rigorous and time-consuming analysis. The idea is to determine which metric is more appropriate to understand whether the network of learners is performing well and whether the service supplied by the platform is useful for the network. The usage of the platform is a straightforward value which allows to understand the reception of the platform itself.

In this paper, we want to translate the KP model to a scholastic environment as such some modifications are needed because of the different purposes and means between a school and a company. For a company the aim of an internal training is to achieve a monetary/strategic gain, while for schools the overall growth of the learners is the final goal, and within it an increased education is of primary importance. In an educational context, the problems presented in [18] could be overcome: (1) social learning creates community discussion forums and group-based projects which encourage collaboration by reducing the learner's isolation, (2) for learners: personal qualities are achieved by considering standard grades, social skills are obtained, for example, by monitoring the learner's activities in the Social LMS as explained in Sects. 4 and 5, and (3) the data quality can be evaluated with standard metrics by teachers (e.g., tests, grades, . . .), and social evaluation metrics by peers (e.g., social grades on materials, comments, liking, . . .).

The novel interpretation of the KP model for the educational context is defined as follows (see Fig. 2):

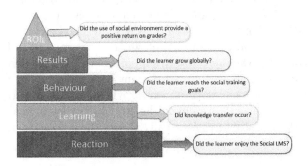

Fig. 2. The Kirkpatrick-Phillips model for a Social LMS.

- *Reaction*: "Did the learner enjoy the Social LMS?" The learner's satisfaction can be analyzed by two main techniques: explicit and implicit [5,17]. With the explicit approach users must openly specify their preferences to the system. This can be achieved by filling in questionnaires and/or by providing short textual descriptions in order to specify the topics of interests. With the implicit approach, the user's preferences are automatically gathered by monitoring the user's actions. To this regard, the defined techniques range from click-through data analysis to query and log analysis, sentiment analysis, etc. We prefer to use an implicit approach, where it would be possible to estimate the actual activity of an engaged learner, and use that as a target.
- *Learning*: "Did knowledge transfer occur?" *Learning* refers to the idea of assessing how much of the information has been understood and retained by the learners of the training. In a Social LMS for educational context, a formal assessment can be defined by the teacher's judgment or by official tests before and after the use of the system for analyzing if an improvement of the academic performance of the student is occurred; an informal assessment can be obtained by analyzing how peers and teachers have evaluated the learner's materials such as new multimedia documents added in the Social LMS, comments on specific subjects, messages from chat and forum, etc.
- *Behavior*: "Did the learner reach the social training goals?" Nowadays most of the schools have personalized learning programs, thus the percentage of completion of these programs can be thought as a measure of how the previous knowledge has changed the student's present knowledge allowing her/him to take further steps forward. The gamification of the learning programs is an important aspect which helps to keep interest in the objectives.
The learner's attitude can be measured by specific behavior indicators: are learners increasing the social activity? The metrics are related to the usage of collaborative tools which reflect on the quality and quantity of the relationships that they establish with their peers.
- *Results*: "Did the learner grow globally?" This level includes in a global statement if the learner's has improved his/her academic performance during the usage of the Social LMS. The data obtained from the indicators considered in the previous levels are gathered, elaborated, and then visualized in customized dashboards. The learning analytic uses data analysis to inform the progress related to teaching and learning. The global evaluation is a mush-up of data collected ranging from formal evaluations (e.g., tests, grades, etc.) to informal evaluations (e.g., social evaluations, judgment of peers, etc.). The heterogeneous types of data available allow to support more adaptive and personalized forms of learning enabling enhanced student performance.
- *ROI*: "Did the use of social environment provide a positive return on grades?" In an educational context, we define the ROI_L in order to evaluate whether the effort in the social interaction really affects the scholastic performance of a learner, L.

4   Definition of KPI for Social LMS

The failure or success of a student is determined by many factors. In the present paragraph, we want to analyze key performance indicators which take into account various aspects of the student life associated with the levels of the KP model presented in Sect. 3. The students grades are an obvious starting point, in particular at the end of the academic year the average grade provides a good insight about the student success. However, the grades do not provide the information regarding the general development of a student. For example, social skills prove to be essential in modern life; for this reason, we want make quantitative and qualitative assessments on them. The sources which provide the data needed for the indicators are the following:

- The "normal" student information (grades, absences, etc.) usually collected by the teachers, and in this case in digital format.
- The data inserted by the students in the social network (forum, chat, evaluations, comments, etc.).
- Feedback from special instructors, families, etc.
- The multimedia material uploaded by teachers and students (for formal and informal learning).

There are many aspects which are important for determining the growth of a student and it is beyond the scope of this paper to take all of them into account. Since this study is devoted to social e-learning we will concentrate on criteria dealing with the learning and social parts of the platform. In particular, the novelty of this paper is to try to combine the social aspect of modern technology with the standard learning practice.

The integration of social aspects in an educational learning environment can be defined as an exchange of information and ideas, supplemented by interactions with a personal or professional network of students. An educational social network (i.e., EduSN) is then created in a natural way by analyzing the interactions of students. The EduSN's effectiveness can be measured by adopting the revised KP model of Sect. 3 by providing metrics for each level as reported in this section.

A prime model of an EduSN is given by a graph where users are vertexes and the links among them are the edges. In this case, we choose to restrict the network to the students; in future works, we plan to create more complex networks including teachers, parents, etc. In order to represent the network we need to establish how the students are connected. Grouping all the students of the same class is a simple idea which does not take into account all the possible scenarios. Groups of students can be defined according to several factors: classes, sharing of a same subject, organization of flipped classrooms across classes or institutions, sharing of a same scholastic interest, etc. For this reason, the concept of *learning groups* might be more appropriate than classes, so the definition of the network must be flexible. The network that we propose is an undirected graph where the edges convey the strength of the relationship between two students. With this model the visual representation of the graph can give insights about

Group 1   Group 3   Group 2

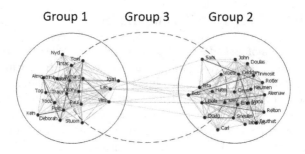

Fig. 3. An example of an EduSN.

the activity of the learners. Figure 3 shows a portion of an EduSN where three groups are defined: (1) Group 1 identifies a class of 19 students, (2) Group 2 identifies a class of 23 students, and (3) Group 3 identifies a *virtual* class of 18 students defined by social interactions on the same topic. Formally, an *EduSN* is the union of non-disjoint groups defined as $EduSN = \bigcup_i^g \mathbf{G}_i$ (g is the total number of groups present in EduSN), where:

- $\mathbf{G}_i = (V, L)$ is a group;
- V is a set of nodes, representing the students S;
- L is a set of links (or edges) between nodes in V, representing the interaction among the students.

Finally, we can define the metrics for each KP's level as follows:

Reaction. As indicated in Sect. 3, explicit and implicit methods can be used to establish the level of student's satisfaction in the usage of the Social LMS. At this level, we consider an implicit approach assuming that the usage of the Social LMS is related to the acceptance of the LMS platform. We define a quantity, called *total connection* (explained in detail in Sect. 5), which is a function of two students, and it represents the strength of the social bond between two students in a given day. At this level, we are not interested in the quality of the bond, but only to the amount of information in the academic time course.

Learning. The student's academic performance is evaluated according to formal and informal methods. The formal approach considers the standard evaluation obtained by grades (i.e., proof, tests, etc.), whereas the informal one considers social aspects based on the student's social behavior when he/she uses the Social LMS. In this last case, we define the *social contribution* indicator - which gives an idea of how the other participants of the network judge the interactions of the student with the social educational network. The social contribution of a student has to take into account two measures: appropriateness, and quality of the material added (by material, we also take into account the posts on the forum and on the chat room). Formally, the *social contribution* measure is defined by:

- *Appropriateness*: each material can be rated with four different levels of appropriateness (material which has no appropriateness rating counts as appropriate since we expect that normal material does not trigger the need for an appropriateness rating). For example, posting the timetables of the movie theater on the math forum would be considered inappropriate, while asking a clarification of a theorem would be appropriate. Highly inappropriate material adds −3 point to the appropriateness score of the student who posted it and also to the material itself (in such a way the teachers can easily ban particularly inappropriate material), inappropriate adds −1 points, appropriate adds 1 point and very appropriate adds 3 points. The average appropriateness score of a material is always visible to the other users in order to allow the instructors to ban any material which should not be visible.
- *Quality*: the quality of all the material published is divided in four different levels:
  - the material adds wrong content (−1 points)
  - the material adds no content to the discussion (for example if the material contains only information already present in the discussion) (0 points)
  - the material adds some content (a good question is also associated to this category) (1 point)
  - the material provides a complete answer to a problem (3 points)

For every published material the peers and the teachers of the network can assign a value of *appropriateness* and *quality*. Since we expect the teachers to be more qualified than the students, teachers' ratings will weigh three times more than those of the students. At the end of the year, it is possible to calculate the average value for these quantities. The overall score of the *social contribution* is obtained for each learner L and it is the sum of all the contributions (for this reason it comprises quantitative and qualitative aspects of the production of each learner):

$$SC_L = \sum_i \left(\alpha App_L^i + (1 - \alpha) Q_L^i \right) \tag{1}$$

where i runs on all the materials produced by learner L; the quantities App_L^i, Q_L^i are the average appropriateness and quality of item i by learner L; $0 \leq \alpha \leq 1$. When α has a value of 0, the *appropriateness* value, App_L, is not considered, and the final weight is equivalent to the weight obtained by analyzing the *quality* value, Q_L. If α has a value of 1, the *quality* value is ignored and only the *appropriateness* value is considered. The importance of *appropriateness* value with respect to *quality* value can be balanced by varying the value of parameter α. In order to evaluate the whole network activity it is reasonable to consider the average Social Contribution of the participants of the network.

Behavior. At this stage (see Sect. 3 for more details), we take into account how the learner's attitude can be measured by specific indicators: are learners increasing their social activity (such as forum posts, participation to school tennis team, theater activities, etc.)? The *attitude* level is measured by analyzing the topological structure of EduSN, and by considering the *informal success* indicator. In the former case, once the EduSN graph is established we can calculate various centrality measures which allow to determine, for example, who are the reference students, who tends to connect participants, etc.; here, we will focus on the betweenness centrality since it perfectly reflects the social feature of the Social LMS and on an indicator called compactness (defined in the next section) which measures the amount of interaction in the group. *Informal success* is associated with the awards attributed to: (1) the student about all the informal activities he/she was involved in, and (2) the achievements by a gamification approach to track the progresses of the student's academic path. In the former case, the idea is to also consider extra educational activities that are an indicator on how a student is socially active in the school; for example, a student can be a member of the: scholastic journal, soccer school team, etc. When considering the metric associated to the whole Social LMS the average value of the all the students should be taken into account.

Results. The two KPI we use to determine the student's final result are: (1) the *average final grade* obtained during the academic year in order to assess the formal performance for each subject (e.g., math, history, science, etc.), and (2) the *influence factor*. The *influence factor* represents the student's influence by considering his/her interactions with the EduSN's participants, his/her ability to drive discussions, to share materials, to attend to extra activities, etc. The more influential you are, the higher your *influence factor* is. The influence factor is a quantity we determine as the aggregation of three indicators previously described:

- attitude level (comprising of betweenness centrality - which is a measure of the activity of the person in the network and compactness).
- informal success - which takes into account how active a person is in the informal context.
- social contribution - which gives an idea of how the other people of the network judge the interaction of the person with the social network.

Since these quantities are measured with different indicators, we first transform them in percentages, and then we calculate the average value for calculating the *influence factor* in percentage. In an environment with a large population we expect that there should be learners who excel in each of the three indicators whose scores can be used to calculate percentages.

- *Betweenness.* We first group the scores of all the students of a school in a list and we find the highest value. At this point, we can re-evaluate the score of each learner as a percentage of the best result. Let us consider an example of a school where the highest value of betweenness is 0.88, if a student has a betweenness equal to 0.72; this value is transformed into $82\% = 0.72/0.88 \times 100$.
- *Informal success.* It is associated with the awards attributed to the student about all the informal activities he/she was involved in. By associating a number to each activity (for example 50 points for being part of the soccer

team who arrived second in the school tournament, etc.) each person will obtain a total number of points. Also in this case we can make an ordered list of results, where the percentage score of a single student is given by the ratio of his/her result and the best one (times 100).

- *social contribution*. Its value is transformed in percentage in the standard way (see Eq. 1).

The *influence factor* for a student, A_L, is calculated with a linear combination of the *betweenness, informal success* and *social contribution* (where the weights are positive and sum to unity, and allow to emphasize each of the three factors):

ROI. The definition of Sect. 3 requires some discussions since it leads to differences in respect to the original formula defined in [12]. It might be useful to notice that we are trying to evaluate the success of a Social LMS, this is obtained by observing how much of the (average) success of the students is related to the characteristics of the Social LMS. If one wants to stick with the formal definition of the ROI [12] it should be considered the difference between the benefits and the costs divided by the costs. For this reason, it is important to carefully understand and define what the benefits and the costs are. The general benefit for a student due to the usage of a Social LMS should include both of the formal grades and the social success (which is generally accounted for as "soft skills"). What is the cost in the case of a Social LMS? It seems meaningful to consider the usage of the social platform as an additional cost on the normal routine of a student life. Although students are prone to using ICT, it takes time and effort to produce material for a Social LMS and to interact with the others. The usage of a Social LMS can thus be naturally considered as a cost. In order to quantify it, one should take into account the quality and quantity of the interactions. However, this is essentially identical to consider the social success of a learner. This quantity has been summarized by the influence factor previously defined. This leads to a rather simple (but surprising) conclusion which is that the difference between benefits and costs due to the Social LMS reduce to the average formal grades. According to this reasoning a natural definition of the ROI (for learner L), following the original one would be:

$$ROI_L = \frac{benefits - costs}{costs} = \frac{\Delta G_L}{\Delta A_L} \tag{2}$$

Where ideally one considers the time variation (Δ) of the grades (G_L) in respect to the variation of the influence factor A_L (in the case where the initial values of G_L and A_L are not known this quantity reduces to a normal ratio). It should be noticed that this quantity can be calculated for each student (L), while at the end it is interesting to know whether the Social LMS was successful, this can be obtained by considering the average value of the ROI on all the students. This formula however can be problematic in the case of those students who have a very limited interaction with the Social LMS (A_L close to zero) since this quantity appears at the denominator, and as such, the ROI of those cases tends to explode.

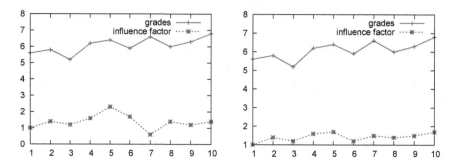

Fig. 4. (left) Simulated cases of low (0.08) correlation, and (right) high correlation (0.81) between the grades and influence factor.

A possible way out of this problem is to calculate the correlation between the grades and the social activity of each student, see Fig. 4 for an example. It is possible to monitor the time evolution of the grades and the influence factor; thus, obtaining a time series for the grades and one for the social activity for each student. It is clear that the elements of those time series do not happen at the same time and a calculation of a correlation might pose problems. However, one can consider fixed time intervals (for example one month) where the average grades and the average social activity can be calculated. By using the correlation, the problems connected with very limited social activity is naturally overcame (in the limit of constant activity the correlation with the grades drops to zero). It is fair to say that correlation and causation are different concepts and supposing that changes in the social usage imply a change in the grades of a learner might be too strong. In practice, it would require a deep textual analysis of the interaction among the peers to understand whether important concepts have passed from a student to another one. On the other hand, since this analysis is carried out at the level of whole learning institutions it seems reasonable to consider the grades - social activity correlation as a good hint of the ROI.

Since the ROI can be calculated for each learner it is also possible to use it as an tool for understanding the social activity of the single students. This can be carried out by breaking down the various parts of the social interaction (by defining a threshold good/bad quality). For example, let us consider the case of two students one which has a very high grades - social activity correlation in this case one expects that most of the social activity was high quality since it helped him/her in the academic success. If for the other student there is anti-correlation among the academic and social activity it might be interesting to consider how much of the latter was of good quality or was based solely on fruitless activities.

5   Evaluating Forum and Chat

In this section, we provide the specific metrics related to the forum and chat room which are needed to comply to the definitions of Sect. 4, while other indicators (e.g. the grades) are not collected from the information derived by the forum and chat room, and for this reason they are not addressed here. In detail, we describe how to formally obtain an EduSN from the forum and the chat room and how to exploit its properties. In a Social LMS, chat room and forum play a key role, and they are some of the social modules most accessed by students. The idea is thus to inter-correlate the social aspect and the academic success by exploiting the natural tendency of human beings (and in particular of students) to be connected. Forum and chat room are different social elements: the forum is expected to be divided in topics to deepen the study subjects, whereas the chat room is intended as a more free space where also informal topics can be discussed (such as sports, movies, vacations, etc.).

Reaction. A *connection* represents how many interactions between two students occur in a given time. At the beginning, the *connection* C is set equal to 0, i.e., $C(a, b; n) = 0$, where: $a, b \in S$ are two students, n is a time factor (e.g., a day). In the case of the forum, each time a person contributes to a conversation, a quantity "f" is added to all the participants of the conversation. Let us suppose that a student $a \in S$ asks a question about math on day 0, and a student $b \in S$ replies on day 1, then:

$$C_F(a, b; 1) = C_F(a, b; 1) + f$$

While student $d \in S$ replies on day 3:

$$C_F(a, d; 3) = C_F(a, d; 3) + f$$
$$C_F(d, b; 3) = C_F(d, b; 3) + f$$

Since, at this level, we are not interested in the directionality of the interaction, this graph is undirected $C_F(a, b; n) = C_F(b, a; n)$, we also want a simple graph [4], that is $C_F(a, a; n) = 0$.

In a school environment there are also informal interests (friendships, feelings, sports, tournaments, etc.). These aspects are better represented by the chat room which does not have the same kind of moderation as the forum, and it allows for more freedom. Due to the synchronous nature of the chat room there is no obvious thread which allows to distinguish easily whether a student is talking to another, so we suggest to adopt a time approach, that is, if two students write messages in the chat room within a given time interval (Δt), then we can consider that those students are connected (for example because of a common interest or by personal reasons). At the beginning of each day the chat connection is zero. If students a, b and d write on the chat room within "Δt" their chat-connections will be enhanced by a factor "c" (similar to the case of the forum).

$$C_C(a, b; 1) = C_C(a, b; 1) + c$$
$$C_C(a, d; 1) = C_C(a, d; 1) + c$$
$$C_C(d, b; 1) = C_C(d, b; 1) + c$$

Unfortunately time distance is not a very good form of correlation (and we plan to use better indicators in further publications), moreover we have to take into account that the bond between two people develops with time. We define a quantity called total forum (chat) connection which takes into account the history of connections. The total forum connection is a function of two students (for all the participants of the network) and of time, in the following definition we omit the student dependency for simplicity, and we keep only the time dependency starting from day 0:

$$
\begin{aligned}
K_0^F &= C_F(0) \\
K_1^F &= C_F(1) + \lambda_F C_F(0) \\
K_2^F &= C_F(2) + \lambda_F C_F(1) + \lambda_F^2 C_F(0) \\
&\quad \dots \\
K_n^F &= C_F(n) + \lambda_F K_{n-1}^F
\end{aligned}
\tag{3}
$$

In practice the total forum connection is an exponentially weighed average of the connections from a given day 0. The weights take into account the time dependency of the interpersonal relationships.

Similarly to the forum, the total chat connection is defined recursively based on the connection derived from the chat room as:

$$
K_n^C = C_C(n) + \lambda_C K_{n-1}^C
\tag{4}
$$

Once we have the "total connection" value among two students we can define the length of the edge in the corresponding graph by aggregating forum and chat room values. The length of the edge between student a and student b $E(a, b)$ at time n is set to be a the inverse of the sum of the total forum and total chat connections, and it is defined as follows:

$$
E(a, b; n) = \frac{1}{K_n^F(a, b) + K_n^C(a, b)}
\tag{5}
$$

A proper choice of the parameters $(\lambda_F, \lambda_C, f, c)$ has to take into account that the connection established via the forum is less frequent but requires more effort than the rapid and more casual connections of the chat room. For this reason the quantity $f \gg c$. The weights λ_F and λ_C give more importance to the most recent interaction in respect to the old ones, as an example we show the connection between two people in a 200 days period with two different lambda parameters: $\lambda_F = 0.99$ and $\lambda_F = 0.999$ (see Fig. 5).

Learning. The *social contribution* indicator has to be used in a different way according to the usage of forum and/or chat room. The *appropriateness* is the metric used to evaluate posts on the chat room, whereas the *quality* metric is taken into account to evaluate the posts on the forum.

In the chat room informal topics can be discussed (such as sports, movies, vacations, etc.). For example, posting the timetables of the movie theater on the math forum would be considered inappropriate, while asking a clarification of a theorem would be appropriate. In the case of the chat room the moderation is

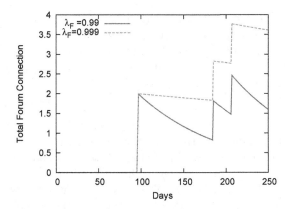

Fig. 5. Simulated total forum connection according to different weights.

limited (due to the very nature of this form of communication), and thus more power is given to the students themselves. All the people accessing the chat room can signal if a material is particularly inappropriate with a red flag which immediately sends a warning to the teachers who can connect and change/censor the material. Also the value of inappropriateness of the author of the post increases (once a teacher confirms the bad quality of the content). A teacher is in fact a super-user having the possibility to censor particularly inappropriate content. Instead, the forum is expected to be divided in topics, and the teacher of the related subject can judge the quality of the students added material with the use of specific labels (i.e., *wrong, neutral, complete, comprehensive*). To each label a point is associated as defined in Sect. 4. In case of a wrong post, a teacher can ban the material in order to guarantee a high quality level of discussions. Peers can also judge the material added in the forum sessions.

Finally, the *social contribution* value is calculated by given the following balancing of its two indicators, i.e., $SC_L = 0.4 * App_L + 0.6 * Q_L$ with $\alpha = 0.4$. In this case, a greater importance is given to the *quality* indicator as the information from the forum are considered most significant than the ones obtained from the chat room where the information is only evaluated by the *appropriateness* indicator. This difference is given by the importance of the *formal* role of a forum with respect to the *informal* role of a chat room. In addition, the posts in a forum are less frequent than the ones of a chat room: a different score can reduce the possibility or overestimate the quantity of interactions obtained by the use of the chat room.

Behavior. Once the EduSN's graph is established we can calculate various centrality measures (e.g., the betweenness as described in Sect. 4) which allow to determine for example, who are the reference students, who tend to connect people, etc. A non-secondary aspect of the graph representation is that we can have an idea of how compact a group is. For example, a graph where students share a lot of edges and these edges are short is expected to tell us that strong relationships have been established in the group. Collaboration is thus expected to be more present than in a group with looser bonds; this in turn has an impact on social learning [2]. A measure of this quantity is the sum of all the total connections between the students of a group (G), divided by the number of students of the group:

$$comp = \frac{1}{|G|} \sum_{i<j \in G} \left(K_n^F(i,j) + K_n^C(i,j) \right) \tag{6}$$

(where $|G|$ is the number of students belonging to the group G). A similar quantity can be obtained for the single student, but in this case it is enough to sum the total connections of a student to his/her "alters" (we will refer to this quantity as the *single compactness*). We emphasize that with this approach the graph associated with social network is based solely on the amount of interactions without taking into account its quality.

Results. For each student, the forum and chat room are overall evaluated by the *influence factor* indicator that is calculated as the average values of: (1) the *single compactness* value (see Eq. 6, and (2) the *social contribution* value as calculated in the *Learning* level. The *informal success* indicator (which is related to the gamification of the informal activities) does not pertain directly to the forum and chat room and for this reason it is not analyzed here. An indirect way to deduce the informal success would be to discover the achievements through a textual analysis of the topics on extra activities such as sports, theater, etc. but we reserve it for future works.

ROI. The ROI is calculated as defined in Sect. 4, by utilizing the metrics previously defined.

6   Dashboard

A dashboard representation is used to control the status of a learning process over a series of reports since the former has a strong visual and descriptive impact. Furthermore, evaluations have shown the importance of a dashboard for the learners to get insights about their performances [6,15]. Within a dashboard one can group a series of analytic tools in a single page and obtain a general overview of the desired study case [15]. As described in [19], learning dashboards are classified into three basic classes that are: (1) dashboards to support the traditional face-to-face lectures, (2) dashboards to support face-to-face group work and classroom orchestration, and (3) dashboards to support blended or online-learning settings. Our work refers to learning dashboard classified in the third group.

For a correct interpretation of the information of a dashboard, both a synchronous and a diachronous approach should be considered. In detail, how a KPI performs in respect with the others at a given time is a synchronous approach to understanding information. Let us consider a case of a student who has good grades in English. Is this student also a good communicator (i.e., he/she also has a very good social network) or not? In the former case, it means that good language skills are not only confined to the academic environment but also to the "real world", while sometimes people do not develop good interdisciplinary abilities.

One has to take into account that within a dashboard representation not only the time evolution of the indicators is automatically updated but it is one of its most important aspects. A diachronous approach to information allows a person accessing to the dashboard to grasp a more deep understanding of information. Knowing that student's grades are at the minimum sufficient level in a subject does not tell us whether this person is doing a good job by improving form very low grades or is in a difficult period dropping from very high performances.

The dashboard of this project is to be thought as modular. This means that we can associate to each different KPI a different module which can be interfaced with the dashboard. This approach allows for great flexibility. In fact, the data representation has to take into account the background of each person accessing it and a modular approach to the dashboard allows for simple re-ordination of the data in this sense. Students and teachers require many KPIs, some of them are common while some of them should be provided only to the appropriate roles, thus leading to dashboards differences. For example, the graph representation of the social network is more important for the teachers, who need to understand better the connections within groups. By using this graph it is immediate to have an idea of how the students are active on the network, if there are subgroups and how the groups are compact. This tool is fundamental since it gives to the instructor a global overview which is sometimes difficult without quantitative evaluations. On the other hand, providing the same tool to students might lead to unnecessary competition and unnatural modification of the network itself. The students' dashboard is thought to give a general overview to the student about his/her current status of achievements. We will focus on it in Fig. 6 as

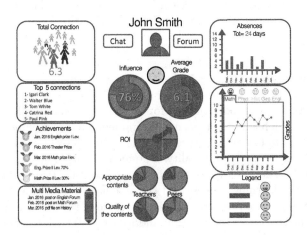

Fig. 6. Student's dashboard.

an example. At the center of the dashboard there are two main indicators (see Sect. 4), that are the *influence factor* and the *average grade*. The smiley, which can have 4 possible results, gives a general idea of the student status. At the top center of the dashboard there is a picture of the student and his name, at its sides there are two buttons, one to access to the chat room and one to access to the forum. The absence days (divided by month and total) are reported at the top right of the dashboard. Right below it there is the indicator of the grades, where a label can be clicked to access each single subject. Multiple labels can be clicked (unclicked) at the same time in order to have comparative overview on subjects over time. On top of the subject name there is a smiley which reports how the student likes the subject, which can been obtained, for example, with a sentiment analysis of the comments of the students on the forum and chat room. At the bottom center there are the degrees of appropriateness and quality of the material posted by the student on the social network, these include the posts on the forum, the multimedia material attached, the comments on the chat. The assessments are divided by the students (peers) and teachers. At the top left of the dashboard there is in indicator of the *total connection* (see Sect. 4), this indicator is a box containing many small images of people and the actual value of total connection. The number of people not shaded increases proportionally with the value of the connection. Below the indicator of the total connection there is the list of the top 5 connections. Below the top 5 connections there are the medals which include both formal and informal activities. These awards are assigned by the teachers. At the very bottom left we list the material published by the student and by clicking over it the student can access it and its ratings.

Although the dashboard here introduced is not yet functional it is a guideline for the implementation phase. The prototype aspect is a crucial preliminary step in user interface design to obtain an immediate feedback with the goal to minimize costs in the software development [15]. This has to take into account that our goal is to create interfaceable modules which allow to personalize the final dashboards according to the relevant needs of the considered LMS. In this respect our proposal is centered around the modules rather than a single instance of dashboard.

7   Conclusions

Although a LMS is an important tool for learning institutions the introduction of the social aspect has not yet been fully implemented and exploited. For this reason, it is useful to have a structure which helps to control the dynamics within a Social LMS and to assess its quality. The idea of this article is to exploit the big amount of data deriving from the educational social network to make precise statements on the learning process and the associated Social LMS.

At the beginning, we presented the architecture of the social learning analytics where all the sources of information and interaction within a Social LMS are taken into account. At this point, we borrowed the structure of the KP model, which is a paradigmatic tool for evaluating the quality of company trainings, in order to provide an evaluation scheme. Following the KP model the assessment of a training should be composed of five steps (reaction, learning, behavior, results and ROI), which we adapted to a school environment with a strong emphasis on the ICT social aspects. Once these steps are clear, it is essential to choose and define new metrics which are able to describe the most important aspects of the educational environment.

In order to obtain quantitative assessments, we focused on the interaction among students; at this level we analyzed only the amount of interaction, neglecting any semantic analysis or its directionality.

We modeled the new metrics defined on two standard modules of the ICT, i.e. the forum and chat room. These are complementary tools which allow to exchange information, the forum returns higher quality and less rapid information while the chat room is devoted to more informal and quick exchanges of ideas. By considering the amount of information exchanged as a bond between two learners it is possible to build an educational social network. The graph associated with the network comprises of nodes (the students) and edges (the amount of information exchanged among the students). The metric that we proposed takes into account the time evolution of personal interaction and the different efforts required by the forum or the chat room. With the help of the graph it is possible to make quantitative estimates of factors which are usually left to a more intuitive level, such as: Who is the leader of the group? Who are the connecting people? Who is generating more discussions? Some of the metrics that we proposed return insights about the performance of each student while others are related to global quantities of the whole network. This introduces a novel vision related to the management of the learning process where the bottom/up view (from the student to the class) is complemented with a top-down (from the group to the student) analysis. The extension of this model beyond a single learning institution up to include many schools in a region or state is very natural and it opens the possibility to have more quantitative assessment about the importance of social aspects in relation with the academic success.

The complex mechanisms of a scholastic environment supplemented by a social network require a wide range of KPI, for this reason we provided examples of modular dashboards for their control. A dashboard is optimal for this purpose since the typical information related to the school activity (such as grades, absences, etc.) can easily be enriched with the information stemming from the social network (compactness of the graph, centrality measures, informal activities, etc.). The importance of modular dashboards is related to the ease of customization in relation to the different needs of the subjects of the scholastic institution (e.g., students, teachers, parents). In fact, some of the proposed metrics should be common to all of these players (e.g., grades) while others should be accessible only to restrict groups. For example, it might be better to not share the information about the global properties of the educational social network with the students in order to not trigger competitive behaviors which can modify the natural scores of the network.

In future works, we plan to modify the EduSN in order to include also teachers, parents, school managers, etc. This would lead to a multi-network view of the school. The interaction among the different social structures (students' network, teachers' network, ...) should provide a more complete vision of the challenging problem of learning management. Furthermore, the utilization of metrics able to detect the directionality of the interaction between the students could allow to better understand the knowledge flow within the scholastic institution.

In summary, this work paves the way to a quantitative estimate and control of the interactions within an educational social network from both a local and global perspective.

References

1. Almosallam, E.A., Ouertani, H.C.: Learning analytics: definitions, applications and related fields. In: Herawan, T., Deris, M.M., Abawajy, J. (eds.) DaEng 2013. LNEE, vol. 285, pp. 721–730. Springer, Heidelberg (2014). doi:10.1007/978-981-4585-18-7_81
2. Bandura, A.: Social Learning and Personality Development. Holt, Rinehart, and Winston, London (1963)
3. Buckingham Shum, S., Ferguson, R.: Social learning analytics. Educ. Technol. Soc. **15**, 3–26 (2012)
4. Butts, C.T.: Social network analysis: a methodological introduction. Asian J. Soc. Psychol. **11**, 13–41 (2008)
5. Claypool, M., Brown, D., Le, P., Waseda, M.: Inferring user interest. IEEE Internet Comput. **5**, 32–39 (2001)
6. Corrin, L., de Barba, P.: How do students interpret feedback delivered via dashboards? In: Proceedings of the Fifth International Conference on Learning Analytics and Knowledge, LAK 2015, pp. 430–431. ACM, New York (2015)
7. Kautz, H., Selman, B., Shah, M.: Referral Web: combining social networks and collaborative filtering. Commun. ACM **40**(3), 63–65 (1997)
8. Kirkpatrick, D.L., Kirkpatrick, J.D.: Evaluating training programs: the four levels. ReadHowYouWant.com/Berrett-Koehler Publishers, Sydney/San Francisco (2010)
9. Nespereira, C.G., Dai, K., Redondo, R.P.D., Vilas, A.F.: Is the LMS access frequency a sign of students' success in face-to-face higher education? In: Proceedings of the Second International Conference on Technological Ecosystems for Enhancing Multiculturality, TEEM 2014, pp. 283–290. ACM, New York (2014)
10. Newstrom, J.W.: Evaluating Training Programs: The Four Levels, by Donald L. Kirkpatrick. Hum. Resour. Dev. Q. **6**(3), 317–320 (1994, 1995)
11. Osguthorpe, R.T., Graham, C.R.: Blended learning environments: definitions and directions. Q. Rev. Distance Educ. **4**(3), 227–233 (2003)
12. Phillips, J.J., Phillips, P.P.: Using action plans to measure ROI. Perform. Improv. **42**(1), 24–33 (2003)
13. Rössling, G., Joy, M., Moreno, A., Radenski, A., Malmi, L., Kerren, A., Naps, T., Ross, R.J., Clancy, M., Korhonen, A., Oechsle, R., Iturbide, J.A.V.: Enhancing learning management systems to better support computer science education. SIGCSE Bull. **40**(4), 142–166 (2008)
14. Sakarya, S., Bodur, M., Yildirim-Öktem, Ö., Selekler-Göksen, N.: Social alliances: business and social enterprise collaboration for social transformation. J. Bus. Res. **65**(12), 1710–1720 (2012)
15. Santos, J.L., Govaerts, S., Verbert, K., Duval, E.: A case study with engineering students. In: Proceedings of the 2nd International Conference on Learning Analytics and Knowledge, LAK 2012, pp. 143–152. ACM, New York (2012)

16. Siemens, G., Baker, R.S.J.D.: Towards communication and collaboration. In: Proceedings of the 2nd International Conference on Learning Analytics and Knowledge, LAK 2012, pp. 252–254. ACM, New York (2012)

17. Teevan, J., Dumais, S.T., Horvitz, E.: Personalizing search via automated analysis of interests and activities. In: Proceedings of the 28th Annual International ACM SIGIR Conference on Research and Development in Information Retrieval, SIGIR 2005, pp. 449–456. ACM, New York (2005)

18. Tour, F., Ameur, E., Dalkir, K.: AM2O - an efficient approach for managing training in enterprise. In: Proceedings of the International Conference on Knowledge Management and Information Sharing, pp. 405–412 (2014)

19. Verbert, K., Govaerts, S., Duval, E., Santos, J.L., Assche, F., Parra, G., Klerkx, J.: Learning dashboards: an overview and future research opportunities. Pers. Ubiquit. Comput. **18**(6), 1499–1514 (2014)

20. Sclater, N., Berg, A., Webb, M.: Developing an open architecture for learning analytics. In: Proceedings of the 21st Congress on Information Systems for Higher Education, EUNIS 2015, Abertay University, Dundee, 10–12 June 2015, pp. 303–313 (2015)

Extending the BPMN Specification to Support Cost-Centric Simulations of Business Processes

Vincenzo Cartelli, Giuseppe Di Modica, and Orazio Tomarchio[✉]

Department of Electrical, Electronic and Computer Engineering,
University of Catania, Viale A. Doria 6, 95125 Catania, Italy
{Vincenzo.Cartelli,Giuseppe.DiModica,Orazio.Tomarchio}@dieei.unict.it

Abstract. Business Process Simulation is considered by many a very useful technique to analyze the impact of some important choices designers take at process design or optimization time, right before processes are actually implemented and deployed. In order for the simulation to provide accurate and reliable results, process models need to take into account not just the workflow dynamics, but also many other important factors that may impact on the overall performance of process execution, and that form what we refer to as the Context of a process. In this paper we formalize a new Business Process Model that encompasses all the features of a business process in terms of workflow and execution Context respectively. The model allows designers to build a cost-centric perspective of a business process. Also, we propose an extension to the Business Process Model and Notation (BPMN) specification with the aim of enhancing the power of the BPMN to also model resources and the process execution environment. In the paper we provide some details of the implementation of a novel Business Process Simulator capable of simulating the newly introduced process model. To prove the overall approach's viability, a case study is finally discussed.

1 Introduction

Many organizations operating in both the global and local markets are recognising the value of Business Process Management (BPM) as a strategic business methodology. BPM combines established management methods with information technology to measure performance, identify bottlenecks, (re)design and deploy processes in a quicker and easier way [18]. *Business Process Simulation* (BPS) is one of the most powerful techniques that supports re-designing of processes. Through simulation, the impact that design choices are likely to have on the overall process performance (measured through KPIs) may be quantitatively estimated. BPS involves steps such as the identification of sub-processes and activities and the definition of the process control flow. But one of the most interesting aspects that simulation should not neglect is the *process context*, i.e., all factors that during the process execution may consistently impact on the process dynamics, and thus, on the process KPIs. The accuracy of the final estimate produced by the simulation depends, among others, on how accurate the process context model is.

© Springer International Publishing AG 2016
A. Fred et al. (Eds.): IC3K 2015, CCIS 631, pp. 492–514, 2016.
DOI: 10.1007/978-3-319-52758-1_26

A literature review revealed that many works deal with the modeling of the process context, but none proposes a really comprehensive view. The purpose of the work presented in this paper, which extends the one presented in [5], is to define a novel business process model, capable of representing all the main aspects of the business processes and their runtime context under a *cost-sensitive* perspective. In order to build that view, our approach leverages and integrates the well-know Business Process Modeling and Notation (BPMN) specification and the Activity Based Costing (ABC) methodology. Further, we developed a business process simulator to simulate business processes defined according to the proposed model. A simple but well structured business process was also designed and used to test the simulator.

The paper is structured in the following way. In Sect. 2 the literature is reviewed. In Sect. 3 a novel cost-sensitive process model is introduced. Section 4 discusses the proposal of a business process simulator; architectural and implementation details are provided along. In Sect. 5 we present the definition of a use-case process and discuss the result obtained from simulating the process. Concluding remarks can be found in Sect. 6.

2   Literature Review

Process simulation plays an important role in the context of the Business Process Management. The purpose of simulation is to give business analysts a clue on what the performance and the cost of processes could be according to the actual process design, and to provide hints on the corrective actions to take in order to improve the overall process performance.

Process simulation is a key features provided by many commercial Business Process Management Suites (BPMS). Most BPMS offer process analysts a tool to simulate process workflows and to customize the process context scenario. A simulation tool's strategy and implementation is specific to the suite it is embedded in. In particular, it is tightly coupled with the process modeling approach and language (be it proprietary or open) adopted by the BPMS. ARIS Simulation is an extension of the well known ARIS Architect suite[1]; it is based on the L-SIM simulation engine and allows to simulate process modeled by using Event-driven Process Chain notation [17]. Recently the BPMN support has been added. iGrafx Process[2] is a process analysis and simulation tool enabling dynamic "what-if" process analysis. It may be used as a standalone tool or within the iGrafx suite of process modeling and analysis. Oracle BPM Suite[3] has a simulation module too, which allow to define simulation scenario with various configurable parameters. Besides full fledged BPMS suites, specialized simulation products exist, such as SimProcess[4]. It is a hierarchical modeling tool that combines process mapping, discrete-event simulation, and the ABC;

---

[1] www.softwareag.com.
[2] www.igrafx.com/products/process-modeling-analysis/process.
[3] www.oracle.com/us/technologies/bpm.
[4] www.simprocess.com.

it allows users to carry on powerful analysis such as "what-if" scenarios, but it does not use any standard notation for mod-eling processes. An interesting survey on some commercial simulation tools is proposed in [9].

The majority of literature works addressing the business process simulation calls on Petri Nets. Petri Nets are preferred to other modeling techniques because of their rigor and clear formalism. Also, there are many easily available and freely accessible tools which analyse Petri Nets and evaluate their "soundness" [11], i.e., whether the model is free of potential deadlocks and livelocks. The basic approach adopted by researchers is to transform the business process model (usually defined in the BPMN language) into its "equivalent" Petri Net model and to feed the model to a simulator engine [7,12]. The Protos2CPN tool's strategy [8] is to transform a Protos model into a Colored Petri Net (CPNet), and simulate it in standard CPNet tools. CPNets [10] is a very powerful discrete-event modeling language combining the capabilities of Petri Nets with the capabilities of a high-level programming language.

Petri Nets are rather a good technique which is proved to be particularly effective in the simulation of workflows. Unfortunately, business processes are not just workflows. Though the activity flow may be considered the skeleton of a business process, there are some important process elements which complement the workflow and may have a strong impact on the process dynamics and performance. In [1] authors discuss the limitations of traditional simulation approaches and identify three specific perspectives that need to be defined in order to simulate business processes in a structured and effective manner: the control flow, the data/rules and the resource/organization. Focus is put on modeling at the right abstraction level the business data influencing the process dynamics and the resources to be allocated to process activities. Similarly, authors in [2] argue that a business processes simulator can not disregard the *environment* where processes are executed and the *resources* required to carry on the process activities. They also propose to use an ad-hoc workflow language (YAWL [3]) to model resources involved in the process dynamics. In [15] authors propose the definition of a conceptual resource model which covers all the types of resource classes and categories that may be involved in the process execution. The resulting model is expressed in the Object Relation Model (ORM) notation as they believe it fits the need to define resources and their mutual relationships in a way that can be easily understood by a non-technical audience. In that paper, a concrete example of resource modeling through ORM is provided and integrated to a workflow model which, in turn, is expressed in YAWL.

The idea proposed in this paper differs from the mentioned works in that it melts two consolidated and broadly adopted techniques to build an integrated business process perspective. The basic principles of ABC and the expressiveness of the BPMN are combined to devise a *cost-sensitive* business process model that can be used to characterize both the workflow and the execution context of a process.

3   A Cost-Centric Business Process Model

In this section we propose the definition of a novel Business Process (BP) model. The purpose of the model is to enable process designers to represent both the (static) structural features of BPs and all external entities and factors that may potentially affect the dynamics of BPs at execution time. For what concerns the structure of a BP, the model will have to support the representation of process' activities and tasks and their control flow, or in a word, the representation of the *process workflow*. Instead, we will refer to the "Context" of a process as the set of factors, external to the process logic, yet capable of influencing the process behaviour and performance at execution time. This set includes, for instance, the human resources responsible for carrying out process tasks, the non-human resources consumed by process tasks (such as goods and services), the business rules enforced on decisor elements, and in general any impromptu event that may occur at process execution time.

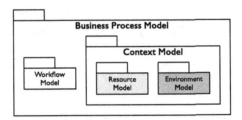

Fig. 1. Business process model.

In our design, the process *Context Model* is defined to be the union of the *Resource Model*, that can be used to characterize both human and non-human resources, and the *Environment Model*, used to represent the rest of external factors that build up the process environment. In Fig. 1 a package-view of the overall process model is depicted. In the remainder of the section, first some key concepts of the Activity Based Costing methodology are briefly recalled. Then we discuss in details the Resource Model, the integration of the Resource Model with a well known process workflow specification, and the Environment Model.

3.1   Basic ABC Concepts

An accurate tracking of costs (and revenues) is one of the most fundamental internal procedures an organization can utilize. "Analytical accounting" is an umbrella term for the financial component of business management. It relies on financial data to make determinations about how, when and why a business spends (and receives) money.

For the sake of cost-flow tracking we draw inspiration from the approach used in the ABC analytical accounting system [6]. The basic assumption of

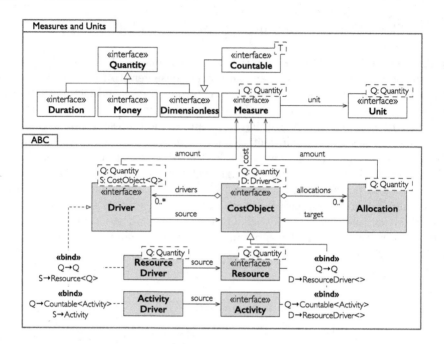

Fig. 2. Activity-based-costing accounting system model.

ABC is that all enterprise costs are generated whenever an activity consumes a resource. The strength of ABC is that all costs generated in the system from the consumption of resources by activities may be directly allocated to these activities by means of appropriate *resource drivers*; in other words, all costs (including overhead costs produced by enterprise's support activities [16]) are treated as *direct costs* when allocated to activities that are indeed precise *costs carriers*.

Given the full activity cost configuration, it is afterward possible to allocate these costs to the *final cost-objects* through their *activity drivers* that define how each activity is "consumed" by the final cost-object. By "final cost-object" we refer to every business entity whose cost has to be computed for the analysis, such as products, customers and channels. Although activity drivers can be represented by any meaningful "demand of activity" function for the final cost-object, they are generally defined by the ratio $\frac{A_{i,j}}{\sum_k A_{i,k}}$ or $\frac{D_{i,j}}{\sum_k D_{i,k}}$ being $A_{i,j}$ the number of instances and $D_{i,j}$ the work-times of the i-th activity dedicated to the j-ht cost-object.

The ABC model depicted in Fig. 2 defines the *CostObject* as the base entity for all subsequent cost-oriented concepts. Each *CostObject* has an unit of measure and can be "driven" by a set of *Drivers* that defines its requirements as the amounts of different cost-objects demanded for one unit of it. In addition, every *CostObject* can target in the run a set of other cost-objects through *Allocations* of its quantities to them over different dimensions such as cost, units and time.

An implementation of the Java Units of Measurement API[5], namely JSR-363 which is currently under review for approval, has been used as measures and units framework (package *Measures and Units* in Fig. 2). The JSR-363 specification defines the concepts of Quantity, Unit and Measure as follows:

- **Quantity**. Any type of quantitative properties or attributes of thing. No unit is needed to express a quantity, nor to do arithmetic on one. Units are needed to represent measurable quantities. Mass, time, distance, heat, and angular separation are among the familiar examples of quantitative properties.
- **Unit**. A quantity adopted as a standard, in terms of which the magnitude of other quantities of the same dimension can be stated. Units can be created from some quantity taken as reference.
- **Measure**. The result of a measurement, expressed as the combination of a numerical value and a unit. A measure is the scalar magnitude of a quantity.

3.2 Resource Model

We focus on the definition of a cost-oriented Resource Model capable of capturing the resource categories that BPs may need. The Resource Model aims at capturing and representing costs and other process related information regarding any kind of "resource" that has to be "consumed" by a process task, be it a *human* or a *non-human* resource, and more specifically puts the basis for an ABC analysis of processes.

The **Resource Model** defines the resource concept as a cost-object extension itself. Figure 3 presents its relevant classes and their relationships in the UML notation [13]. A Resource, no matter the kind, produces a cost whenever it is allocated and actually consumed by some business operation. As better explained in the Sect. 3.3, the CostObject concept depicted in the diagram represents the element that, in our proposed view, bridges the domain of resources and the domain of all the business operations that actually make use of resources.

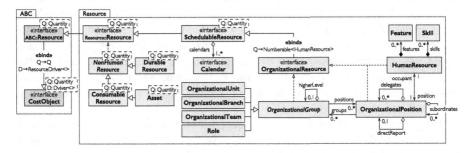

Fig. 3. Resource model.

The *NonHumanResource* class may be used to define resources such as goods and services. *ConsumableResource* extends it by introducing the concept of "residual amount" and it is suited for available-if-in-stock resources.

---

[5] http://www.unitsofmeasurement.org.

SchedulableResource represents a generic resource whose availability is defined by one or more calendars where each *Calendar* represents a set of time intervals in which the resource is available to the whole system; the most common calendar based resource is the *HumanResource* (HR) which represents the physical person who, in turn, occupies an *OrganizationalPosition* within an *OrganizationalGroup* structure such as an *OrganizationalUnit* structure (organizational chart).

Each *OrganizationalPosition* can be occupied by one HR at a time and is related to other organizational positions through the *directReport* and the *delegates* relations. The *directReport* relation points to the HR which is accountable for the position (usually a manager) whereas the *delegates* relation allows the **Resource Model** to identify a set of alternative HRs when the requested one is not readily available.

3.3   Integrating the Resource Model with BPMN

There is a lot of specifications available to define BP models. Some provide features to represent basic activities and the control flow (UML's activity diagrams), others integrate the representation of *Events* that trigger activities and/or that are triggered by activities [17]. To our knowledge there is no open specification capable of representing all the features of the Business Process model that have been so far discussed. The well known OMG's Business Process Model and Notation (BPMN) [14] standard is though a good candidate for our purpose.

Our choice is in part motivated by the fact that BPMN has reached a good maturity level, and is adopted and continuously supported by many vendors. Besides new semantic elements, such as callable elements and event-triggered sub-processes, the BPMN 2.0 introduced the complete Collaboration, Process and Choreography meta-models, the BPMN Diagram Interchange meta-model, the BPMN Execution Semantics, the Basic and Extended Mapping of BPMN Models to WS-BPEL and the BPMN Exchange Formats with its XMI and the XSD files for XML serialization. Most important, the BPMN 2.0 *"..introduces an extensibility mechanism that allows the extension of standard BPMN elements with additional attributes. It can be used by modelers and modeling tools to add non-standard elements or artifacts to satisfy specific needs, such as the unique requirements of a vertical domain, and still have valid BPMN Core."* This appealing features drove us to choose the BPMN as the specification on which to ground our modeling needs.

The BPMN provides for a comprehensive notation that allows to represent many features of a business process. Basically, in BPMN it is possible to define a process workflow articulated in sub-processes, call activities, tasks, and to also model the control flow of activities. Various type of external and internal events may also be defined. Further, the organization units responsible for a given process, a sub-process or even a single task may be modeled by means of concepts like pools and lanes. Like other specifications, though, BPMN lacks of notation to model the "contextualization" of a process into its execution environment. We intend to tackle this deficiency by integrating the discussed *Context model* to the process model defined in the BPMN specification.

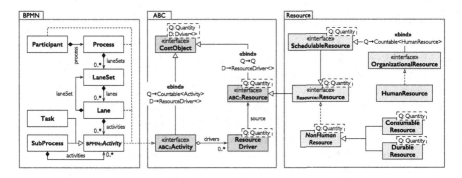

Fig. 4. Integration between the BPMN and the Resource model.

On the left of the class diagram in Fig. 4 relevant BPMN elements have been reported (white boxes), while on the right end the Resource Model package is depicted (light gray boxes). In order to integrate the two models, it is of primary importance to bring the elements of the BPMN domain under a *cost-sensitive* perspective. The bridge between the two models is realized by the ABC package, whose basic concepts are depicted in dark gray boxes.

The BPMN model is integrated into the cost perspective by means of the *BPMN::Activity* class. This class implements the behaviour of the *ABC::Activity* interface, which represents the Activity concept in the ABC sense. Basically, we let the *Task*, the *SubProcess* and the *Process* behave like an *ABC::Activity*.

According to this representation, the BPMN entities in Fig. 4 are cost objects as well. Further, their classes are bound to the Resource class (which again, is a cost object) by means of the *ResourceDriver*. This is to model that, at the lowest level, a *Task* is an elementary business operation that basically needs to consume Resources for its execution: the cost produced by the task execution is exactly the aggregated cost of the consumed resources. Further, *Participant*, *Process*, *Lane* and *SubProcess* are *ABC::Activity* as well, but in this perspective they act as cost aggregators. It is easy to build a cost "chain" where a Process aggregates the costs of it's Lanes, which in turn aggregate the costs produced by their Tasks, which in turn aggregate the costs produced for consumption of their associated Resources. The aggregation is realized by means of an Allocation scheme that sums up the measures (costs, number of instances and times) to the higher level elements in a bottom-up fashion, starting from tasks.

In Fig. 4 both the Resource and the business operations (process, sub-process, task) are CostObjects; but with the slight difference that Resources are *cost generators*, whilst the business operations are *cost aggregators*.

The granularity of the business elements provided by the BPMN standard well matches the ABC philosophy. By associating costs to the most atomic business element (the resource), by means of direct-cost allocation it is easy to derive the costs of every operation at any level in the hierarchy of business operations, from the simplest *task* to the most complex *process*.

3.4   Environment Model

We discuss the factors that are external to the process logic but that are anyway capable of affecting the process execution. In the literature, researchers [1] have identified numerous aspects that designers either tend to neglect or are not able to account for at design time, and that have a negative impact when the process executes. Some of those are the employment of oversimplified (or even incorrect) models, the discontinuous availability of data and resources, the inhomogeneous skill of the employed human resources (which leads to non deterministic human tasks' duration), the lack of knowledge about the exact arrival time of process' external stimuli, and so on.

Our objective is to provide the process designers with a tool to simulate the effects that external factors may have on the process dynamics and performance. Basically, we introduce a model which provides for the representation of non-deterministic behaviours of the process elements. The proposed model is an extension of the one proposed in an earlier work [4]. Figure 5 shows the BPMN elements which have been associated a statistical behaviour, and the categories of statistical behaviours that have been modeled.

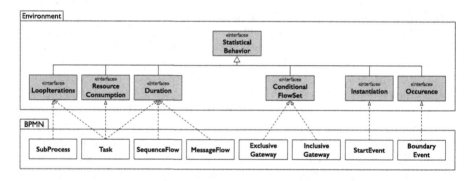

Fig. 5. Process environment model.

StatisticalBehavior is the root class representing a generic behaviour which is affected by non-deterministic deviations. It includes several probability distribution models (uniform, normal, binomial to cite a few) that can be used and adequately combined to build specific behaviours. *Duration* models the temporal length a certain event is expected to last. This concept will be used to specify the expected duration of the following BPMN elements: *tasks, sequence flows* and *message flows*. Tasks, in fact, may have a variable duration in respect, for instance, to the skills of the specific person in charge of it or, in the case of a non-human task, to the capability of a machine to work it out. A sequence flow may have a variable duration too (e.g. "transportation time"). It is not rare, in fact, that time may pass since the end of a preceding task to the beginning of the next one (think about a very simple case when physical documents need to be transported from one desk to another desk in a different room). Finally, message flows fall in this category too since they are not "instant" messages, and the time to reach the destination may vary from case to case. *ResourceConsumption* models the uncertainty regarding the unpredictability of the consumption of a *Resource* by a given *Task*. The implemented statistical behavior concerns the *amount* of a given resource type that can be consumed by a task. For instance, in the case of human type of resource, it is possible to model a task consuming a discrete number of resource units which is computed by a statistical function; whereas in the case of a non-human resource type, it will be possible to specify the statistical amount of the resource expressed in its unit of measure (kilowatts, kilograms, meters, liters, etc.). *LoopIteration* models the uncertainty of the number of iterations for a specific looped activity. This is the case of both BPMN's *standard loop* and *multi-instance loop activities,* i.e., tasks or sub-processes that have to be executed a certain number of times which either is prefixed or depends on a condition. *ConditionalFlowSet* models the uncertainty introduced by the conditional gateways (both inclusive and exclusive). It will be used to select which of the gateway's available output flow(s) are to be taken. The *Instantiation* concept models the rate at which a specific event responsible for instantiating a process may occur. In particular, it will be used to model the behaviour of the BPMN's *Start Event* elements that are not triggered by the flow. The *Occurrence* models the delay after which an event attached to an activity may occur. The event is triggered exactly when the activity processing-time exceeds the delay.

4   Design of a Business Process Simulator

In this Section we discuss the design of a novel simulator of business processes defined according to the Business Process Model presented in the Sect. 3. The simulator was designed to help process designers understand the real dynamics of enterprise processes once they are deployed in their execution context, and estimate the performance of processes in terms of execution time and incurred costs.

The basic design requirement is the capability of simulating the workflow of context-aware business processes. The adopted strategy was to exploit a well known and robust technique for the simulation of workflows, and concentrate our efforts on the integration of the process context with the workflow dynamics.

As for the simulation of the workflow, we opted on the Colored Petri Nets (CPNets) [10]. The interesting thing of CPNets is that they preserve useful properties of plain Petri nets, and at the same time allow tokens to have attached data (which is said the "token color) and enable generic data manipulation. The idea is then to use this feature in order to integrate the context data with the workflow, and let the CPNet run *context-aware* simulations.

In the proposed system, all a process designer has to do is supply a *Business Process Definition* (BPD) file, which embeds all the specific definitions of the process to be simulated in terms of workflow and execution context respectively. Examples of what a BPD file may look like are provided in the Sect. 5.

Figure 6 depicts the system's component architecture. The *BPSim Manager* is the component responsible for managing the system initialization and the

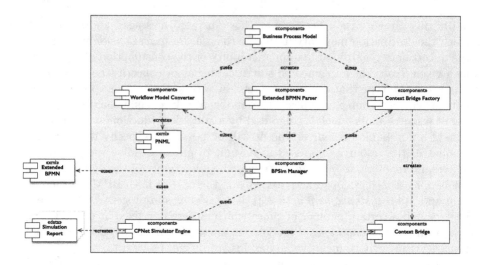

Fig. 6. System architecture.

actual simulation. It acts as a front-end interface to the designer (accepting BPD files), while on the back-end it coordinates the system components' activities. The *Extended BPMN Parser* component receives in input the BPD file and produces the *Business Process Model*. The *Workflow Model Converter* is the component responsible for converting process workflow contained in the *Business Process Model* into a Petri Net file (*PNML*), in a form which is ready to be processed by the *CPNet Simulator Engine*[6](the Simulator). The *Context Bridge Factory* provides a functionality to extract from the *Business Process Model* the information concerning the context scenario(s) to be simulated and to generate a *Context Bridge* for each defined scenario[7]. A *Context Bridge* embeds the information concerning a specific scenario and will act as a process' Context manager at simulation time. It is called into play in two different phases. During the system initialization (before the simulation is actually triggered), it is responsible for "coloring the plain Petri Net's created by the *Workflow Model Converter*. During the simulation, a *Context Bridge* will act as a context manager to the Simulator, in the sense that it will serve any request originated by the Simulator aiming at either inspecting the Environment's runtime status or at reserving/releasing Resources. Finally, the *CPNet Simulator Engine* elaborates the PNML and orchestrates the simulation process.

4.1 Workflow Model Converter

In this section we disclose some implementation details of the *Workflow Model Converter* component of the Fig. 6. The Workflow Model Converter is

---

[6] For the purpose of this work the engine of the Renew tool was used: www.renew.de.

[7] It is up to the designer to define as many context scenarios as they like. The system is capable of simulating many scenarios and producing the relative reports.

responsible for extracting the process workflow from the Business Process Model and converting it into its equivalent plain Petri Net (the *PNML* file). The approach used for the conversion grounds on the idea that all basic BPMN elements (sub-process, activity, task, gateways, etc.) may have equivalent Petri Net representations, which we are going to refer to as *PNML patterns*. A generic workflow model can be considered as a particular combination of BPMN elements. Therefore, building the workflow's equivalent Petri Net translates into a work of patterns' "composition". So our first effort was to build a repository of *PNML patterns*. The actual conversion is then carried out by composing the basic building blocks retrieved from the repository. In the component diagram represented in Fig. 7 the sub-components responsible for the conversion are depicted. In the following, a description of the work carried out by the sub-components is given.

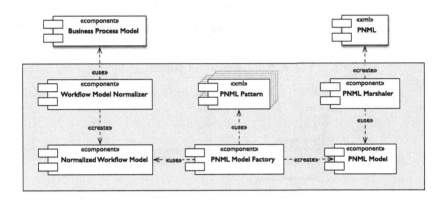

Fig. 7. Workflow model converter.

Workflow Model Normalizer. In order to keep the number of possible patterns as small as possible, a model normalization must be applied. In fact, in some cases the BPMN specification allows the designer to model some elements of a workflow in different, yet equivalent, graphical notations. It is responsibility of the Normalizer to pre-process the BPMN model in order to normalize those elements.

PNML Patterns. PNML Patterns are PNML files describing the Petri Nets that represent BPMN elements. In the Figs. 8 and 9 we reported a graphical representation[8] of the PNML patterns for the *None Start Event* and the *Task*. Each PNML pattern is a non-executable Colored Petri Net (CPNet) with Java as an inscription language. The Java inscription language allows a simple integration of the CPNet with the *ContextBridge* instance at runtime. The ContextBridge instance is used in the simulation initialization phase to fill the PNML key places with colored tokens representing the Java instances of the Process Model Context

---

[8] The graphical notation adopted to depict the Petri Net is taken from the Renew tool.

elements that are to be queried and manipulated in the successive run phase. In some cases the PNML Pattern uses one or more synchronous channels to interact with other CPNet instances that implement common complex logic, such as the *start* and *activity* CPNet instances used in the patterns in Figs. 8 and 9.

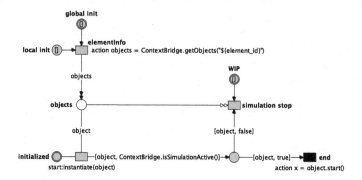

Fig. 8. None start event pattern.

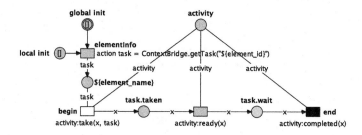

Fig. 9. Task pattern.

PNML Model Factory. It is the component charged for the creation of the *PNML model*. Here the overall Petri Net model is built by inspecting the *Normalized Workflow Model*, identifying all the workflow elements, retrieving from the repository the PNML patterns relative to the identified elements, and adequately composing the patterns. The obtained model is then pushed to the *PNML Marshaler* component which finally will produce the Petri Net instance of the original workflow (*PNML*) to be eventually simulated by the *CPNet Simulator Engine* (see Fig. 6).

5   Case Study Example

In this section a simple yet comprehensive case study is presented which will guide the reader through the steps designers are required to take in order to use the business process simulator, and will provide an example of what the output of a process simulation looks like.

First of all, designers have to supply a *Business Process Definition* (BPD) file, which is an instance of their process model embedding all the specific features of the process in terms of workflow and execution context respectively. A BPD is an XML file structured according to a schema that we are going to briefly describe. In order to define the process workflow, plain BPMN standard in its XML serialized form will be used. The standard envisions some points of extensions [14] that, instead, we are going to exploit to represent the process context features. In the end, the BPD file is still compliant to the BPMN standard.

The basic schema for the overall process representation, then, is that of the BPMN standard. Listing 1.1 reports the BPMN *baseElement* schema definition which is the abstract base class for all the BPMN elements and introduces the *extensionElements* reference:

Listing 1.1. XSD excerpt of the BPMN BaseElement and ExtensionElements.

```xml
<xsd:schema xmlns="http://www.omg.org/spec/BPMN/20100524/MODEL"
    ...>

<xsd:element name="baseElement" type="tBaseElement"/>
<xsd:complexType name="tBaseElement" abstract="true">
 <xsd:sequence>
  <xsd:element ref="documentation" minOccurs="0" maxOccurs="unbounded"/>
  <xsd:element ref="extensionElements" minOccurs="0" maxOccurs="1" />
 </xsd:sequence>
 <xsd:attribute name="id" type="xsd:ID" use="optional"/>
 <xsd:anyAttribute namespace="##other" processContents="lax"/>
</xsd:complexType>

<xsd:element name="extensionElements" type="tExtensionElements" />
<xsd:complexType name="tExtensionElements">
 <xsd:sequence>
  <xsd:any namespace="##other" processContents="lax" minOccurs="0"
     maxOccurs="unbounded" />
 </xsd:sequence>
</xsd:complexType>
```

The BPMN *extensionsElements* tag identifies a section that can be used to "*...attach additional attributes and elements to standard and existing BPMN elements*" [14]. There it is where we are going to provide extra information concerning the process context. In particular, using that tag we can add the environmental features discussed in Sect. 3.4 to *Subprocess, Task, SequenceFlow, MessageFlow, ExclusiveGateway, InclusiveGateway, StartEvent* and *BoundaryEvent* respectively. Moreover, being the BPMN *resource* an extension of the *tRootElement*, which in turn extends the *tBaseElement*, extra information concerning the newly introduced Resource::Resource concept (Fig. 3 in Sect. 3.2) can be added through that extension point. In Listing 1.2 the schema excerpt describing the Resource::Resource element has been reported.

Listing 1.2. XSD excerpt of the BPSIM resource.

```
<xsd:schema xmlns=" http://www.unict.it/bpmn/bpsim/2.0" ...>

<xsd:element name=" resource" type=" Resource" />
<xsd:complexType name=" Resource">
 <xsd:sequence>
  <xsd:element name=" unit" type=" Unit" minOccurs=" 1"
     maxOccurs=" 1" />
  <xsd:element name=" timeUnit" type=" TimeUnit" minOccurs=" 1"
     maxOccurs=" 1" />
  <xsd:element name=" moneyUnit" type=" MoneyUnit" minOccurs=" 1"
     maxOccurs=" 1" />
  <xsd:element name=" usage" minOccurs=" 1"
      maxOccurs=" unbounded">
   <xsd:complexType>
    <xsd:sequence>
     <xsd:element name=" availability" type=" Amount"
         maxOccurs=" 1" />
     <xsd:element name=" usageCost" type=" Amount" minOccurs=" 0"
         maxOccurs=" unbounded" /> <!-- EUR -->
     <xsd:element name=" unitCost" type=" TimeableAmount"
         minOccurs=" 0" maxOccurs=" unbounded" />
    </xsd:sequence>
    <xsd:attribute name=" scenarioRef" type=" xsd:IDREF"
        use=" required" />
   </xsd:complexType>
  </xsd:element>
 </xsd:sequence>
 <xsd:attribute name=" type" type=" xsd:string" />
</xsd:complexType>
```

Finally, the standard *category* tag is another element of the BPMN schema that we are going to use to supply other specific information regarding the definition of the simulation scenarios, the calendars and the cost objects.

In the following we provide an example of how to define a BPD file for a case study process. A BPMN description of the investigated case study process is shown in Fig. 10. It represents the process of **releasing a construction permit** by the Building Authority of a Municipality. The whole process involves different actors who interacts to each other exchanging information and/or documents (represented in the model as specific messages). The involved actors are: *Applicant*, the private citizen/company who needs to obtain the building permit and triggers the whole process; *Clerk*, the front-office employee of the Building Authority who receives the Application and is in charge of (a) checking the documentation of the application, (b) interacting with the applicant in order to obtain required documents and (c) sending back the result of the application;

Senior Clerk, the back-office employee of the Building Authority who evaluates and decides on the application; *Expert*, an external expert who may be called upon by the Senior Clerk whenever specific technical issues arise and the final decision may not be autonomously taken.

The business process spans four swimlanes, one for each actor. Both the Applicant and the Expert are external entities, i.e., are not part of the enterprise's business process dynamics. While there was no reason to represent the Applicant in the resource model, the Expert was modeled as a *DurableResource* (paid by the hour). The Clerk and the Senior Clerk were modeled as *OrganizationalPosition*. Other considered resources in the scenario are energy, modeled as *DurableResource*, paper and stamps, both modeled as *ConsumableResource*.

The BPMN diagram depicted in Fig. 10, which represents just the process workflow, has an equivalent *Xml* representation that we are not going to show for space reason. Instead, we deem interesting to report what the *Xml* representation of the process context will look like. Suppose we want to set up a context scenario (say *scenario1*) for which we intend to employ all the above mentioned resources. Also, we need a calendar (*official-calendar*) to state that for human resources the working days are Monday through Friday and the working hours follow the pattern [8:00 AM to 12:00 AM, 1:00 PM to 5:00 PM]. We then specify four different types of application ("cost object", in the ABC terminology) that potential applicants may submit. Further, in the specific scenario we require 1 unit of the Clerk resource type and 2 units of Senior Clerk resource type, whose hourly costs are 10€ and 20€ respectively. Listing 1.3 reports an Xml excerpt of the definition of the scenario, the cost objects and the Clerk. Note that all the new elements describing the process context, and that are out of the BPMN standard, have been assigned a *bpsim* prefix, while BPMN elements shows no prefix.

Listing 1.3. XML excerpt defining the scenario, the cost objects and the Senior Clerk resource.

```
<definitions id="sid-6cc2411a-..."
    xmlns="http://www.omg.org/spec/BPMN/20100524/MODEL"
    xmlns:bpsim="http://www.unict.it/bpmn/bpsim/2.0" ...>
<import location="BPSIM20.xsd"
    namespace="http://www.unict.it/bpmn/bpsim/2.0"
    importType="http://www.w3.org/2001/XMLSchema" />
<extension definition="bpsim:BPMNExtension" mustUnderstand="true" />
<category id="bpsim-scenarios" name="BpSim_Scenarios">
 <categoryValue id="scenario1" value="Scenario_1">
  <extensionElements>
   <bpsim:scenario default="true">
    <bpsim:timeUnit>month</bpsim:timeUnit>
    <bpsim:moneyUnit>EUR</bpsim:moneyUnit>
    <bpsim:maxInstances>1000000</bpsim:maxInstances>
    <bpsim:startInstant>2015-01-01T00:00:00.000</bpsim:startInstant>
    <bpsim:endInstant>2018-01-01T00:00:00.000</bpsim:endInstant>
   </bpsim:scenario>
  </extensionElements>
 </categoryValue>
</category>
```

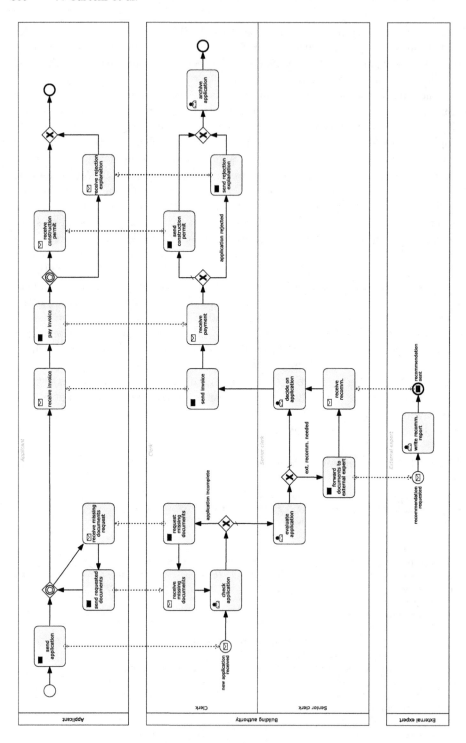

Fig. 10. BPMN model of the example process.

```xml
<category id="bpsim-cost-objects" name="BpSim_Cost_Objects">
 <categoryValue id="application" value="Application">
  <extensionElements>
   <bpsim:costObjectType id="application-maintenance"
       name="Maintenance" />
   <bpsim:costObjectType id="application-building-renovation"
       name="Building_Renovation" />
   <bpsim:costObjectType
       id="application-preservation-and-restoration"
       name="Preservation_and_Restoration" />
   <bpsim:costObjectType id="application-urban-restructuring"
       name="Urban_Restructuring" />
  </extensionElements>
 </categoryValue>
</category>
<resource id="senior-clerk" name="Senior_Clerk">
 <extensionElements>
  <bpsim:resource type="OrganizationalPosition">
   <bpsim:unit>#HR</bpsim:unit>
   <bpsim:timeUnit>h</bpsim:timeUnit>
   <bpsim:moneyUnit>EUR</bpsim:moneyUnit>
   <bpsim:usage scenarioRef="scenario1">
    <bpsim:availability calendarRef="official-calendar">2
    </bpsim:availability> <!-- 2 #HR -->
    <bpsim:unitCost timeBound="true">20.00
    </bpsim:unitCost> <!-- 20 EUR/h/#HR -->
   </bpsim:usage>
  </bpsim:resource>
 </extensionElements>
</resource>
```

In Listing 1.4 we show how to assign resources to a specific task. The *bpsim:resourceConsumption* tag models the statistical behaviour "ResourceConsumption" of Fig. 5 discussed in Sect. 3.4. Specifically, the *send invoice* task is assigned exactly one Clerk, an amount of energy that is distributed according to the Normal law, a number of paper sheets that is Bernoulli distributed over a minimum of 4 sheets and exactly 2 stamps. Further, the duration of *send invoice* task (which models the statistical behaviour "Duration" of Fig. 3) has a Beta (PERT parametrized) distribution (*bpsim:duration* section). Finally, the excerpt reports the parameters of the *autonomous decision* gateway. In particular, it is stated that in the case that at the input flow arrives a cost object of type *application-maintenance, application-building-renovation* or *application-urban-restructuring*, one of the outgoing flow has 20 % probability of being selected, the other takes the remaining 80 %; instead, in the case that an *application-preservation-and-restoration* cost object arrives, a specific output flow is alway selected. In our cost-based view, Gateways are one of the many elements where the type of the processed cost object affects the workflow.

Listing 1.4. XML excerpt defining the "send invoice" task's resource assignement and the exclusive gateway statistical behaviour.

```xml
<sendTask id="sid-7B9D15A6-..." name="send_invoice"
  startQuantity="1" completionQuantity="1" ...>
 <extensionElements>
  <bpsim:task scenarioRef="scenario1">
   <bpsim:resourceConsumption resourceRef="clerk" unit="#HR">
    <bpsim:countable type="Constant">
     <bpsim:parameter name="value">1</bpsim:parameter>
    </bpsim:countable>
   </bpsim:resourceConsumption>
   <bpsim:resourceConsumption resourceRef="energy" unit="W">
    <bpsim:fractional type="Normal">
     <bpsim:parameter name="mu">500</bpsim:parameter>
     <bpsim:parameter name="sigma">3.33</bpsim:parameter>
    </bpsim:fractional>
   </bpsim:resourceConsumption>
   <bpsim:resourceConsumption resourceRef="paper" unit="#paper">
    <bpsim:countable type="Bernoulli" shift="4">
     <bpsim:parameter name="p">0.12</bpsim:parameter>
    </bpsim:countable>
   </bpsim:resourceConsumption>
   <bpsim:resourceConsumption resourceRef="stamp" unit="#stamp">
    <bpsim:countable type="Constant">
     <bpsim:parameter name="value">2</bpsim:parameter>
    </bpsim:countable>
   </bpsim:resourceConsumption>
   <bpsim:duration unit="h" type="Pert">
    <bpsim:parameter name="a">10</bpsim:parameter>
    <bpsim:parameter name="m">15</bpsim:parameter>
    <bpsim:parameter name="b">30</bpsim:parameter>
   </bpsim:duration>
  </bpsim:task>
 </extensionElements>
 <incoming>sid-8982DB9F-...</incoming>
 <outgoing>sid-8E50B134-...</outgoing>
</sendTask>

<exclusiveGateway id="sid-4A21D4F4-..."
    default="sid-59595B8C-..." gatewayDirection="Diverging"
    name="autonomous_decision?">
 <extensionElements>
  <bpsim:exclusiveGateway scenarioRef="scenario1">
   <bpsim:conditionalFlowSet flowRef="sid-A1FD99EE-...">
    <bpsim:probabilities>
     <bpsim:probability costObjectTypeRef=
        "application-maintenance">0.2
     </bpsim:probability>
     <bpsim:probability costObjectTypeRef=
        "application-building-renovation">0.2
     </bpsim:probability>
     <bpsim:probability costObjectTypeRef=
        "application-preservation-and-restoration">1.0
     </bpsim:probability>
```

```
<bpsim:probability costObjectTypeRef=
    "application−urban−restructuring">0.2
</bpsim:probability>
</bpsim:probabilities>
</bpsim:conditionalFlowSet>
</bpsim:exclusiveGateway>
</extensionElements>
<incoming>sid−D15F5375−...</incoming>
<outgoing>sid−59595B8C−...</outgoing>
<outgoing>sid−A1FD99EE−...</outgoing>
</exclusiveGateway>
```

The cost-sensitive approach adopted to design the Business Process Model eventually allowed us to gather data produced by the simulation and easily put them in a form that facilitates ABC analysis. Focus is put on the application. As explained earlier, each submitted application is a *cost-object*, in the sense that it accumulates the costs produced by every activity that is going to process the application itself. Also, when traversing the activities an application is also capable of recording both the time spent in every activity's queue and the time for being processed.

	resources	activities										TOTAL
		check application	evaluate application	decide on application	send invoice	receive payment	send constr. permit	archive application	...	write rec. report	...	send rejection explanation	
Building Authority	Clerk	3.025,72			676,42	312,73	611,59	1.062,48	...		...	104,88	6.360,32
	Senior Clerk		267.660,75	706,67					...		...		268.450,76
	Energy	27,23	481,79		6,09		5,50		...		...	0,94	526,13
	Paper				2,12		3,84		...		...	0,40	7,64
	Stamp				339,20		1.536,00		...		...	64,00	2.144,00
	TOTAL	3.052,96	268.142,54	706,67	1.023,83	312,73	2.156,94	1.062,48	...		...	170,22	277.488,85
	time unit cost [€/h]	10,09	20,04	20,00	15,14	10,00	35,27	10,00	...		...	16,23	19,74
	unit cost [€/#activity]	11,06	1.264,82	3,33	4,83	1,48	11,23	5,01	...		...	8,51	19,74
External Expert	External Expert								...	61.771,13	...		61.771,13
	TOTALS								...	61.771,13	...		61.771,13
	time unit cost [€/h]								...	30,00	...		30,00
	unit cost [€/#activity]								...	908,40	...		908,40

(a)

	activities	cost-objects				TOTAL
		Maintenance	Building Renovation	Preservation & Restoration	Urban Restructuring	
Building Authority	check application	2.444,56	371,33	214,46	22,60	3.052,96
	evaluate application	214.532,66	37.760,23	13.413,99	2.435,66	268.142,54
	decide on application	566,67	100,00	33,33	6,67	706,67
	send invoice	818,94	148,31	46,55	10,03	1.023,83
	receive payment	250,95	43,72	14,75	3,30	312,73
	send constr. permit	1.718,95	304,58	110,64	22,77	2.156,94
	archive application	848,09	152,69	52,06	9,65	1.062,48
	request missing documents	557,08	42,33	77,69		677,10
	receive missing documents	82,33	6,41	10,93		99,67
	forward docs to ext. expert	36,88	5,33	7,11		49,32
	receive recomm.	25,75	3,53	5,12		34,40
	send rejection explanation	145,18	25,04			170,22
	TOTAL	222.028,06	38.963,48	13.986,64	2.510,68	277.488,85
	time unit cost [€/h]	17,34	17,80	14,00	19,80	17,22
	unit cost [€/#cost-objects]	1.306,05	1.298,78	1.398,66	1.255,34	1.308,91
External Expert	write rec. report	46.625,99	6.550,08	8.595,05		61.771,13
	TOTAL	46.625,99	6.550,08	8.595,05		61.771,13
	time unit cost [€/h]	30,00	30,00	30,00		30,00
	unit cost [€/#cost-objects]	274,27	218,34	859,51		291,37

(b)

Fig. 11. Cost reports for scenario1: (a) resource/activity; (b) activity/cost-object.

activities	work times					queue times				
	cost-objects				TOTAL	cost-objects				TOTAL
	Maintenance	Building Renovation	Preservation & Restoration	Urban Restructuring		Maintenance	Building Renovation	Preservation & Restoration	Urban Restructuring	
Building Authority										
check application	242,3	36,8	21,3	2,2	302,6	3,8	2,0			5,8
evaluate application	10.707,4	1.884,6	669,5	121,6	13.383,0	11.863,4	1.271,2	166,0	17,7	13.318,3
decide on application	28,3	5,0	1,7	0,3	35,3	11.765,5	261,4	654,3	85,6	12.766,9
send invoice	54,0	9,9	3,0	0,7	67,6	16,8	4,1	1,7	0,2	22,9
receive payment	25,1	4,4	1,5	0,3	31,3	3.384,9	242,5	371,0		3.998,3
send constr. permit	48,8	8,7	3,0	0,7	61,2	3,5	0,5			4,0
archive application	84,8	15,3	5,2	1,0	106,2	3,8				3,8
request missing documents	38,3	2,9	5,5		46,7					
receive missing documents	8,2	0,6	1,1		10,0	12.870,4	2.262,9	772,8	140,6	16.046,7
forward docs to ext. expert	1,8	0,3	0,4		2,5	232,8	23,2	389,1		645,0
receive recomm.	1,3	0,2	0,3		1,7	1.578,8	42,3	21,5		1.642,6
send rejection explanation	9,0	1,5			10,5	0,2				0,2
TOTAL	11.249,3	1.970,2	712,3	126,8	14.058,6	41.724,0	4.110,0	2.376,4	244,2	48.454,6
time unit duration [h/h]	100%	100%	100%	100%	100%	371%	209%	334%	193%	345%
unit durations [h/#instance]	66,2	65,7	71,2	63,4	66,3	245,4	137,0	237,6	122,1	228,6
External Expert										
write rec. report	1.554,2	218,3	286,5		2.059,0					
TOTAL	1.554,2	218,3	286,5		2.059,0					
time unit duration [h/h]	100%	100%	100%		100%					
unit durations [h/#instance]	9,1	7,3	28,7		9,7					

Fig. 12. Work and queue time reports for scenario1.

Three simple process KPIs are going to be assessed through the simulation: the *monetary costs* incurred by the Building Authority to attend the requests, the *work times* spent by the activities to process applications and the *queue times* applications had to wait before being processed by the activities. Figure 11 depicts a report of analytical costs computed for *scenario1*. In particular, costs incurred for the allocation of resources to activities are reported in the section "resource/activity cost" of the report; activity costs absorbed by the four types of cost-object (applications) are instead shown in the section "activity/cost-object cost" of the same report. In the Fig. 12 times spent by cost-objects for traversing the process are shown. In particular, work times and queue times have been reported respectively.

The obtained data provide a picture of the performance of the process in terms of costs and times. Process designers may use those data to make typical ABC analysis, and guess which part of the process (workflow, resource assignments) is susceptible of improvements. New scenarios may be easily defined by fine-tuning the basic scenario.

6 Conclusion

In the past decade there has been a growing interest of enterprises towards the Business Process Management's methodology and techniques. Field experts agree that one of the most critical step in the management of business processes is process analysis and modeling. Modelling a business process is not merely defining its *workflows*. In order to precisely estimate the cost and the performance of a process, the context where the process will execute must be carefully identified and modeled at design time. In the this paper we proposed the definition of a novel process context model, which grounds on the BPMN notation and on the ABC accounting principles, which process designers can leverage to define cost-sensitive representations of business processes. In the paper, we discussed some architectural and implementation details of a business process simulator capable of making simulations of such context-aware business processes, and of producing estimates of both the costs they would incur and their execution times. A simple yet comprehensive use case example was also discussed in the paper. In the future, the just proposed *cost* model will be enhanced to encompass new aspects that goes beyond the mere "monetary" aspect, the objective being to make multidimensional analysis of the process performance. Also, a study will be conducted to get the simulator exploit the semantics of the BPMN choreography model.

References

1. Van der Aalst, W.M.P.: Business process simulation revisited. In: Barjis, J. (ed.) EOMAS 2010. LNBIP, vol. 63, pp. 1–14. Springer, Heidelberg (2010). doi:10.1007/978-3-642-15723-3_1
2. Van der Aalst, W.M.P., Nakatumba, J., Rozinat, A., Russell, N.: Business process simulation: how to get it right. In: vom Brocke, J., Rosemann, M. (eds.) International Handbook on Business Process Management. Springer, Berlin (2008)
3. Van der Aalst, W.M.P., Ter Hofstede, A.H.M.: YAWL: yet another workflow language. Inf. Syst. **30**(4), 245–275 (2005)
4. Cartelli, V., Di Modica, G., Tomarchio, O.: A Resource-aware simulation tool for Business Processes. In: ICE-B 2014 - Proceedings of the 4th International Conference on e-Business, pp. 123–133, Vienna (Austria), August 2014
5. Cartelli, V., Di Modica, G., Tomarchio, O.: A Cost-centric model for context-aware simulations of business processes. In: IC3K 2015 - Proceedings of the 7th International Joint Conference on Knowledge Discovery, Knowledge Engineering and Knowledge Management, pp. 303–314, Lisbon (Portugal), November 2015
6. Cooper, R., Kaplan, R.S.: Activity-based systems: measuring the costs of resource usage. Account. Horiz. **6**, 1–12 (1992)
7. Dijkman, R.M., Dumas, M., Ouyang, C.: Semantics and analysis of business process models in BPMN. Inf. Softw. Technol. **50**(12), 1281–1294 (2008)
8. Gottschalk, F., Van der Aalst, W., Jansen-Vullers, M., Verbeek, H.: Protos2CPN: using colored Petri nets for configuring and testing business processes. Int. J. Softw. Tools Technol. Transfer **10**(1), 95–110 (2008)
9. Jansen-vullers, M.H., Netjes, M.: Business process simulation a tool survey. In: Workshop and Tutorial on Practical Use of Coloured Petri Nets and the CPN (2006)
10. Jensen, K.: Coloured Petri Nets: Basic Concepts, Analysis Methods and Practical Use. Springer, London (1996)
11. Kherbouche, O., Ahmad, A., Basson, H.: Using model checking to control the structural errors in BPMN models. In: 2013 IEEE Seventh International Conference on Research Challenges in Information Science (RCIS), pp. 1–12, May 2013
12. Koizumi, S., Koyama, K.: Workload-aware business process simulation with statistical service analysis and Timed Petri Net. In: Proceedings - 2007 IEEE International Conference on Web Services, ICWS 2007, pp. 70–77 (2007)
13. OMG: The UML Specification, July 2005. http://www.omg.org/spec/UML/2.0/
14. OMG: Business Process Model and Notation (BPMN 2.0), January 2011. http://www.omg.org/spec/BPMN/2.0/

15. Ouyang, C., Wynn, M., Fidge, C., ter Hofstede, A., Kuhr, J.C.: Modelling complex resource requirements in business process management systems. In: ACIS 2010 Proceedings - 21st Australasian Conference on Information Systems (2010)
16. Porter, M.E.: Competitive Advantage: Creating and Sustaining Superior Performance. Free Press, New York (1985)
17. Scheer, A.-W., Nüttgens, M.: ARIS architecture and reference models for business process management. In: Aalst, W., Desel, J., Oberweis, A. (eds.) Business Process Management. LNCS, vol. 1806, pp. 376–389. Springer, Heidelberg (2000). doi:10.1007/3-540-45594-9_24
18. Weske, M.: Business Process Management: Concepts, Languages, Architectures, 2nd edn. Springer, Heidelberg (2012)

Supporting Semantic Annotation in Collaborative Workspaces with Knowledge Based on Linked Open Data

Anna Goy[(⊠)], Diego Magro, Giovanna Petrone,
Marco Rovera, and Marino Segnan

Dipartimento di Informatica, Università di Torino, Turin, Italy
{annamaria.goy, diego.magro, giovanna.petrone,
marco.rovera, marino.segnan}@unito.it

Abstract. The management of shared resources on the Web has become one of the most pervasive activities in everyday life, but the heterogeneity of tools and resource types (documents, emails, Web sites, etc.) usually causes users to be lost and to spend a lot of time in organizing resources and tasks. Structured semantic annotation can provide a smart support to collaborative resource organization, but, as demonstrated by our user studies, users have often to deal with ambiguous or unknown expressions, suggested by the system or by other users. As a consequence, it is important to provide them with an "explanation" of unclear annotations, which can be based on formally encoded domain knowledge, retrieved from the LOD Cloud. We chose commonsense geospatial knowledge to implement a proof-of-concept prototype providing such "explanations". After a brief presentation of the background, represented by the SemT++ project, we describe the approach and present a user evaluation of it.

Keywords: Collaborative workspaces · Semantic annotation · Linked Open Data · Semantic Web · Ontology-driven applications · Geospatial knowledge

1 Introduction

The collaborative management of shared resources on the Web has become one of the most pervasive activities, not only for knowledge workers, but for everybody in everyday life. However, due to the pervasive nature of this activity, which is required by almost every task – from buying a ticket for a concert to organizing a holiday, from writing a scientific paper to managing children activities – the number of different tools that users have to use is very large, and useful resources belong to very heterogeneous types (documents, web sites, images, email conversations, posts, etc.). In this scenario, users are often lost and spend a lot of time in trying to organize resources and tasks.

In order to take on this challenge, a mechanism to handle heterogeneous Web resources in a uniform way is required. Moreover, such a mechanism should enable sharing and collaboration among users, and should help them by providing a smart support to resource organization. Semantic technologies can be the key factor to face this challenge. Ontologies and semantic representations of the handled resources, in

© Springer International Publishing AG 2016
A. Fred et al. (Eds.): IC3K 2015, CCIS 631, pp. 515–531, 2016.
DOI: 10.1007/978-3-319-52758-1_27

fact, can endow collaborative tools with some "expertise" about the objects they are managing, and this can turn such tools into smart companions. In particular, the annotation of resources, based on shared semantic vocabularies, can enable collaborative applications to help users in both resource organization and retrieval.

However, semantic annotation typically imposes a great overload on the users: In order to gain a future advantage in organization and retrieval, users are required an extra-work to annotate resources. To alleviate such an overload, tools for the collaborative management of Web resources should provide automatic annotations or, at least, suggestions. We started the design and implementation of a (semi-)automatic support to semantic annotation of resources (exploiting HTML parsing and Named Entity Recognition); some preliminary results of this work can be found in [1]. As demonstrated by our user studies, in a system that provides users with suggestions for semantic annotation, and where annotations are collaboratively defined, users have often to deal with ambiguous or unknown expressions, suggested by the system or by other users for the annotation of resources. As a consequence, it is of paramount importance providing them with an easy and quick access to the meaning of such unclear annotations. In order to reach this goal, collaborative tools should be equipped with formally encoded domain knowledge and the Linked Open Data (LOD) Cloud (lod-cloud.net) represents a very rich source of knowledge about a wide range of domains, in the form of data and semantic models that can be exploited within collaborative applications. However, a collaborative environment can be used to carry out very different activities, each one referring to a different domain: Which are the most suited datasets? Two considerations can help us answer this question.

First, datasets based on Linked Data best practices typically provide links to related datasets: This means that from a selected set of data referring to a specific domain, other datasets are usually reachable; in particular, many datasets contains links to cross-domain resources such as DBpedia (wiki.dbpedia.org).

Second, there are some types of knowledge that, due to their intrinsic nature, can be considered (almost) universal: *Geospatial knowledge* is one of the most popular type of such a cross-domain knowledge. As demonstrated by the pervasive presence of services based on geolocation, maps, and directions functionalities, geospatial knowledge is involved in a lot of different specific domains and is used by everyone to carry on everyday activities: Geospatial concepts and relations are used when planning a journey, when taking care of environmental issues, when arranging an appointment, when organizing a conference, etc. This knowledge does not represent a scientific perspective, but instead a *commonsense* one (i.e. a perspective enabling people to distinguish different geospatial entities, to identify and to georeference them); in fact, it does not provide a formal precision degree in geographic descriptions, but instead a model enabling people to describe, and ultimately organize, representations of real-world entities, like mountains, cities, or streets.

The cross-domain nature of geospatial information is further confirmed by a recent report by the LOD work team, where geography appears as one of the nine thematic categories the whole LOD cloud is divided into [2]. In particular, GeoNames (www. geonames.org) has assumed, together with DBpedia, a role of hub (see: lod-cloud.net), becoming the de facto reference geospatial dataset in the LOD Cloud. Moreover, geospatial knowledge can act as a "glue" in integrating and linking different datasets [3].

These considerations led us to choose *commonsense geospatial knowledge* as the first testbed for our approach, aimed at endowing shared workspaces with domain knowledge. However, the proposed architecture (see Sect. 4.3) has a more general validity, and can be used to include different types of domain knowledge within collaborative environments.

Ultimately, the main contribution described in this paper shows the role played by (geospatial) ontologies and data retrieved from the LOD Cloud in the implementation of the previously mentioned "explanation functionality" within a collaborative environment for Web resource management.

The rest of the paper is organized as follows. In Sect. 2 we discuss the main related work; in Sect. 3 we briefly summarize the main characteristics of the SemT++ project, representing the background of the proposal presented in this paper. In Sect. 4, which describes the contribution with respect to our earlier work, we present the motivations of the presented approach, we sketch a simple usage scenario, and we describe the implementation of the "explanation functionality", together with a user evaluation of it. Section 5 concludes the paper and outlines future developments.

2   Related Work

Several research fields have to be taken into account in order to outline the background reference work for the approach presented in this paper.

As far as the original idea underlying SemT++ is concerned, an interesting reference work is represented by [4], presenting the problems of the *desktop metaphor* and several approaches trying to replace it. In particular, an interesting model presented in the mentioned book is Haystack [5], a flexible and personalized system enabling users to define and manage workspaces referred to specific tasks. Another interesting set of approaches are those grounded into Activity-Based Computing, where the core concept structuring the interaction model is that of *user activity* [6, 7]. The main enhancement of SemT++ with respect to these approaches is the explicit geospatial knowledge model and the exploitation of LOD sets, discussed in this paper.

Another relevant research field that is worth to be considered is represented by research about social tagging systems, where resources can be tagged with meta-data referring to different aspects (*facets*); the user-centered, bottom-up tagging process leads to the creation of multi-facets classifications called *folksonomies* [8]. Interesting semantic enhancements of tagging systems have been developed [9], with particular attention to knowledge workers [10]. With respect to social tagging systems, SemT++ shifts its focus from mass social communities to (small) collaborative groups of people sharing specific activities.

In general, the idea of exploiting semantic technologies to support collaborative resource management is not new. For example, a new research area has recently emerged, the *Social Semantic Web* [8], an approach relying on the idea that semantic technologies can support the creation of machine readable interlinked representations of social objects (people, contents, resources, tags, etc.) enabling different social "islands" (i.e., isolated communities of users and data) to be connected and integrated. The approach presented in this paper can be seen as part of this project, since it aims at enhancing a collaborative environment for resource management with semantics, in order to provide users with a smarter support to resource management.

Another project aimed at coupling desktop-based user interfaces and Semantic Web is the *Semantic Desktop* [11]. In particular, the NEPOMUK project (nepomuk. semanticdesktop.org) defined an open source framework for implementing semantic desktops that rely on a set of ontologies and integrate existing applications to support collaboration among knowledge workers. Drăgan and colleagues [12] propose an approach to connect the Semantic Desktop to the Web of Data: This enables the system to "bring Web data to the user", thus supporting the exploitation of external data within the user personal context. The proposal by Drăgan and colleagues is one of a great number of recent semantic approaches trying to use LOD to enhance services for the users. In the same direction, the LinkZoo tool [13] propose a collaborative annotation platform based on LOD: Semantic annotations are stored as RDF triples and they enable LinkZoo to couple standard keyword search with property-based filtering. [14] contains a survey of the approaches to exploit LOD in metadata for multimedia content, while Linkify [15] is an add-on for major browsers that adds a link to Named Entities recognized in online texts, pointing to a mashup of information items extracted from LOD sources. Passant and Laublet [16] present MOAT (Meaning Of A Tag), a semantic framework for the definition of machine-readable meanings of tags: Tags are represented as quadruples (<*User, Resource, Tag, Meaning*>), their meaning is linked to well-known LOD sets (such as DBpedia and GeoNames) and can be shared with other uses.

Lots of work has also been done, within different research communities, in the field of collaborative semantic annotation. In the NLP field, tools have been implemented aimed at supporting collaborative annotation of textual corpora [17]. NLP-oriented annotation tools enable users to associate "semantic" labels to phrases within a text and usually refer to an annotation schema that can be formally encoded as an ontology (e.g., [18, 19] among many others).

In the Knowledge Management field, a similar notion of "semantic annotation" has been used, where annotations link words or phrases within a document to instances in a semantic knowledge base (and indirectly to classes of a domain ontology); see, for instance, [20], which also contains an interesting survey of annotation frameworks. A good survey of ontology-based annotation environments can also be found in [21]. Usually, ontologies provide the metadata structure, and describe document properties, such as author, date, format, etc. (e.g., Dublin Core: www.dublincore. org). In some cases, annotation systems can rely on more domain-dependent semantic resources (e.g., the Getty Thesaurus of Geographic Names: www.getty.edu/research/ tools/vocabularies/tgn in the geographic domain). Annotations have also been widely used in e-learning [22] and in the so-called "semantic wikis"; for instance, Buffa and colleagues [23] describe SweetWiki, a wiki tool supporting a structured, semantic annotation of resources.

Finally, given that geospatial knowledge plays a major role in the proposal presented in this paper, we dedicate the final part of this related work overview to it. The importance of geospatial knowledge, especially in information retrieval and knowledge organization, is claimed in the literature (see, for instance, [24]), and is demonstrated by the leading role that geography acquired in the Web of Data during the last ten years.

In particular, the growth of Web 2.0 and its related practices, like crowdsourcing, found in Geographic Information a preferential knowledge domain. Goodchild termed the so gathered information *volunteered geographic information*, and considered it an interesting example of user-generated content [25]. This phenomenon emphasizes the major role played by a commonsense perspective over geographic knowledge: Services like OpenStreetMap, WikiMapia, Google Earth, Google Maps were contributing to change the way people interact with the Web, turning them into information prosumers, rather than mere information consumers. Moreover, the recent mobile revolution, the availability of social networks like Foursquare, and the pervasive trend of geolocation and resource geo-tagging, increased the role of geospatial knowledge in our everyday life. Within this scenario, ElGindy and Abdelmoty propose a framework for analyzing folksonomies derived from geo-tagging activity and discovering place-related semantics (e.g., events, activities, personal opinions, and so on). The results of such an analysis reveals "a much richer structure of concepts and relationships than those defined in a formal data source produced by experts" ([26], p. 222). Moreover, the synergy among semantic technologies, Web of Data and Geographic Information resulted in the establishment of the *Semantic Geospatial Web*, a Semantic Web extension based on a set of spatial ontologies that can be exploited in geography-based retrieval systems, leading to better quality results [27]. In conclusion, it is worth mentioning the geospatial ontology *Space*, based on GeoNames, WordNet and MultiWordNet [24]. *Space* is aimed at representing geographic and spatial concepts and relations from the commonsense point of view, an aspect which is shared by our perspective.

3 Shared Workspaces as "Round Tables" and Resources as "Information Objects"

The background of the work described in this paper is represented by the *Semantic Table Plus Plus* (SemT++) project. SemT++ started from the idea that shared workspaces could be seen as "round tables", where people sit together in order to collaboratively carry on an activity (such as planning a journey, organizing children care, participating in the social activities of a NGO, write a scientific paper for a conference) [28]. Table participants typically use different types of resources to perform the tasks required by the specific activity: They get information from Web sites, read papers, write documents, watch videos or photo galleries, write emails and posts, and so on. The resources useful to carry on the activity the table is devoted to are typically encoded in different formats, handled by different applications, and stored in different places, although they typically refer to the same semantic context. We thus designed an interaction model aimed at providing an *abstract view* over table resources by handling them as *information objects*, lying on a table, collaboratively managed (added, deleted, modified, annotated) by table participants.

In SemT++, workspace awareness is guaranteed on each table by standard mechanisms such as a presence panel (showing the list of table participants currently sitting at the table); icon highlighting (to notify users about table events); notification messages (filtered on the basis of the topic context represented by the active table [29]).

One of the most important features of SemT++ is that each table is endowed with semantic representations of the resources lying on it. Such representations are based on the *Table Ontology*, a semantic model grounded in O-CREAM-v2 [30], a core reference ontology for the Customer Relationship Management domain developed within the framework provided by the foundational ontology DOLCE (Descriptive Ontology for Linguistic and Cognitive Engineering) [31] and one of its extensions, namely the Ontology of Information Objects (OIO) [32]. The Table Ontology enables SemT++ to represent resources (documents, Web pages, email threads, images, etc.) as *information objects*, with properties and relations. Figure 1 shows a simplified example: The semantic representation of a Web page encoded in UTF-8/HTML5, written in Spanish by Carlos, and containing two parts – an image and a link (to a pdf document); moreover, the represented Web page has a main *topic* (*Cuba music tour*) and it refers to a set of entities (called *objects of discourse*: *Havana, Santiago de Cuba, salsa, Camagüey*).

SemT++ also includes a Reasoner that, on the basis of the semantic representations of the resources lying on a table, can infer possible properties of related resources; for example, if a document contains a hyperlink to a resource written in Spanish, the Reasoner infers that probably the document itself is written in Spanish. A detailed description of the Table Ontology, including classes, relations and axioms supporting the reasoning mechanisms can be found in [33].

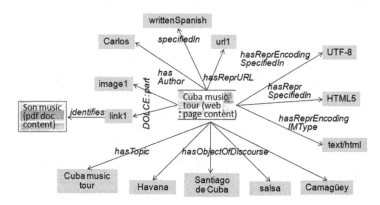

Fig. 1. Semantic representation of a resource (Web page) – simplified example.

The semantic representations of table resources is a kind of *semantic annotation*, in which the structure is provided by the Table Ontology. These representations are collaboratively built by users, with a significant support provided by the system. In fact, when a table participant adds a new resource to a table, a new semantic representation is set up on the basis of contributions from the system and from table participants. In particular, the system (actually, the *Smart Object Analyzer* module) analyzes the resources and, on the basis of the results of the analysis, it performs the following tasks:

- It defines some properties: Typically, the *format* (e.g., UTF-8, HTML) and the *parts* (e.g., images included in the analyzed resource). Parts are proposed to users, who can select the interesting ones and add them to the table.
- It proposes candidate values for other properties: The *author* of the resource, the *language* the resource is expressed in, the main *topic* and the *objects of discourse* (representing the resource *content*). These suggested values are identified as follows:

Authors, from meta information of the document itself;

Language, from meta information of the document itself or from the reasoning mechanisms;

Topic and *objects of discourse*, from meta information about the document (e.g., HTML meta-tags), from the results of a Named Entity Recognition (NER) service; from the Reasoner, which – on the basis of the Table Ontology – infers suggested values.

Users can confirm or discard the suggestions, or they can add new values from scratch. The activity performed by table participants in the definition of the semantic representation of resources can be seen as a *collaborative annotation* task: Values for object properties can be added, deleted, or modified, according to the collaboration policy defined on the table: In case of a *consensual* policy, all participants can always edit semantic representations, while in case of an *authored* or *supervised* policy the final decision about the semantic annotation of a resource is taken by its creator or by the table supervisor; a detailed account of collaboration policies in SemT++ can be found in [34]. Semantic representations represent a *shared view* over table resources; however, each table participant can also keep a *personal view* over resources, containing "private" annotations; see [35] for details.

The current SemT++ proof-of-concept prototype is a Java Web application accessible through a Web browser. Backend core functionality is provided by Java components while services relying on heterogeneous technologies are accessible through a RESTful interface: For example, a Python Parser Service provides the HTML analysis, while a Node.js module is in charge of interfacing with the NER Service, based on TextRazor (www.textrazor.com). Files corresponding to table objects are managed through Dropbox, Google Drive, and Google Mail APIs. The User Interface (UI) is based on Bootstrap (getbootstrap.com), which guarantees responsiveness and thus availability on different devices.

The Table Ontology, as well as the knowledge base containing the semantic representations of table resources, are expressed in OWL (www.w3.org/TR/owl2-overview), and the OWL API library (owlapi.sourceforge.net) is used to interact with them. The current Reasoner implementation is based on Fact++ (owl.cs.manchester.ac.uk/tools/fact).

We performed some user evaluations of the system, at different development stages. The first evaluation is presented and discussed in [28], and demonstrated that communication among users, resource sharing, and resources retrieval is significantly faster when using SemT++ with respect to performing the same tasks without it. The second evaluation is presented and discussed in [33], and told us that users appreciate the functionality of SemT++ User Interface enabling the exploitation of multiple criteria (in particular, resource content) to perform object selection.

Moreover, we carried on a qualitative user study with the goal of defining the model that supports collaborative semantic annotation of table resources, and in particular the suitable collaboration policies [34]. The main results of this user study are the implementation of the mentioned collaboration policies (*consensual, authored, supervised*) that can be set on each table, and the implementation of the *personal views* functionality, clearly requested by users.

4 Providing Explanations Based on Commonsense Geospatial Knowledge in Collaborative Annotation

4.1 The Need for Knowledge About Resource Content

The "expertise" of SemT++ represented by the Table Ontology mainly refers to resources (documents, Web pages, images, email threads, etc.) as *information objects*, i.e., it includes knowing that they are encoded in specific formats, they usually have one or more authors, and so on. This knowledge also includes knowing that information objects typically have a *content*, structured in a main *topic* and a set of entities the resource "talks about" (the *objects of discourse*). However, the Table Ontology does not provide any knowledge about the content itself: If a document talks about "New York", no semantic representation is associated to such a value, which is simply a string.

In the evaluation aimed at testing the availability of multiple selection criteria [33], as well as in the qualitative user study used to define the model handling collaborative semantic annotation of table objects [34], many users pointed out that the meaning of some topics and objects of discourse were unclear and that a sort of explanation would have been very helpful. The examples of unclear values mentioned by users are system suggestions (see Sect. 3) or annotations provided by other table participants. Since system suggestions and collaboration among users are two core aspects of SemT++, it is clear that the system needs to be enhanced with the required "explanation functionality".

The analysis of the users' answers in the mentioned evaluation and user study also showed that a significant number of examples of unclear values refer to places (villages, monuments, regions, mountains, etc.), and users would like to know their nature (is Saint Barthélemy a village or a valley?) or their geolocation (where is Saint Barthélemy? How far is it from Aosta?). These feedbacks from users suggested us two things:

- Besides the semantic knowledge representing resources as information objects (encoded in the Table Ontology), SemT++ tables need to be endowed with specific domain knowledge, aimed at providing a semantic characterization of the entities representing resource *content*; this knowledge would enable the system to offer "explanations" of the unclear property values.

– A significant aspect of the knowledge about resource content is represented by *commonsense geospatial knowledge*, enabling the system to provide information about places (villages, monuments, regions, mountains, etc.). Obviously, the relevance of such a knowledge depends on the specific activity a single table is devoted to: It is intuitively very important on a table devoted to the organization of a journey, while it seems definitely less useful on a table set up to write a paper about neural networks. However, as already claimed in the Sects. 1 and 2, commonsense geospatial knowledge represents a very important cross-domain knowledge, and it can play a major role in SemT++, in particular as far as the "explanation functionality" is concerned.

On the basis of the just presented motivations, we designed and implemented a new module, the *Geospatial Knowledge Manager*, in charge of managing commonsense geospatial knowledge on SemT++ tables. The *Geospatial Knowledge Manager* will exploit geospatial information retrieved from GeoNames, the most popular geographic dataset in the LOD Cloud. Before describing how the *Geospatial Knowledge Manager* works (Sect. 4.3), we sketch a very simple usage scenario (Sect. 4.2), showing how geospatial knowledge can provide table participants with "explanations" of topics and objects of discourse representing resource content.

4.2 Usage Scenario

Imagine that John, together with a group of friends, participates in a table devoted to the organization of a journey to Cuba. John and his friends are particularly interested in cultural and sustainable tourism. The discussion about the itinerary started a few days earlier, and the table is currently populated by some bookmarks to Web sites describing travels in Cuba. Browsing the Web, John finds another interesting site, proposing a music tour of the island, thus he decides to add it to the table. When the new object is dropped on the table, the system starts its analysis, finding that it is an HTML document (encoded in UTF-8) and it is probably written in Spanish. Moreover, the Smart Object Analyzer module discovers that it contains several images and hyperlinks, which represent its parts; parts are proposed to John: he selects an image (showing the tour steps on a map) and an e-book about Son music (linked in the Web page) and adds them to the table. The system also suggests some candidate topics and a set of candidate objects of discourse (among which: *Cuba, Havana, Santiago de Cuba, Moncada Barracks, Camagüey, Music of Cuba, salsa*).

John confirms the language (*Spanish*), provides *Cuba music tour* as main topic, and looks at the candidate objects of discourse, in order to see if some of them could well represent the Web page content. John is uncertain about one of them, which seems to be interesting, namely *Camagüey*: Is it a city, a small village, or a beach? Is it relevant for the music tour and, in general, for a cultural travel through Cuba? Should it be mentioned as objects of discourse of the selected Web page? John thus clicks on the linked item (*Camagüey*) to get an explanation: The system displays a pop-up window, shown in Fig. 2, providing information about Camagüey, namely the kind of place (a *City*), a short description, and its position on a map. On the basis of this information, John decides to add it as an object of discourse.

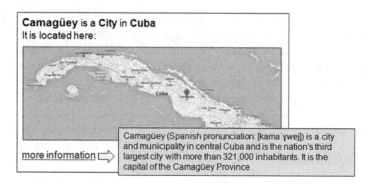

Fig. 2. "Explanation" of an object of discourse on a SemT++ table.

Later on, Mary sits at the table to have a look to the new items: She is notified about the new Web page added by John, and she takes a look at the semantic description (annotations), in order to have a view at-a-glance of its content. Intrigued by *Camagüey*, she clicks on it, gets the explanation (Fig. 2), and starts looking for further information about the Cuban city.

4.3   The Geospatial Knowledge Manager

A preliminary version of the Geospatial Knowledge Manager and its role is described in [36]. In the following we describe its current architecture and present a user evaluation of it (Sect. 4.4). The main components of the Geospatial Knowledge Manager module are shown in Fig. 3.

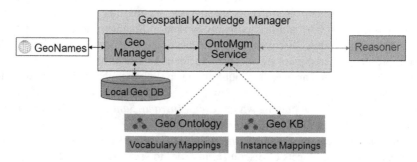

Fig. 3. Architecture of the geospatial knowledge manager.

We describe the role of the different components in the following.

Geo Ontology. The semantic model representing the commonsense geospatial knowledge is provided in the Geo Ontology. This component represents the system "expertise" about geospatial issues, from a commonsense perspective, as described in Sect. 1 and 2. It provides a vocabulary to describe the content of table resources, as far as the geospatial aspects are concerned, and thus it enables the system to "interpret" geospatial data belonging to potentially heterogeneous sources. This role is one of major importance within SemT++: The Geo Ontology, in fact, provides the conceptual framework needed to integrate data coming from different datasets and possibly originally characterized by means of different ontologies.

The Geo Ontology is a lightweight, application ontology, containing classes (about 240) and properties mainly reflecting the properties used by GeoNames to describe geographic features (latitude, longitude, population, altitude, etc.). The top layer of the taxonomic structure of the Geo Ontology is represented by two classes: *GeoSocialEntity* – that includes all the geospatial entities created by people's activities: For example, infrastructures, human settlements, administrative and political institutions, but also concepts used to partition the geographic space (such as regions or borders) – and *GeoPhysicalEntity* – that includes natural or geophysical entities like valleys, rivers, deserts, mountains, and so on.

It is worth underlining that, although the Geo Ontology partially reflects the GeoNames ontology (see below), it is an independent semantic model, aimed at representing a conceptual vocabulary useful to integrate data from heterogeneous sources. This choice also ensures SemT++ not to be committed to any specific external semantic model, and thus to any specific dataset.

Geo KB. The Geo KB is the knowledge base containing all the semantic assertions, i.e. the "facts", about geospatial instances, expressed according to the vocabulary provided by the Geo Ontology. In particular, each geospatial instance (e.g., the instance representing Camagüey) is classified with respect to at least a class of the Geo Ontology (e.g., as an instance of the *City* class).

GeoNames. GeoNames is an open geospatial gazetteer, released in 2006, gathering different data sources provided by governmental organizations, institutes of geography and statistics, as well as users' contribution. The GeoNames dataset contains over 10 millions of toponyms and 9 millions of *features*, uniquely identified by URIs, and classified according to the GeoNames *ontology*, a taxonomy including 9 high-level classes, called *feature classes*, and 650 subclasses, called *feature codes*. GeoNames offers a number of RESTful Web Services (www.geonames.org/export/ws-overview. html) enabling different types of search: A general purpose string-based search, a search for closest toponyms, for the altitude of a geographic point, for cities and toponyms within a user specified bounding box, for postal codes, for earthquakes, and so on. The most part of GeoNames services return XML or JSON objects, while only in some cases (for example for the general search service) RDF results are available. In SemT++, we used the *search* service, i.e., the general purpose search service returning a list of results in JSON format.

Vocabulary Mappings. Vocabulary Mappings represent the alignment between SemT++ Geo Ontology and the GeoNames ontology, and enables GeoNames entities to be classified into classes of the Geo Ontology. Vocabulary Mappings rely on two relations, *conceptual equivalence* and *subsumption*; both these relations can involve a GeoNames *feature code* and a class of the SemT++ Geo Ontology, or they can involve two properties, one belonging to the GeoNames ontology and the other to the SemT++ Geo Ontology.

The *conceptual equivalence* relation is used to state that a GeoNames *feature code* and a Geo Ontology class (or a GeoNames property and a Geo Ontology property) are equivalent; the *subsumption* relation is used to state that a GeoNames *feature code* represents a subclass of a Geo Ontology class (or that a GeoNames property represents a subproperty of a Geo Ontology property). For example, the following is the RDF/XML serialization of the axiom stating the subsumption relationship between the class representing all individuals having H.STMH as *feature code* value in the Geo-Names ontology and the class *WaterSpring* in the Geo Ontology:

```
<owl:Restriction>
  <rdfs:subClassOf
    rdf:resource="http://www.di.unito.it/ontologies/SemTppOntologies/
                  SemTppGeographicOntology#WaterSpring"/>
  <owl:onProperty
    rdf:resource="http://www.geonames.org/ontology#featureCode"/>
  <owl:hasValue
    rdf:resource="http://www.geonames.org/ontology#H.STMH"/>
</owl:Restriction>
```

Current Vocabulary Mappings mention 192 classes belonging to the Geo Ontology and 233 *feature codes* from GeoNames, defining 186 conceptual equivalence relations and 31 subsumption relations. Obviously, new Vocabulary Mappings have to be defined if a new dataset, relying on a different ontology, has to be integrated within the system.

Instance Mappings. Instance Mapping represent the correspondences between SemT++ and GeoNames individuals (e.g. the instance representing the city of Camagüey in SemT++ and the instance representing it in the GeoNames dataset).

GeoManager. The GeoManager submodule interacts with GeoNames to retrieve information about geospatial entities. The current version of the GeoManager relies on the asynchronous Web framework Node.js and uses a local database (Local Geo DB), implemented in MongoDB (www.mongodb.org), to locally store information retrieved from GeoNames.

OntoMgmService. The OntoMgmService manages all the interactions with the Geo Ontology and the Geo KB and invokes the Reasoner (see Sect. 3) to classify Geo-Names entities in the Geo Ontology, on the basis of the Vocabulary Mappings described above. The current version of the OntoMgmService exploits Java Servlets and the OWL API library, and offers a RESTful service interface, providing results in JSON format.

The Geospatial Knowledge Manager implements the "explanation functionality" depicted in the usage scenario above (Sect. 4.2). In the following we detail the steps performed by the Geospatial Knowledge Manager in order to achieve this goal.

When a new topic or object of discourse is added to the semantic representation of a table resource by a user, or when a candidate topic/object of discourse is proposed by the system (as the case of *Camagüey* mentioned in the usage scenario above), the corresponding string – together with the IRI referring to the instance created by the system for that topic/object of discourse – is passed to the Geospatial Knowledge Manager, more specifically to the GeoManager submodule.

The GeoManager checks in the Local Geo DB if the information about that item had already been retrieved from the LOD, and, if not, it invokes the GeoNames *search* service, getting a JSON object containing a list of entities, along with their descriptions. The GeoManager tries to select the relevant entry using simple heuristics (e.g., the presence of the searched string in the name of the GeoNames entity, the position in the results); in some cases (e.g., the case of *Camagüey* in our usage scenario) this lead to a complete disambiguation, while in other cases the user will be presented with a list of alternative possible "explanations", among which s/he can choose the suited one.

At this point, the GeoManager invokes the OntoMgmService, in order to have the instance (identified by a system IRI) classified with respect to the Geo Ontology (e.g., classifying Camagüey as an instance of *City*). Such a classification enables the system to provide the main information item within the explanaion, i.e., the entity type (allowing users to know, for example, that *Camagüey* is a city and not a beach). Moreover, the knowledge available in LOD sets (GeoNames in our prototype) is brought into the system, linked to the semantic description of table resources (as depicted in Fig. 4), and available to table users: When a table user clicks on that (candidate) topic/object of discourse, the result of the instance classification, together with other relevant GeoNames data (e.g., localization on a map, description usually from linked DBpedia data), are displayed (see Fig. 2, where the information about Camagüey is shown).

As shown in the usage scenario (Sect. 4.2), this knowledge provides table participants with an "explanation" of the meaning of the (candidate) topics/objects of discourse, which can be useful when annotating table resources with semantic properties representing their content. Furthermore, these "explanations" can also support users in resource selection.

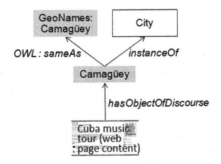

Fig. 4. The semantic representation of a Web page including geospatial knowledge.

In the following section we presents the results of a user evaluation aimed at testing the usefulness of this functionality.

4.4   Evaluation

Following a user-centered design approach, after having implemented a functionality based on the user feedback from the first evaluation round (see Sect. 4.1), we contacted again 8 of the users involved in our previous studies, in order to test the new "explanation functionality". We set up the table of our usage scenario (Sect. 4.2), i.e. a shared workspace were users could organize a cultural and sustainable travel to Cuba. The table was configured with a *consensual* policy for the management of collaborative annotation (see Sect. 3). We invited our participants to imagine that the discussion is now focused on the itinerary and we provided them with a set of bookmarks proposing several tours, some of them being thematic ones (like a music and dance tour, or a horse riding journey).

We left participants some time to take a look at the available resources, in order to become familiar with the context. Then we asked three of them to select a resource each, to be added to the table. This triggers the collaborative annotation process of the added resources, in which the system suggests property values (as described in Sect. 3) and users can select system suggestions or add new values. We asked users to concentrate on *topics* and *objects of discourse* (even though other properties – such as *language*, *author*, etc. – were also available). When the participants reached a stable agreement about the annotation, we stopped the annotation activity.

We recorded the number of times users used the "explanation functionality" available; see Table 1. Moreover, after the test, we asked them to rate, in a 1 to 5 scale, the usefulness of the functionality (1 meaning totally useless, 5 definitely useful); see Table 2. Finally, we collected all free comments they had about the experience.

Table 1. Number of times users used the "explanation functionality".

user1	user2	user3	user4	user5	user6	user7	user8	# users	# times
3	1	2	0	0	3	1	0	5	10

Table 2. Users' evaluation of the "explanation functionality".

user1	user2	user3	user4	user5	user6	user7	user8	mean	st.dev.
5	4	4	3	5	5	5	2	4.125	1.126

From results shown in Table 1 we can see that 5 users out of 8 used the "explanation functionality", while 3 of them did not use it at all. Quite interestingly, one of them (*user5*) rate the usefulness of the functionality at the higher degree (see Table 2), despite that s/he never used it during the test.

In Table 2 we can see that the average rate is 4.125 (on a 1 to 5 scale), indicating that the new feature was appreciated by users; the quite low standard deviation (1.126) tells us that users tend to agree on it (in fact, nobody rated it as totally useless).

Analyzing free comments, we can find interesting suggestions to improve the functionality. Three users said that the functionality would be more interesting if not only geospatial issues were supported; two users pointed out that in the cases in which more than one explanation were available, reading the explanations was quite annoying: Since this derives from the fact that often results from GeoNames search are not unique, it implies that a greater effort should be devoted to the disambiguation phase.

5   Conclusions and Future Work

In this paper we presented the "explanation functionality" implemented within the SemT++ system, providing users with information – retrieved from the LOD cloud – about the entities referred to when describing resource content in a collaborative semantic annotation environment. In particular, we claimed the central, cross-domain role played by commonsense geospatial knowledge in such a context and described the current implementation of the mentioned functionality, together with a user evaluation.

Some open issues clearly emerged from the presented approach. For example, the connection of new datasets, different from GeoNames and in general from geography-oriented datasets, currently requires the manual definition of a local semantic model (taking the role of the Geo Ontology) and of the corresponding Vocabulary Mappings. The investigation of semi-automatic approaches to ontology integration would be interesting; see, for instance, [37].

Moreover, knowledge retrieved from LOD sets could be used to provide users with suggestions about possibly related resources (for example, if a document, lying on a table concerning the organization of travel to Cuba, talks about Son music, a link to DBpedia could provide suggestions for adding resources about Caribbean music on the table).

References

1. Goy, A., Magro, D., Petrone, G., Picardi, C., Rovera, M., Segnan, M.: Semi-automatic support to semantic annotation of web resources in SemT++. Technical report #2015-15, Departent of Computer Science, University of Torino (2015)
2. Schmachtenberg, M., Bizer, C., Paulheim, H.: Adoption of the linked data best practices in different topical domains. In: Mika, P., et al. (eds.) ISWC 2014. LNCS, vol. 8796, pp. 245–260. Springer, Heidelberg (2014). doi:10.1007/978-3-319-11964-9_16
3. Hart, G., Dolbear, C.: Linked Data: A Geographic Perspective. CRC Press, London (2013)
4. Kaptelinin, V., Czerwinski, M. (eds.): Beyond the Desktop Metaphor. MIT Press, Cambridge (2007)

5. Karger, D.R.: Haystack: per-user information environments based on semistructured data. In: Kaptelinin, V., Czerwinski, M. (eds.) Beyond the Desktop Metaphor, pp. 49–100. MIT Press, Cambridge (2007)

6. Bardram, J.E.: From desktop task management to ubiquitous activity-based computing. In: Kaptelinin, V., Czerwinski, M. (eds.) Beyond the Desktop Metaphor, pp. 223–260. MIT Press, Cambridge (2007)

7. Voida, S., Mynatt, E.D., Edwards, W.K.: Re-framing the desktop interface around the activities of knowledge work. In: UIST 2008, pp. 211–220. ACM Press, New York (2008)

8. Breslin, J.G., Passant, A., Decker, S.: The Social Semantic Web. Springer, Heidelberg (2009)

9. Abel, F., Henze, N., Krause, D., Kriesell, M.: Semantic enhancement of social tagging systems. In: Devedžić, V., Gašević, D. (eds.) Web 2.0 & Semantic Web, pp. 25–56. Springer, Heidelberg (2010)

10. Kim, H., Breslin, J.G., Decker, S., Choi, J., Kim, H.: Personal knowledge management for knowledge workers using social semantic technologies. Int. J. Intell. Inf. Database Syst. 3(1), 28–43 (2009)

11. Sauermann, L., Bernardi, A., Dengel, A.: Overview and outlook on the semantic desktop. In: 1st Workshop on The Semantic Desktop at ISWC 2005, vol. 175. CEUR-WS (2005)

12. Drăgan, L., Delbru, R., Groza, T., Handschuh, S., Decker, S.: Linking semantic desktop data to the web of data. In: Aroyo, L., Welty, C., Alani, H., Taylor, J., Bernstein, A., Kagal, L., Noy, N., Blomqvist, E. (eds.) ISWC 2011. LNCS, vol. 7032, pp. 33–48. Springer, Heidelberg (2011). doi:10.1007/978-3-642-25093-4_3

13. Meimaris, M., Alexiou, G., Papastefanatos, G.: LinkZoo: a linked data platform for collaborative management of heterogeneous resources. In: Presutti, V., d'Amato, C., Gandon, F., d'Aquin, M., Staab, S., Tordai, A. (eds.) The Semantic Web: Trends and Challenges. LNCS, vol. 8465, pp. 407–412. Springer, Heidelberg (2014)

14. Schandl, B., Haslhofer, B., Bürger, T., Langegger, A., Halb, W.: Linked data and multimedia: the state of affairs. Multimed. Tools Appl. 59(2), 523–556 (2012)

15. Yamada, I., Ito, T., Usami, S., Takagi, S., Toyoda, T., Takeda, H., Takefuji, Y.: Linkify: enhanced reading experience by augmenting text using linked open data. In: ISWC 2014 Semantic Web Challenge (2014). challenge.semanticweb.org

16. Passant, A., Laublet, P.: Meaning of a tag: a collaborative approach to bridge the gap between tagging and linked data. In: Bizer, C., Heath, T., Idehen, K., Berners-Lee, T. (eds.). Linked Data on the Web (LDOW 2008), vol. 369. CEUR (2008)

17. Bontcheva, K., Cunningham, H., Roberts, I., Tablan, V.: Web-based collaborative corpus annotation: requirements and a framework implementation. In: Witte, R., Cunningham, H., Patrick, J., Beisswanger, E., Buyko, E., Hahn, U., Verspoor, K., Coden, A.R. (eds.) LREC 2010 workshop on New Challenges for NLP Frameworks, pp. 20–27 (2010)

18. Cunningham, H., Maynard, D., Bontcheva, K., Tablan, V., Aswani, N., Roberts, I., Gorrell, G., Funk, A., Roberts, A., Damljanovic, D., Heitz, T., Greenwood, M. A., Saggion, H., Petrak, J., Li, Y., Peters, W.: Text Processing with GATE (Version 6): (2011). gate.ac.uk

19. Fragkou, P., Petasis, G., Theodorakos, A., Karkaletsis, V., Spyropoulos, C.: Boemie ontology-based text annotation tool. In: Calzolari, N., Choukri, K., Maegaard, B., Mariani, J., Odjik, J., Piperidis, S., Tapias D. (eds.) Proceedings of the International Conference on Language Resources and Evaluation (LREC 2008), European Language Resources Association (ELRA) (2008)

20. Uren, V., Cimiano, P., Iria, J., Handschuh, S., Vargas-Vera, M., Motta, E., Ciravegna, F.: Semantic annotation for knowledge management: requirements and a survey of the state of the art. J. Web Semant. 4(1), 14–28 (2006)

21. Corcho, O.: Ontology based document annotation: trends and open research problems. Int. J. Metadata Semant. Ontol. **1**(1), 47–57 (2006)
22. Su, A.Y.S., Yang, S.J.H., Hwang, W.Y., Zhang, J.: A Web 2.0-based collaborative annotation system for enhancing knowledge sharing in collaborative learning environments. Comput. Educ. **55**, 752–766 (2010)
23. Buffa, M., Gandon, F., Ereteo, G., Sander, P., Faron, C.: SweetWiki: a semantic wiki. Web Semant. **6**(1), 84–97 (2008)
24. Giunchiglia, F., Dutta, B., Maltese, V.: Feroz, F: A facet-based methodology for the construction of a large-scale geospatial ontology. J. Data Semant. **1**(1), 57–73 (2012)
25. Goodchild, M.F.: Citizens as sensors the world of volunteered geography. GeoJournal **69**(4), 211–221 (2007)
26. ElGindy, E., Abdelmoty, A.: Capturing place semantics on the geosocial web. J. Data Semant. **3**(4), 207–223 (2014)
27. Ballatore, A., Wilson, D.C., Bertolotto, M.: A survey of volunteered open geo-knowledge bases in the semantic web. In: Pasi, G., Bordogna, G., Jain, L.C. (eds.) Quality issues in the management of Web information, pp. 93–120. Springer, Heidelberg (2013)
28. Goy, A., Petrone, G., Segnan, M.: A cloud-based environment for collaborative resources management. Int. J. Cloud Appl. Comput. **4**(4), 7–31 (2014)
29. Ardissono, L., Bosio, G., Goy, A., Petrone, G.: Context-aware notification management in an integrated collaborative environment. In: UMAP 2009 Workshop on Adaptation and Personalization for Web2.0, pp. 23–39. CEUR (2010)
30. Magro, D., Goy, A.: A core reference ontology for the customer relationship domain. Appl. Ontol. **7**(1), 1–48 (2012)
31. Borgo, S., Masolo, C.: Foundational choices in DOLCE. In: Staab, S., Studer, R. (eds.) Handbook on Ontologies, 2nd edn, pp. 361–381. Springer, Heidelberg (2009)
32. Gangemi, A., Borgo, S., Catenacci, C., Lehmann, J.: Task Taxonomies for Knowledge Content. Metokis Deliverable D07 (2005)
33. Goy, A., Magro, D., Petrone, G., Segnan, M.: Semantic representation of information objects for digital resources management. Intelligenza Artificiale **8**(2), 145–161 (2014)
34. Goy, A., Magro, D., Petrone, G., Picardi, C., Segnan, M.: Ontology-driven collaborative annotation in shared workspaces. Future Gener. Comput. Syst. Spec. Issue Semant. Technol. Collab. Web **54**, 435–449 (2016)
35. Goy, A., Magro, D., Petrone, G., Picardi, C., Segnan, M.: Shared and personal views on collaborative semantic tables. In: Molli, P., Breslin, John, G., Vidal, M.-E. (eds.) SWCS 2013-2014. LNCS, vol. 9507, pp. 13–32. Springer, Heidelberg (2016). doi:10.1007/978-3-319-32667-2_2
36. Goy, A., Magro, D., Petrone, M., Rovera, C., Segnan, M.: A semantic framework to enrich collaborative semantic tables with domain knowledge. In: Proceedings of IC3K 2015, KMIS, vol. 3, pp. 371–381. SciTePress (2015)
37. Zhao, L., Ichise, R.: Ontology integration for linked data. J. Data Semant. **3**(4), 237–254 (2014)

Author Index

Printed in the United States
By Bookmasters